T0262970

Self on Audio

Self on Audio: The Collected Audio Design Articles of Douglas Self, Third Edition is the most comprehensive collection of significant articles in the technical audio press. This third edition features 45 articles that first appeared in *Elektor*, *Linear Audio*, and *Electronics World*.

Including expanded prefaces for each article, the author provides background information and circuit commentary. The articles cover both discrete and opamp preamplifier design, mixing console design, and power amplifier design. The preamplifier designs are illuminated by the very latest research on low noise and RIAA equalization. The famous series of 1993 articles on power amplifier distortion is included, with an extensive commentary reflecting the latest research on compensation and ultra-low distortion techniques. This book addresses the widened scope of technology that has become available to the audio designer over the last 35 years.

New materials include:

- Prefaces that explain the historical background of the articles, why they were written, and the best use of the technology of the day

- Extensive details, including schematics, of designs that preceded or followed the design in each article, giving an enormous amount of extra information and a comprehensive overview of how the author's design approaches have evolved

- New directions for the technology, describing new lines of thought such as curvilinear Class-A.

Douglas Self studied engineering at Cambridge University, then psychoacoustics at Sussex University. He has spent many years working at the top level of design in both the professional audio and hifi industries, and has taken out a number of patents in the field of audio technology. He currently acts as a consultant engineer in the field of audio design.

Self on Audio

The Collected Audio Design Articles of Douglas Self

Third Edition

Douglas Self

Focal Press
Taylor & Francis Group

NEW YORK AND LONDON

First published 2016
by Focal Press
70 Blanchard Road, Suite 402, Burlington, MA 01803

and by Focal Press
2 Park Square, Milton Park, Abingdon, Oxon OX14 4RN

Focal Press is an imprint of the Taylor & Francis Group, an informa business

© 2016 Taylor & Francis

The right of Douglas Self to be identified as author of this work has been asserted by him in accordance with sections 77 and 78 of the Copyright, Designs and Patents Act 1988.

All rights reserved. No part of this book may be reprinted or reproduced or utilised in any form or by any electronic, mechanical, or other means, now known or hereafter invented, including photocopying and recording, or in any information storage or retrieval system, without permission in writing from the publishers.

Notices
Knowledge and best practice in this field are constantly changing. As new research and experience broaden our understanding, changes in research methods, professional practices, or medical treatment may become necessary.

Practitioners and researchers must always rely on their own experience and knowledge in evaluating and using any information, methods, compounds, or experiments described herein. In using such information or methods they should be mindful of their own safety and the safety of others, including parties for whom they have a professional responsibility.

Product or corporate names may be trademarks or registered trademarks, and are used only for identification and explanation without intent to infringe.

Library of Congress Cataloging-in-Publication Data
Self, Douglas.
[Works. Selections]
   Self on audio : the collected audio design articles of Douglas Self / by Douglas Self. — Third edition.
      pages cm
   Includes index.
   Collection of articles written for the journal Wireless world (now Electronics world), 1976–2012.
   1. Audio amplifiers—Design and construction.   2. Power amplifiers—Design and construction.   I. Wireless world (London, England : 1932)   II. Electronics & wireless world.   III. Title.
   TK7871.58.A9S456 2006
   621.3815′35—dc23
   2015014158

ISBN: 978-1-138-85446-8 (pbk)
ISBN: 978-1-138-85445-1 (hbk)
ISBN: 978-1-315-72109-5 (ebk)

Typeset in Times New Roman
By Apex CoVantage, LLC   Cover design by Douglas Self

Contents

Preface

This book is a newly expanded collection of the articles I wrote for various journals between the years 1974 and 2014. It is however much more than just an expanded collection. Each article has a newly written preface which gives the historical background to the project, why it was written, relevant anecdotes, and how it tried to make the best use of the technology of the day through new circuit techniques. It maps out the technical progress that has been made since the article was written, evaluated from a forty-year perspective. It goes from history to the cutting edge. The first edition of *Self on Audio* had no prefaces at all, the second edition had some short prefaces, and this third edition has much-extended prefaces for every chapter.

This volume contains a considerable amount of hitherto unpublished research, and indicates new directions to be explored, describing new lines of thought such as curvilinear Class-A. I have added a lot of information, including schematics of designs that preceded or followed the design in the article, giving a comprehensive overview of how my design approaches have evolved. Any errors in the original article are corrected in the corresponding Preface. In some places, names have been redacted to make sure that historical technical problems do not become current legal ones.

All the articles in this book are reproduced just as they were published, with no editing; corrections or afterthoughts are to be found in the relevant Preface. Attempting to polish the text in the light of later knowledge would, I think, be cheating, would produce something much less useful than the latest editions of my other books, and would destroy whatever value the articles might have as historical documents. The only exception to this is Chapter 19, a very recent article on new developments in RIAA equalisation. A few typos crept into a rather closely reasoned article, and leaving them uncorrected could only cause confusion. The corrections are listed in the associated preface.

The earlier articles were all published in *Wireless World* (now *Electronics World*), which was then the premier publication for new ideas and innovative designs in audio electronics. During my long career of writing for them, I was in the happy position that *Wireless World/ Electronics World* (WW/EW) would publish pretty much everything I sent them, and often very quickly. I only once had an article turned down; that was back in 1978. It was, I admit, a rather dry piece on minimising the distortion in emitter-followers; perhaps crucially it was devoid of illustrations. However I never throw anything away, and the principles I set out in that article have been incorporated in my book *Small Signal Audio Design*, now in its second edition. I worked pretty much in perfect harmony with WW/EW over this period, the only point of disagreement being the enthusiasm of the various editors for changing the article title, all too often to incorporate an eye-rollingly awful pun. In feeble belated protest I have given in each Preface the title that I intended.

I stopped writing for WW/EW when they stopped paying their authors. The remuneration was always tiny compared with the work involved in preparing an article, but that was accepted because it was then an influential and much-loved magazine. Working for free to make profits for someone else, on the other hand, is a line I do not cross. Their place has been admirably filled by Jan Didden's *Linear Audio* journal [1], which I strongly recommend to anyone interested in audio design.

The technology of writing has changed drastically since the early articles were written. My writing process consisted of getting down a draft on paper, revising it with much eraser work and crossing out, and then hammering out the final version on a portable typewriter handed down from my aunt Winifred, god bless her, using carbon paper to make a simultaneous copy, and effacing mistakes with Tippex. This is a very different business to using a word processor; if you need to shift a paragraph, you will have to type the whole page all over again. This is a powerful incentive to do as much thinking ahead as possible.

Working with a word processor is completely different. I find myself diving in and out of the text, fixing a sentence here and a sentence there. This can easily lead to bumpy text that does not flow well, and so the last stage is what I call 'annealing', in which it is checked for smooth reading. My word processing career began with Wordwise [2] on the BBC Model-B computer, which was remarkable for its day but not exactly WYSIWYG. PCs then appeared, and I tried Wordstar 2000 [3], which utterly failed to impress. Wordperfect 5.1 and 6.0 [4] were much better and were used to write most of the articles in this book. As time moved on, resistance was futile, and capacitance wasn't much better; I was assimilated by Word 2003 and I'm still using it.

I have never been sure if word-processing has made for better writing or worse. I likewise am unsure if, at the end of the day, it is quicker or slower overall. It is certainly more convenient, and electronic off-continent backups do make an author feel more secure.

When the earlier articles were written, CAD was not to be found outside the aerospace industry. My design approach was to mull over the requirements for a while, then put down a first cut of the schematic on a big A2 sheet of paper. It was considered cool to do this with Rotring pens, and I still have all mine. When the big sheet of paper was suitably amended in the light of development, it was passed on to a draughtsman/woman who would produce a neat version drawn in black ink on plastic film, using, yes, more Rotring pens. This translucent original could be copied as required by the diazo process, involving ultra-violet light and a strong smell of ammonia. These draughtspersons have now marched into history, and the last one I knew ended up as a postman. He told me he preferred it.

In the last forty years, the breadth of technology available to the audio designer has increased relatively slowly. At the beginning of the period covered by this collection, the only choice in preamp design was between discrete transistor stages and opamps. The former were usually two or three transistor stages, though a few 'discrete opamp' designs were produced, by me for one in the MRP1 preamplifier design. Truly usable IC opamps like the LM741 were still relatively new, and had dubious characteristics as regards noise and distortion, particularly crossover distortion. This most unloved of audio defects was tolerated in power amplifiers because it had to be, but very properly, there was considerable opposition to incorporating it in preamplifiers. The appearance of the NE5534 in 1978 revolutionised noise and distortion, but for a long time, it was an expensive part to be used sparingly. Now, as the 5532 dual version, it is in most circumstances the cheapest opamp available, and still extremely useful for small-signal design.

In the power amplifier field there were by now some much-improved complementary pairs of bipolar power transistors available, but power FETs were still three or four years away from wide availability. At this time (say 1974) using valve circuitry in a new design would have been considered absolutely ridiculous. Despite much initial enthusiasm, it was found that power FETs did not solve all amplifier design problems overnight, and in fact introduced some new ones such as low gm and unpredictable Vgs. I still firmly believe that if FETs had been invented first, the appearance of BJTs would have been heralded as a great step forward. The biggest technological change in the period has been the appearance of Class-D amplifiers with acceptable performance and reliability; that technology is not examined here.

My first article published by WW/EW was a FET compressor/limiter; its design was my third-year project at Cambridge, and it was of course aimed at WW/EW from the start. This debut was followed by what I called 'An Advanced Preamplifier' in 1976. The Advanced Preamplifier certainly gave (and gives—I still have the prototype) very low distortion indeed for its day, obtained by making each stage a discrete-component operational amplifier. This required dual supply rails, using dual IC regulators to produce ±24 V; at the time the cost of this power-supply scheme was significant.

As a reaction to this complexity, I decided to try my hand at what might be called 'traditional' discrete circuitry in a preamplifier, and this became the 'High Performance Preamplifier' published in 1979, though actually designed nearly two years earlier. It was conceived in an era when opamps were still regarded with considerable suspicion by designers seeking the best possible audio performance. In the search for simplicity, a single supply rail was used, without regulation, but with a simple RC filter after the reservoir capacitor to reduce ripple to a manageable 50 mV or so. Combined with a well-filtered V/2 rail to bias the various stages, plus suitable stage decoupling, experiment proved that this minimal-cost arrangement could give hum and noise results that were as good as those yielded by the dual-IC-regulator approach.

In contrast, the Precision Preamplifier of 1983 was designed at a time when the remarkable 5534/5532 opamps had become available at reasonable prices. Since they delivered very low noise with then almost unmeasurable distortion, it was clearly time to try a 'third way' as regards preamp design. Having explored discrete opamps, and conventional discrete circuitry, an IC opamp solution was an obvious next step. The return to opamps meant a return to dual power supplies, but this was a small price to pay for the convenience of dual rails. This design later gained a moving-coil head amplifier. I had designed several of these stages before, at least two of which made their way into commercial production, but this was the first version that got both noise and distortion down to what I considered to be satisfyingly low levels. The salient features are the discrete transistor input devices which then, and indeed now, provide the best possible noise figure. At the time many head amps were outboard units, often relying on battery power, presumably to sidestep intractable ground-loop problems. However, no difficulties were found in grafting this design onto existing preamplifiers.

Some years later, having devoted much time in between to power amplifier design, I felt the call to take another look at preamplifiers. My last design was twelve years old, and it seemed likely that some significant improvements could be made. Much thought and a lot of calculation and simulation led to the 'Precision Preamp 96' articles, including in-depth mathematical modelling of the noise generated by the RIAA stage. This allowed each noise contribution to be studied independently, and permitted comparison between the actual noise and the theoretical minimum. The latter is rather sensitive to the exact assumptions made. It was also possible to discover why opamps that appeared to be quieter than the 5534 in theory, were actually noisier in practice. The answer was that opamp bias-cancellation networks maybe great for DC precision, but the extra common-mode noise they generate in audio circuitry is just an embarrassment. The '96 preamp was in fact not so much an updated version as a thorough re-design, with only the moving-coil input amp remaining essentially unaltered. It demonstrated, amongst other things, that obtaining an interchannel separation of 100 dB on a stereo PCB is perfectly possible with careful component and track layout.

In 1990, I had again turned my attention to power amplifiers. For many years I had felt that the output stages of power amplifiers presented very great possibilities for creative design, and so I explored some of them, the first being a hybrid output stage with BJT drivers and FET output devices (see Chapter 20). One of the first difficulties I met with was the problem of determining how much of the overall distortion was produced in the small-signal sections, and how much was generated by the output stage. Traditionally, the latter was regarded as the major source of distortion, but there was very little published research to back this up, and so I attacked the problem myself. When I began it was not clear if there were two, twenty, or two hundred significant distortion mechanisms, but after a good deal of study it suddenly became clear that seven or eight were sufficient to explain all the observed distortion. This number has grown until today I identify twelve (see the Preface to Chapter 31), and it looks as though there is at least one more mechanism to be identified and rendered harmless. There are almost certainly others, but the non-linearities they produce are currently below the level of practical measurement with THD analysers. When output stage distortion has been completely eliminated forever (and it has to be said there is no sign of this happening in the immediate future) then it may be time to dig into the deeper levels of non-linearity.

It quickly became clear that by taking a few simple circuit precautions, it was possible to design amplifiers with very much lower distortion than the norm. Such amplifiers, with their very low THD figures, are rather distinct from average designs, and so I looked around for a suitable name. Once the critical factors are identified, designing an low-distortion amplifier becomes more a matter of avoiding mistakes rather than being brilliant, so I decided to call them 'Blameless' amplifiers, rather than 'hyper-linear' or something similarly pretentious, to emphasise this point. In a Blameless Class-B amplifier, all the distortions that are easily fixed have been fixed; this does not mean we have a distortionless amplifier, because the crossover distortion and LSN from the output stage are *not* easily fixed. However, the crossover distortion is all that we do get. The results and conclusions of this major investigation were published in eight parts as 'Distortion in Power Amplifiers'. 'Distortion Residuals' followed this up, providing a visual guide to the appearance of the various distortion mechanisms on the oscilloscope screen.

These endeavours built up to a substantial body of information on just how to minimise amplifier distortion, and two designs that exemplify this are included in the 'Distortion In Power Amplifiers' articles, one working in Class-B and the other in Class-A. This foundation of knowledge simply begged to be put to further use, and so two major power amplifier projects were created; the Trimodal amplifier based on Class-A, and the Load-Invariant amplifier in relatively conventional Class-B.

The Trimodal amplifier demonstrated how to make a Class-A amplifier that coped gracefully with varying load impedances. This project had its roots in an insistent demand for a PCB for the Class-A power amplifier presented in the last part of 'Distortion in Power Amplifiers'. I find it goes against the grain to reproduce a design without trying to improve it, and the Trimodal article was the result. My first intention was to demonstrate how a Class-A amplifier could, with appropriate design, move gracefully into a relatively linear version of Class-AB when the load impedance became too low for Class-A operation to be maintained, rather than clipping horribly as some configurations do. Improvements were also made in the noise and DC offset performance of the basic amplifier. I was concerned to guard against catastrophic currents flowing if there were errors in building the quiescent-current controller, so a safety network was added to set an upper limit on the bias voltage. It was simple to make the amplifier switchable between A and B by making the limiting value of this second bias circuit preset-adjustable, and the Trimodal was born.

The Load-Invariant power amplifier project was a direct development of the work done on amplifier distortion. Power amplifiers always give worse distortion into lower load impedances. For bipolar output devices, as the load value drops from 100 Ω to about 8 Ω, the crossover distortion increases steadily and predictably. However, at about 6 Ω (depending on transistor characteristics) an extra low-order distortion appears that can easily double the THD at 4 Ω compared with 8 Ω. I decided to see to what extent I could thwart this extra distortion, aiming to produce the first semiconductor amplifier that gave exactly the same THD at 4 Ω as it did at 8 Ω. While it did not prove possible to quite attain this, I did get reasonably close. This design seems to have generated a lot of interest.

A few of my articles have been written in reaction to contributions to *Electronics/Wireless World* that suggested promising new approaches, or that I simply found intriguing. Investigating a particular amplifier topology takes a lot of time and effort, but preparing the results for publication does not add a great deal to this, so some more articles resulted. They may not have advanced the art of audio greatly, if at all, but they did explore a few paths which would otherwise have remained untrod. Two examples are given in this book: 'Common-Emitter Amplifiers' and 'Two-Stage Amplifiers'. In neither case were the results sufficiently encouraging for me to proceed further with the concepts involved. When the idea that loudspeakers could, in certain circumstances, draw much more current from an amplifier than its impedance curve suggested first came to my notice in the mid-1980s, I must confess I felt a degree of scepticism. I was wrong; the effect is real, though its relevance to real-life signals rather than artificial stimulus waveforms is extremely doubtful. The abnormally high currents that flow are provoked by using the stored energy in the circuit elements, such as the inertia of the speaker cone. This requires a rectangular stimulus waveform with rapid full-amplitude transitions, carefully timed to catch the speaker resonance at its worst moment, and the difficulty is that real waveforms do not have these features, or anything remotely resembling them. Eventually, I got around to putting the idea to the test, using an electrical analogue of a speaker system. The article 'Excess Speaker Currents' describes the effect and how to produce it.

When semiconductors were first applied to audio amplification, the choice of operating mode was simple: Class-A, Class-AB, or Class-B. It took several years before proper complementary pairs of output devices were available, but it was clear that they allowed a good deal more flexibility in the design of output stages, and variations on the standard configurations began to appear, becoming more radical as time went on and the technology of audio developed. The 1960s gave us Class-D, though the rudimentary versions available then were not much of a gift. The 1970s saw the advent of the Blomley concept, current-dumping, and, significantly, Class-G. In this situation, it was inevitable that extra letters of the alphabet would be called in to describe new methods of operation, though it was clear that calling something, say, 'Class-Y' said nothing about how it operated, and the prospect of trying to remember what an alphabet soup of 26 (or more) class letters actually represented was not enticing. With this in mind, I produced the article 'Class Distinction', which attempts to simplify amplifier classification into simple combinations of A, B, C, and D in such a way that at least some information about the mode of action is given. I will not pretend that I expect my classification system to sweep the world overnight. Should it fail to sweep the world at all, I think the article is still useful because it allows the generation of a matrix of amplifier types, some of which no-one has got around to inventing yet. Give me time.

Power FETs first began to reach the market in the mid-1970s, and as so often with new technology, they were claimed to be superior to existing methods in just about every possible way. However, experience soon showed that FETs were not dramatically more linear than bipolar transistors, nor were they inherently short-circuit proof. There appeared, however, to be definite advantages in their high bandwidth and freedom from carrier-storage effects. Initially, I was intrigued by the possibility that power FETs, with their much-advertised speed and bandwidth, would allow the implementation of various ingenious output stages involving local feedback that in bipolar format had proved difficult or impossible to stabilise, apparently due to the slowness of the output devices. The results were not encouraging: any increase in stability due to the faster devices was more than outweighed by their tendency to parasitic oscillation when used in anything other than the simplest of configurations.

Most serious power amplifiers are fitted with a muting relay that disconnects the electronics from the loudspeaker at turn-on and turn-off, to reduce thumps and bangs. The same relay is used to protect the loudspeaker from incineration if the amplifier suffers a fault which puts a large DC voltage on the output. Because of the safety implications, this relay must operate promptly and reliably, and designing its control circuitry is not trivial. My article on relay control delved deeply into the control circuit design, with special emphasis on a rapid relay response to dangerous conditions or intrusive transients; this is not, as it might appear, merely a matter for the relay designer. Apparently tiny details of circuit design have a major effect.

Chapters 40 and 41 in this book introduced new ways of displaying amplifier efficiency and power dissipation that I call 'power partition diagrams'. The first one dealt with many different kinds of amplifier, and studied how they disposed of the power involved in driving resistive and reactive loads with sine waves. This produced the interesting conclusion that a proper appreciation of peak transistor dissipations into real loads with phase-shift could be the most crucial factor in determining amplifier reliability. Whenever sine waves are used for testing, and there are many good reasons why they should be used most of the time, the criticism is likely to be levelled that this is unrealistic, which of course it is. However, the lack of realism, expressed as a very different peak-mean ratio (PMR) from music, means if an amplifier can handle sinewaves without thermal embarrassment, it will work fine on music, with a healthy safety margin. Music is less demanding than sinewaves. The second article looked at the statistics of music (which were surprisingly obscure) and ways of calculating true power dissipations from the results, for comparison with the sinewave data. This showed, amongst other things, that Class-A amplifiers are in reality not more than 1% efficient. This raises serious questions about their desirability in an energy-conscious world.

In the course of the investigations that led to these articles, I found over and over again that the conventional wisdom on power amplifiers was more conventional than wise. Some examples are given here, though they may not make much sense until you have read the relevant article. Some statements turned out to be half-true: an example is the widespread assumption that a current-source loaded Voltage Amplifier Stage (VAS) gives current drive to the output devices. The reality is much more complex; the impedance might be high at low frequencies, but the Miller capacitor around the VAS causes the impedance to fall with frequency until it is a few kilo ohms at the top of the audio band. This hardly counts as a current source. The drive point is also loaded by the non-linear input impedance of the output stage, which complicates matters further. There is much more to most questions about amplifiers than at first meets the eye.

Some long-held beliefs turned out to be completely wrong, though plausible in theory and workable in practice. The best illustration of this is the universal belief that the crucial parameter in biasing a Class-B output stage to minimise crossover distortion is the quiescent current. In actual fact, the critical factor is the voltage across the emitter resistors. If the value of these are changed, then the quiescent current can be radically altered although the amplifier remains at the same optimal bias point, because the voltage drop is unchanged. However, emitter resistor values are rarely changed in this way, so setting up for a given current is the same as setting up for a given voltage. The difference is unimportant if you are simply repairing or adjusting amplifiers, but vital if you seek to understand how they work.

In some cases, there was no argument about the distortion mechanism operating, but very little, if any, published information quantifying the size of the effect. This applied to most of the amplifier distortions examined, and it took a little thought to develop ways of measuring each one separately, wondering in each case why the relatively simple test had apparently never been done before. It may be, of course, that parts of this work have also been done by various audio manufacturers, who have every reason for keeping their private research to themselves.

You might want to know what amplifiers I was listening to over the period in which these articles were written. The first was a Metrosound ST20 integrated amplifier, bought in 1971 along with a pair of Wharfedale Super 8 full-range loudspeakers from a chap in Trinity Hall, for a modest price. The ST20 worked very well for the price, and proved remarkably resistant to short-circuiting considering it had no overload protection system.

I later built a powered mixer with two Henry's 50W amplifier modules, for a disco operation of which I was one half. When not out earning money, this made a very acceptable hi-fi amplifier. The loudspeakers were upgraded to Wharfedale Unit-5's treated with formica to make them look like giant liquorice allsorts; I can't remember if that came about accidentally or on purpose.

For a long time, an example of the quasi-complementary amplifier described in Chapter 31 was used for the hi-fi, and was perfectly acceptable. I moved on to a Quad 303 power amp, used with the MRP1 preamplifier design described in Chapter 1. Right at this moment, I am using a Cambridge Audio 340A, one of my early designs for Audio Partnership.

This collection of articles obviously cannot be a totally exhaustive exposition of audio design, as various topics that have already been fully expounded in *The Audio Power Amplifier Design Handbook* and *Small-Signal Audio Design* are not represented here. There is some overlap in the material here and there, but that is unavoidable if each book is to stand alone. The articles are in general in chronological order within the two categories of Preamplifier and Power amplifier material, but with some exceptions to maintain the flow of ideas.

The articles reproduced here are an account of a stimulating intellectual journey. I hope that reading them will share some of the sense of discovery that I felt, and that this collection will be both useful and entertaining to all those concerning themselves with audio electronics.

To the best of my knowledge, no supernatural assistance was received in the making of this book.

All suggestions for its improvement that do not involve combustion will be gratefully received. You can find my email address on the front page of my website at douglas-self.com.

References

1. http://www.linearaudio.net/ (accessed Feb. 2015)

2. http://en.wikipedia.org/wiki/Wordwise (accessed Jan. 2015)

3. http://en.wikipedia.org/wiki/WordStar#WordStar_2000 (accessed Feb. 2015)

4. http://en.wikipedia.org/wiki/WordPerfect (accessed Feb. 2015)

Preamplifiers and related matters

Chapter I

Advanced preamplifier design

A no-compromise circuit with noise gating
(Wireless World, November 1976)

This was my first preamplifier design, conceived in 1975, the first year that I worked in the audio industry. The two gain-control solution to the dynamic-range problem was inspired by what was thought of as advanced industry practice at the time—notably in products by Radford and Cambridge Audio. I thought the rumble-gate concept was rather clever, but I seem to have been in a minority of one, as it was never even mentioned in the correspondence that followed. That was rather a disappointment.

The major part of the circuitry consisted of two discrete opamps with differential inputs, which performed the amplification and tone-control functions. The ±24V supply rails, which permitted a maximum signal level of about 16 Vrms, were provided by two 7824 IC regulators. The Moving Magnet (MM) input stage, however was a simple three–transistor stage running between +24 V and 0V. The reasoning was that in the MM stage low noise was the priority, the low signal levels making distortion much less of an issue, and so it was better to have a single input transistor rather than a differential pair; I had a vague notion that the pair would be 3 dB noisier. This is not actually the case, as the transistor of the pair that handles the negative feedback works under much better conditions for low noise than the input transistor, which is faced with the highly inductive source impedance of an MM cartridge. I have not yet got around to quantifying the difference, but clearly two devices must be noisier than one to some extent; having said that, the 5534 opamp with its differential inputs gives a very acceptable noise performance as an MM input stage. I later developed a more advanced discrete MM stage that had still only one input transistor, but could have been run between the +24V and −24V rails, giving a theoretical 6 increase in headroom. This approach is fully described in Chapter 2.

As shown in the block diagram (Preface Figure 1.2) the RIAA equalisation was divided into two halves. The HF roll-off at 2.1 kHz was implemented in the first stage, and the feedback around the normalisation amplifier implemented the boost starting at 500 Hz and levelling off at 50 Hz. Doing it this way round avoids a massive loss of headroom in the first stage [1], though at the cost on some compromise on noise. Splitting up the equalisation of course makes the RIAA

The MRP1 Advanced Preamplifier.

Preface Figure 1.1 The Advanced Preamplifier neatly boxed: 1976

component calculations very much easier, and unless I'm very much mistaken that is the inglorious and only reason that the idea has the popularity it does. That was *not* the reason I did it here. The main motivation was to make life easier for the MM input stage, which was a relatively simple three–transistor Walker configuration [2] with my addition of second-transistor collector load bootstrapping, what I called a BootWalker at the time, though it is perhaps not the most euphonious of terms. I did not want to implement the LF boost in this stage as its rather limited open-loop gain would have led to excessive distortion.

During the development period, the need for an HF correction pole to counteract the levelling-out of the response as unity gain is approached in the first stage was appreciated; this is implemented by R1, C1 in Preface Figure 1.2. The extra 6 dB/octave roll-off *exactly* compensates in level and phase for the levelling-out in the first stage so long as you get its frequency right; this is not a case of an approximate bodge.

While giving excellent results considering its simplicity, the BootWalker did not have the capabilities of a discrete opamp; therefore it was not required to implement the boost in the LF region, where it might have had inadequate negative-feedback to give good linearity. The normalisation amplifier is a discrete opamp with nine transistors rather than three, and better fitted to giving high gain at low distortion. The components implementing the LF part of the RIAA were switched out of the feedback network of the normalisation amplifier when a line input was in use; all my later opamp designs do the whole of the RIAA equalisation in one stage, thanks to the Lipshitz equations. I have heard the discrete opamp approach disrespected on the grounds that ultra-low distortion occurs almost automatically; such statements show woeful ignorance, as there are many design issues to that need to be dealt with properly. These are described in [3]. Be aware that the discrete opamps used in the article here are more start-of-art than state-of-art and could be much improved in the light of today's knowledge; for example, now I would put current mirrors in the collectors of the input pairs to enforce equality of collector current, and so prevent the generation of second-order distortion. For full details of how to make a state-of-art discrete opamp, see [3].

Absolute phase was an unheard of concept in 1976, and the phase-inversion produced here by the tone-control was ignored. It is difficult to see how it could have been fixed without adding another stage, which did nothing but invert to get the phase correct again; not exactly an economical design approach.

You may be wondering why the coupling capacitors are in most cases rather small non-electrolytics. This is because electrolytics were not as reliable then as they are now, and so I wanted to use as few as possible. Looking back, this was not my brightest idea, although I was right in the sense that the lifetime of a piece of audio equipment is still primarily limited by electrolytics drying out and falling in value. In top-end mixing consoles where internal temperatures are relatively high and 24/7 operation is the norm, the electrolytics need replacing (an extremely specialised job) long before the pots and switches are worn out.

You will note the 25 kΩ level pots and the 100 kΩ pot in the tone-control network; these values were typical of the era. In succeeding designs, I used electrolytic coupling capacitors, often of 47 uF, because this allowed the circuit impedances to be an order of magnitude lower, with corresponding reductions in Johnson noise and the effects of device input current noise. Pot impedances were reduced to 10 kΩ; for an example see Chapters 5 and 6. Since there were always electrolytic reservoir capacitors in the power supply, it seemed pointless to worry too much about a few coupling capacitors working under rather more benign conditions from the point of view of reliability.

As my thoughts on preamp design progressed, I reduced the circuit impedances in my high-end designs by another order of magnitude, so that pots were 1 kΩ, and coupling capacitors of 470 uF or 1000 uF were used whenever DC had to be blocked. The limits on this process are to an extent the increased drive requirements, though this can be easily overcome by using multiple opamps in parallel, which in itself also reduces noise by partial cancellation; see Chapters 14 and 15. A more final limit is the unavailability of two- and four-gang pots in very low values, but this too can be circumvented by using multiway switches and resistor ladders.

Let us see how far this idea can be pushed. If the pots/switched-resistor-networks are reduced to 100 Ω then driving them with paralleled opamps starts to get clumsy, as you would probably need to use eight of them for a good distortion performance. Using opamps combined with discrete output transistors may cut the component count but is likely to impact the THD for the worse. If you decide to get really radical and replace the opamps with power amplifiers, I suppose the ultimate limit would be overall power dissipation; imagine driving a 10 Ω volume control at 10 Vrms (roughly maximum opamp level). You would be dissipating 10 W in the pot, or 20 W for stereo; if you insist on a pot rather than switched-resistors, it's going to be a wirewound rheostat (I actually have some of them in my component collection, but unfortunately none are double gang for stereo.) Good luck getting a rheostat with a log law. Admittedly 10 Vrms is a high level, and for a more likely 2 Vrms, the pot dissipation is 400 mW, and more doable, but this will be repeated at many places in the circuitry. Apart from the problems of sourcing dual-gang log rheostats, there are likely to be crosstalk and offness problems if the ground connections are not equally reduced in impedance, which won't be easy.

Pushing the idea a little further, the use of power amplifiers removes the level limitations of opamps. With ±45 V rails, we could sustain 28 Vrms, 9 dB above 10 Vrms, which could be used to either increase headroom or signal/noise ratio (by raising the internal level), but we would then be dissipating 80 W in our 10 Ω volume control. This is obviously well beyond the line of sensible, but in hi-fi, that is a very blurred line. If you want to take the idea even further, I will merely observe that there is no difficulty in designing an amplifier to drive a 2 Ω load, and leave you to it.

If you're wondering if this sort of thing has been tried for real, the answer is yes, sort of. I recall a microphone preamplifier with a very low impedance negative feedback network intended to reduce noise; this had to be driven by a power amplifier.

The LED biasing of current-sources (to minimise current changes with temperature) was an elegant method but, as I strongly suspected at the time, quite unnecessary, as the discrete op-amps used are really not that sensitive to their internal operating conditions. Nevertheless, as LEDs were pretty new at the time, it was generally regarded as well cool. A PCB was available from *Wireless World* for the princely sum of £3.50.

The level monitoring circuitry was perhaps a bit over the top, with a big pair of VU meters originally intended for a mixing desk. There were also '1V Peak' LEDs to indicate nominal level, and bipolar 'Clip' LEDs to alert the user that the ±24V rails had been hit. I have to admit that I never once saw the 'Clip' LEDs come on in real use. The bipolar clip-detect circuit proved extremely useful in other applications however, and I have been using it ever since. There is also a bit of an infelicity in the '1V Peak' driver circuit, where the full output swing of one of the discrete op-amps is potentially applied to the trigger input of a 555 timer. Still, it never failed.

The Advanced Preamplifier was inspired by an article written by Daniel Meyer in *Wireless World* for July 1972. It was called: 'Audio Pre-amplifier Using Operational Amplifier Techniques', and still makes interesting reading, though some of the statements made are questionable. I have not been able to confirm it, but this is almost certainly the same Daniel Meyer that founded SWTP and designed the 'Tiger' series of power amplifiers [4].

When I measured the distortion performance of Meyer's version of a discrete opamp in April 1975, I found the distortion performance was less than inspiring, because the output is connected directly to the VAS collector, and loading on this point is very bad for linearity; I measured 0.008% at 1 Vrms out, with a flat closed-loop gain of 16 times and ±15V supply rails (The supply rails used by Meyer). Although I was but a lad, it still seemed to me that this was rather high for what was supposed to be an advanced configuration, and I set out to improve it. In the course of this, I taught myself some of the rules of good amplifier design, such as do not connect your main gain stage directly to the load, but instead, buffer it with some kind of emitter-follower output stage. I used a constant-current emitter follower, increased the supply rails to ±30V, and *voila!* Less than 0.003% at 20 Vrms out. This pleasurably straightforward process started off an interest in reducing distortion which has stayed with me ever since.

A follow-up article was published called 'Additions to the Advanced Preamplifier' in October 1977. This added a second-order subsonic filter and a switched-frequency second-order scratch filter to the signal path. Also added was a virtual earth summing amp to give simultaneous mono and stereo outputs. (The mono output was to drive a sound-to-light converter, or light show, as we called them in those heady days.) Rather than add another complete discrete opamp, I used a simpler four-transistor circuit with the gain provided by a cascode stage using current-injection (this technique is described in the preface to Chapter 2) and a constant-current Class-A output. Other features that were described were meter suppression (to stop the VU needles smacking the end-stops at switch-on) and a remote muting facility. The latter was a switch on the end of a piece of cable rather than an IR system.

In designing these later add-ons, I was deliberately following the path of other preamp designers in *Wireless World.* It was definitely the done thing to come up extra features at a later date, and somehow gave the project an aura of greater import. I was danged to heck if I was going to be left out of that.

This was the first preamplifier that I developed into a completely worked-out and neatly boxed entity. The initial design work on the discrete opamps was done with the Sound Technology 1700A in 1975, using one of the first examples to enter Britain. In its day it was a major advance in audio measurement. The prototype Advanced Preamplifier was built on stripboard, but it could have been laid out as a PCB and put into production. This is therefore preamplifier No. 1 in my personal numbering system, and I described it as the MRP1. The RP naturally stands for Reference Preamplifier, and the M comes from Meadow House in Swaffham Prior, Cambridgeshire, where I was living when I did the design work.

References

1. Self, D., *Small-Signal Audio Design,* 2nd edition, Chapter 9, pp. 252–256. Focal Press (Taylor & Francis) 2014. ISBN: 978-0-415-70974-3 Hardback. 978-0-415-70973-6 Paperback. 978-0-315-88537-7 ebook.

2. Ibid., pp. 301–303.

3. Ibid., pp. 110–115.

4. http://en.wikipedia.org/wiki/Daniel_Meyer_(engineer) (accessed Jan. 2015)

   (The Wikipedia entry claims that Meyer published an article in *Electronics World* in May 1960: this must refer to a USA magazine of that name as the British journal was called *Wireless World* at the time, and its May 1960 issue does not contain an article by Meyer.)

ADVANCED PREAMPLIFIER DESIGN

November 1976

This preamplifier design offers a distortion figure of below 0.002%, an overload margin of around 47 dB, and a signal-to-noise ratio of about 71 dB for the disc amplifier. A novel noise gate mutes the output when no signal is presented to the disc input and conversely, by using the subsonic information present on record pressings, eliminates the problem of muting low-level signals.

This article describes a stereo pre-amplifier that equals or exceeds the performance of many of those available. The circuit incorporates a novel method of muting the signal path, when the disc input is quiescent, by using a noise gate that never mutes a wanted low-level signal.

Many of the important performance factors, such as signal-to-noise ratio, overload margin, and accuracy of the RIAA equalization, are essentially defined by the design of the disc input circuitry. This therefore merits close attention. The best attainable s/n ratio for a magnetic cartridge feeding a bipolar transistor stage with series feedback is about 71 dB with respect to a 2 mV r.m.s. input at 1 kHz, after RIAA equalization. This has been clearly demonstrated by Walker.[1] The equivalent amplifier stage with shunt feedback gives an inferior noise performance over most of the audio band due to the rise in cartridge source impedance with frequency. This limits the maximum s/n ratio after equalization to about 58 dB. These facts represent a limit to what the most advanced disc input stage can achieve.

Overload margin appears to be receiving little attention. The maximum velocities recorded on disc seem to be steadily increasing and this, coupled with improved cartridges, means that very high peak voltages are reaching disc inputs. Several writers have shown that short-term voltages of around 60–80 mV r.m.s. are possible from modern disc and cartridges, and higher values are to be expected.[2,3] This implies that to cater for signal maxima, a minimum overload margin of 32 dB with respect to 2 mV r.m.s. at 1 kHz is essential. Obviously a safety factor on top of this is desirable. However, most pre-amplifiers at the top end of the market provide around 35–40 dB only. There are certain honourable exceptions such as the Technics SU9600 control amplifier which achieves an overload margin of 54 dB, mainly by the use of a staggeringly high supply of 136 V in the disc input amplifier. The Cambridge P50/110 series offers a margin in excess of 60 dB by the artifice of providing unity-gain buffering, for correct cartridge loading, but no amplification before the main gain control. This allows the use of an 18 V supply rail, but does limit the maximum s/n ratio.

The overload margin of a pre-amplifier is determined by the supply voltage which sets the maximum voltage swing available, and by the amount of amplification that can be backed-off to prevent overload of subsequent stages. Most preamplifiers use a relatively high-gain disc input amplifier that raises the signal from cartridge level to the nominal operating level in one jump. Low supply voltages are normally used which reduce static dissipation and allow the use of inexpensive semiconductors. The gain control is usually placed late in the signal path to ensure low-noise output at low volume settings. Given these constraints, the overload performance is bound to be mediocre, and in medium-priced equipment the margin rarely exceeds 30 dB. If these constraints are rejected, the overload margin of the system can be improved.

Two separate gain controls remove the most difficult compromise, which is the placement of the volume control. This approach is exemplified in the Radford ZD22 and the Cambridge P60 circuitry. One gain control is placed early in the signal path, preceded by a modest amount of gain. Cartridges of high output can be accommodated by the use of this first control. The second is placed late in the pre-amplifier and is used as a conventional volume control, see Figure 1.1.

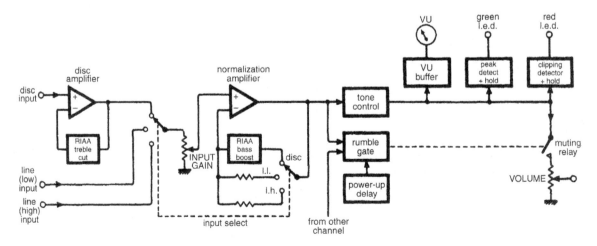

Figure 1.1 Block diagram of the complete circuit. Two gain controls are used in the signal path to allow a substantial increase in overload margin.

The other performance criterion which is largely defined by the disc input circuitry is frequency response, as defined by the accuracy of the RIAA equalization. Assuming that the relevant amplifying stage has sufficient open-loop gain to cope with the bass boost required, the accuracy of the equalization depends entirely on the time constants within the feedback loop. Careful design, and the use of close-tolerance components can assure an accurate response to within ± 0.2 dB from 30 Hz to 20 kHz.

Pre-amplifier distortion seems to have received little attention compared with that generated by power amplifiers, perhaps because the former has traditionally been much lower. However, power amplifiers, with such low THD that the residual harmonics can no longer be extracted from the noise at normal listening levels, are now commonplace, particularly with the advent of techniques such as current dumping. This desirable state of affairs unfortunately does not extend to pre-amps, which in general produce detectable distortion at nominal operating levels, usually between 0.02% and 0.2%. In this design the THD at 1 kHz is less than 0.002% even at 25 dB above the nominal operating level of 0 dBm. A Sound Technology 1700 A distortion measurement system was used during development.

At this point it is convenient to consider the noise gate principle. When the pre-amplifier is being used for disc reproduction the output from each channel is continuously sampled to determine if a signal is present; if nothing is detected within a specified time interval, dependent on the previous signal levels received, the pre-amplifier is muted by the opening of a reed relay in series with the output signal path. This allows only power amplifier noise to reach the loudspeakers and considerably reduces the perceived noise generated by a quiescent sound system. Noise in the quiescent state is particularly noticeable when headphones are in use. The reed relay is also used to prevent switch on transients from reaching an external power amplifier. So far this circuit appears to be a fairly conventional noise gate. The crucial difference is that signals from disc that have not been subjected to rumble filtering are always accompanied by very low frequency signals generated by record ripples and small-scale warps. Even disc pressings of the highest quality produce this subsonic information, at a surprisingly high level, partly due to the RIAA bass boosting. The l.f. component is often less than 20 dB below the total programme level but this is quite sufficient to keep the pre-amplifier unmuted for the duration of a l.p. side. The preamplifier is unmuted as soon as the stylus touches the disc, and muted about a second after it

has been raised from the run-out groove. This delay can be made short because the relative quiet at the start of the run-out groove is sensed and stored. The rumble performance of the record deck is largely irrelevant because virtually all of the subsonic information is generated by disc irregularities.

Audio circuitry

A detailed block diagram of the pre-amplifier is shown in Figure 1.1, and Figure 1.2 shows the main signal path. The disc input amplifier uses a configuration made popular by Walker, but the collector load of the second transistor is bootstrapped. This increases the open-loop gain and hence improves the closed-loop distortion performance by a factor of about three to produce less than 0.002% at an output of 6.5 V r.m.s. (1 kHz). This stage gives a s/n ratio (ref 2 mV) of about 70 dB and a gain of 15 at 1 kHz. This is sufficient to ensure that the noise performance is not degraded by subsequent stages of amplification. The maximum output of this stage before clipping is about 6.5 V r.m.s. and the nominal output is 30 mV r.m.s. Because this is the only stage before the input gain control, these two figures set the overload margin at 47 dB. To ensure that this overload margin is maintained at high frequencies, the treble-cut RIAA time-constant is incorporated in the feedback loop. This leads to slightly insufficient cut at frequencies above 10 kHz because the gain of the stage cannot fall below unity, and hence fails to maintain the required 6 dB/octave fall at the top of the audio spectrum. This is exactly compensated for outside the feedback loop by the low-pass filter $R_1 C_1$, which also helps to reject high frequencies above the audio band.

For convenience I have referred to the next stage of the circuit as the normalization amplifier because signals leaving this should be at the nominal operating level of 0 dBm by manipulation of the input gain controls. Separate controls are provided for each channel to allow stereo balance. A later ganged control is used for volume setting and causes no operational inconvenience. In the disc replay mode, the normalization amplifier provides the RIAA bass boost, by the feedback components R_{2+3} and C_2. Two line inputs are also provided; line low requiring 30 mV and line high 100 mV to give 0 dBm from the normalization stage with the input gain control fully advanced. When these inputs are selected, the feedback networks are altered to adjust the gain and give a flat frequency response. Ultrasonic filters are incorporated to ensure stability and aid r.f. rejection. Capacitor C_3 in the feedback arm

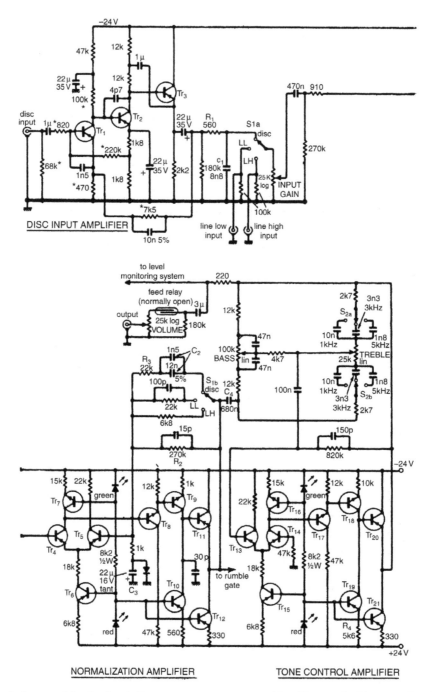

Figure 1.2 Circuit diagram of the signal path. Constant-current sources are biased from an l.e.d./resistor chain for improved thermal stability.

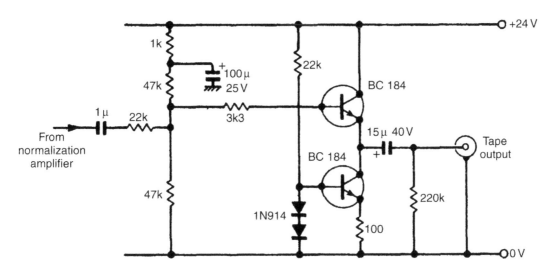

Figure 1.3 Tape output circuit. The smallest allowable load impedance for an undistorted output is about 2.2kΩ. Line inputs of the preamplifier are suitable for playback purposes.

reduces the gain to unity at d.c. for good d.c. stability. If a fault causes the amplifier output to saturate positively the capacitor is protected by a diode which has no effect on the distortion performance.

The circuitry of the normalization amplifier is complicated because its performance is required to be extremely high. The harmonic distortion is far below 0.002% at the maximum output of 14.5 V r.m.s. which is 25 dB above nominal operating level. This large amount of preamplifier headroom allows gross preamplifier overload before clipping. The input stage of the amplifier is a differential pair with a constant-current source for good common-mode rejection. The operating currents are optimized for good noise performance, and the output is buffered by an emitter-follower. The main voltage amplifier, Tr_9 has a constant-current collector load so that high voltage gain at low distortion can be obtained. This performance is only possible if the stage has very little loading so it is buffered by the active-load emitter-follower. The various current sources are biased by a l.e.d.-resistor chain because the forward voltage drop of an l.e.d. has a negative temperature coefficient that approximates closely to that of a silicon transistor V_{be} drop. Hence, this method provides exceptionally stable d.c. conditions over a very wide temperature range.

After the normalization stage the signal is applied to a tone-control circuit based on the Baxandall network. The main limitation of the Baxandall system

is that the turnover frequency of the treble control is fixed. In contrast, the bass control has a turnover frequency that decreases as the control nears the flat position. This allows a small amount of boost at the low end of the audio spectrum to correct for transducer shortcomings. The equivalent adjustment at the high end of the treble spectrum is not possible because boost occurs fairly uniformly above the turnover frequency for treble control settings close to flat. In this circuit the treble turnover frequency has been given three switched values which have proved useful in practice. Switch 2 selects the capacitors that determine the turnover point. The maximum boost/cut curves are arranged to shelve gently, in line with current commercial practice, rather than to continue rising or falling outside the audio range. In addition, the coupling capacitor C_4 has a significant impedance at 10 Hz so that the maximum bass boost curve not only shelves but begins to fall. Full boost gives +15 dB at 30 Hz but only +8 dB at 10 Hz. The tone control system has a maximum effect of ±14 dB at 50 Hz and ±12.5 dB at 10 kHz.

The tone-control amplifier uses the same low distortion configuration as the normalization stage, but it is used in a virtual-earth mode. The main difference is that the open-loop gain has been traded for open-loop linearity by increasing the emitter resistor of the main voltage amplifier from 1 to 10 kΩ thus increasing local feedback. Resistor R_4 has been increased to 5.6 k to maintain appropriate d.c. conditions. This modification makes it much easier to compensate for stability in the unity-gain condition that occurs when treble-cut is applied.

Level detection circuitry

From the tone-control section the signal is fed to the final volume control via the muting reed-relay. Note that this arrangement allows the volume control to load the input of the external power amplifier even when the relay contacts are open, thus minimising noise. The signal level leaving the tone-control stage is comprehensively monitored by the circuitry shown in Figure 1.4. Each channel is provided with two peak-detection systems, one lights a green l.e.d. for a pre-determined period if the signal level exceeds 1 V peak, and the other lights a red l.e.d. if the tone-control stage is on the verge of clipping. Each channel is also provided with a VU meter driver circuit. Transistor Tr_{22} forms a simple amplifying stage which also acts as a buffer. Voltage feedback is used to ensure a low-impedance drive for the meter circuitry. The first peak detector is formed by IC_1 and

its associated components. When the voltage at pin 2 goes negative of its quiescent level by 1 V, the timer is triggered and the l.e.d. turns on for a defined time. The relatively heavy l.e.d. current is drawn from an unstabilized supply to avoid inducing transients into any of the stabilized supplies.

The clipping detector continuously monitors the difference in voltage between the tone-control amplifier output and both supply rails. If the instantaneous voltage approaches either rail, this information is held in a peak-storage system. Normally Tr_{24} and Tr_{25} conduct continuously but if the junction of D_1 and R_5 approaches the +24 V rail then Tr_{24} and hence Tr_{25} turn off. This allows C_5 to charge and turn on Tr_{26}, and Tr_{27} and hence the l.e.d. until the charge on C_5 has been drained off through emitter-follower Tr_{26}. If the measured voltage nears the—24 V rail, then D_1 conducts to pull up the junction of R_6 and R_7, which once again turns off Tr_{25}. In this way both positive and negative approaches to clipping are indicated. This comprehensive level indication does of course add significantly to the task of building and testing the preamplifier. If desired, any or all of the three sections may be omitted.

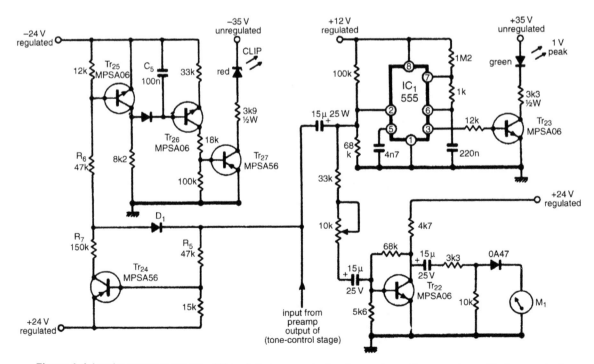

Figure 1.4 Level monitoring circuitry. Although three separate circuits are shown, these may be omitted as required.

Noise gate

The final section controls the muting reed-relay. At switch-on, the +12 V rail rises rapidly until stabilized by the zener diode. Pin 2 on IC_4 is, however, briefly held low by C_6, and the 555 is therefore immediately triggered to send pin 3 high. This saturates Tr_{28} which prevents Tr_{29} from turning on. At the end of the time delay, pin 3 goes low and relay driver Tr_{29} is no longer disabled (Figure 1.5).

The noise gate uses two amplifiers with gains of about 100. These sample both channels at the output of the normalization stage and the inputs are clamped with diodes so that the normalization amplifiers may use their full voltage swing capability without damaging the 741s. Due to their high gain, under normal signal conditions the op-amp outputs move continuously between positive and negative saturation which keeps the storage capacitor C_7 fully charged. In the silent passages between l.p. tracks the l.f. signal is not normally of sufficient amplitude to cause saturation

but will usually produce at least +3 to +4 V across C_7 which gives a large margin of safety against unwanted muting. To facilitate this the response of the amplifiers is deliberately extended below the audio band. When the stylus leaves the record surface and the l.f. signals cease, C_7 slowly discharges until the non-inverting input of comparator IC_3 falls below the voltage set on the inverting input. At this point the 741 switches and its output goes low to cut off the base drive to Tr_{29}, and switch off the relay. When the stylus is replaced on a record, the process takes place in reverse, the main difference being that C_7 charges at once due to the low forward impedance of D_2. To prevent the relay sporadically operating when the preamplifier is handling signals presented through the line inputs, an extra wafer on the source-select switch is arranged to override the rumble-sensing circuit, and provide permanent unmute. This is achieved by pulling the inverting input of comparator IC_3 negative of the +15 V rail by the 10 kΩ resistor so that even when C_7 is fully discharged,

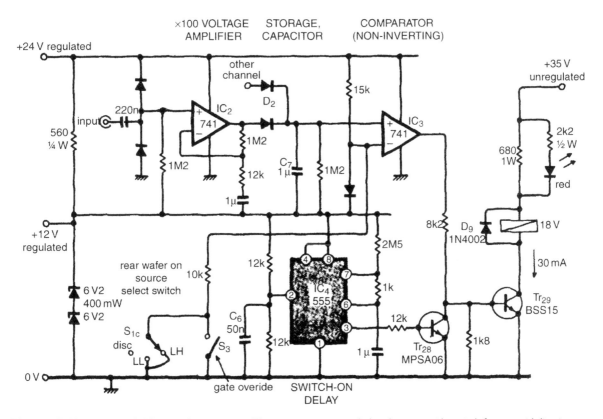

Figure 1.5 Noise gate and delay switch on circuitry. The noise gate is provided with an override switch for use with line input signals. The delay switch-on overrides all of the circuitry. Amplifier IC2 is repeated for a stereo system.

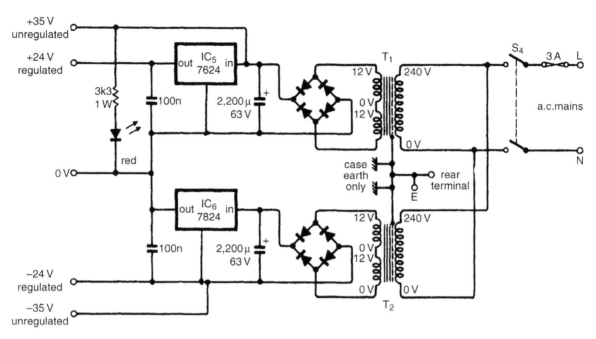

Figure 1.6 Power supply. Two regulator integrated circuits are used which should be mounted on heat sinks.

IC_3 will not switch. In addition, S_3 provides a manual override for testing and comparison purposes.

The power supply is shown in Figure 1.6. Regulators are used to provide stabilized ±24 V rails. The unregulated supply rests at about ±35 V. The signal circuitry has been designed to withstand ±35 V appearing on the supply rails, so that even in the unlikely event of both regulators failing, no further destruction will arise. Each regulator IC requires about 7 cm² of heat sink area.

Physical layout of the preamplifier is no more critical than that of any other piece of audio equipment. In general it is wise to use a layout that places the disc input amplifier as close as possible to its input socket, and as far as possible from the mains transformer. Screened cable should be used between the disc input stage and its input socket, and between the final volume control and the output socket. The earthing requirements are straightforward and the circuit common 0 V rail is led from the input sockets through the signal path to the output volume control, and finally to the 0 V terminal of the power supply. This arrangement minimises the possibility of spurious e.m.fs arising between stages. The only problem likely to be encountered is the formation of an earth loop when the preamplifier is connected to a power amplifier. Therefore, it may be satisfactory in a permanent installation to have the preamplifier circuitry connected to mains earth only through the signal lead to the power amplifier. The preamplifier case must of course be connected to the mains earth for safety reasons. It is preferable to define the potential of the preamplifier even if the power amplifier is disconnected. In the prototype the 0 V rail was connected to the mains earth via a 22 Ω resistor which stops the formation of an earth loop and prevents the signal circuitry from taking up a potential above earth due to leakage currents etc. Testing is relatively straightforward, providing the preamplifier is constructed and checked stage by stage. Dynamic parameters such as THD are not accurately measurable without expensive test gear, but it has been found in the course of experimentation that if the d.c. conditions are correct then the various signal stages almost always show the desired a.c. performance. The non-signal circuitry should be relatively simple to fault-find. No problems should be encountered with the noise gate section which has proved to be very reliable throughout a protracted period of testing. The only preamplifier adjustment is for the VU meter calibration. This should be set to IV r.m.s. = 0VU, which is completely non-standard but very useful in terms of the dynamic range of the signal path. For normal operation the input gain controls should be set so that the meter indications do not exceed 0VU, to preserve a safety margin in the later stages. This completes the preamplifier design.

Component notes

All unmarked diodes are 1N914 or equivalent.

Red bias l.e.ds are TIL209 or equivalent.

Green bias l.e.ds are TIL211 or equivalent.

Resistors marked with an asterisk should be metal oxide types

Tr1 to Tr6 and Tr13 to Tr15 are BCY71

Tr7,8,9,16,17,18,22,23,25,26,28 are MPS A06.

Tr10,11,12,19,20,21,24,27 are MPS A56

Tr9 is BFX85 or equivalent.

The muting reed relay should be a two pole make type with an 18V coil. If a different coil voltage is used,

---

the value of the dropper resistor should be adjusted.

The VU meter should have a 1 mA movement.

If an internal diode and series resistor are fitted, the external components should be omitted.

Switch 1 (source select) is a five pole 3 way.

Switch 2 (treble frequency) is a four pole 3 way.

References

1. Walker, H.P. 'Low-noise Audio Amplifiers', *Wireless World*, May 1972.

2. King, Gordon J. *The Audio Handbook*, Newnes-Butterworth, 1975.

3. Heidenstrom, P.N. 'Amplifier Overload', *Hi-Fi News*, December 1974.

Chapter 2

High-performance preamplifier

Low-cost design with active gain control
(Wireless World, February 1979)

Intended title: *An Economical High-Performance Preamplifier*

This was my first preamplifier design to be published that might be called 'conventional' in its use of discrete transistors. It was my own reaction to the relative complexity of the discrete opamp based Advanced Preamplifier just described; I set out to produce a preamplifier that was relatively conventional, in that its stages used only two or three transistors, to see just how good it could be. At that time, the available opamps were looked at with entirely justified suspicion; they were relatively noisy and prone to crossover distortion in their output stages. Crossover might be inescapable in a power amplifier, but it was definitely not a good thing to have in a preamp. Hence the use of discrete Class-A circuitry throughout. (The 5534 opamp was just becoming available at the time, but was ferociously expensive.)

The basic philosophy was the use of simple two or three-transistor stages, enhanced with current-source outputs when required, running from a single and rather high-voltage supply rail to increase headroom and reduce distortion at a given signal level. For simple transistor stages, increasing the supply rail, with suitable adjustment to the DC conditions, is a guaranteed and virtually foolproof way to reduce distortion. It has its limits, as power dissipation is naturally increased, and turn-on transients may become large enough to cause damage further along the signal path.

A single supply rail of +38V was used for simplicity, allowing a maximum signal amplitude of 13 Vrms with appropriate biasing. It was without regulation but with an extra RC filter to reduce ripple to about 5 mV peak-to-peak. This minimal-cost arrangement gave hum and noise figures as good as those from the dual-rail, IC regulator method. Where this approach fell down a bit was that the absence of DC regulation made the audio analyser residual signal heave up and down like a rough sea when the rather low noise and distortion were being measured, making accurate measurements difficult, even when a 400 Hz high-pass filter was used in the testgear.

The single rail saved a little on power-supply parts, but a more important consideration is that the simple transistor stages used were better suited to single-rail operation, because both the volume-control-stage and the tone-control had the lower rail (ground) as a signal and biasing reference. You will note a certain concern for the number of transistors used to accomplish a given function. At the time, discrete transistors were significantly more expensive than their accompanying resistors and capacitors, and this is reflected in some of the design decisions taken.

Note the transistor equivalent of the White cathode-follower at the disc-stage output, giving push-pull Class-A operation, doubling the peak output current available for the same quiescent. Its beautiful simplicity depends on having a reasonably low rail ripple so its operation is not disrupted; it worked nicely with the filtered supply rail used here. The RIAA accuracy is unimpressive by modern standards, having maximum errors of −0.6 dB at 20 Hz, and +0.6 dB at 20 kHz. The latter error is not due to the absence of an HF correction pole that is present in the form of R17, C11, but is largely due to resistors being restricted to E12 values in those days. The input transistor TR1 is a BCY71 simply because I had lots of them.

The subsonic filter is a single-stage third-order design based on the constant-current emitter-follower TR6, TR5. If you are wondering what R21 does, it is a 'base-stopper' that prevents the emitter-follower from going into oscillation in the VHF band if it sees a whiff of source inductance. This is a bit of a high value to stick right in the signal path, and I wonder what the effects were of its Johnson noise, and the current noise of TR6 flowing through it. Nowadays I would aim for a more elegant solution.

This was the first preamplifier I designed with an active-volume-control, implemented by using a log pot as a variable resistor in a shunt-feedback stage. This is definitely a rough edge in the design, because channel balance depends on the track resistance as well as the uncertainties of a dual-slope log law. If the tolerance of the track resistance is ±20%, you could theoretically have a channel balance error of 4 dB with the volume fully up, though statistically that is extremely unlikely. Peter Baxandall did not publish his most ingenious active-volume-control until the end of 1980;[1] it cancels out almost all pot inaccuracies and I never used the simple-minded approach shown here again in a published design. The active-volume stage gives a phase-inversion that cancels out the inversion in the tone-control. This was therefore my first preamplifier design that preserved absolute phase; in other words the output was in phase with the input.

The line-inputs feed the volume-control stage through series resistors that define the gain of each input in conjunction with shunt feedback through the volume control. I thought this was highly convenient as it meant there was no need to switch the feedback of the volume stage to obtain different gains for different line inputs. Thus the output of the MM stage (50 mVrms) is applied through a 10 kΩ resistor which with the 100 kΩ pot track resistance gives a maximum gain of ten times and an output of 500 mVrms. The 100 mV Line Input 1 is applied through two 10 kΩ resistors, R27 and R29, so the maximum gain is five times. All very neat, but inflexible and unsatisfactory when examined more closely. If the minimum acceptable input impedance is taken as 10 kΩ, which is a good general rule, the volume control track has to be 100 kΩ to get a gain of ten times. For unity gain, the input resistor also has to be 100 kΩ, and the active volume stage is working at a high impedance and will be much noisier than if unity gain was achieved by using a 10 kΩ input resistor and pot track resistance. Noise measurements are sparsely distributed in my 1977 notes, but I find that the active volume stage (set to six times gain) and the tone-control (set flat) measured together gave a noise output on the Left channel of −79. 8 dBu and the Right channel of −77.5 dBu, measured in a 80 kHz bandwidth because that was the only option on the Sound Technology 1700A. Assuming it was white noise, that converts to −85.4 dBu and −83.1 dBu in a 22 kHz bandwidth. This is definitely noisy by modern standards; (the Elektor 2012 preamp has a noise output of −115 dBu at middling volume settings), and I am very sure that most of the noise here is coming from the active volume stage. The fact that there is a 2.3 dB difference between the two channels is also rather worrying.

The volume-control stage has an interesting feature which has never been discussed before. The voltage-gain stage TR7, TR8 is a cascode configuration and has what I called a current-injection resistor R33. This allows the gain transistor TR7 to run at a higher current than the cascode transistor TR8. This idea was taken from an article called 'Transistor Wide-Band Cascade Amplifiers' by F Butler in *Wireless World*.[2] The resistor is shown in a circuit diagram but not mentioned in Butler's text; I thought it looked interesting and soon discovered that it much improved the distortion performance of the stage. My explanation was (and is, because I haven't looked into the matter again since) that running TR7 at a higher I_c increased its transconductance, and because of the fixed voltage on the emitter of TR8, the current through the injection resistor would be constant. This meant that the varying part of I_c all went through TR8 to the relatively high-impedance collector load. If memory serves, this article also prompted me to introduce the transistor version of the White cathode-follower mentioned above; see the description of the MRP2 preamplifier below.

The distortion performance of the active-volume-control stage was a major issue because it had to give a closed-loop gain of up to ten times, at which the negative feedback was much reduced. Using a standard two-transistor stage like the tone-control of the MRP4 gave 0.037% THD at 1 kHz and 0.30% at 10 kHz, both at 8 Vrms out. That really wasn't good enough, especially remembering the much better performance of the MRP1, but I did not want to go to the complexity of a full discrete opamp stage. The three-transistor stage with current injection gave 0.0046% at 1 kHz and 0.0088% at 10 kHz, (both 8 Vrms out) and was a very acceptable compromise. Without the current injection the THD was 0.011% at 1 kHz at 8 Vrms out, more than twice as much.

The tone-control stage had a less arduous task, only having to give a gain of three times at maximum, and the straightforward two-transistor configuration with bootstrapped collector load proved quite adequate. It was probably a mistake to put the tone-control stage last in the signal path, after the active volume-control stage, because you get the full noise of the tone-control stage at the output even at zero volume. The balance control was supposed to be implemented using coaxial volume control pots. It was a cop-out, and I admit it.

While I set out to make this design relatively conventional, it was my version of conventional. With its push-pull emitter-followers, current-injected cascode, and active-volume-control, it turned out to be not so very conventional after all. It aroused much interest and copious correspondence. The

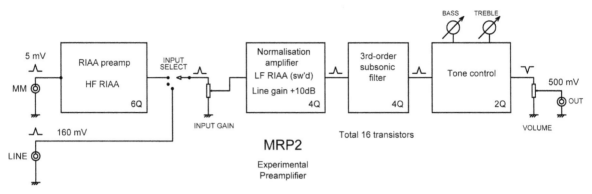

Preface Figure 2.1 The block diagram of the MRP2: 1978. The Q figures show the number of transistors in each stage.

noise performance of discrete preamplifiers like this is in theory better than that of a design based on opamps because in each stage the input element that determines the noise performance is a single discrete transistor, rather than a differential pair of integrated transistors, which could be up to 3 dB noisier, and possibly worse if you made the negative feedback network of too high an impedance.

This was preamplifier MRP4 in my numbering system. A PCB carrying two channels was available from *Wireless World* for £5.00.

You are probably wondering what happened to the MRP2 and MRP3. The MRP2 was an experimental design that proved the concept of using relatively simple 'conventional' stages with two or three transistors, and a single supply-rail, though it used a +35V regulated supply rather than the simple rail filtering used in the MRP4. BC182 (NPN) and BC212 (PNP) transistors were used throughout. The block diagram is shown in Preface Figure 2.1. The use of separate input gain and volume controls was continued, so the level in the signal path could be optimised, though some sort of level indication would have been necessary to do this effectively; at this time I would have been considering an LED bar-graph meter, probably based on the SN16880 stereo bar-graph IC.

Concern about absolute phase still lay in the future then, and the phase-inversion produced by the tone-control was ignored. In the absence of a balanced input or output, there is no way to correct the phase without adding another active stage that would contribute its own share of noise and distortion.

As in the MRP1, the RIAA equalisation was performed in two goes, with the LF part of the RIAA being switched out of the feedback network of the normalisation amplifier when a line input was in use. The HF roll-off at 2.1 kHz was done in the first stage, while the feedback around the normalisation amplifier provided LF boost starting at 500 Hz and levelling off at 50 Hz. The original Moving Magnet (MM) stage used a standard Walker three-transistor configuration [3] without the addition of collector bootstrapping to the second transistor. During the development process this was replaced by a six-transistor circuit which had an input stage, a cascoded gain stage, and that followed by a current-source emitter follower. This was the first MM input stage I designed with a this feature; since it is the most sophisticated version of the

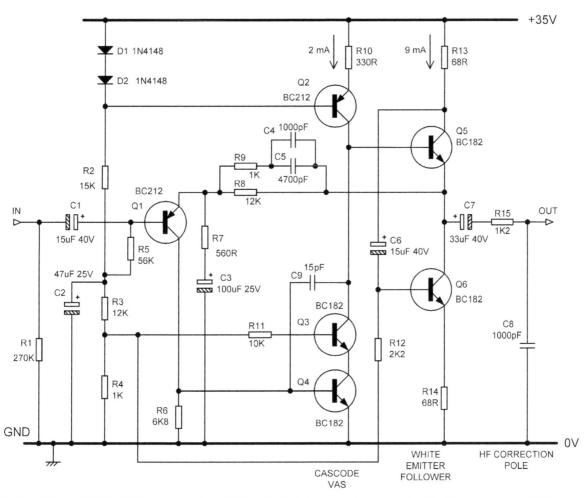

Preface Figure 2.2 The MM input stage of the MRP2, with single input transistor, cascade VAS with current-source, White push-pull output stage, and HF correction pole

circuit I have used (the MRP4 has a bootstrapped VAS load rather than a current-source), it is reproduced in Preface Figure 2.2.

The configuration is essentially that of the conventional solid-state power amplifiers of the time, which tended to use a single input transistor to perform the feedback subtraction, the great advantages of the differential pair for low distortion as well as DC precision not being much appreciated at this point. Here a single input transistor was used to minimise noise, as two transistors must be noisier than one, though in this case not 3 dB worse. The input transistor has its collector current defined at about 88 uA by the 0.6V established across R6; this apparently low value gives a better noise performance than higher currents with the highly inductive source resistance of an MM cartridge. Q1 passes its output current to the Voltage Amplifier Stage, or VAS, and its transconductance combined with the value of the dominant-pole Miller capacitor C9 sets the open-loop gain at high frequencies.

The main sources of distortion in the VAS are Early effect and the non-linear variation of Cbc with Vce. A cascode VAS reduces both effectively, though not as well as putting an emitter-follower inside the Miller loop would have done. (We live and learn.) At this distance of time, I have no idea why R11 was set as high as 10 kΩ; it seems unlikely it was a stability issue. In testing the current-source emitter follower still showed premature negative clipping at high frequencies, due to the heavy loading of the HF RIAA capacitor in the feedback network, and also loading by the HF correction pole; this was pretty much eliminated by converting it to a push-pull Class-A White output structure as in Preface Figure 2.2. Everything is biased from a single divider D1-D2-R2-R3-R4, which is heavily filtered by C2 to remove supply-rail noise. The voltage across R4 is 1.2V, and biases both the cascode transistor Q3 and the push-pull output current-source Q6; why I used a resistor there rather than another pair of diodes, I have no idea. So far as I'm concerned, this configuration may have six transistors, but it is not a 'discrete opamp' because it does not have a differential input. However, unlike the other discrete transistor configurations I have used, for this one there should be no difficulty in converting it to dual-rail operation, because the biasing system is not referenced to the bottom rail. Because a single input transistor has not the DC precision of a differential pair, and there is a voltage drop across feedback resistor R8, there will be a significant offset voltage at the output, which will need to be DC blocked.

The input impedance is set by the parallel combination of R1 and R5, which is 46.4 kΩ. This is a little low compared with the nominal value of 47 kΩ, representing an error of 1.3%, but it is the best you can do with two E12 resistor values; bear in mind that R5 should not be too big (say 100 kΩ or less) as it carries the base current for Q1. At the time of the design E24 resistors were relatively rare and expensive, and no consideration was given to using them to get quantities like this more accurate. Nowadays R1 = 300 kΩ and R5 = 56 kΩ would give an input impedance of 47.2 kΩ, an error of only 0.4%, and even greater accuracy could be achieved if required by using three E24 resistors in parallel. For example R1 = 560 kΩ in parallel with 620 kΩ, and R5 = 56 kΩ, reduces the error in nominal value to 0.10% high, which is likely to be much less than the resistor tolerances. There are many combinations of three resistors that give a combined value very close to 47 kΩ, and adding the extra constraint that they should be as nearly equal as possible allows significant improvements in accuracy when tolerances are taken into account, as random errors partially cancel; this is explained in detail in Chapter 19. The three resistors just mentioned of 560 kΩ, 620 kΩ, and 56 kΩ show this effect to some extent; if 1% resistors are used the tolerance of the combination is improved to 0.85%. A better choice would be R1 = 160 kΩ in parallel with 200 kΩ, and R5 = 100 kΩ, which has a nominal value only 0.12% high, and a combined tolerance of 0.60%, significantly better.

A third-order subsonic filter was fitted as standard and worked on the line inputs as well as the MM input. The normalisation amplifier was a three-transistor Walker configuration with my addition of bootstrapping to the second transistor collector load—what I called a BootWalker. The volume control was again right at the end of the signal path, so noise from all the previous stages would be effectively attenuated at low volume levels. The phase-inversion produced by the tone-control was ignored, so absolute phase was not preserved. The design was never taken as far as a fully built, nicely boxed project, but it did lay out most of the groundwork for the MRP4. It was never published, but I have all the original notes, so it could be if there was enough interest.

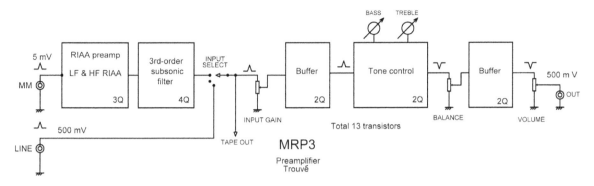

Preface Figure 2.3 The block diagram of the MRP3: 1976. The Q figures show the number of transistors in each stage.

The MRP3 was what you might call 'preamp trouvé'; at the time I was designing for Electrosonic, and part of my work was a series of audio modules which could be connected together to build almost any sort of custom audio system. One module was an MM input stage plus a third-order Butterworth subsonic filter, and another was a tone-control stage with balance-control facilities and an output buffer. You don't have to be a genius to realise that connecting these two modules together with a volume control gives you Instant Hi-fi Preamplifier. The block diagram is shown in Preface Figure 2.3. Separate input gain and volume controls were still retained.

The MM input stage was designed in July 1976 and was basically the Walker configuration, without bootstrapping on the second transistor, rather than the more complex stage used in the MRP2. The tone-control module was designed in November 1976, and was a standard two-transistor stage with collector bootstrapping. The balance control used a pull-up resistor (not shown) to minimise the attenuation when it was set to central. There were also two unity-gain buffers in the shape of emitter-followers with current-source emitter loads; this was advanced technology for its day. The volume control was once more at the very end of the signal path, so low volume settings would attenuate the noise from all the previous stages. Again, the phase-inversion done by the tone-control was ignored and absolute phase not maintained.

The supply rail was +24V regulated, a standard established long before I joined the company; a higher voltage like +30V or +35V would have given more headroom, though possibly with a slight cost increase on some of the electrolytic capacitors. The +24V rail limited the maximum signal level to about 8 Vrms, the lowest figure for any of the preamplifiers described in this book.

References

1. Baxandall, P., Audio Gain Controls, *Wireless World,* Nov. 1980, p. 80.

2. Butler, F., Transistor Wide-Band Cascade Amplifiers, *Wireless World,* March 1965, p. 124.

3. Walker, H.P., Low-noise Audio Amplifiers, *Wireless World,* May 1972, p. 236.

HIGH-PERFORMANCE PREAMPLIFIER

February 1979

Some years ago Doug Self described a no-compromise preamplifier which was designed using high voltage transistors to give exceptional performance. This new design sacrifices very little of that performance and uses a small number of low-cost transistors to significantly reduce the cost. A novel active gain control makes best use of the dynamic range and removes the problem of volume control placement.

This preamplifier offers a similar performance to that of the advanced preamplifier published previously, [1] but with a simpler design that reduces the parts count and hence cost. In normal use, the signal levels are kept around 50 mV by exchanging the normal potentiometer volume control, which acts as an attenuation control, for an active gain control. Therefore, the signal receives only the amplification required for a given output and so makes best use of the amplifier's dynamic range.

The distortion performance is also improved because unwanted gain will be used to give higher negative feedback and thus greater linearity. The active gain control uses a shunt feedback circuit where the volume control varies the resistance of a feedback arm as shown in Figure 2.1. The disc input stage has a relatively low gain of +20 dB at 1 kHz which allows a very high input overload margin. This is followed by a third-order high-pass filter which removes subsonic signals while they are still at a low level.

Both bass boost and treble cut portions of the RIAA equalisation take place in the first stage. The gain control stage is positioned after the input switching and has a maximum voltage gain of +20 dB. This is followed by a Baxandall tone control which has unity gain at 1 kHz.

The use of an active gain control eliminates the problems associated with a normal volume control. If all of the gain is placed before the control, the supply voltage limits the overload margin. If some gain occurs after the volume control, then the signal-to-noise ratio is degraded because noise generated in the later stages does not undergo attenuation. The use of two controls, one early and one late in the signal chain, is one method of avoiding this compromise [1] but a true gain control is considered to be a more elegant solution.

Because a low-cost, single unregulated power supply is used with first-order RC smoothing to reduce ripple, all sections of the preamplifier are designed with high ripple-rejection performance.

Disc input stage

The most difficult stage to design in a preamplifier is the disc input, and the problems are compounded if, as in this case, the gain of the stage is low to allow a high overload margin. A low voltage gain at 1 kHz means that the feedback network which defines the gain and RIAA equalisation will have a relatively low impedance, and

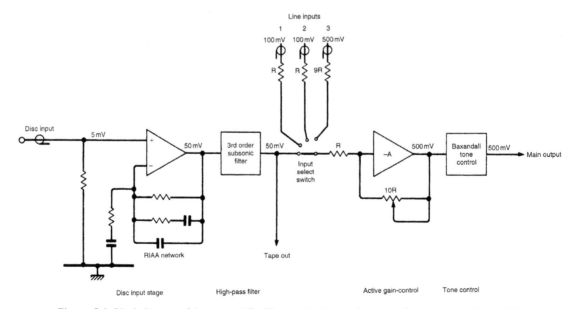

Figure 2.1 Block diagram of the preamplifier. The signal voltages shown are for maximum gain at 1 kHz.

thus appear as a heavy load to be driven by the disc amplifier. This situation becomes worse at higher frequencies when the reactance of the equalisation components falls. Therefore, as a large voltage swing at the output is desirable, a large amount of current must be able to flow into and out of the feedback network at high frequencies. A second, and related problem, is that if the gain at 1 kHz is low, the gain at 20 kHz must be 19.3 dB lower due to the RIAA equalisation, which makes it close to unity. Therefore, it becomes more difficult to set the top end of the RIAA curve accurately. For this reason an extra low-pass section, with a -3 dB frequency of about 22 kHz, is added after the disc amplifier to ensure that the high-frequency gain continues to fall at a steady rate. It should be noted that if the correct turn-over frequency is chosen for the final low-pass network, the RIAA amplitude and phase curves are obtained exactly.

Preamplifier specification
Input sensitivity
for 500 mV output
Disc 5 mV at 47 kΩ
Line I 100 mV at 20 kΩ
Line 2 100 mV at 20 kΩ
Line 3 500 mV at 100 kΩ

Disc input overload level
1.1 V r.m.s. at 1 kHz
3.8 V r.m.s. at 10 kHz

Outputs
Main output
500 mV r.m.s. at 100 Ω source impedance
Tape output
50 mV r.m.s. at 1 kΩ source impedance
Maximum possible level from main output
8.5 V r.m.s

Frequency response
Disc input (RIAA equalisation)
±1.0 dB 20 Hz to 20 kHz
±0.5 dB 100 Hz to 20 kHz
Line inputs (flat)
+0, to −0.5 dB 20 Hz to 20 kHz

Total harmonic distortion
From disc input to main output, at 1 kHz with the gain control set to × 6
less than .008% at 8 V r.m.s.
less than .005% at 5 V r.m.s.
Because the output signal level will normally be around 500 mV the THD level will be much lower.

Noise
Disc input better than 68 dB below 5 mV r.m.s.
Line inputs below −75 dBm at full gain
Residual below −90 dBm at zero gain
Tone controls
Bass ±14 dB at 50 Hz
Treble ±10 dB at 10 kHz

Power consumption
Approx. 78 mA each channel from a +38 V supply.

Another consequence of the fall in closed-loop gain at high frequencies is that the compensation for Nyquist stability is more difficult, and in this design it was necessary to add a conventional RC step-network to the normal dominant-pole compensation. The dominant-pole capacitor is kept as small as possible to preserve the slew-rate capability of the stage.

The basic disc input stage is shown in Figure 2.2. In this series-feedback configuration almost all of the voltage gain is provided by the second transistor, which has a bootstrapped collector load for high open-loop gain and linearity. The final transistor is an emitter-follower for unitygain voltage buffering. This configuration allows the use of a p-n-p input transistor for optimum noise performance, but it also means that the collector current must flow through the feedback resistance R_F. This places another constraint on the design of the feedback network because an excessive voltage drop must be avoided.

As the disc input amplifier must be capable of sourcing or sinking large peak values of current to drive the capacitive feedback loop at high frequencies, the conventional emitter-following output circuit in Figure 2.2 is not suitable because the sink current causes a voltage drop in R_E. Lowering the value of R_E reduces the effect, but this is a poor solution as it leads to a high quiescent power dissipation. Replacing R_E with a constant-current source is more effective because the maximum sink current becomes equal to the standing current of the stage. However, this would still limit the output of the disc stage at high audio frequencies due to an inability to sink sufficient current. For this reason, the push-pull class A configuration in Figure 2.3 was chosen. The bottom transistor is a current-source which is modulated in anti-phase to the top emitter-follower, via the current-sensing resistor R_A and a capacitor. This can also be considered as a negative-feedback loop that attempts to keep the current in R_A constant. However, the open-loop gain is only unity and so with 100% negative feedback the current variations in the top transistor are reduced to one half by the capacitor. Due to the anti-phase drive of the

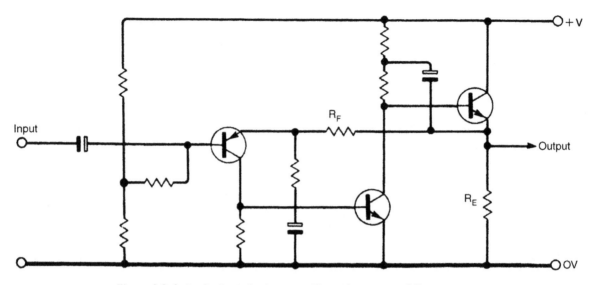

Figure 2.2 Series-feedback disc input amplifier with an emitter-follower output.

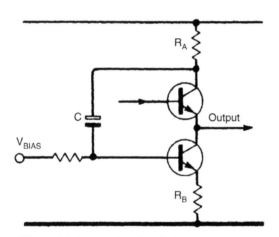

Figure 2.3 Improved push–pull class A output stage.

lower transistor, this stage can sink a peak current of twice the standing current, and therefore give twice the output swing at high frequencies.

A practical circuit of the disc input amplifier and its associated subsonic filter is shown in Figure 2.4. All of the d.c. bias voltages are provided by the potential divider R_2, R_3, R_4, D_1 and D_2. This chain is heavily decoupled by C_2 to prevent supply-rail ripple entering this sensitive part of the circuit. Note that Tr_3 and Tr_5 are isolated from the bias voltage by R_{14} and R_{22} to simplify any fault-finding.

The RIAA equalisation is provided by R_{10}, R_{11}, C_6 and C_7 in the feedback loop, and R_7, C_4 forms a step network that aids h.f. stability. Resistor R_{17} and C_{11} make

up the low-pass section that corrects the top octave of the RIAA curve. The subsonic filter is a 3-pole Butterworth type with an ultimate slope of 18 dB/octave. Although the frequency response shows a loss of only 1.5 dB at 20 Hz, the attenuation is increased to more than 14 dB at 10 Hz. The unity-voltage gain element of the filter is formed by Tr_5 and Tr_6 arranged as an emitter-follower with a current-source load. This configuration was chosen for its excellent linearity. An output of about 50 mV is available for tape recording although the exact voltage will depend on the cartridge sensitivity. Resistor R_{24} prevents damage to Tr_6 if the tape output is shorted to earth, and resistor R_{25} maintains the output of the disc stage at 0 V d.c., and also prevents switching clicks.

The total harmonic distortion from input to tape output at 6 V r.m.s. is below 0.004% from 1 to 10 kHz but because the anticipated signal level here from most cartridges is about 50 mV r.m.s. The distortion during use will be even lower. The disc input will accept more than 1 V r.m.s. at 1 kHz, and about 3.8 V r.m.s. at 10 kHz before overloading. It is felt that the improvement these figures show over conventional methods justifies the complication of a low-gain disc input stage. The accuracy of the RIAA equalisation depends on how closely the RC time-constants can be set. If 5% components are used the deviation should be less than ±0.5 dB from 1 to 15 kHz, and within ±1 dB from 20 Hz to 20 kHz. The signal-to-noise ratio for a 5 mV r.m.s. input at 1 kHz is better than 68 dB.

The remaining part of the preamplifier comprises an active gain-control and the tone-control stage. The input

(a)

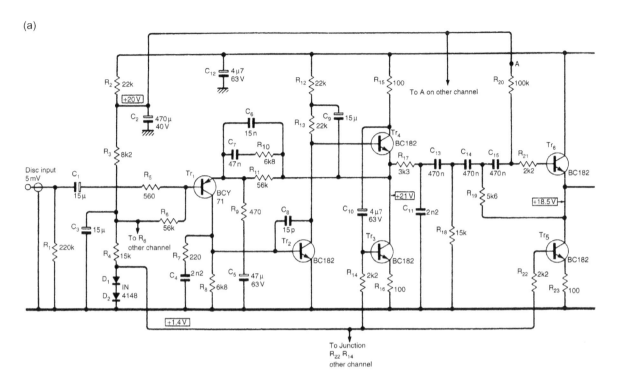

(b)

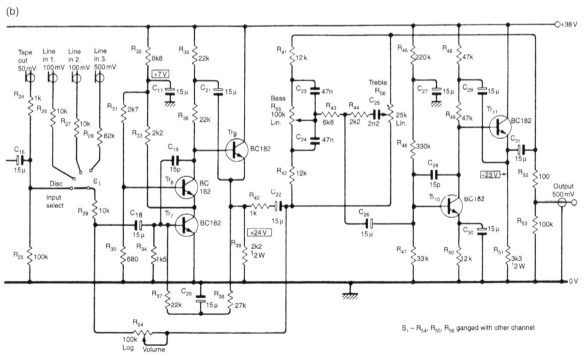

Figure 2.4 One complete channel of the preamplifier. Components in the bias suppy, R2, R3, R4, C2, C3, D1, D2 and also C12 are not repeated in channel 2. All electrolytic capacitors are rated at 40 V and all resistors are ¼ W unless otherwise stated.

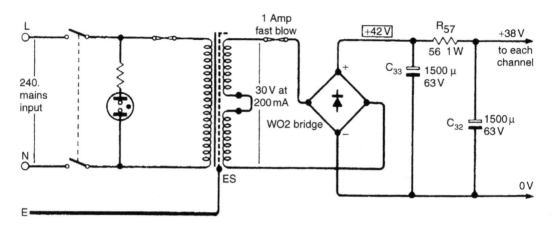

Figure 2.5 Power supply. If a higher voltage transformer is used, R57 should be increased accordingly.

switching is simple and requires only one switch section per channel. Also, any line input of suitable sensitivity can be used as a tape monitor return.

The shunt-feedback configuration of the active gain control enables each line input to have its sensitivity defined by the value of a single series input resistor. The maximum voltage gain available from the stage is the ratio of the feedback resistance to the input resistance, and is +20 dB when the volume control is at maximum resistance. This gain is only used in the disc mode. The most sensitive line input is rated at 100 mV for a 500 mV output and the least sensitive input has unity gain. Any sensitivity between these two limits may be provided by using the appropriate series resistor value.

The gain control comprises Tr_7 and Tr_8 arranged as a cascode voltage amplifier with a bootstrapped collector load and Tr_9 as a conventional emitter-follower. The d.c. conditions are set by negative feedback through R_{37} and R_{38}, and a.c. feedback is applied through the volume control. The linearity of this circuit is increased by a current injected into Tr_7 through R_{33}. The voltage at the top of R_{33} and the potential divider R_{30}, R_{31}, is smoothed by R_{32} and C_{17}. Resistor R_{40} prevents high-frequency instability when the volume control is set to zero gain.

The tone-control is a conventional Baxandall circuit, with Tr_{10} providing a high voltage-gain by its bootstrapped collector load. Transistor Tr_{11} is another emitter-follower which buffers the high impedance at the collector of Tr_{10}. The output is taken through R_{52}, which protects the output against short-circuits. Because the output impedance is low, long cables may be used without loss of high frequencies. The power supply is shown in Figure 2.5.

Construction

Normal precautions should be taken to keep a.c. power away from the disc input stage, and to avoid earth loops. The leads to R_{54} should be kept short to prevent hum pick-up on the virtual-earth point of the gain control. Typical voltages for various parts of the circuit are shown in Figure 2.4. These measurements should be made with a 20 kΩ/V meter.

Several modifications can be made to the preamplifier to suit individual requirements. Firstly, the treble turnover frequency of the tone-control section can be increased from 2 kHz as shown in Figure 2.4, to 5 kHz, for example, by reducing C_{25} to 1000 pF. For variable turnover frequencies C_{25} can be made switchable. Some purists may feel that the provision of a tone-control is unnecessary, and even undesirable. In this case, the output should be taken from the junction of C_{22} and R_{54}, but R_{52} and R_{53} should be retained at the output. Because the current drawn by the preamplifier will now be less, it is advisable to raise the value of R_{57} to keep the supply rail at +38 V.

In the circuit of Figure 2.4, no balance control is included. This function was performed in the prototype by a dual-concentric volume control. If, however, a conventional balance network is required this can be added at the output of the preamplifier although the low output impedance will be sacrificed.

Reference

1. Self, D. 'Advanced preamplifier', *Wireless World*, November 1976, p 41.

Chapter 3

Precision preamplifier

(Wireless World, October 1983)

By the time the Precision Preamplifier of 1983 was conceived, the remarkable 5534/5532 opamps were available at a reasonable price, and delivered very low noise with then almost unmeasurable distortion, given a few simple precautions. It became clear that even if ultimate performance was the goal, it was no longer economical or sensible to assemble eight or more transistors into a home-made discrete opamp. The adoption of opamps meant a return to dual regulated power supplies, but time and progress had made this option less costly than it had been when the Advanced Preamplifier was designed. The supply rails were ±15V rather than the ±17V, which I later made universal for powering 5534/5532 opamps. This allowed maximum signal levels of about 9 Vrms, some 5 dB less than the Advanced Preamplifier described in Chapter 1, with its ±24V rails. It was the last preamp I designed using the Sound Technology 1700A measurement system.

I put a lot of effort into producing the best design I could at the time, taking the 'precision' bit very seriously. I found that getting the RIAA equalisation accurate beyond a certain point by cut-and-try methods was virtually impossible, and very quickly exhausted my limited supplies of patience, so I fired up the BBC Model-B (computers only took a second or so to boot in those days—we have come a long way since then) and wrote some software using the Lipshitz equations to optimize the RIAA component selection. The easiest equations to work with were those for Configuration A, so that configuration was used every time I needed RIAA equalisation until quite recently; this is all explained in Chapter 19 of this book. I also evaluated dozens of different volume-control laws by plotting their design equations; this was a significant project as it was my first in which the computer part of the design process was really central. In bench testing, I did my best to minimise noise at every point. The preamplifier was very well received, so it was worth the effort. As with several of the designs described here, I still use the prototype on a regular basis.

There are some salient technical points. Note the HF correction pole R8, C11 which is required to keep the RIAA equalisation accurate at HF because the midband gain is only +26 dB (1kHz). Nowadays I use +30dB (1kHz), so the HF correction pole has less work to do. The high input-impedance of the bootstrapped buffer A3 prevents operation of the tape-monitor switch from causing small step-changes in signal level and gives a low-impedance drive to the Baxandall tone-control, which is now placed before the volume control, so its noise is attenuated at low volume settings. The tone-control uses the two-capacitor version of bass-control, because I felt at the time (and still do) that having turnover frequencies that increased with the amount of boost or cut made the approximate equalisation of loudspeakers more effective. The active-volume stage gives an inversion that cancels out the phase-inversion in the tone-control, and so absolute phase is now preserved from input to output.

The active balance control is integrated with the tone-control stage; it alters the proportion of the output signal fed back to the Baxandall network and so alters the stage gain. If this was done with a simple pot of reasonable value, its source impedance would be high at the central setting, interfering with the tone-control curves. The network R22, R23, R24 in Preface Figure 3.3 minimises this impedance variation, so it can be approximately compensated for by the addition of the fixed 1 kΩ resistor R17 at the other end of the Baxandall network. This network was designed by 'manual optimisation', that is, trial and error, as this was before the days of affordable simulators. Contemporary simulation shows that the gain of the stage with tone controls flat and balance central is +0.8 dB; with balance hard over it is +5.1 dB on one channel and 0.0 dB on the other, which confirms the range quoted in the article. While I confidently asserted at the time that this was enough balance range, now I'm not so sure. I think an 8 or 10 dB range would be safer, but that would make the design of the impedance-equalising network more difficult.

I have now assessed the variation of the tone control response with balance setting, by SPICE simulation. If we set the treble control to Mark-3 (i.e. with the treble wiper at the 2kΩ–8kΩ point on the track of the 10kΩ pot), then the treble flattens out at a cut of −4.0 dB with balance central. At either extreme balance setting, the amount of cut is increased to −4.3 dB, as the impedance of the control and its associated network are at a minimum and therefore there is more negative feedback and so more treble cut. This is demonstrated in Preface Figure 3.1. The traces have been shifted up and down to eliminate the changes in gain below 100 Hz caused by the balance control.

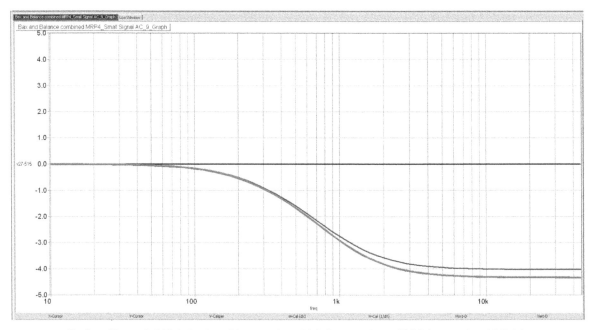

Preface Figure 3.1 Variation in treble cut action with balance settings of full left, central, and full right

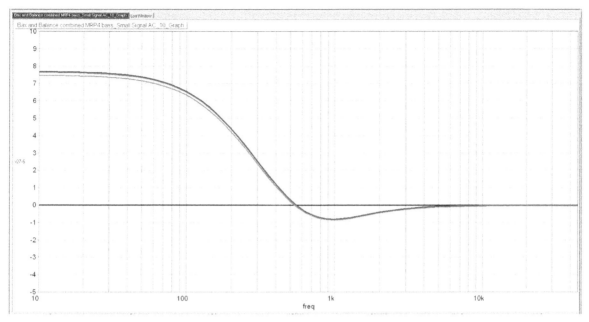

Preface Figure 3.2 Variation in bass boost action with balance settings of full left, central, and full right

Very similar results are obtained with +7.5 dB of bass boost, as shown in Preface Figure 3.2. Here the variation in boost with balance setting is only 0.2 dB. (The 'undershoot' of the response around 1 kHz is normal for the two-capacitor version of the Baxandall bass control.) I think this demonstrates that the impedance-equalising network around the balance control is actually doing a rather good job; a better one than I thought at the time. I can say with absolute certainty that no comments on this issue, negative or otherwise, were ever received.

If you're struck by the large number of 15 uF electrolytic capacitors in the circuit, even in places where they are unnecessarily large, the explanation is simple. Some time previously I had bought a big batch of 15uF 40V capacitors at a most advantageous price, and so I used them everywhere I could.

This was preamplifier MRP10 in my numbering system.

Once again, you are no doubt wondering about MRP5 to MRP9. Between the MRP4 High Performance Preamplifier of Chapter 2 and this, the MRP10, I designed several preamplifiers for experimental use or for consultancy clients, numbered MRP5 to MRP9, but the MRP10 was the next design to be published. For the MRP5 to MRP9, the same technology was used in each case; discrete transistors and a single filtered but unregulated supply rail. These are in no sense 'the lost preamplifiers of Douglas Self' as I have them all fully documented, and they could be properly written up and published if anyone was sufficiently enthusiastic. Now it can be told . . .

The MRP5 was based on the MRP4 described in Chapter 2; it was designed in March and April 1978, with an eye to selling the design to a well-known hi-fi manufacturer of the day; in the end this didn't happen. It too was a discrete transistor design with a single supply rail at +38V. The MM input stage was a four-transistor configuration very similar to that in the MRP4, with a bootstrapped voltage-amplifier-stage (VAS) and a push-pull White output emitter-follower. The major difference was that all the RIAA equalisation was done in this first stage. RIAA accuracy was within ±0.4 dB up to 10 kHz, but worsened to ±0.8 dB at 20 kHz. That was considered pretty good at the time; it was a year later that the famous *Lipshitz* paper with its RIAA design equations appeared [1], and really accurate equalisation became much easier to obtain.

The active-volume-control stage had an improved configuration so that the pot resistance value had much less effect in determining the gain. There was a separate stage after it configured as an active balance control, and there I have to admit that getting equal gain with the control central was to some extent pot resistance-dependent. This stage included a fairly elaborate mono/stereo switching system which properly summed the L and R channels on a virtual-earth basis. You may wonder why this facility was included at a time when music media were universally stereo; there were still old mono LPs about, but the real reason is that preamplifiers were just considered 'incomplete' without them at the time. They were rarely used. The next stage was a variable-slope filter scratch filter, which took a bit of thought, and lots of measuring and hand-drawn graphs, to bring to a successful conclusion. The design approach for this filter, which is still entirely valid, has now been published in my *Small Signal Audio Design* [2]. Finally, there was a conventional two-transistor Baxandall tone-control stage. The block diagram is shown in Preface Figure 3.3. Much of the MRP5 technology was used in the later MRP8.

You will note from Preface Figure 3.3 that this is the first of my preamplifier designs to have the tone-control at the end of the signal path with no volume control after it. Presumably that seemed like a good idea at the time, but it doesn't now. The full noise of the tone-control stage will always be running straight into the power amplifier; if it is a Baxandall tone-control working with shunt feedback, then the noise gain will be at least two times. In the structure of Preface Figure 3.3, all the noise from the active balance control and the scratch filter will also pass to the power amplifier at all volume settings, including zero, and I think I have to admit that, as regards preamplifier architecture, this was not my finest hour. The upside is that the signal levels through the balance control, filter and tone-control are much reduced (assuming the volume is not set to maximum), and

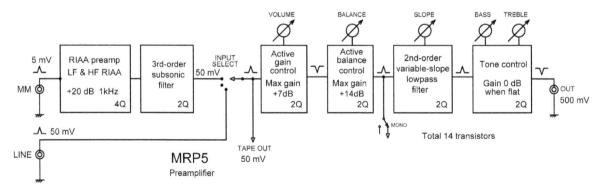

Preface Figure 3.3 The block diagram of the MRP5: 1978. The Q figures show the number of transistors in each stage.

so the distortion in these stages will also be much reduced. I would also point out in my defence that discrete transistor circuitry was and is much quieter than the affordable opamps of the day, essentially the LM741 with its voltage noise density of around 20 nV/√Hz.

On mature reflection, I think the noise issue may be the more important. This is a good example of the difficulty in placing the volume control, and making appropriate noise/distortion/headroom tradeoffs, even when designing preamplifiers with active gain controls.

A major design objective was to get the active gain and balance controls working in a more predictable way, with gain depending more on the ratio between two halves of a pot track rather than the absolute value of that track, as in the MRP4. This worked quite well for the volume control, but there was still something of an issue with the gain being partly set by the ratio of the pot track resistance to a fixed resistor. Down to −50 dB ref full volume the interchannel gain errors stayed below 1.2 dB, rising to 3 dB at −70 dB. I was a bit less successful for the balance control, but since an error in matching the channel gains only gives rise to a fixed displacement of the balance knob rather than image-shift it is of much less importance. The distortion performance of the whole rear section after the input-select switch, at 8 Vrms out and at full volume with balance central, was 0.0084% at 1 kHz and 0.061% at 10 kHz.

The tone-control stage used in the MRP5 was a fairly simple two-transistor circuit having a gain stage Q1 with a bootstrapped collector and an output emitter-follower, as shown in Preface Figure 3.4. The

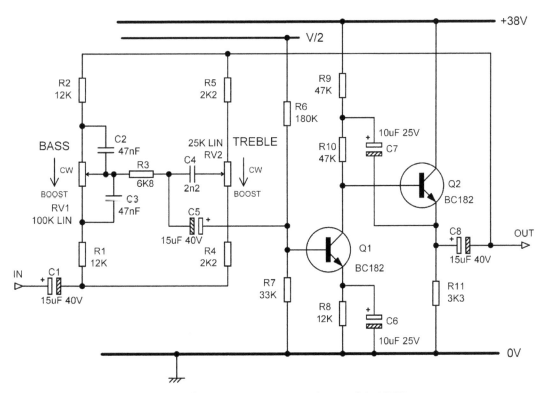

Preface Figure 3.4 The tone control stage of the MRP5.

gain stage is biased from the filtered V/2 bus via R6, R7, its operating current being defined by the DC voltage set up across R8. This voltage should not be too large as it may limit the negative voltage swing. However, the main limitation on this is usually the output stage, its maximum sink current being set by the value of emitter resistor R11.

The MRP6 was designed in July 1978. The MM input stage was as for the MRP4, except the output was a simple emitter-follower with emitter-resistor; hence only three transistors. There was a third-order subsonic filter, this time placed after the input select switch so it worked on all inputs and not just the MM input; this required a two-transistor buffer to be placed in front of it to ensure it had a low-impedance drive and gave a predictable frequency response. There was no scratch filter. The two-transistor Baxandall tone-control stage was this time placed *before* the volume-control, so its noise could be turned down. There was a simple mono/stereo switch. For reasons of which I have no recall whatever, I decided to use a conventional passive volume control, followed by a three-transistor amplifier giving 14 dB of gain, at the end of the signal path. The vernier balance control altered the gain of this stage by altering the amount of negative feedback; there was some dependence on the pot track resistance, but this was reduced by the padding resistors that limited the range of balance control. The block diagram is shown in Preface Figure 3.5.

It was never put into a neat box, but I still have the stripboard prototype, which was thoroughly measured and evaluated; see Preface Figure 3.6.

The MRP7 of 1980 was a consultancy design for a client. It had no features of any real technical interest (the client wanted conventional and he got it). The thing that sticks in the mind is *I never got paid*. I'm not going to point names or name fingers, but: You Know Who You Are. Even the great Peter Baxandall had trouble getting paid by at least one well-known name in the audio business. That's why nowadays I insist that all my consultancy work is paid for in advance. Fortunately, I can do that.

The MRP8 of 1982 was another design for a client, but this time I held out for stage payments, so I did get paid (mostly). This design was also somewhat similar to the MRP4, employing discrete

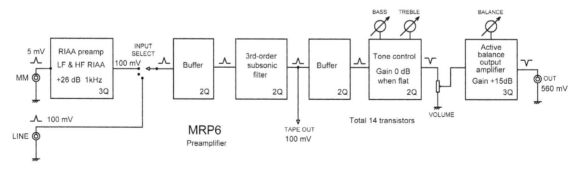

Preface Figure 3.5 The block diagram of the MRP6: 1978. The Q figures show the number of transistors in each stage.

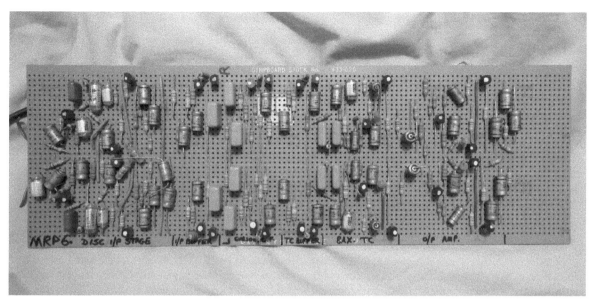

Preface Figure 3.6 The MRP6 stripboard prototype: 1978.

transistor design with a single supply rail at +38V, but used the variable-slope scratch filter from the MRP5. The client was very rightly looking for a Unique Selling Proposition, so I used an improved active volume control. You can never get the noise output of an active volume stage to zero by turning the volume setting to zero, as you can with a passive control. What you can do is gang the active volume control with a passive control put at the end of the signal path, giving the much desired zero-noise at zero-volume, and much improving the rather flat control law you get with the Baxandall active volume configuration. This required a four-gang pot for stereo but there were no problems with sourcing the part. This active-passive approach to volume control is still very much valid when really high performance is required, and I have used it in a design very recently. The principle is described in [3], and there is some new information in the preface to Chapter 18 in this book.

The MRP9 is a mystery, even to me. All I can find are a few notes referring to 'minimalism'. I am sure it did not refer to omitting tone-controls. It looks like this really *is* The Lost Preamplifier of Douglas Self, but I will keep looking.

References

1. Lipshitz, S.P., On RIAA Equalisation Networks, *J. Audio Eng Soc,* June 1979, p. 458 onwards.

2. Self, D., *Small-Signal Audio Design,* 2nd edition, Chapter 9, pp. 281–285. Focal Press (Taylor & Francis) 2014. ISBN: 978-0-415-70974-3 Hardback. 978-0-415-70973-6 Paperback. 978-0-315-88537-7 ebook.

3. Ibid., Chapter 13, pp. 368–369.

PRECISION PREAMPLIFIER

October 1983

Until relatively recently, any audio preamplifier with pretensions to above-average quality had to be built from discrete transistors rather than integrated circuits. The 741 series of op-amps was out of the question for serious audio design, due to slew-rate and other problems, and the TL071/72 types, though in many ways excellent, were still significantly noisier than discrete circuitry. In an article some years ago, [1] I attempted to show that it was still feasible to better the performance of such devices by using simple two or three-transistor configurations.

The appearance of the 5534 low-noise op-amp at a reasonable price, has changed this. It is now difficult or impossible to design a discrete stage that has the performance of the 5534 without quite unacceptable complexity. The major exception to this statement is the design of low-impedance low-noise stages such as electronically-balanced microphone inputs or moving-coil head amplifiers, where special devices are used at the input end.

5534 op-amps are now available from several sources, in a conventional 8-pin d.i.l. format. This version is internally compensated for gains of three or more, but requires a small external capacitor (5–15 pF) for unity-gain stability. The 5532 is a very convenient package of two 5534s in one 8-pin device with internal unity-gain compensation, as there are no spare pins.

The 5534/2 is a low-distortion, low-noise device, and a typical audio stage could be expected to generate less than 0.005% THD over the range 1–20 kHz, leaving the residual distortion lost in the noise of all but the most expensive analysers. Noise performance obviously depends partly on external factors, such as source resistance and measurement bandwidth, but as an example consider the moving-magnet disc input stage shown in Figure 3.3. When prototyped with a TL071, the noise (with a 1 k resistor input load) was −69 dB with reference to a 5 mV r.m.s. 1 kHz input. Substituting a 5534 improved this to −84 dB, a clear superiority of 15 dB.

Another advantage of this device to the audio designer is its ability to drive low-impedance loads (down to 500 Ω in practice) to a full voltage swing, while maintaining low distortion. This property is much appreciated by studio mixer designers, whose output amplifiers are still expected to drive largely fictitious 600 Ω loads. As a comparison, the TL071 is only good for loads down to about 2 kΩ.

Architecture

As explained in a previous article, [1] the most difficult compromise in preamplifier design is the distribution of the required gain (usually at least 40 dB) before and after the volume control. The more gain before the volume control, the lower the headroom available to handle unexpectedly large signals. The more gain after, the more the noise performance deteriorates at low volume settings. Another constraint is that it is desirable to get the signal level up to about 100 mV r.m.s. before reaching the volume control, as tape inputs and outputs must be placed before this. The only really practical way to get the best of both worlds is to use an active gain-control stage—an amplifier that can be smoothly varied in gain from effectively zero up to the required maximum.

If the input to the disc stage is a nominal 5 mV r.m.s. (assumed to be at 1 kHz throughout the avoid confusion due to RIAA equalization) from either moving-magnet cartridge or moving-coil head amp, then 26 dB of gain will be needed to give the 100 mV which is the minimum it is desirable to offer as a tape output. This can easily be got from a single 5534 stage, and taken together with the supply rails (±15 V) this immediately fixes the disc input overload at about 320 mV r.m.s. A figure such as this is quite adequate, and surpasses most commercial equipment.

One must next decide how large an output is needed at maximum volume for the 5 mV nominal input. 1 V r.m.s. is usually ample, but to be certain of being able to drive exotic units to their limits, 2 V r.m.s. is safer. This decision is made easier because using an active gain-control frees us from the fear of having excessive gain permanently amplifying its own noise after the volume control. Raising the 100 mV to this level requires the active gain stage to have another 26 dB of gain available; see the block diagram in Figure 3.1.

The final step in fixing the preamp, architecture is to place the tone-control in the optimum position in the chain. Like most Baxandall stages, this requires a low-impedance drive if the response curves are to be predictable, and so placing it after the active gain-control block (which has the usual very low output impedance) looks superficially attractive. However, further examination shows that (a) the active-gain stage also requires a low-impedance drive, so we are not saving a buffer stage after all, and (b) since it uses shunt feedback the tone-control stage is rather noisier than the others, [2] and should therefore be placed before the gain control so that its noise can be attenuated along with the signal at normal volume settings.

The tone-control is preceded by a unity-gain buffer stage with low output impedance and a very high input

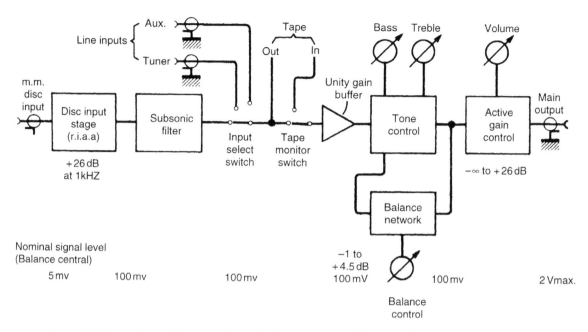

Figure 3.1 Block diagram. Tone-control placed before gain-control block to reduce noise from tone-control.

impedance, so that the load placed on line input devices does not vary significantly when the tape-monitor switch is operated. This brings us to the block diagram in Figure 3.1. Figure 3.3 shows the circuit diagram of the complete preamplifier. The components around A_1 and A_2 make up the moving-magnet disc stage and its associated subsonic filter. Disc preamplifier stage A_1 uses a quite conventional series feedback arrangement to define the gain and provide RIAA equalisation. This provides a clear noise-performance advantage of 13 dB over the shunt feedback equivalent, [2,3] which is sometimes advocated on the rather dubious grounds of 'improved transient response'. The reality behind this rather woolly phrase is that the series configuration cannot give the continuously descending frequency response in the ultrasonic region that the RIAA specification seems to imply, because its minimum gain is unity. Hence sooner or later, as the frequency increases, the gain levels out at unity instead of dropping down towards zero at 6 dB per octave. As described in Refs. 1 and 2, when a low-gain input stage is used to obtain a high overload margin, 'sooner' means within the audio band, and so an additional low-pass time-constant is required to cancel out the unwanted h.f. breakpoint; once more it is necessary to point out that if the low-pass time-constant is correctly chosen, no extra phase or amplitude errors are

introduced. This function is performed in Figure 3.3 by R_8 and C_{11}, which also filter out unwanted ultrasonic rubbish from the cartridge.

It was intended from the outset to make the RIAA network as accurate as possible, but since the measuring system used (Sound Technology 1700 A) has a nominal accuracy of 0.1 dB, 0.2 dB is probably the best that could be hoped for. Designing RIAA networks to this order of accuracy is not a trivial task with this configuration, due to interaction between the time-constants, and attempting it empirically proved most unrewarding. However, Lipshitz, in an exhaustive analysis of the problem, using heroic algebra in quantities not often seen, gives exact but complicated design equations.[4] These should not be confused with the rule-of-thumb time-constants often quoted. The Lipshitz equations were manipulated on an Acorn Atom microcomputer until the desired values emerged. These proved on measurement to be within the 0.2 dB criterion, with such errors as existed being ascribable to component tolerances.

Design aims were that the gain at 1 kHz should be 26 dB, and that the value of R_3 should be as small as feasible to minimize its noise contribution. These two factors mean that the RIAA network has a lower impedance

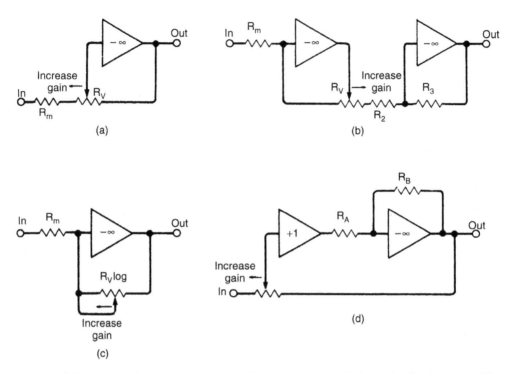

Figure 3.2 Evolution of active gain-control stage. That due to Baxandall, chosen for this design, is at (d).

than usual, and here the load-driving ability of the 5534 is helpful in allowing a full output voltage swing, and hence a good overload margin.

There is a good reason why the RIAA capacitors are made up of several in parallel, when it appears that two larger ones would allow a close approach to the correct value. It is pointless to design an accurate RIAA network if the close-tolerance capacitors cannot be easily obtained, and in general they cannot. The exception to this is the well-known Suflex range, usually sold at 2.5% tolerance. These are cheap and easy to get, the only snag being that 10 nF seems to be the largest value widely available, and so some paralleling is required. This is however a good deal cheaper and easier than any other way of obtaining the desired close-tolerance capacitance.

Metal-oxide resistors are used in the RIAA network and in some other critical places. This is purely to make use of their tight tolerance (1% or 2%), as tests proved, rather unexpectedly, that there was no detectable noise advantage in using them.

The recently updated RIAA specification includes what is known as the 'IEC amendment'. This adds a further 6 dB/octave low-cut time-constant that is −3 dB at 20.02 Hz. It is intended to provide some discrimination against subsonic rumbles originating from record warps, etc., and in a design such as this, with a proper subsonic filter, it is rather redundant. Nonetheless the time-constant has been included, in order to keep the bottom octave of the RIAA accurate. The time-constant is *not* provided by R_3C_3 (which is no doubt what the IEC intended) but by the subsonic filter itself, a rather over-damped third-order Butterworth type designed so that its slow initial roll-off simulates the 20.02 Hz time-constant, while below 16 Hz the reponse drops very rapidly. Implementing the IEC roll-off by reducing C_3 is not good enough for an accurate design due to the large tolerances of electrolytic capacitors. However, the R_3, C_3 combination is arranged to roll-off lower down (−3 dB at about 5 Hz) to give additional subsonic attenuation.

Capacitor C_1 defines the input capacitance and provides some r.f. rejection. A compromise value was chosen, and this may be freely modified to suit particular cartridges.

The noise produced by the disc input stage alone, with its input terminated with a 1 k resistor to simulate roughly a moving-magnet cartridge, is −84.5 dB

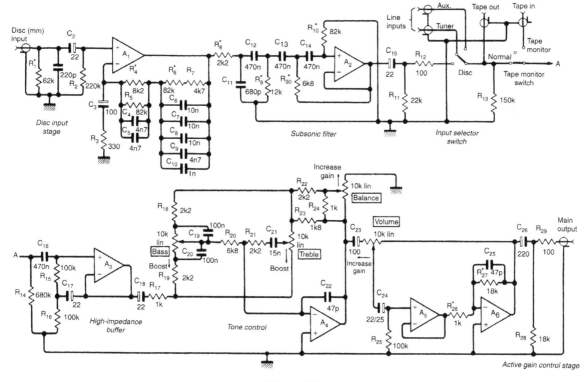

Figure 3.3a

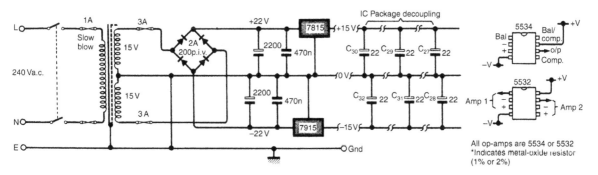

Figure 3.3b Complete circuit diagram. Decoupling capacitors for i.cs must be close to packages.

with reference to a 5 mV r.m.s. 1 kHz input (i.e. 100 mV r.m.s. out) for a typical 5534 A sample. The suffix A denotes selection for low noise by the manufacturer. When the 1 k termination is replaced by a short circuit, the level drops to −86 dB, indicating that in real life the Johnson noise generated by the cartridge resistance is significant, and so that stage is really as quiet as it is sensible to make it.

Subsonic filter

As described above, this stage not only rejects the subsonic garbage that is produced in copious amounts by even the flattest disc, but also implements the IEC roll-off. Below 16 Hz the slope increases rapidly, the attenuation typically increasing by 10 dB before 10 Hz is reached. The filter therefore gives good protection against subsonic rumbles, that tend to peak in the 4–5 Hz region.

This filter obviously affects the RIAA accuracy of the lowest octave, and so C_{12}, C_{13}, C_{14} should be good-quality components. A 10% tolerance should in practice give a deviation at 20 Hz that does not exceed 0.7 dB, rapidly reducing to an insignificant level at higher frequencies. The tape output is taken from the subsonic filter, with R_{12} ensuring that long capacitive cables do not cause h.f. instability. If it really is desirable to drive a 600 Ω load, then C_{15} must be increased to 220 μF to maintain the base response.

High-impedance buffer

This buffer stage is required because the following tone-control stage demands a low-impedance drive, to ensure that operating the tape monitor switch S_2 does not affect the tape-output level. If the input selector switch S_1 was set to accept an input from a medium impedance source (say 5 k), and the buffer had a relatively low input impedance (say 15 k), then every time the tape-monitor switch was operated there would be a step change in level due to the change of loading on the source. This is avoided in this design by making the buffer input impedance very high by conventional bootstrapping of R_{15}, R_{16} via C_{17}. This is so effective that the input impedance is defined only by R_{14}. Unlike discrete-transistor equivalents, this stage retains its good distortion performance even when fed from a high source resistance, e.g. 100 k.

Tone-control stage

Purists may throw up their hands in horror at the inclusion of this, but it remains a very useful facility to have. The range of action is restricted to ±8 dB at 10 kHz and ±9 dB at 50 Hz, anything greater being out of the realm of hi-fi. The stage is based on the conventional Baxandall network with two slight differences. Firstly the network operates at a lower impedance level than is usual, to keep the noise as low as possible. The common values of 100 k for the bass control and 22 k for the treble control give a noise figure about 2.5 dB worse. Even with the values shown, the tone stage is about 6 dB noisier than the buffer that precedes it. Both potentiometers are 10 k linear, which allows all the preamplifier controls to be the same value, making getting them a little easier. The low network impedance also reduces the likelihood of capacitive interchannel crosstalk. Once again,

implementing it is only possible because of the 5534's ability to drive low-value loads.

Secondly, the tone-control stage incorporates a vernier balance facility. This is also designed as an active gain-control, with the same benefit of avoiding even small compromises on noise and headroom. The balance control works by varying the amount of negative feedback to the Baxandall network, and therefore some careful design is needed to ensure that the source resistance of the balance section remains substantially constant as the control is altered, or the frequency response may become uneven. Resistors R_{22}, R_{23}, R_{24} define this source resistance as 1 k, which is cancelled out by R_{17} on the input side. The balance control has a range of +4.5 to −1.0 dB on each channel, which is more than enough to swing the stereo image completely from side to side. If you need a greater range than this, perhaps you should consider siting your speakers properly.

Active gain-control stage

An active gain-control stage must fulfil several requirements. Firstly, the gain must be smoothly variable from maximum down to effectively zero. Secondly, the law relating control rotation and gain should be a reasonable approximation to logarithmic, for ease of use. Finally, the use of an active stage allows various methods to be used to obtain a better stereo channel balance than the usual log. pot. offers.

All the configurations shown in Figure 3.2 meet the first condition, and to a large extent, the second. Figures 3.2(a) and 3.2(b) use linear controls and generate a quasi-logarithmic law by varying both the input and feedback arms of a shunt-feedback stage. The arrangement of Figure 3.2(c), as used in the previous article, offers simplicity but relies entirely on the accuracy of a log. pot. While 2(a) and 2(b) avoid the tolerances inherent in the fabrication of a log. track, they also have imperfect tracking of gain, as the maximum gain in each case is fixed by the ratio of a fixed resistor R_m to the control track resistance, which is not usually tightly controlled. This leads to imbalance at high gain settings.

Peter Baxandall solved the problem very elegantly, [1] by the configuration in 2(d). Here the maximum gain of the stage is set not by a fixed-resistor/track-resistance ratio, but by the ratio of the two fixed resistors R_a, R_b. A buffer is required to drive R from the pot. wiper, because in a practical circuit this tends to have a low value. It can be readily shown by simple algebra that the control track

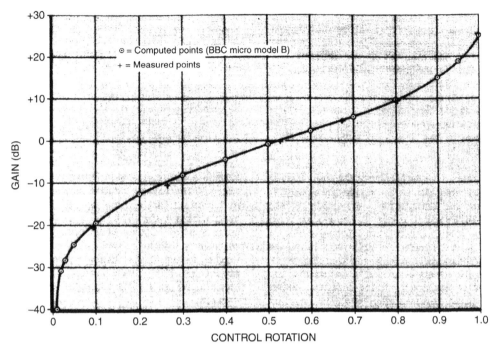

Figure 3.4 Law of gain-control pot., approximately linear over main part of range.

resistance now has no effect on the gain law, and hence the channel balance of such a system depends only on the mechanical alignment of the two halves of a dual linear pot. The resulting gain law is shown in Figure 3.4, where it can be seen that a good approximation to the ideal log (i.e. linear in dB) law exists over the central and most used part of the control range.

A practical version of this is shown in Figure 3.3. A_5 is a unity-gain buffer biased via R_{25}, and R_{26}, R_{27} set the maximum gain to the desired +26 dB. Capacitor C_{25} ensures h.f. stability, and the output capacitor C_{26} is chosen to allow 600 Ω loads to be driven. A number of outwardly identical Radiohm 20 mm dual-gang linear pots were tested in the volume control position, and it was found that channel balance was almost always within ±0.3 dB over the gain range −20 to +26 dB, with occasional excursions to 0.6 dB. In short, this is a good way of wringing the maximum performance from inexpensive controls, and all credit must go to Mr Baxandall for the concept.

At the time of writing there is no consensus as to whether the absolute polarity of the audio signal is subjectively important. In case it is, all the preamplifier inputs and outputs are in phase, as the inversion in the tone stage is reversed again by the active-gain stage.

Power supply

The power supply is completely conventional, using complementary i.c. regulators to provide ±15 V. Since the total current drain (both channels) is less than 50 mA, they only require small heatsinks. A toriodal mains transformer is recommended for its low external field, but it should still be placed as far as possible from the disc input end of the preamplifier. Distance is cheaper (and usually more effective) than Mu-Metal. Since the 5534 is rated up to ±20 V supplies, it would be feasible to use ±18 V to get the last drop of extra headroom. In my view, however, the headroom already available is ample.

Construction

The preamplifier may be built using either 5534 op-amps or the 5532 dual type. The latter are more convenient (requiring no external compensation) and usually cheaper per op-amp, but can be difficult to obtain. To compensate each 5534 for unit gain, necessary for each one, connect 15 pF between pins 5 and 8. Note that the rail decoupling capacitors should be placed as close as possible to the op-amp packages—this is one case in which it really does matter, as otherwise this i.c. type is prone to h.f. oscillation that is not visible on a scope, but which results in a very poor distortion performance. It must also be borne in mind that both the 5534 and 5532 have their inputs tied together with back-to-back parallel diodes, presumably for voltage protection, and this can make fault-finding with a voltmeter very confusing.

Only 2.5% capacitors should be used in the RIAA networks if the specified accuracy is to be obtained. Resistors in Figure 3.3 marked * should be metal oxide 1% or 2%, for reasons of tolerance only. Each of these resistors sets a critical parameter, such as RIAA equalization or channel balance, and no improvement, audible or otherwise, will result from using metal oxide in other positions.

Several preamplifier prototypes were built on Veroboard, the two channels in separate but parallel sections. The ground was run through in a straight line from input to output. Initially the controls were connected with unscreened wire, and even this gave acceptable crosstalk figures of about −80 dB at 10 kHz, due to the low circuit impedances. Screening the balance and volume connections improved this to −90 dB at 10 kHz, which was considered adequate. It must be appreciated that the crosstalk performance depends almost entirely on keeping the two channels physically separated.

Some enthusiasts will be anxious to (a) use gold-plated connectors; (b) by-pass all electrolytics with non-polarized types; or (c) remove all coupling capacitors altogether, in the pursuit of an undefinable musicality. Options (a) and (b) are pointless and expensive, and (c) while cheap, may be dangerous to the health of your loudspeakers. Anyone wishing to dispute these points should arm themselves with objective evidence and a stamped, addressed envelope.

Specification
(Based on measurements made on three prototypes, with Sound Technology 1710A.)

Moving-magnet

noise ref. 5 mV r.m.s., 1 kHz input	−81 dB
RIAA accuracy	±0.2 dB
input overload point (1 kHz)	300 mV r.m.s.

Line inputs

noise ref. 100 mV r.m.s. i/p	−85 dB
maximum input	9 V r.m.s.
maximum gain	+26 dB
treble control range	±8 dB
bass control range	±9 dB
vernier balance control	−1 dB to +4.5 dB
volume control channel balance	±0.3 dB
distortion (1 kHz–20 kHz)	0.005%
maximum output	9.5 V r.m.s.

References

1. Self, D. 'High-Performance Preamplifier', *Wireless World*, February 1979.

2. Walker, H.P. 'Low noise audio amplifiers', *Wireless World*, May 1972, pp 233–237.

3. Linsley-Hood, J. 'Modular preamplifier', *Wireless World*, October 1982, p 32 onwards.

4. Lipshitz, S.P. 'On RIAA equalisation networks', *J. Audio Eng. Soc.*, June 1979, p 458 onwards.

5. Baxandall, P. 'Audio gain controls', *Wireless World*, November 1980, pp 79–81.

Chapter 4

Design of moving-coil head amplifiers

(Wireless World, December 1987)

Moving Coil (MC) cartridges are generally considered to have superior performance to the moving magnet kind, but they have a very low output voltage. Only the accompanying very low source impedance allows reasonable signal-to-noise ratios to be achieved. A further complication is that the output voltage varies over a very wide range between different brands. In some ways a step-up transformer to get the signal up to moving-magnet level is the most elegant solution, as no power supply is required, but it is also an expensive one, because of the need for highly effective magnetic screening to keep hum out of the tiny signals. There are also issues with the frequency response of transformers, though non-linearity should not be a problem at the very low levels involved. A further consideration is Johnson noise from the winding resistances; this is about the only place where the use of silver wire has some sort of rational basis, as the conductivity of silver is 5% higher than that of copper. However, Johnson noise varies as the square root of resistance, so the noise reduction is only 0.22 dB. Silver is much more costly; it's not exactly a brilliant technical fix.

For this reason, electronic amplification of the signal is the more popular approach, and here the major challenge is to design an amplifier that can exploit the very low source impedance to give very low levels of noise. For many years, moving-coil head amps were exotic devices of highly specialised design. They were usually all-discrete, to minimise noise, and because of this, and the very low feedback impedances to be driven, linearity was poor, and only acceptable because the signal levels were so low.

This design is a hybrid discrete/op-amp configuration, which gives very low distortion even at signal levels monstrously above the normal operating conditions. It sidesteps several problems by having much more gain than is normally required; this would normally be a very bad move, severely curtailing headroom, but here it works as it can be assumed that the MC amp is always followed by a moving-magnet disc stage with substantial gain, so clipping will occur there first.

When the article appeared, the 2N4403 transistor was the usual choice for low noise from a low source impedance. I have been unable to confirm this from modern datasheets, but as I recall it the base resistance rbb was 40Ω typical, significantly lower than that of most small-signal transistors. The 2SB737 transistor, with its magnificently low rbb of only 2Ω, existed, but it was expensive and not that easy to obtain, so I didn't use it. In a year or two, this situation changed for the better. Now it has changed again, for the worse.

The 2SB737, with its uniquely low base spreading resistance (rbb) of 2Ω was a wonderful transistor, and I say "was" because it has now been obsolete for some time. The question is why? I have never found anybody able to give any sort of answer to this question. It was a device with unique properties, and since MC cartridges show no sign of going away, you would think there would be a secure if not enormous market for it. They are a strictly limited resource though there are probably more of them out there than there are nuvistors. They can still be obtained for a price, for example from The Signal Transfer Company, but there are only so many in existence. The big question is what can replace it? Other types of transistors which might be suitable are also obsolete or threatened with obsolescence, once again for no obvious reason.

In the previous version of this preface I wrote, 'Nobody now would consider devices like the 2N4403 for this application'. It looks like I might have spoken too soon. However, paralleling twenty 2N4403's to get the effective rbb down to the level of one single 2SB737 is not exactly elegant design, and I imagine you might run into trouble with a build-up of device capacitances. Later versions of the MC preamp used three 2SB737's in parallel, giving a very handy noise reduction. Using sixty 2N4403's to try and emulate this is not practical politics, though you could argue anything goes in the wonderful world of hi-end hi-fi.

It looks like it might be time to explore the use of low-noise JFETs in MC headamps. The voltage noise is higher but the current noise is lower, so they are normally thought of as being best matched to medium-impedance sources rather than the very low values seen in MC use. One of the quietest amplifiers I know of is a design by Samuel Groner which uses eight JFETs in parallel to obtain a noise density of 0.39 nV√Hz [1].

There is more on MC preamp stages in the second article on Preamplifier 2012.

Reference

1. Groner, S., A Low-Noise Laboratory-Grade Measurement Preamplifier, *Linear Audio Volume* 3, April 2012, p. 143.

DESIGN OF MOVING-COIL HEAD AMPLIFIERS

December 1987

In recent years, moving-coil cartridges have increased greatly in popularity. This is not the place to try and determine if their extra cost is justified by an audible improved performance; suffice it to say that a preamplifier now needs a capable moving-coil cartridge input if it is to be considered complete. The head-amplifier design presented here as an example was originally intended to be retrofitted to the precision preamplifier previously published in *Wireless World*, [1] feeding the existing moving-magnet disc input. However, it is adaptable to almost any preamplifier and cartridge as the gain range available is very wide; it should therefore be of interest to any engineer working in this field. Hereafter 'moving coil' is abbreviated to m.c., and 'moving-magnet' to m.m.

Traditionally, moving-coil cartridges were matched to moving-magnet inputs by special transformers, which give 'free gain'—in a sense—and are capable of a good noise performance if the windings are carefully designed for very low series resistance. However, the inescapable problems of low-frequency distortion, high-frequency transient overshoots and the need for obsessive screening to avoid 50 Hz mains pickup render them unattractive and expensive.

The requirements for a high-quality m.c. head-amplifier are as follows. The overwhelming need is for a good noise performance, as the signals generated by m.c. cartridges are, in general, very low. However, this sensitivity is also much more variable than that of m.m. cartridges, where one can take a nominal output of 5 mV r.m.s. for 5 cm/s at 1 kHz as being virtually standard. In contrast, a survey of the available m.c. cartridges gave a range from 2.35 mV (Dynavector DV10X IV) to 0.03 mV (Audionote 102 vdH), though these are both exceptional and the great majority fell between 0.2mV and 0.4 mV. Figure 4.1 shows the output levels of a number of current m.c. cartridges plotted on a scale of dBu (i.e. referred to 775 mV) and m.m. cartridges are included on the right for comparison. It is notable how these bunch together in a range of less than 7 dB.

A representative m.c. cartridge used both as a basis for design, and for testing, is the Ortofon MC10 Super, which has an output of 0.3 mV for 5 cm/s, and an internal resistance of 3 Ω. There is general agreement that this is a good-sounding component.

As detailed above, there is a need for easily variable gain over a wide range. This can be quite adequately provided in switched steps, avoiding the problems of uncertain stereo balance on dual potentiometers. From the above output figures, a gain range of 6 dB to 46 dB appears necessary to cater for all possible cartridges. It would seem, at the low-gain end, that the amplifier is virtually redundant, and so a minimum gain of 20 dB was chosen.

Moving-coil cartridges are very tolerant of the loading they see at an amplifier's input, as a result of their own very low internal impedance. For example, Ortofon, who might be reckoned to know a thing or two about m.c. cartridges, simply state that the recommended load for most of their wide range of cartridges is 'greater than 10 Ω'. Nonetheless, since experimenting with cartridge loading is a harmless enough pastime, provision for changing the input loading resistor over a wide range has been made in this design.

The preamplifier should have the ability to drive a normal m.m. cartridge input at sufficient level to ensure that the head amplifier does not limit the disc headroom. Any figure here over about 300 mV r.m.s. should be satisfactory. A less obvious point is that the input impedance, apart from the nominal 47 k resistive component, usually includes a fair amount of capacitance, either to adjust cartridge frequency response or to exclude r.f. This can cause head-amplifier instability unless it is dealt with.

Finally, a head amplifier should meet the usual requirements for frequency response, crosstalk, and linearity. Capacitive crosstalk is usually not a problem, due

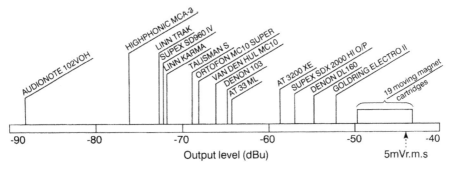

Figure 4.1 Output levels of representative moving-coil cartridges plotted on a scale of decibels relative to 0.775 V (1 mW in 600 Ω), with the outputs of a number of moving-magnet cartridges as a comparison.

to the very low impedances involved, but for the same reason, linearity can present problems despite the low signal levels.

Design problems

The theoretical noise characteristics of amplifiers have been dealt with very competently in other articles, [2] and there is no need to repeat the various mathematical derivations here. The designer's options are usually limited to choosing a suitable input device, operating it at roughly the right current, (not usually critical due to the flat bottoms of the noise curves) and then making sure that the surrounding circuitry doesn't mess things up too much. M.c. head amplifiers are almost always built around discrete devices, with or without the addition of an accompanying op-amp (for an exception see Ref. 3). Figure 4.2 shows the reason why: when source resistances are low (say below 1 k) even advanced op-amps are easily outperformed by discrete devices, due to the inevitable compromises in i.c. fabrication. The values of equivalent input noise (e.i.n.) in Figure 4.2 were taken from five samples of each device, using a source resistance of 3R3, and the general circuit configuration in Figure 4.3. The rather non-standard measurement bandwidth is due to the use of the internal filters on a Sound Technology measuring system; adding a third-order 20 kHz active filter at the ST input would be very difficult, as the levels of noise being measured are so low. To convert to 20 kHz upper bandwidth limit, subtract 1.5 dB. One of the prerequisites for good performance in this role is a low value for R_b, and this has led to a fine miscellany of devices being applied to a job they were never intended for: medium power devices, print-hammer drivers, (a lot of

transistors seem to have been designed as print-hammer drivers) and so on.

Apart from careful device selection, the other classical way of reducing noise with low source impedances is to use multiple devices. The assumption here is that m.c. amplifier noise will swamp the miniscule Johnson noise inherent in the source (this is usually all too true) and therefore, if two input devices have their outputs summed, the signals will simply add, giving a 6 dB gain, while the two uncorrelated device noise contributions will partially cancel, giving only 3 dB.

Thus, there is a theoretical gain of 3 dB in noise performance every time the number of input devices is doubled. There are, of course, clear economic limits to the amount of doubling you can go in for; eight parallel devices is the most that I have seen. It also seems difficult in practice to get the full theoretical benefit.

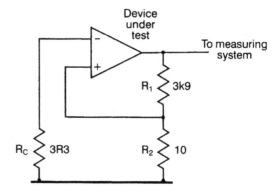

Figure 4.3 Circuit used to obtain the measurements shown in Figure 4.2.

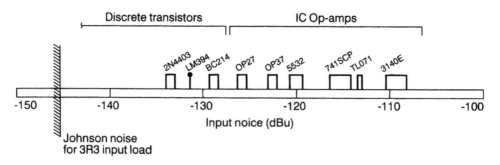

Figure 4.2 Discrete transistors still, in the main, provide a better noise performance than op-amps at low source resistances, as shown here for five examples of each type.

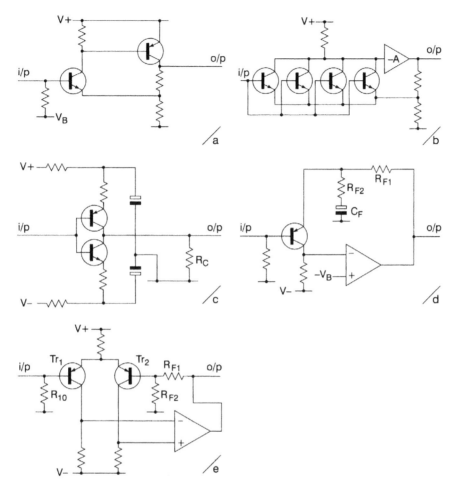

Figure 4.4 Some head-amplifier configurations. A fairly low open-loop gain in the circuit at (a) results in poor linearity. At (b), the gain is provided by multiple transistors, which theoretically gives an improvement of 3 dB in noise performance for twice the number of transistors, but can also present current-sharing problems. The arrangement at (c) provides the 3 dB improvement without current sharing: linearity is not of the highest order. Circuit (d) uses one input device, the gain being provided by an ip-amp: the necessity for Cf presents problems, which are overcome in the (e) configuration at the expense of a lowered noise performance.

M.c. head-amplifiers in use today can be roughly divided into three common topologies, as shown in Figure 4.4. That shown in 4(a) relies on a single device with low R_b, and the combination of limited open-loop gain and the heavy loading of the low-impedance of the feedback network on the final transistor means that both linearity and maximum output level tend to be uninspiring. Given the technical resources that electronics can deploy, there seems no need to ask the paying customers to put up with any measurable distortion at all. An amplifier of this type is analysed in Ref. 4.

Figure 4.4(b) shows the classic multiple-parallel-transistor configuration; the amplifier block A is traditionally one or two discrete devices, that usually have difficulty in driving the low-impedance feedback network. Effort is usually expended in ensuring proper current-sharing between the input devices.

This can be done by adding small emitter resistors to swamp V_{be} variations, but these will effectively appear in series with the source resistance, and compromise the noise performance unless they are individually decoupled with a row of very large electrolytics. Alternatively, each transistor can be given its own d.c. feedback loop to set up its collector current, but this tends to be even more prodigal of components. Having said this, experiment proved that the problem of current-sharing was not as serious as conventional wisdom holds; this is explained below. For examples of circuitry see Ref. 5.

Figure 4.4(c) shows the series-pair scheme. This simple arrangement allows two input devices to give the normal 3 dB noise improvement without current-sharing problems as substantially the same collector current goes through each device. The collector signal currents are summed in R_c, which must be reasonably low in value to absorb any current imbalance. This configuration has its adherents but it also has its difficulties, such as indifferent linearity.

It was therefore originally decided to base the design presented here on a single well-chosen device, with the spadework of providing open-loop gain and output drive capability left to an op-amp. This leads to the configuration in Figure 4.4(d), which gives excellent linearity, and less than 0.002% THD at full output may be confidently expected. The first problem to be dealt with is the very low value of R_{f2}; this must be as low as possible (say 10 Ω) as it is effectively in series with the input source resistance and will degrade the noise performance accordingly. This means that C_f must be very large, of the order of 2200 μF, to preserve the l.f. response. A 3R3 resistor in the R_{f2} position demands 4700 μF to give –3 dB at 10 Hz; this is not elegant. The capacitance C_f cannot be dispensed with, since there is a d.c. level of +0.6 V on the emitter of the input device, leading to a wholly impossible offset at the output of the op-amp.

One solution to this is the use of a differential pair, as in Figure 4.4(e). This cancels out the V_{be} of the input transistor Tr_1, at the cost of some degradation in the noise performance of the circuit, and hopefully the d.c. offset is so much smaller that, if C_f is omitted and the offset is amplified by the full a.c. gain, it will not seriously reduce the output voltage swing. In effect, the second transistor Tr_2 is an emitter follower transferring the feedback signal to the emitter of Tr_1, and such a circuit element introduces a small but inescapable amount of extra noise. In this case, with the component values shown, the degradation is about 2.8 dB.

A possibly more serious objection to this circuit is that the offset at the output is non-negligible, about 1 V, much of which is due to the base bias current flowing through R_n. A d.c.-blocking capacitor on the output is essential, and if it is an electrolytic there may be some doubt as to which way round to put it, as the exact level of input pair balance is unpredictable.

After practical trials, it was decided that a 3 dB noise penalty was too great, and that a way had to be found to use a single-ended input.

A new approach

The new method evolved is shown in the block diagram Figure 4.5. There is no C_f in the feedback loop, and indeed no overall d.c. feedback at all. The two halves of the circuit, the input transistor and the op-amp, each have their own d.c., feedback systems. The transistor relies on simple shunt negative feedback via d.c. loop 1, while the op-amp has its output held precisely to a d.c. level of 0 V by the integrator A_2. This senses the mean output level, and sets up a voltage on the non-inverting input of A_1 that is very close to the level set on Tr_1 collector, such that the output stays firmly at zero; its time-constant is made large enough to ensure that an ample amount of open-loop gain exists at the lowest audio frequencies. Failure to do this results in a rapid rise of distortion as the frequency is lowered. Any changes in the direct voltage on Tr_1 collector are completely uncoupled from the output. However, a.c. feedback passes through R_{f1} as usual and ensures that the linearity of the compound arrangement is near-perfect, as is often the case with transistor op-amp hybrid circuits. Due to the high open-loop gain of A the a.c. level on Tr_1 collector is very small and so a.c. feedback through d.c. loop I does not significantly affect the input impedance of the amplifier, which is about 8 kΩ.

The device chosen for the input transistor was the 2N4403, a type that has been acknowledged as superior for low-noise applications for some years. The R_b is quoted as

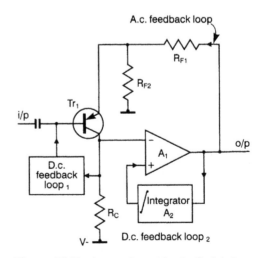

Figure 4.5 The layout adopted for the final design.

about 40 Ω.[5,6] More modern purpose-designed devices such as the 2SB737 will improve the noise performance by up to 1 dB, but the extra cost is significant.

A single device used in the circuit of Figure 4.6 gives an e.i.n. of −138 dB with a 4 mA collector current, which is certainly not bad, but it was consistently found that putting devices in parallel without any current-sharing precautions whatever always resulted in a significant improvement in noise performance. On average, adding a second transistor reduced noise by 1.2 dB, and adding a third reduces it by another 0.5 dB. Beyond this the law of diminishing returns sets in and, since further multiplication was judged unprofitable, a triple-device input was settled on. The current-sharing under these conditions was checked by measuring the voltage across 100 Ω resistors inserted in the collector paths. Using 3.4 mA as the total current for the array it was found after much device-swapping that the worst case of imbalance was 0.97 mA in one transistor and 1.26 mA in another. No attempt was made to ensure that all the devices came from the same batch. It therefore appears that, for this device at least, matching is good enough to make simple paralleling worthwhile, and it was therefore decided to use three devices in parallel in the final circuit.

There now remains the problem of setting the gain. Usually it would be simple enough to alter R_{f1} or R_{f2}, but here it is not quite so simple. The resistance R_{f2} is not amenable to alteration, as it must be kept to the lowest practicable value of 3.3 Ω, and R_{f1} must be kept up to a reasonable value so that it can be driven to a full voltage swing by an op-amp output. This means a minimum of 500 Ω if the op-amp is

to be of an easily obtainable type such as the 5534. (It is paradoxical that amplifiers whose output is measured in millivolts are required to chuck around so much current.)

These two values fix a minimum closed-loop gain of about 44 dB, which is far too high for all but the most insensitive cartridges. The only solution is to use a ladder output attenuator to reduce the overall gain; this would be anathema in a conventional signal path, because of the loss of headroom involved, but since an output of 300 mV r.m.s. would be enough to overload virtually all m.m. inputs, we can afford to be prodigal with it. If the gain of the head amplifier is set to be a convenient 200 × (+46 dB) then attenuation to reduce overall gain to a more useful +20 dB still allows a maximum output of 480 mV r.m.s.; this comfortably exceeds the input capability of the intended host preamplifier, though one previous design would accept it all and come back for more.[7] Smaller degrees of attenuation to provide intermediate gains allow greater outputs, and these are summarized in the specification. The Ortofon MC10 was used with +26 dB of gain, to give similar output levels to m.m. cartridges driving the precision preamplifier RIAA stage direct.

The last constraint is the need to provide a low output impedance to the succeeding m.m. input stage, so that it can give a good noise performance; it is likely to have been optimized to give of its best with a source impedance of 500 Ω or less. This implies that the ladder attenuator will need low resistor values, imposing yet more loading on the unfortunate op-amp, so this problem has been side-stepped by making the ladder an integral part of the a.c. feedback loop, as shown in Figure 4.6. This

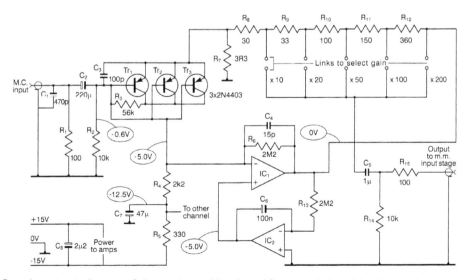

Figure 4.6 Complete circuit diagram of the moving-coil head amplifier, intended to drive the moving-magnet input of a preamplifier.

is only practicable because it is known that the load resistance presented by the next stage will be too high at 47 kΩ to cause any significant gain variations.

The final circuit

This is shown in Figure 4.6, and most closely follows the configuration of Figure 4.4(d), with the exception that the input devices have suddenly multiplied themselves by three. Capacitor C_1 is soldered on the back of the m.c. input phono sockets and is intended for r.f. filtering rather than modification of the cartridge response. If the need for more capacitive or resistive loading is felt, then extra components may be freely connected in parallel with R_1. If R_1 is raised in value, then load resistances of up to 5 kΩ are possible, as the impedance looking into C_2 is about 8 kΩ. Capacitor C_2 is large to give the input devices the full benefit of the low source impedance, and its value should not be altered. Resistors R_2, R_3 make up d.c. loop 1, setting the d.c. operating conditions of $Tr_{1,2,3}$, while R_4 is the collector load, decoupled from the supply rail by C_9 and R_5, which are shared between the two channels. Opamp IC_1 provides the main a.c. open-loop gain, and is stabilized at h.f. by C_4 : R_6 has no real effect on normal operation, but is included to give IC_1 a modicum of negative feedback and hence tidy behaviour at power-up, when this would otherwise be lacking due to the charging time of C_2 the other op-amp, IC_2, is the integrator that makes up d.c. loop 2, its time-constant carefully chosen to provide plenty of open-loop gain from IC_1 at low frequencies, and to avoid a peaking in the l.f. response that can occur due to the second time-constant of C_2.

The ladder resistors R_8–R_{12} make up the combined feedback-network and output-divider, overall gain being selected by a push-on link in the prototype. A rotary switch could be used instead, but this will produce loud clicks when moved with the volume up, since the emitter current of Tr_1–Tr_3 flows through R_7, and a small current therefore flows down the divider chain. The output resistor R_{15} ensures stability when driving long screened cables, and C_5 is included to eliminate any trace of d.c. offset from the output because the stage might find itself driving a horribly vulnerable 'esoteric' input stage with direct coupling and possibly substantial gain at d.c. Anything is possible these days.

Comparing performance parameters

These are given in the specification, and I think there will be few opportunities to quibble. On the vital question of noise it would be instructive to compare it with other

preamplifiers—not easy because the noise performance of m.c. head amplifiers is specified in so many different ways it is virtually impossible to reduce them all to a similar form, particularly without knowing the spectral distribution of the noise. Noise performances are specified with and without CCIR-ARM weighting, over different band-widths, and with different source impedances. This article has dealt throughout with unweighted noise referred to the input, over a 400 Hz–30 kHz bandwidth, and with RIAA equalisation *not* taken into account. Without getting bogged down in invidious comparisons, I can only say that it is my belief that the design given here is quieter than most current designs, being within 6 dB of the theoretical minimum.

When using this design with the precision preamplifier, it was noted with some surprise that it was so quiet that the m.m. RIAA stage actually caused the noise performance to deteriorate by about 3 dB. Since the RIAA stage is itself very quiet (s/n ratio –81 dB referred to 5mV r.m.s. input) it is considered that the design goals were met.

Specification

Careful earthing is needed if the noise and cross talk performance quoted is to be obtained.

Gain	Gain (dB)	Max output (r.m.s.)
10×	+20 dB	480 mV
20×	+26 dB	960 mV
50×	+34 dB	2.4 V
100×	+40 dB	4.6 V
200×	+46 dB	10 V

Input overload level. 48 mV r.m.s.

Equivalent input noise. –139.5 dBu, unweighted, no RIAA.

THD Less than 0.002% at 7 V r.m.s. output, (maximum gain) at 1 kHz. Less than 0.004% 40 Hz–20 kHz.

Frequency response. +0, –2 dB, 20 Hz–20 kHz.

Crostalk. Less than –90 dB 1 kHz–20 kHz (layout dependent).

Power requirements. 20 mA at ±15 V, for both channels.

Practice

P.c.b. layouts of require some care if the full performance is to be realised. First, the grounding should be carefully planned, as it must be realised that with such low impedances as R_7 (3R3) playing a vital role, the resistance of tracks can be significant. It is suggested that a single star ground point be chosen on the p.c.b., and critical paths (input ground, R_1, R_7) all connected to this, to prevent signal currents causing voltage drops where they are least wanted. It is vital to avoid making loops in the input path that will pick up 50 Hz magnetic fields.

It is essential to place the decoupling capacitor C_8 next to IC_1 to prevent insidious h.f. oscillation which makes its presence known only by severely impaired linearity. When interfacing the head amplifier to an existing design, note that about 8 mA flows down the ground connection.

References

1. Self, D. 'A precision preamplifier', *Wireless World*, October 1983, pp 31–34.

2. Walker, H.P. 'Low-noise audio amplifiers', *Wireless World*, May 1972.

3. Duncan, B. 'AMP-01', *HiFi News*, October 1984, pp 67–73.

4. Nordholt and Van Vierzen, 'Ultra Low Noise Preamp For Moving-Coil Phono Cartridges', *JAES*, April 1980, pp 219–223.

5. Barleycorn, J. (a.k.a. S. Curtis) *HiFi For Pleasure*, August 1978, pp 105–106.

6. Foord, A. 'Introduction to low-noise amplifier design', *Wireless World*, April 1981, p 71.

7. Self, D. 'High-quality preamplifier', *Wireless World*, February 1979, p 40.

Chapter 5

Precision preamplifier '96, Part I

(Electronics World, July/August 1996)

Sometimes you just feel that the world needs another preamplifier. Having put a lot of effort into power amplifiers and their problems in the meantime, I thought it would be interesting to see if the original Precision Preamplifier (Chapter 3) could be significantly improved. It could. The improvements were due to a bit more thinking being done, rather than any new and exciting parts becoming available.

The 5532-based moving-magnet disc stage was retained with a few changes (most importantly, a cost-effective way to tighten up the RIAA accuracy) as attempts to make a hybrid stage that was quieter were not encouraging; the plan was single bipolar input transistor followed by an opamp to provide open-loop gain and load-driving capability. The noise performance was disappointing, and this is a matter I still feel I have not got quite to the bottom of.

There was a major upgrade to the tone-control section, with the standard Baxandall stage being replaced by a non-Baxandall configuration that gave fully variable bass and treble turnover frequencies, free of interaction with the amounts of boost and cut demanded. This was the application of mixing console technology to preamplifiers. The tone-control is fully described in Chapter 6, the second part of the Precision Preamplifier '96 article, and its further development and application to the Linear Audio Low-Noise Preamplifier is comprehensively covered in Chapter 18.

Naturally this design is showing its age a little as my design techniques have moved on. If I was revisiting the design tomorrow, I would:

1) Update the MC head amp stage for lower noise as described in Chapter 15, the second article on the Preamplifier 2012.

2) Convert the MM preamp stage from RIAA Configuration A to Configuration C, to reduce the amount of precision capacitance required. This is fully explained in Chapter 19 on RIAA network optimisation.

3) In several places, a particular resistance value is obtained by putting a small value in series with a large one. This allows very precise nominal values to be obtained, but the tolerance is essentially that of the larger resistor. If, on the other hand, it is possible to make up the required value with sufficient accuracy by using two resistors of the same value, either in series or

parallel, then the variation in total resistance is reduced by a factor of √2. If three equal resistors are used the variation is reduced by a factor of √3, and so on. Therefore a better design would use two or perhaps three near-equal resistors to set the critical parameters of the RIAA network. Three resistors give a much greater number of combinations and allow a nominal value to be approached much more closely than do two, and the extra cost is very small; however the design procedure is, as you might imagine, rather more complex. This is all explained in some detail in Chapter 19 on RIAA network optimisation.

4) The IEC amendment is an extra 6 dB/octave roll-off at 22 Hz, introduced without much if any consultation, apparently to make life easier for those promoting disk noise-reduction systems like dbx [1]. At the time of the design work, there seemed to be a vague notion about that it ought to be included to give an accurate RIAA response. I therefore duly included it, by integrating the extra roll-off with the response of the subsonic filter, for economy of components; the result is not mathematically exact, but the errors involved are tiny. As time went by, opinion hardened against the IEC amendment and the general conclusion was that it should be made switchable in/out, or if that was too expensive just left out altogether. With my integrated approach, switching would have been distinctly clumsy, and I would change that now. My later designs have a subsonic filter with a standard Butterworth response, and the IEC amendment is implemented as a separate CR network that can be switched in/out easily. You've gotta ask yourself one question, punk, does the IEC even matter at all if you have a third-order subsonic filter cutting in around 20 Hz?

5) The third-order subsonic filter in the Precision Preamplifier '96 is of the single-stage Sallen & Key type [2], where all the time constants are in front of the opamp. This has the merit that there is no time-constant after the opamp so it can drive a low impedance unhindered, and so an extra buffer opamp is avoided. Unfortunately, while I did not appreciate it at the time, this sort of configuration may amplify the distortion of the opamp by a greater factor than a conventional multi-stage Sallen & Key filter [3], and I would now use the multi-stage configuration as in the Preamplifier 2012, where the third time-constant comes after the opamp and does interact with the other two. The subsonic filter in Preamplifier 2012 is not switchable in/out, but it could be easily made so.

There is one statement in the article that raises my eyebrow. In the text box 'Filtering Subsonics', I say that 100V polyester caps generate ten times less capacitor distortion than those rated at 63V. After more investigation, I now have grave doubts as to whether that general statement is actually true; if distortion is an issue, then polystyrene or polypropylene capacitors eliminate it. Trying to pick and choose between different types of polyester capacitor is probably not a good route to go. There is always the possibility of variation in non-linearity between different batches of the same polyester part because this parameter is never specified on data sheets.

6) When this preamplifier was designed, the 5532 was the pre-eminent audio opamp, though it was expensive compared with the TL072. Nowadays, the 5532 is usually the cheapest opamp you can buy, and a better but commensurately more expensive opamp is available in the shape of the LM4562. I would now certainly put one of those in the subsonic filter, to reduce noise and distortion, but I would retain the 5534A in the MM stage because its combination of voltage

and current noise is better suited to MM cartridge impedances. When allocating opamps in your design you want to avoid sharing the same package between left and right channels, as this inevitably brings their circuitry close together and worsens interchannel crosstalk.

This was Preamplifier MRP11 in my numbering system.

References

1. http://en.wikipedia.org/wiki/Dbx_%28noise_reduction%29 (accessed Feb. 2015)

2. Self, D., *The Design of Active Crossovers,* pp. 218–219. Focal Press 2011. ISBN 978-0-240-81738-5.

3. Billam, P.J., Harmonic Distortion in a Class of Linear Active Filter Networks, *Journ Audio Eng Soc,* Volume 26, Number 6, June 1978, pp. 426–429.

PRECISION PREAMPLIFIER '96, PART I

July/August 1996

A new preamp design is timely. There is more variation in audio equipment than ever before, so to a greater extent preamps are required to be all things to all persons. High source resistance outputs and low-impedance inputs must be catered for, as well as ill-considered and exotic cabling with excessive shunt capacitance. The last preamp design I placed before the public was in 1983, [1] extended in facilities by the moving-coil head amp stage published in 1987.[2]

In the last ten years, small-signal analogue electronics has undergone few changes. Most circuitry is still made from *TL072s*, with resort to 5532s when noise and drive capability are important. In this period many new op-amps have appeared, but few have had any impact on audio design; this is largely a chicken/egg problem, for until they are used in large numbers the price will not come down low enough for them to be used in large numbers. Significant advantage over the old faithfuls is required.

This new design uses the architecture established in Ref. 1, which has not been improved upon so far. The already low noise levels have been further reduced. The tone controls were fixed-frequency, and proved inflexible compared with the switched-turnover versions in my previous designs, [3,4] so these frequencies are now fully variable, and a non-interrupting tone-cancel facility provided.

This preamplifier is designed to my usual philosophy of making it work as well as possible, by the considered choice of circuit configurations etc., rather than the alternative approach of specifying exotic components and hoping for the best.

The evolution of preamplifiers

Adding tape facilities and tone control

There are two basic architectures for tape record/replay handling. The simpler, in Figure 5.1(d), adds a tape output and a tape monitor switch for off-tape monitoring on triple-head machines.

The more complex version in Figure 5.1(e) allows any input to be listened to while any input is being recorded, though how many people actually do this is rather doubtful. This method demands very high standards of crosstalk inside the preamp. There is usually no tape return input or tape monitor switch as there is now no guarantee that the main path signal comes from the same original source as the tape output.

The final step is to add tone controls. They need a low-impedance drive for predictable equalisation curves, and a vital point is that most types—including the Baxandall—phase-invert. Since the maintenance of absolute polarity is required, this inversion can conveniently be undone by the active gain control, which also uses shunt feedback and phase-inverts. The tone-control can be placed before or after the volume control, but if afterwards it generates noise that cannot be turned down. Putting it before the volume control reduces headroom if boost is in use, but since maximum boost is only +10 dB, the preamp inputs will not overload before 3 V r.m.s is applied; domestic equipment can rarely generate such levels. Figure 5.1(f) shows the final architecture.

Minimal requirements are source selection and level control, as in Figure 5.1(a); an RIAA disc preamp stage is one input option. This sort of 'passive preamplifier' (a nice oxymoron) is only practical if the main music source is a low-impedance high-level output like CD.

The only parameter to decide is the resistance of the volume pot; it cannot be too high because the output impedance, which reaches a maximum of one quarter the track resistance at −6 dB, will cause high-frequency roll-off with the cable capacitance. On the other hand, if the pot resistance is too low, the source equipment will be unduly loaded. If the source is valve equipment, which does not respond well to even moderate loading, the problem starts to look insoluble.

Adding a unity-gain buffer stage after the selector switch, Figure 5.1(b), means the volume control can be reduced to 10 kΩ, without loading the sources. This still gives a maximal output impedance of 2.5 kΩ, which allows you only 5.4m of 300 pF/m cable before the response is 1.0 dB down at 20 kHz. For 0.1 dB down at 20 kHz, only 1.6m is permissible.

The input *RC* filters found on so many power-amps as a gesture against transient intermodulation distortion add extra shunt capacitance ranging from 100 pF to 1000 pF, and can cause additional unwanted h.f. roll off.

Unfortunately only a CD source can fully drive a power amplifier. Output levels for tuners, phono amps and domestic tape machines are of the order of 150 mV rms, while power amplifiers rarely have sensitivities lower than 500 mV. Both output impedance and level problems are solved by adding a second amplifier stage as Figure 5.1(c), this time with gain. The output level can be increased and the output impedance kept down to 100 Ω or lower.

This amplifier stage introduces its own difficulties. Nominal output level must be at least 1 V r.m.s. (for 150 mV in) to drive most power amps, so a gain of 16.5 dB is needed. If you increase the full-gain output level to 2 V r.m.s., to be sure of driving exotica to its limits, this becomes 22.5 dB, amplifying the input noise of the gain stage at all volume settings. Noise performance thus deteriorates markedly at low volume levels—the ones most of us use most of the time.

One answer is to split the gain before and after the volume control, so that there is less gain amplifying the internal noise. This inevitably reduces headroom before the volume control. Another solution is double gain controls—an input-gain control to set the internal level appropriately, then an output volume control that requires no gain after it.

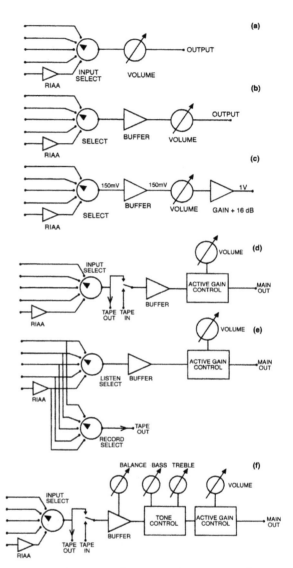

Figure 5.1 The course of preamp evolution, as impedance and level matching problems are dealt with.

Input gain controls can be separate for each channel, doubling as a balance facility.[3] However this makes operation rather awkward. No matter how attenuation and fixed amplification are arranged, there are going to be trade-offs on noise and headroom.

All compromise is avoided by an active gain stage, i.e. an amplifier stage whose gain is variable from near-zero to the required maximum. You get lower noise at gain settings below maximum, and the ability to generate a quasilogarithmic law from a linear pot. This gives excellent channel balance as it depends only on mechanical alignment.

Design philosophy

There is great freedom of design in small-signal circuitry, compared with the intractable problems of power amplification. Hence there is little excuse for a preamp that is not virtually transparent, with very low noise, crosstalk and THD.

Requirements for the RIAA network

- The RIAA network must use series feedback, as shunt feedback is 14 dB noisier.
- Correct gain at 1 kHz. Sounds elementary, but you try calculating it.
- Accuracy. The 1983 model was designed for ±0.2 dB accuracy 20–20 kHz, which was the limit of the test gear I had access to at the time. This is tightened to ±0.05 dB without using rare parts.
- It must use obtainable components. Resistors will be E24 series and capacitors E12 at best, so intermediate values must be made by series or parallel combinations.
- Ro (Figure 5.2), must be as low as possible as its Johnson noise is effectively in series with the input signal. This is most important in moving-coil mode.
- The feedback network impedance to be driven must not be low enough to increase distortion or limit output swing—especially at high frequencies.
- The resistive path through the feedback arm should ideally have the same d.c. resistance as input bias resistor R18 (Figure 5.8), to minimise offsets at A1 output.

The circuitry here meets all these requirements.

Once all the performance imperatives are addressed, the extra degrees of freedom can be used to, say, make components the same value for ease of procurement. Opamp circuitry is used here, apart from the hybrid moving-coil stage. The great advantage is that all the tricky details of distortion-free amplification are confined within the small black carapace of a 5532.

One route to low noise is low-impedance design. By minimising circuit resistances the contribution of

Johnson noise is reduced, and hopefully conditions set for best semiconductor noise performance. This notion is not exactly new—as some manufacturers would have you believe—but has been used explicitly in audio circuitry for at least 15 years.

In the equalisation and AGS stages, gains of much less than one are sometimes required. In these cases, avoiding the evils of attenuation-then-amplification (increased noise) and amplification-then-attenuation (reduced headroom) requires the use of a shunt feedback configuration. In the classic unity-gain stage, the shunt amplifier works at a noise gain of ×2, as opposed to unity, so using shunt feed-back introduces a noise compromise at a very fundamental level.

Absolute phase is preserved for all input and outputs.

The preamp gain structure

Compared with Ref. 1, the moving-magnet disc amplifier gain has been increased from +26 to +29 dB (all levels are at 1 kHz) to bring the line-out level up to 150 mV nominal. This is done to match equipment levels that appear to have reached some sort of consensus on this value. The input buffer has a gain of +1.0 dB with balance central.

The maximum gain of the AGS is therefore reduced from +26 to +22 dB, to retain the same maximum output of 2 V. This affects only the upper part of the gain characteristic.

Disc input

While vinyl as a music-delivery medium is almost as obsolete as wax cylinders, there remain many sizable album collections that it is impractical to either replace with CDs or transfer to digital tape. Disc inputs must therefore remain part of the designer's repertoire for the foreseeable future.

The disc stage here accepts a moving-coil cartridge input of 0.1 or 0.5 mV, or a moving-magnet input of 5 mV. It also includes a third-order subsonic filter and the capability to drive low impedances. The moving-coil stage simply provides flat gain, of either 10 or 50 times, while the moving-magnet stage performs the full RIAA equalisation for both modes.

Moving-coil input criteria

This stage was described in detail in Ref. 2. The prime requirement is a good noise figure from a very low source impedance—here 3.3 Ω to comply with, for example, the Ortofon *MC10* cartridge. The circuit features

- triple low-r_b input transistors
- two separate d.c. feedback loops
- combined feedback-network and output-attenuator.

The very low value of R_6 means that a series capacitor to reduce the gain to unity at d.c. is impracticable; there is no d.c. feedback through R_7, R_{10} around the global loop. Local d.c. negative feedback via R_2, R_3 sets input transistor conditions, and dc servo IC_2 applies whatever is needed to IC_1 non-inverting input to bring IC_1 output to 0 V.

The two gains provided are 10× and 50×, so inputs of 0.5 mV and 0.1 mV will give 5m V r.m.s. out. The equivalent input noise of the moving-coil stage alone is −141 dBu, with no RIAA. Johnson noise from a 3.3 Ω resistor is −147 dBu, so the noise figure is a rather good 6 dB. Resistor R_6 is also 3.3 Ω. This component generates the same amount of noise as the source impedance, which only degrades the noise figure by 1.4 dB, rather than 3 dB, as transistor noise is significant.

If discrete transistors seem like too much trouble, remember a 5532 stage here would be at least 15 dB noisier.

The moving-magnet input stage

The first half of Morgan Jones's excellent preamp article [5] appeared just after this preamp design was finalised. While I thoroughly endorse most of his conclusions on RIAA equalisation, we part company on two points. Firstly, I am sure that 'all-in-one-go' RIAA equalisation as in Figure 5.2(a) is definitely the best method, for IC op-amp designs at least. In my design the resultant loss of high-frequency headroom is only 0.5 dB at 20 kHz, which I think I can live with.

Secondly, I do not accept that the difficulties of driving feedback networks with low-impedance at h.f. are

insoluble. I quite agree that 'very few preamps of any age' meet a +28 dB ref 5 mV overload margin, but some exceptions are Ref. 1 with +36 dB, Ref. 3 with +39 dB, and Ref. 4 with a tour-de-force +47 dB. My design here gives +36 dB across most of the audio band, falling to +33 dB at 20 kHz (due to h.f. pole-correction) and +31 dB at 10 Hz (due to the IEC rolloff being done in the second stage).

Many contemporary disc inputs use an architecture that separates the high and low RIAA sections. Typically there is a low-frequency RIAA stage followed by a passive h.f. cut beginning at 2 kHz, Figure 5.2(b). The values shown give a correct RIAA curve.

Amplification followed by attenuation always implies a headroom bottleneck, and passive h.f. cut is no exception. Signals direct from disc have their highest amplitudes at high frequencies so this passive configuration gives poor h.f. headroom. Overload occurs at A1 output before passive h.f. cut can reduce the level.

Figure 5.3 shows how the level at A1 output (Trace B) is higher at h.f. than the output signal (Trace A). Trace C shows the difference, i.e. the headroom loss; from 1 dB at 1 kHz this rises to 14 dB at 10 kHz and continues to increase in the ultrasonic region. The passive circuit was driven from an inverse RIAA network. Using this, a totally accurate disc stage would give a straight line just below the +30 dB mark.

A related problem is that A1 in the passive version must handle a signal with much more h.f. content than A1 in Figure 5.2(a). This worsens any difficulties with slew-limiting and h.f. distortion. The passive version uses two amplifier stages rather than one, and more precision components.

Another difficulty is that A1 is more likely to run out of open-loop gain at h.f. This is because the response plateaus above 1 kHz, rather than being steadily reduced by increasing negative feedback. Passive RIAA is not an attractive option.

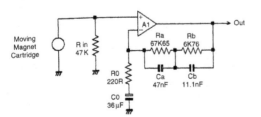
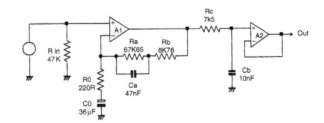

Figure 5.2 The basic RIAA configurations. Figure 5.2(a) is the standard 'all-in-one-go' series feedback configuration; the values shown do not give accurate RIAA equalisation. Figure 5.2(b) is the most common type of passive RIAA, with a headroom penalty of 14 dB at 10 kHz.

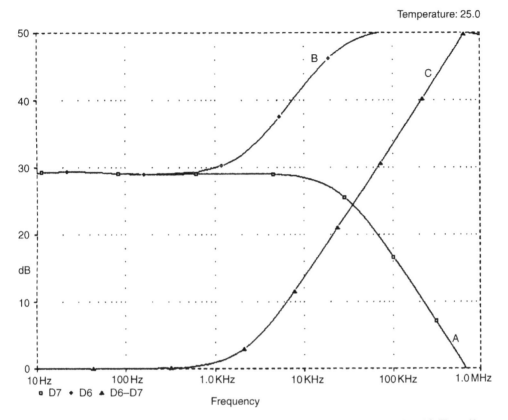

Figure 5.3 Headroom loss with passive RIAA equalisation. The signal at A1 (Trace B) is greater than A2 (Trace A) so overload occurs there. The headroom loss is plotted as Trace C.

Alternatively there may be a flat input stage followed by a passive h.f. cut and then another stage to give the l.f. boost, which has even more headroom problems and uses yet more bits. The 'all-in-one-go' series feedback configuration in Figure 5.2(a) avoids unnecessary headroom restrictions and has the minimum number of stages.

In search of accurate RIAA

I have a deep suspicion that such popularity as passive RIAA has is due to the design being much easier. The time-constants are separate and non-interactive; only the simplest of calculations are required.

In contrast the series-feedback system in Figure 5.2(a) has serious interactions between its time-constants and design by calculation is complex. The values shown in Figure 5.2(a) are what you get if you ignore the interactions and simply implement the time-constants as $R_a \times C_a$

equals 3180 μs, $R_b \times C_a$ equals 318 μs, and $R_b \times C_b$ equals 75 μs. The resulting errors are ±0.5 dB ref 1 kHz.

Empirical approaches (cut-and-try) are effective if great accuracy is not required, but attempting to reach even ±0.2 dB by this route becomes very tedious and frustrating. Hence the Lipshitz equations[6] have been converted to a spreadsheet, and used to synthesise the design in Figure 5.8.

A great deal of rubbish has been talked about RIAA equalisation and transient response, in perverse attempts to render the shunt RIAA configuration acceptable despite its crippling 14 dB noise disadvantage. The heart of the matter is that the RIAA replay characteristic apparently requires the h.f. gain to fall at a steady 6 dB/octave forever.

A series-feedback disc stage-with relatively low gain cannot make its gain fall below one, and so the 6 dB/octave fall tends to level out at unity early enough to cause errors in the audio band. Adding a high-frequency correction pole—i.e. low-pass time constant—just after the input stage makes the simulated and measured frequency response identical to a shunt-feedback version, and retains the noise advantage.

At this level of accuracy, the finite gain open-loop gain of even a 5534 at h.f. begins to be important, and the frequency of the h.f. pole is trimmed to allow for this.

What RIAA accuracy is possible without spending a fortune on precision parts? The best tolerance readily available for resistors and capacitors is ±1%, so at first it appears that anything better than ±0.1 dB accuracy is impossible. Not so. The component-sensitivity plots in Figures 5.4, 5.5 show the effect of 1% deviations in the value of R_a, R_b; the response errors never exceed 0.05 dB, as there are always at least two components contributing to the RIAA response.

Sensitivity of the RIAA capacitors is shown in Figures 5.6, 5.7 and you can see that tighter tolerances are needed for C_a and C_b, than for R_a and R_b to produce the same 0.05 dB accuracy. The capacitors have more effect on the response than the resistors.

Finding affordable close-tolerance capacitors is not easy; the best solution seems to be, as in 1983, axial polystyrene, available at 1% tolerance. These only go up to 10 nF, so some parallelling is required, and indeed turns out to be highly desirable. The resistors are all 1%, which is no longer expensive or exotic, though anything more accurate certainly would be.

For C_a, the five 10 nF capacitors in parallel reduce the tolerance of the combination to 0.44%. This statistical trick works because the variance of equal summed components is the sum of the individual variances. Thus for five 10 nF capacitors, the standard deviation (square root of variance) increases only by the square root of five, while total capacitance has increased five times. This produces an otherwise unobtainable 0.44% close-tolerance 50 nF capacitor.

Similarly, C_b is mainly composed of three 4n7 components and its tolerance is improved by root-three, to 0.58%.

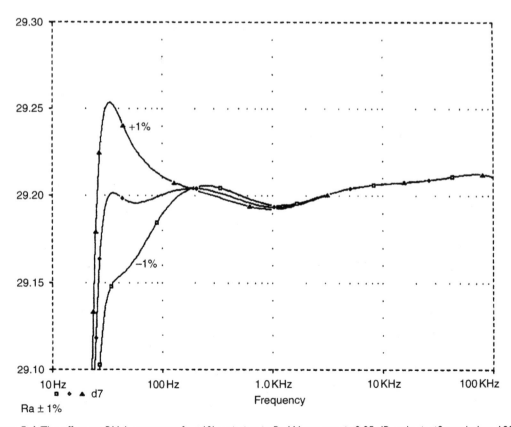

Figure 5.4 The effect on RIAA accuracy of a ±1% variation in Ra. Worst-case is 0.05 dB, only significant below 100 Hz.

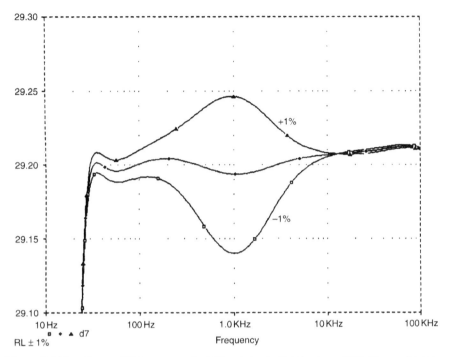

Figure 5.5 The effect on RIAA accuracy of a ±1% variation in Rb. Worst-case 0.05 dB around 1 kHz.

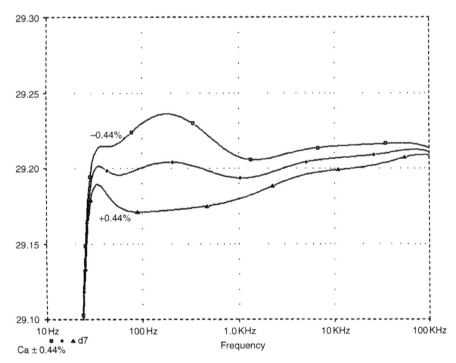

Figure 5.6 The effect on RIAA accuracy of a ± 0.44% variation in Ca. Effect is less than ± 0.05 dB at low frequencies, with a small effect on the upper audio band.

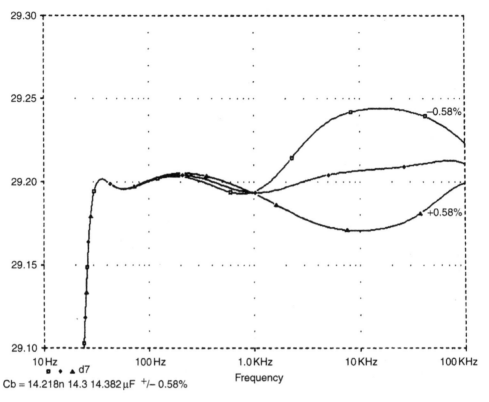

Figure 5.7 The effect on RIAA accuracy of a ±0.58% variation in Cb. Effect is less than ±0.05 dB on top four octaves. Smaller variation is permissible in the capacitors for the same RIAA error.

Noise considerations

The noise performance of any input stage is ultimately limited by Johnson noise from the input source resistance. The best possible equivalent input noise data for resistive sources, for example microphones with a 200 Ω source resistance, i.e. –129.6 dBu, is well-known, but the same figures for moving-magnet inputs are not.

It is particularly difficult to calculate equivalent input noise for moving magnet stages as a highly inductive source is combined with the complications of RIAA equalisation.[7] The amount by which a real amplifier falls short of the theoretical minimum equivalent input noise is the noise figure, NF. I often wonder why noise figures are used so little in audio; perhaps they are a bit too revealing.

The noise performance of disc input stages depends on the input source impedance, the cartridge inductance having the greatest influence. It is vital to realise that no value of resistive input loading will give realistic noise measurements.

A 1 kΩ load models the resistive part of the cartridge impedance. But it ignores the fact that the 'noiseless' inductive reactance makes the impedance seen at the preamp input rise very strongly with frequency, so that at higher frequencies most of the input noise actually comes from the 47 kΩ loading resistance. I am grateful to Marcel van de Gevel [8] for drawing my attention to this point.

Hence, for the lowest noise you must design for a higher impedance than you might think, and it is fortunate that the RIAA provides a treble roll-off, or the noise problem would be even worse than it is. This is not why it was introduced. The real reason for pre-emphasis/de-emphasis was to discriminate against record surface noise. Table 5.1 shows the two most common audio op-amps, the 5532 being definitely the best and quieter by 5 dB.

To calculate appropriate EINs, I built a spreadsheet mathematical model of the cartridge input, called MAG-NOISE. The basic method is as in Ref. 9. The audio band

50–22 kHz is divided into nine octaves, allowing RIAA equalisation to be applied, and the equivalent generators of voltage noise (e_n) and current noise (i_n) to be varied with frequency.

Filtering subsonics

This stage is a third-order Butterworth high-pass filter, modified for a slow initial rolloff that implements the IEC amendment. This is done by reducing the value of R27 + R28 below that for maximal flatness. The stage also buffers the high-frequency correction pole, and gives the capability to drive a 600 Ω load, if you can find one.

Capacitor distortion10 in electrolytics is—or should be—by now a well-known phenomenon. It is perhaps less well known that non-electrolytics can also generate distortion in filters like these. This has nothing to do with Subjectivist musicality, but is very real and measurable.

The only answer appears to be using the highest-voltage capacitors possible; 100 V polyester generates ten times less distortion than the 63 V version.

Table 5.1 Measured noise results, showing the 5532's superiority

Z_{source}	TL072	5532	5532	5532
			benefit	EIN
1k	−88.0	−97.2 dBu	+9.8 dB	−126.7 dBu
Shure M75ED	−87.2	−92.3 dBu	+5.1 dB	−121.8 dBu

(Preamp gain +29.55 dB at 1 kHz. Bandwidth 400–22 kHz, r.m.s. sensing)

Noise generated by the 47 kΩ resistor R_{in} is modelled separately from its loading effects so its effect can be clearly seen. I switched off the bottom three octaves to make the results comparable with real cartridge measurements that require a 400 Hz high-pass filter to eliminate hum, and 1/f effects are therefore neglected. No psychoacoustic weighting was used, and cartridge parameters were set to 610 Ω + 470 mH, the measured values for the Shure *M75ED*.

The results match well with my 5532 and *TL072* measurements, and I think the model is a usable tool. Table 5.2 shows some interesting cases; output noise is calculated for gain of +29.55 dB at 1 kHz, and signal-to-noise ratio for a 5 mV r.m.s. input at 1 kHz.

I draw the following conclusions. The minimum equivalent input noise from this particular cartridge, without the extra thermal noise from the 47 kΩ input loading, is −133.5 dBu, no less than 7 dB quieter than the loaded cartridge. (Case 1) It is the quietest possible condition. The noise difference between 10 MΩ and 1 MΩ loading is still 0.2 dB, but as loading resistance is increased further to 1000 MΩ the EIN asymptotes to −133.5 dBu. A 47 kΩ loading is essential for correct cartridge response.

With 47 kΩ load, the minimum EIN from this cartridge is −126.5 dBu. (Case 5) All other noise sources, including R_0, are ignored. This is the appropriate noise reference for this preamp design.

Resistor R_0, the 220 Ω resistor in the bottom arm of negative feedback network, adds little noise. The difference between Case 5 and Case 7 is only 0.3 dB.

A disc preamp stage using a good discrete bipolar device such as the remarkable *2SB737* transistor (r_b only 2 Ω typ) is potentially 2.8 dB quieter than a *5532*, when the noise from R_0 and the input load are included. Compare Cases 11 and 16.

The calculated noise figure for a 5532 is 4.5 dB. Measured noise output of the moving magnet stage is −92.3 dBu (1 kHz gain +29.5 dB) and so the equivalent

Table 5.2 Calculated minimum noise results

Case	e_n	i_n	R_{in}	R_o (R)	Output	S/N ref (dB)	EIN
	nV/√Hz	pA/√Hz			dBu	5 mV	dBu
1 Noiseless amp	0	0	1000 *M*	0	−104.0	−89.7	−133.5A
5 Noiseless amp	0	0	47 k	0	−97.1	−82.8	−126.5C
7 Noiseless amp	0	0	47 k	220	−96.7	−82.4	−126.2
11 *2SB737*, I_c = 70 μA	1.7	0.4	47 k	220	−95.3	−81.0	−124.8
16 5532	5	0.7	47 k	220	−92.5	−78.2	−122.0
18 TL072	18	0.01	47 k	220	−86.9	−72.6	−116.5

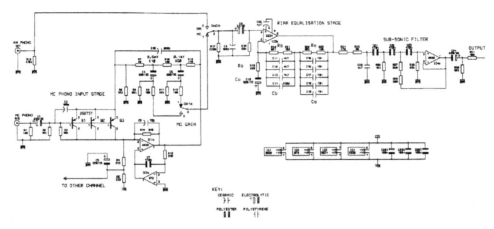

Figure 5.8 Switchable for moving coil or moving-magnet type cartridges, the disc amplifier includes a subsonic filter to reduce cone excursions and distortion due to warped vinyl.

input noise is −121.8 dBu, and the real noise figure is 4.7 dB, which is not too bad. Noise from the subsonic filter is negligible.

Taking e_n and i_n from data books, it looks as though the 5534/5532 is the best op-amp possible for this job. Other types—such as *OP-27*—give slightly lower calculated noise, but measure slightly higher. This is probably due to extra noise generated by bias current-cancellation circuitry.[8]

There is an odd number of half-5532s, so the single 5534 is placed in the moving-magnet stage, where its slightly lower noise is best used. The RIAA-equalised noise output from the disc stage in moving-coil mode is −93.9 dBu for 10× times gain, and −85.8 dBu for 50× times. In the 10× case the moving-coil noise is actually 1.7 dB lower than moving-magnet mode.

Circuit details

The complete circuit of the disc amplifier and subsonic filter is Figure 5.8. Circuit operation is largely described above, but a few practical details are added here. Resistors R_9 and R_{12} ensure stability of the moving-coil stage when faced with moving-magnet input capacitance C_8, while R_8 and R_{11} are dc drains.

The 5534 moving-magnet stage has a minimum gain of about 3×, so compensation should not be required; if it is, a position is provided (C_{26}) for external capacitance to be added; 4.7 pF should be ample. The moving magnet stage feedback arm $R_{20–23}$ has almost exactly the same d.c. resistance as the input bias resistor R_{18}, minimising the offset at the output of IC_3. The h.f. correction pole is $R_{24} + R_{25}$ and C_{20}.

Capacitor C_{24} is deliberately oversized so low loads can be driven. Resistor R_{31} ensures stability into high-capacitance cables.

References

1. Self, D. 'A precision preamplifier', *Wireless World*, October 1983, p 31.

2. Self, D. 'Design of moving-coil head amplifiers', *Electronics & Wireless World*, December 1987, p 1206.

3. Self, D. 'An advanced preamplifier design', *Wireless World*, November 1976, p 41.

4. Self, D. 'High-performance preamplifier', *Wireless World*, February 1979, p 43.

5. Jones, M. 'Designing valve preamps', *Electronics World*, March 1996, p 192.

6. Lipshitz, S. 'On RIAA equalisation networks', *Journ Audio Eng Soc*, June 1979, p 458.

7. Walker, H.P. 'Low noise amplifiers', *Wireless World*, May 1972, p 233.

8. van de Gevel, M. Private communication, February 1996.

9. Sherwin, J. 'Noise specs confusing?', National Semiconductor Application Note AN-104, Linear Applications Handbook 1991.

10. Self, D. 'Capacitor distortion, Views', *HiFi News & RR*, November 1985, p 23.

Chapter 6

Precision preamplifier '96, Part II

(Electronics World, September 1996)

The input switching was a radical new design that gave exceptionally good isolation between the inputs. You could argue that this was pointless as no one sensible would be running sources into the amplifier apart from the one they were actually listening to, and you might be right. Morgan Jones raised this particular issue with a recommendation that alternate positions on rotary input-select switches should have every other switch position grounded to reduce inter-source crosstalk [1], and so I decided to see what could be done with push-switches instead. The answer is −95 dB at 10 kHz, which I would suggest is enough for most of us.

The following line input stage, like that in the original Precision Preamplifier (1983) was given a very high input impedance by preamplifier standards, achieved not by bootstrapping this time but simply using a high-value bias resistor R33; this works fine so long as you can accommodate the offset resulting from the opamp bias current flowing through the resistor. The stage also implemented an active vernier-balance control.

There was a major upgrade to the tone-control section that gave the usual bass and treble controls continuously variable turnover frequencies. A lot of work went into using 10 kΩ linear pots everywhere, with suitable control-laws obtained by various devious means; the reasoning behind this was not so much to make ordering the parts easier (though that was a consideration), but to avoid the channel imbalances inherent in log and anti-log pots. The variable-frequency tone-controls worked very well indeed, although this approach went dead against the prevailing fashion for omitting tone-controls altogether; this did not bother me greatly.

This tone-control design clearly had more impact than I thought. In 2012, I casually mentioned that I had developed an improved version of it, and there was quite a clamour on for me to write it up for *Linear Audio.* Since a tone-control by itself was of limited utility, I added a line input, balance, and volume control, and *voila,* the *Linear Audio* Low-noise Preamplifier of 2013 was born. This was published in *Linear Audio,* Volume 5, April 2013, and that article is included later in this book as Chapter 18.

The active volume control section of the Precision Preamplifier '96 design also shows its age a little. If I was updating the design tomorrow, and looking to improve the performance, I would use one of the new and quieter versions of the volume control described in Chapters 14 and 18, or at least make use of the same principles for reducing noise, as follows:

1) The impedance of the volume pot would be reduced from 10 kΩ to 5 kΩ, 2 kΩ, or 1 kΩ. Lower values than this are difficult to obtain as two-gang pots, though you can go as low as you like by using by using multiway switches and resistor ladders. The lower value pots mentioned require special circuitry to drive their low impedances.

2) The basic active volume-control uses a buffer, to isolate the rest of the circuitry from the pot wiper, and an inverting amplifier to provide the gain. Multiplying either amplifier in number reduces noise by averaging when the outputs are joined together by low-value resistors. More noise reduction is obtained by multiplying the inverting amplifiers (which have gain) rather than the buffers.

3) The gain of the volume-control can be set to zero, or at any rate very close to it, this being determined by the end-resistance of the volume pot track. However, the noise output of the stage does not drop to zero; there is residual output noise that originates in the voltage-noise of the opamps. This will be very low (it is about −121 dBu in the *Linear Audio* Low-noise Preamplifier), but it is not zero as it would be with a passive volume pot. In one of my recent preamplifier designs (not for publication), a passive pot was added after the active volume stage; if it is an unloaded linear pot then the volume still only depends on the angular control setting and good channel balance is preserved. This approach requires a four-gang pot but naturally it really does give zero noise at zero volume setting [2]. That will of course be compromised if there are further stages downstream, such as a balanced output or a ground-cancelling output, but these are normally unity-gain and can be designed to be very quiet.

These improvements can be seen demonstrated in the *Elektor* Preamp 2012 (Chapter 14), which uses a 1 kΩ pot, four buffers, and four inverting amplifiers, and the *Linear Audio* Low-noise Preamplifier (Chapter 18), which uses a 5 kΩ pot, a single buffer, and two inverting amplifiers.

This was Preamplifier MRP11 in my numbering system.

References

1. Jones, M, Designing Valve Preamps, *Electronics World,* March 1996, p. 193.
2. Self, D, *Small-Signal Audio Design,* 2nd edition, Chapter 9, pp. 368–369. Focal Press (Taylor & Francis) 2014. ISBN: 978-0-415-70974-3. Hardback. 978-0-415-70973-6 Paperback. 978-0-315-88537-7 ebook.

PRECISION PREAMPLIFIER '96, PART II

September 1996

Morgan Jones raised the excellent point of crosstalk in the input-select switching in a recent article (Valve Preamplifiers, *Electronics World*, March/April 1996). If the source impedance is significant then this may be a serious problem.

While I agree that Morgan's rotary switch with every other contact grounded may be slightly superior to conventional rotary switches, measuring a popular Lorlin switch type showed the improvement to be only 5 dB. I am also unhappy with all those redundant 'mute' positions between input selections, so I instead chose interlocked push-switches rather than a rotary. A four-pole-changeover format can then be used to reduce crosstalk.

The problem with conventional input select systems like Figure 6.1(a) is that the various input tracks necessarily come into close proximity, with significant crosstalk through capacitance C_{stray} to the common side of the switch, i.e. from A to B. Using two changeovers per input side—i.e. four for stereo—allows the intermediate connection B-C to be grounded by the NC contact of the first switch section. This keeps the 'hot' input A much further away from the common input line D, as shown in Figure 6.1(b).

Crosstalk data in Table 6.1 was gathered at 10 kHz, with 10 kΩ source impedances. The emphasis here is on minimising inter-source crosstalk, as interchannel (L–R) crosstalk is benign by comparison. Interchannel

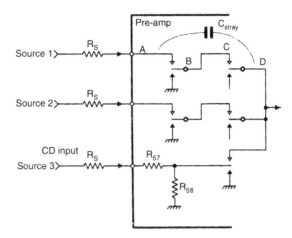

Figure 6.1b Improved input selection using four-pole switching to reduce capacitance between the different sources. The CD input attenuator can be grounded when not selected, so two-pole switching is sufficient for high isolation of Source 3.

Table 6.1 Crosstalk exhibited by four switch arrangements using a 10 kHz test signal

Simple rotary:	71 dB
Morgan-Jones rotary:	76 dB
2 c/o switch	74 dB
4 c/o switch	95 dB

isolation is limited by the placement of left and right on the same switch, with the contact rows parallel, and limits L–R isolation to −66 dB at 10 kHz with 10 kΩ source impedance.

With lower source impedances, both intersource and interchannel crosstalk is proportionally reduced. In this case, a more probable 1 kΩ source gives 115 dB of inter-source rejection at 10 kHz for the four-pole-changeover method.

Line input criteria

Nowadays the input impedance of a preamp must be high to allow for interfacing to anachronistic valve equipment, whose output may be taken from a valve anode. Even light loading compromises distortion and available output swing. A minimum input impedance is 100 kΩ, which many preamp designs fall well short of.

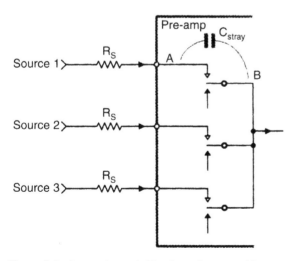

Figure 6.1a Input-select switching for audio preamplifiers—the conventional method, with poor rejection of unselected sources due to Cstray.

The CD input stands out from other line sources in that its nominal level is usually 1 V rather than 150 mV. This is perfectly reasonable, since digital sources have rigidly defined maximum output levels, and these might as well be high to reduce noise troubles. There is no danger of the analogue output section clipping. However, this means a direct line input cannot be used without the trouble of resetting volume and recording-level controls whenever the CD source is selected.

This problem is addressed here by adding a 16 dB passive attenuator, as shown for Source 3 in Figure 6.1(b). The assumption is that a CD output has a low impedance, and that a 10 kΩ input impedance will not embarrass it. As a result, resistance values can be kept low to minimise the noise degradation. Output impedance of this attenuator is 1.4 kΩ, which generates −120.9 dBu of Johnson noise as opposed to −135.2 dBu from a direct 50 Ω source. This is still much less than the preamp internal noise and so the noise floor is not degraded. It is now possible to improve inter-source crosstalk simply by grounding the CD attenuator output when it is not in use, so only a two-pole switch is required for good isolation of this source.

The tape-monitor switch allows the replay signal from the tape deck to be compared with the source signal. With three-head machines, this provides a real-time quality check. But with the much more common two-head appliances, where the input signal is looped straight back to the amplifier in RECORD mode, it only provides confirmation that the signal has actually got there and back.

Line input buffering

This stage has to provide a high input impedance and variable gain for the balance control. My last preamp [1] had the balance control incorporated in the tone stage, but this does not appear to be practical with the more complex tone system here.

The vernier balance control alters the relative stage gain by +4.5, −1.1 dB—a difference of 6 dB—which is sufficient to swing the image wholly from one side to the other. Since the minimum gain of this non-inverting stage is unity, the nominal gain with balance control central is 1.1 dB. Maximum gain of the active gain stage, or AGS, is reduced to allow for this. The active nature of this balance control means that the signal never receives unwanted attenuation that must be undone later with noisy amplification. The

gain law is modified by R_{34} to give as little gain as possible in the centre. Maximum gain is set by R_{35}, Figure 6.11(b).

A high input impedance is obtained simply by using a high-value biasing resistor R_{33}, accepting that the bias current through this will give some negative output offset; at −180 mV this is not large enough to reduce headroom. Input impedance is therefore 470 kΩ, high enough to prevent loading problems with any conceivable source equipment.

In discussing noise there are fundamental limits that lend perspective to the process. If the external source impedance is 50 Ω, which is about as low as is plausible, the inherent thermal noise from it is −135.2 dBu in 22 kHz bandwidth. This is well below the measuring equipment, (AP System 1) which has an input noise floor I measured at −116.8 dBu, 50 Ω source again.

The noise output of the buffer/balance stage is of the same order and cannot be measured directly—a good way would be to use the flat moving-coil cartridge stage as a preamp for the testgear [2]. Calculated noise output is −116 dBu with balance central.

Controlling tone

I plan to ignore convention once again. I think tone controls are absolutely necessary, and it is a startling situation when, as frequently happens, anxious inquirers to hi-fi advice columns are advised to change their loudspeakers to correct excess or lack of bass or treble. This is an extremely expensive way of avoiding tone controls.

This design is not a conventional Baxandall tone control. The break frequencies are variable over a ten to one range, because this makes the facility infinitely more useful for correcting speaker deficiencies. This enhancement flies in the face of Subjectivist thinking, but I can live with that. Variable boost/cut and frequency enables any error at top or bottom end to be corrected to at least a first approximation. It makes a major difference, as anyone who has used a mixing console with comprehensive EQ will tell you.

Middle controls are quite useless on a preamplifier. They are no good for acoustic correction: after all, even a third-octave graphic equaliser isn't that much use. Variable frequency mid controls are standard on mixing consoles because their function is voicing—i.e. giving a sound a particular character—rather than correcting response anomalies.

Certain features of the tone control may make it more acceptable to those with doubts about its sonic correctness. The tone control range is restricted to ±10 dB, rather than the ±15 dB which is standard in mixing consoles. The response is built entirely from simple 6 dB/octave circuitry, with inherently gentle slopes. The stage is naturally minimum-phase, and so the amplitude curves uniquely define the phase response. This will be shown later, where the maximum phase-shift does not exceed 40° at full boost.

This is a return-to-flat tone control. Its curves do not plateau or shelve at their boosted or cut level, but smoothly return to unity gain outside the audio band. Boosting 10 kHz is one thing, but boosting 200 kHz is quite another, and can lead to some interesting stability problems. The fixed return-to-flat time-constants mean that the boost/cut range is necessarily less at the frequency extremes, where the effect of return-to-flat begins to overlap the variable boost/cut frequencies.

The basic principle is shown in Figure 6.2. The stage gives a unity-gain inversion, except when the selective response of the side-chain paths allow signal through. In the treble and bass frequency ranges, where the sidechain does pass signal, boost/cut potentiometers $VR_{2,4}$ can give either gain or attenuation. When a wiper is central, there is a null at the middle of the boost/cut potentiometer, no signal through that side-chain, and gain is unity.

If the potentiometer is set so the side-chain is fed from the input then there is a partial cancellation of the forward signal; if the side-chain is fed from the output then there is a partial negative-feedback cancellation. To put it another way, positive feedback is introduced to counteract part of the negative feedback through R_{37}.

This apparently ramshackle process actually gives boost/cut curves of perfect symmetry. In fact this symmetry is pure cosmetics, because you can't use both sides of the curve at once, so it hardly matters if they are exact mirror-images.

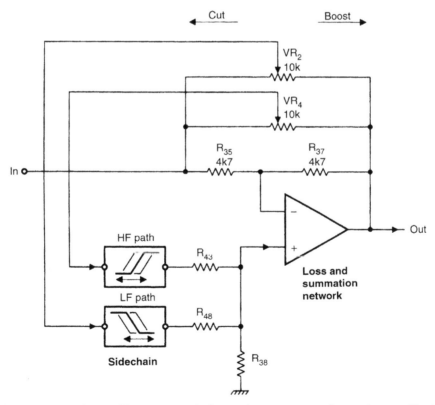

Figure 6.2 The basic tone control circuit. The response only deviates from unity gain at frequencies passed by the h.f. or if side-chain paths.

Bass and treble

The tone control stage acts in separate bands for bass and treble, so there are two parallel selective paths in the side-chain. These are simple *RC* time-constants, the bass path being a variable-frequency first-order low-pass filter, and the associated bass control only acting on the frequencies this lets through.

Similarly, the treble path is a variable high-pass filter. The filtered signals are summed and returned to the main path via the non-inverting input, and some attenuation must be introduced to limit cut and boost.

Assuming a unity-gain side-chain, this loss is 9 dB if cut and boost are to be limited to ±10 dB. This is implemented by R_{43}, R_{48} and R_{38}, Figures 6.2, 6.3 and 6.4. The side-chain is unity-gain, and so has no problems with clipping before the main path does. As a result, it is highly desirable to put the loss after the sidechain, where it attenuates side-chain noise.

The loss attenuator is made up of the lowest value resistors that can be driven without distortion. This minimises both the Johnson noise therein and noise generated by op-amp IC_{7b}.

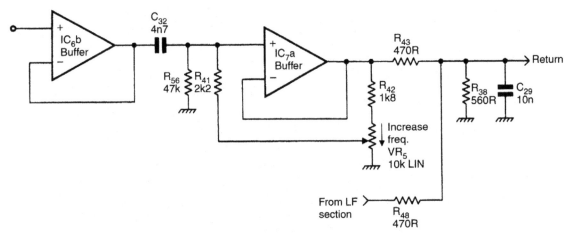

Figure 6.3 The treble frequency control circuit, with a range of 1 to 10 kHz. The variable bootstrapping of R41 via VR5 renders the control law approximately logarithmic.

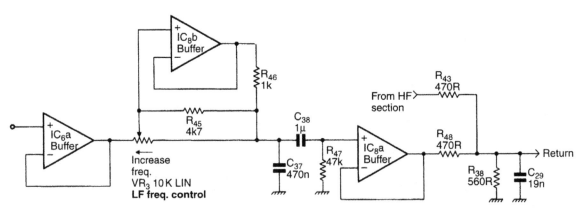

Figure 6.4 The bass frequency control circuit for 100 Hz to 1 kHz. R46 aids VR3 in driving C37 by an amount that varies with the control setting.

The tone cancel switch disconnects the entire sidechain, i.e. five out of six op-amps, from any contribution to the main path, and usefully reduces the stage output noise by about 4 dB, depending on the h.f. frequency setting. It leaves only IC_{7b} in circuit, which is required anyway to undo the gain-control phase-inversion.

Unlike configurations where the entire stage is by-passed, the signal does not briefly disappear as the switch moves between two contacts. This minimises transients due to suddenly chopping the waveform and makes valid tone in/out comparisons much easier.

Having all potentiometers identical is very convenient. I have used linear 10 kΩ controls, so the tolerances inherent in a two-slope approximation to a logarithmic law can be eliminated. This only presents problems in the tone stage frequency controls, as linear potentiometers require thoughtful circuit design to give the logarithmic action that fits our perceptual processes.

Basics of the treble path are shown in Figure 6.3. Components C_{32}, R_{41} are the high-pass time-constant, driven at low-impedance by unity-gain buffer IC_{6b}. This is needed to prevent the frequency from altering with the boost/cut setting. The effective value of R_{41} is altered over a 10:1 range by varying the amount of boot-strapping it receives-from IC_{7a}, the potential divider effect and the rise in source resistance of VR_5 in the centre combining to give a reasonable approximation to a logarithmic frequency/rotation law, Figure 6.5.

Resistor R_{42} is the frequency end-stop resistor. It limits the maximum effective value of R_{41}. Capacitor C_{29} is the treble return-to-flat capacitor. At frequencies above the audio band it shunts all the sidechain signal to ground, preventing the treble control from having any further effect.

The treble side-chain does degrade the noise performance of the tone control stage by 2–3 dB when connected. This is because it must be able to make a contribution at the h.f. end of the audio band. As you

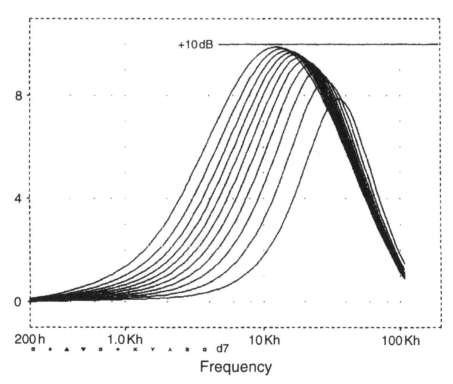

Figure 6.5 Treble frequency control law for constant increments of rotation. The curves approximate to linear spacing on the log frequency axis. PSpice simulation.

would expect, the noise contribution is greatest when the h.f. frequency is set to minimum, and so a wider bandwidth from the side-chain contributes to the main path.

The simplified bass path is shown in Figure 6.4. Op-amp IC_{6a} buffers VR_2 to prevent boost/frequency interaction. The low-pass time-constant capacitor is C_{37}, and the resistance is a combination of VR_3 and $R_{45,46}$.

Capacitors $C_{38,39}$ with R_{47} make up the return-to-flat time-constant for the bass path, which blocks very low frequencies, limiting the lower extent of bass control action. The bass frequency law is made approximately logarithmic by IC_{8b}; for minimum frequency VR_3 is set fully counter-clockwise, so the input of buffer IC_{8b} is the same as the C_{37} end of R_{46}, which is thus bootstrapped and has no effect.

Turnover

When VR_3 is fully clockwise, $R_{45,46}$ are effectively in parallel with VR_3 and the turnover frequency is at a maximum. Resistor R_{45} provides some extra law-bending,

Figure 6.6. Sadly, an extra op-amp is required. However, despite its three op-amps, the bass side-chain contributes very little extra noise to the tone stage. This is because most of its output is inherently rolled off by the low-pass action of C_{37} at high frequencies, almost eliminating its noise contribution.

Once the active elements have been chosen—here *5532s*—and the architecture made sensible in terms of avoiding attenuation-then-amplification, keeping noise-gain to a minimum, and so on, there remains one further means of improving noise performance. This is to reduce the impedance of the circuitry.

The resistances are lowered in value, with capacitances scaled up to suit, by a factor that is limited only by op-amp drive capability. This is another good reason to use the *5534/2*.

Two examples of this process as applied to the tone stage are given here. In each case the noise improvement is for the stage in isolation, set flat with high frequency set at minimum:

Firstly, in this sort of stage $R_{36,37}$ are conventionally 22 kΩ. This was reduced to 4.7 kΩ, and noise output dropped by 1.3 dB. Second, the summation/loss network

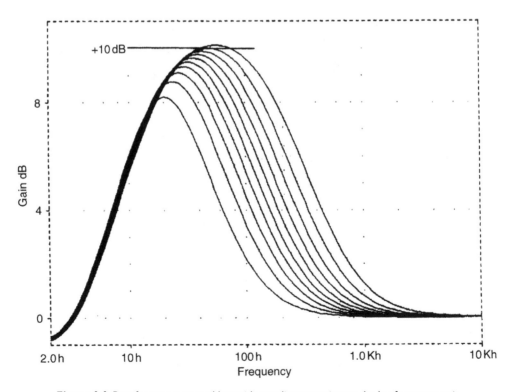

Figure 6.6 Bass frequency control law with near-linear spacing on the log frequency axis.

began with $R_{43,48}$ as 4.7 kΩ, and R_{38} as 5.6 kΩ. Reducing this by a factor of ten to 470 Ω and 560 Ω respectively reduced output noise by 0.6 dB.

With balance control central and tone cancel pressed, noise output of the tone stage, plus the line/balance buffer before it is −107.2 dBu. This is 22 kHz bandwidth. With tone controls active but set flat, noise output at minimum high frequency is −104.7 dBu, and at maximum is −106.7 dBu.

The final tone stage may look rather a mess of pottage, and be afflicted with more buffers than Clapham Junction. This is unavoidable if control interaction is to be wholly eliminated. Sadly, the practical tone circuit is somewhat more complex than Figures 6.2, 6.3 and 6.4, reflecting one of the disadvantages of low-noise opamps. This is that bipolar input stages mean that the bias currents are non-negligible. They must not be allowed to flow through potentiometers if crackling noises are to be avoided when they are moved.

These bias currents also tend to be reflected in significant output offset voltages, as the source resistances for the two op-amp inputs are not normally the same. All gain-variable circuit stages therefore have their gain reduced to unity at d.c. This subject is detailed later.

Figure 6.7 shows the measured extremes of cut and boost at the frequency extremes. Figure 6.8 gives the phase-shift at h.f. while Figure 6.9 shows phase-shift at low frequencies. In both cases it is very modest.

Active gain stage

The active gain stage, or AGS, used here as in, [1] is due to Baxandall.[3] Maximum gain is set to +23 dB by the ratio of $R_{52,53}$, to amplify a 150 mV line input to 2 V with a small safety margin.

An active volume-control stage gives the usual advantages of lower noise at gain settings below maximum,

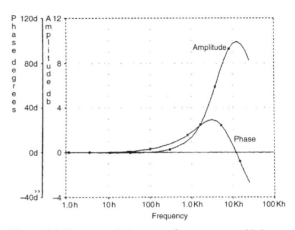

Figure 6.8 Tone control phase curve for maximum treble boost. Maximum phase-shift is 29° at about 4 kHz. PSpice simulation.

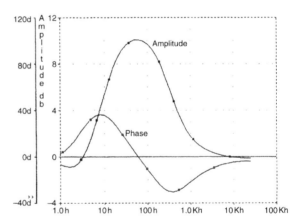

Figure 6.9 Tone-control phase curve for maximum bass boost. Maximum phase-shift is 31° at 40 Hz.

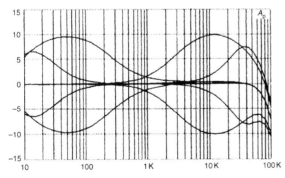

Figure 6.7 Tone-control maximum boost/cut curves (measured).

and for the Baxandall configuration, excellent channel balance that depends solely on the mechanical alignment of the dual linear potentiometer. All mismatches of its electrical characteristics are cancelled out, and there are no quasi-log dual slopes to induce anxiety.

Note that all the potentiometers are 10 kΩ linear types and identical, apart from the question of centre-detents, which are desirable only on the balance, treble and bass boost/cut controls.

Compared with, [1] noise has been reduced by an impedance reduction on the gain-definition network $R_{52,53}$. The limit on this is the ability of buffer IC_{5a} to drive R_{52}, which has a virtual earth at its other end. Figure 6.10 shows the volume control law for different maximum gain settings; only the very top end of the curve alters significantly.

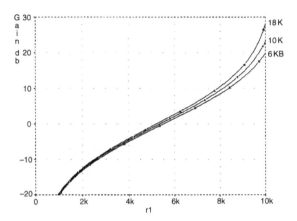

Figure 6.10 Plot of the Active Gain Stage volume control law. Varying the maximum gain has little effect except at the top end; the middle curve is the one used in the preamp.

Table 6.2 Characteristics of the tone-control stage

	Tone cancel (dBu)	Tone flat (dBu)
AGS zero gain	−114.5	−114.5
AGS unity gain	−107.4	−105.3
AGS fully up	−90.2	−86.4

Figure 6.11(a) Part 1 Circuit of input buffer, tone control, Active Gain Stage.

For the rear section of the preamp—i.e. that shown in Figure 6.11(a)—the noise performance depends on control settings. The below gives results for h.f. frequency at minimum, the worst case, Table 6.2.

The figures for maximum gain may look unimpressive, but remember this is with +23 dB of gain; at normal volume settings the noise output is below −100 dBu. I think this is reasonably quiet.

Output muting and relay control

The preamp includes relay muting on the main outputs. This is to prevent thuds and bangs from upstream parts of the audio system from reaching the power amplifiers and speakers at power-up and power down. Most op-amp circuitry, being dual-rail (i.e. outputs at 0 V) does not inherently generate enormous thumps, but it cannot be guaranteed to be completely silent. It may produce a very audible turn-on thud, and often

objectionable turn-off noise. I recall one design that emitted an unnerving screech of fading protest as the rails subsided. . . .

Electronic muting is desirable, but introduces unacceptable compromises in performance. Relay muting, given careful relay selection and control, is virtually foolproof. The relay must be normally-open so the output is passively muted when no power is applied. The control system must:

- Delay relay pull-in at power-up, to mute turn-on transients. A delay of at least 1 second before the relay closes.
- Drop out the relays as fast as possible at power-down, to stop the dying moans of the preamp, etc., from being audible.

My preferred technique is a 2 ms or there-abouts power-gone timer, held in reset by the a.c. on the mains transformer secondary, except for a brief period around the a.c. zero-crossing, too short to allow the timer to trigger. When the a.c. disappears, this near-continuous reset is removed, the timer fires, and relay power is removed within 2 ms. This is over long before the reservoir capacitors in the system can discharge, so turn-off transients are authoritatively suppressed.

However, if the mains switch contacts generate an r.f. burst that is in turn reproduced as a click by the preamplifier, then even this method may not be fast enough to completely mute it.

Figure 6.11(b) shows the practical relay-control circuit. At turn-on, R_{211} slowly charges C_{224} until Tr_{205} and D_{207} are forward biased, i.e. when C_{224} voltage exceeds that set up by $R_{214,215}$. This is the turn-on delay. Transistor Tr_{206} is then turned on via R_{213}, energising the relays, and LD_{201} is brightly lit through D_{208} and R_{216}. This led is dimly lit via R_{217} as soon as power is applied, but only brightens when the initial mute period is over.

As long as mains power is applied, Tr_{203} is kept turned on through $D_{205,206}$ by the a.c. ahead of the bridge rectifier, except during the zero-crossing period every 10 ms, when the voltage is too low for Tr_{203} base to conduct. When Tr_{203} switches off, C_{223} starts to charge through R_{208}, but is quickly discharged through R_{207} when the very brief zero-crossing period ends. If it does not end—in other words mains power has been switched off—C_{223} keeps charging until Tr_{204} turns on, discharging C_{224} rapidly via R_{210}, and removing power from the relays almost instantly.

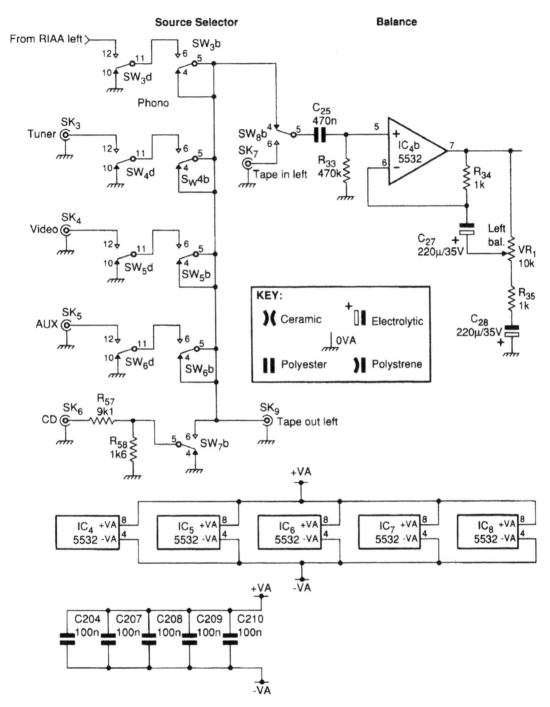

Figure 6.11a Part 2 Circuit of input buffer, tone control, Active Gain Stage.

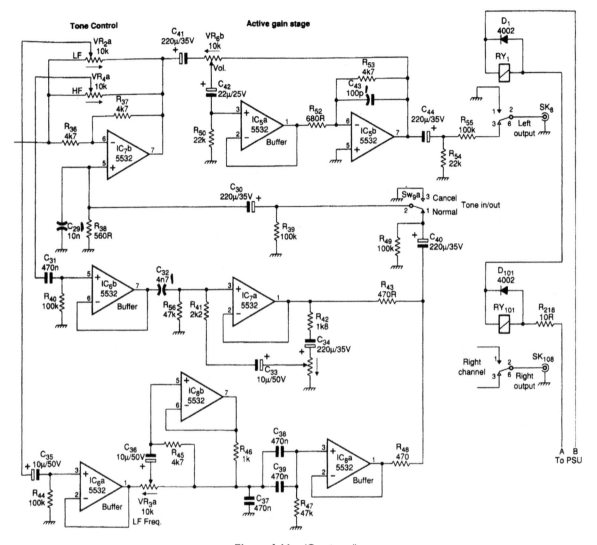

Figure 6.11a (Continued).

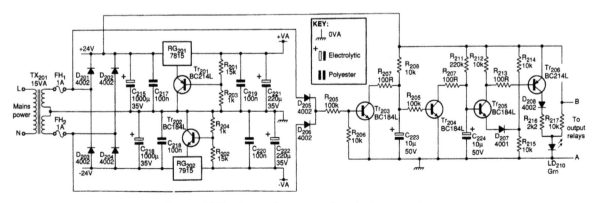

Figure 6.11b Circuit of power-supply and relay controller.

DC blocking and additional details

The preamp circuitry has been described as each stage was dealt with, so this section is confined to d.c. blocking problems and other odd subjects.

The complete circuit of the line section of the preamp is Figure 6.11(a). Bias current is kept out of balance potentiometer VR_1 by C_{27}, and d.c. gain held to unity by C_{28}. Capacitors C_{31} and C_{35} keep bias currents out of $VR_{2,4}$, necessitating bias resistors R_{40}, R_{44}.

The treble frequency law is corrected by bootstrapping through C_{33}, which keeps the bias current of IC_{7a} out of VR_5. Similarly, C_{34} prevents any offset on IC_{7a} output reaching VR_5. In the bass path C_{36} keeps IC_{8b} bias out of VR_3, while return-to-flat components $C_{38,39}$ and R_{47} provide inherent d.c.-blocking.

Final offsets at the side-chain output are blocked by C_{40}, while IC_{7b} bias is blocked by C_{30}. This is essential to prevent the tone-cancel switch clicking due to d.c. potentials. Bear in mind that this switch may still appear to click if it switches in or out a large amount of response-modification of a non-zero signal. This is because the abrupt gain-change generates a step in the waveform that is heard as a click. This is unavoidable with hard audio switching.

Capacitor C_{41} keeps IC_{7b} output offset from volume control VR_6, while C_{42} blocks IC_{5a} bias current from the pot wiper. Capacitor C_{44} gives final d.c.-blocking to protect the following power amplifier.

Many components in this design are the same value; for example, wherever a sizable non-electrolytic is required, 470 nF could usually be made to work. This philosophy has to be abandoned in areas where critical parameters are set, such as the RIAA network and tone control stage.

Supplying power

This is a conventional power supply using IC regulators. I strongly recommend that you use a toroidal mains transformer to minimise the a.c. magnetic field.

Supply rails have been increased from ±15 to ±18 V to maximise headroom. Nonetheless, 15 V regulators are specified as they are easy to obtain. Their output increased to 18 V by means of $R_{201,203}$, Tr_{201} and $R_{202,204}$, Tr_{202}.

It is common to use a potential divider to 'stand-off' the regulator by a fixed proportion of the output voltage. In the improved version here, positive divider $R_{201,203}$ is buffered by emitter-follower Q_{201}. Thus $R_{201,203}$ can be higher in value—saving power—while Tr_{201} absorbs the ill-defined quiescent current from the regulator COM pin.

Choosing the right op-amps

Exotic and expensive op-amps will probably give a disappointing noise performance. The bipolar input of the *5534/32* is well matched to the medium-low impedances used in this preamplifier. For example, an *OP-27* might be expected to be quieter in the moving-magnet cartridge stage; but when measured, or calculated, it is 2 dB noisier.

The performance

Figure 6.12 shows the THD of the flat moving-coil cartridge stage alone, at maximum gain. The rise at extreme if is due to the integrator time-constant. Figures 6.13 and 6.14 give the THD of the moving-magnet cartridge disc input and the entire rear section respectively. Levels involved are ten times those found in real use. Distortion is not a problem here.

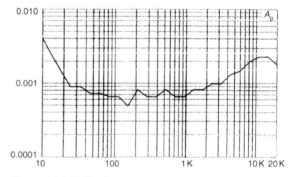

Figure 6.12 THD of the moving coil stage alone, at 2.2 Vrms output. Measurement bandwidth 30 kHz.

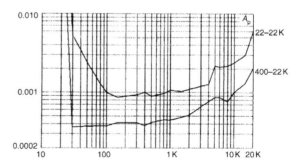

Figure 6.13 THD of disc input stage in moving-magnet mode, at 8Vrms output. Bandwidth 22–22 kHz upper trace and 400–22 kHz lower trace, which gives a more valid result as magnetic hum is excluded. Distortion is very low, but rises at h.f. due to increasing loading.

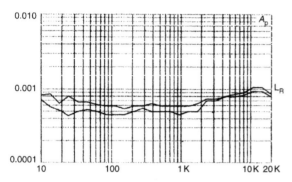

Figure 6.14 THD of rest of preamplifier at 8 Vrms in and out, i.e. volume control set for unity gain. Tone control set flat, bandwidth 80 kHz. Distortion is below 0.001%.

Crosstalk performance attained depends very much on physical layout. Capacitive crosstalk can be minimised by spacing components well apart, or by simple screening. Resistive crosstalk depends on the thickness of the various ground paths.

It would be desirable to specify a grounding topology for optimal results, but this is not so easy. I found that the more tightly the various grounds are tied together with heavy conductors, the better the crosstalk performance. There seemed little scope for subtlety.

As with noise performance, the results depend somewhat on control settings, but under most conditions the prototype gave about −100 dB flat across 20 Hz–20 kHz, with noise contributing to the reading. This was not hard to achieve.

The preamplifier in perspective

In determining what (if anything) has been achieved by this design, we must see if it is capable of any further improvement.

- The moving-magnet stage input noise performance is limited by the electrical characteristics of the cartridge and its loading needs.
- Making the RIAA any more accurate will be expensive.

- Increasing disc input headroom would require the use of higher supply rails, demanding discrete amplifier stages.

Having gone to some effort to make the preamplifier as noise-free and transparent as possible, we should ask how it compares with other parts of the system. The standard Blameless Class B power amplifier [4] output noise is −93.5 dBu, and the Trimodal [5] with the low-impedance feedback network reduces this to −95.4 dBu. In both cases the source impedance is 50 Ω.

Both amplifiers have a closed-loop gain of +27.2 dB, and so the equivalent input noise (EIN) is −120.7 and −122.6 dBu respectively. This can be compared with the source-resistance Johnson noise of 50 Ω, which is −135.2 dBu. The best power-amp noise figure is therefore 12.6 dB, which is some way short of perfection.

In contrast, the noise output from the preamplifier is never less than −114.5 dBu with the volume control at zero. Even in this rather useless condition, the preamplifier increases the total noise output, as it produces 8 dB more than the Trimodal power amplifier input noise. At mid-volume (in-line mode) the preamplifier noise is −105.3 dBu, which is 17 dB worse than the power-amp; clearly as far as preamp design is concerned, history has not yet ended.

Even so, serious thought has been given to whether this may be the quietest preamp yet built. Comments and opinions on this are invited.

References

1. Self, D. 'A Precision Preamplifier', *Wireless World*, October 1983, p 31.

2. Self, D. 'Ultra-Low-Noise Amplifiers and Granularity Distortion', *Journ AES*, November 1987, pp 907–915.

3. Baxandall, P. 'Audio Gain Controls', *Wireless World*, November 1980, pp 79–81.

4. Self, D. 'Distortion in Power Amplifiers: Part 2', *Electronics World*, September 1993, p 736.

5. Self, D. 'Trimodal Audio Power', *Electronics World*, June 1995, p 462.

Chapter 7

Overload matters (phono headroom)

(Electronics World, February 1997)

Intended title: *Signal Levels from Vinyl Discs*

There are usually no gain controls on Moving-Magnet (MM) RIAA inputs. Switched gains are encountered occasionally, but I don't think I have ever seen an input stage with continuously variable gain achieved by altering the negative feedback. This is because it is relatively difficult to design a fixed-gain RIAA network, if you are doing it properly with all the RIAA equalisation performed in one stage. Distributing the RIAA equalisation across two or even more stages makes the mathematics simple, which is why this stupid notion is as popular as it is, but leads inevitably to serious compromises in noise and/or headroom. Implementing variable gain adds another layer of difficulty to obtaining an accurate frequency response. An MM stage with a passive gain control after it is, of course, not the same thing, and quite useless for preventing overload with high inputs.

Because of this the headroom, or 'overload margin', as it is more often called in this context, of an RIAA stage is of considerable importance. The more the better, so long as nothing else is compromised. The issue can get a bit involved, as the frequency-dependant gain is further complicated by a heavy frequency-dependant load in the shape of a feedback network with capacitive paths to ground. This heavy loading was a major cause of distortion and headroom-limitation in conventional RIAA stages that had emitter-follower outputs with highly asymmetrical drive capabilities, and for some reason, it took the preamplifier industry a very long time to wake up to this. Once more, the 5532 op-amp solved that problem.

There are also interesting limitations to the levels which stylus-in-vinyl technology can generate, set by geometrical and mechanical considerations. The highest MM cartridge sensitivity for normal use is 1.6 mV per cm/s. There are then some famous maximal velocity levels to consider. The Pressure Cooker discs by Sheffield Labs were recorded direct to disc and are said to have velocities up to 40 cm/s, giving us $1.6 \times 40 = 64$ mV rms. The jazz record *Hey! Heard the Herd* by Woody Herman (Verve V/V6 8558, 1953) is claimed to have a peak velocity of 104 cm/s at 7.25 kHz, translating to

$1.6 \times 104 = 166$ mV rms. The subject of signal levels from vinyl is dealt with in much more detail in my *Small-Signal Audio Design*, second edition [1].

The Woody Herman figure seems out of line with all other data, but it could credibly be taken as a possible absolute maximum level. If we want an MM stage to accept 166 mV rms at 1 kHz, and assuming a 10 Vrms output is feasible, then we can have no more than 35.6 dB (1kHz) of closed-loop gain. Designing to this figure presents no problems, but these days, I usually go for +30 dB (1kHz) of gain. This gives an extra 5 dB of safety margin on top, and so can very reasonably be regarded as absolutely immune to overload. Such an RIAA stage will accept 316 mV rms at 1 kHz from a cartridge before clipping. With only +30 dB of gain, it will definitely need an HF correction pole to prevent errors at HF where the gain asymptotes towards unity.

If +30 dB (1kHz) of gain sounds over-cautious, bear in mind it has other advantages. This amount of closed-loop gain allows for plenty of negative feedback when opamps like 5534 or 5532 are used, and therefore an accurate realisation of the intended RIAA equalisation. The 5534 can give maximum errors of up to 0.16 dB when configured with +40 dB (1kHz) of gain, and 0.1 dB with +35 dB (1kHz) of gain, simply due to the finite feedback factor available; absolutely accurate passive components are assumed here. At +30 dB (1kHz) of gain, the maximum error drops to 0.04 dB, where it is almost certainly lost in the effects of passive component tolerances in the RIAA network [2].

A gain of +30 dB (1kHz) means that a nominal 5 mV rms input signal will only give an output of 158 mV rms, and so in most cases, more gain will be required downstream to get the level required. This subsequent gain must of course be switchable or otherwise variable so that its gain can be reduced to unity when required, or the headroom in the MM stage becomes meaningless, as unavoidable clipping will be happening later in the signal chain.

References

1. Self, D., *Small-Signal Audio Design,* 2nd edition, Chapter 8, pp. 209–216. Focal Press (Taylor & Francis) 2014. ISBN: 978-0-415-70974-3 Hardback. 978-0-415-70973-6 Paperback. 978-0-315-88537-7 ebook.

2. Ibid., Chapter 8, pp. 243–249.

OVERLOAD MATTERS (PHONO HEADROOM)

February 1997

There was no room in my Preamp '96 article for a proper discussion of the overload behaviour of RIAA preamp stages.[1] Like noise performance, the issue is considerably complicated by both cartridge characteristics and the RIAA equalisation.

There are some inflexible limits to the signal level possible on vinyl disc, and they impose maxima on the signal that a cartridge can reproduce. The absolute value of these limits may not be precisely defined, but they set the way in which maximum levels vary with frequency, and this is perhaps of even greater importance.

Figure 7.1(a) shows the physical groove amplitudes that can be put onto a disc. From subsonic up to about 1 kHz, groove amplitude is the constraint. If the sideways excursion is too great, the spacing will need to be increased to prevent one groove breaking into another, and playing time will be reduced. From about 1 kHz to ultrasonic, the limit is groove velocity rather than amplitude. If the cutter head tries to move sideways too quickly compared with its forward motion, the back facets of the cutter destroy the groove that has just been cut by the forward edges.

At replay time, there is a third restriction—that of stylus acceleration or, to put it another way, groove curvature. This sets a limit on how well a stylus of a given size can track the groove. Allowing for this at cutting time puts an extra limit on signal level, shown by the dotted line in Figure 7.1(a).

The severity of this restriction depends on the stylus shape. An old-fashioned spherical type with a tip diameter of 0.0007 in requires a roll-off of maximum levels from 2 kHz, while a relatively modern elliptical type with 0.0002 in effective diameter postpones the problem to about 8 kHz.[2]

Thus there are at least three limits on the signal level. The distribution of amplitude with frequency for the original signal is unlikely to mimic this, because there is almost always more energy at l.f. than h.f. Therefore the h.f. can be boosted to overcome surface noise without overload problems, and this is done by applying the inverse of the familiar RIAA replay equalisation.

Moving-magnet and moving-coil cartridges both operate by the relative motion of conductors and magnetic field, so the voltage produced is proportional to rate of change of flux. The cartridge is sensitive to velocity rather than to amplitude (and so sensitivity is always expressed in millivolts per cm/s) and this gives a frequency response rising steadily at 6 dB/octave across the whole audio band. Therefore, a maximal signal from disc Figure 7.1(a) would give a cartridge output like Figure 7.1(b)—i.e., 1(a) tilted upwards.

Figure 7.1(c) shows the RIAA replay equalisation curve. The shelf in the middle corresponds with 1(a), while an extra time constant at 50 Hz limits the amount of l.f. boost applied to warps and rumbles. The 'IEC amendment' is an extra roll-off at 20 Hz, (shown dotted) to further reduce subsonics. When RIAA equalisation 1(c) is applied to cartridge output 1(b), the result will look like Figure 7.1(d), with the maximum amplitudes occurring around 1–2 kHz.

Clearly, the overload performance of an RIAA input can only be assessed by driving it with an inverse-RIAA equalised signal, rising at 6 dB/octave except around the middle shelf. My Precision preamp '96 has an input overload margin referred to 5 mV r.m.s. of 36 dB across most of the audio band, i.e., 315 mV r.m.s. at 1 kHz. The margin

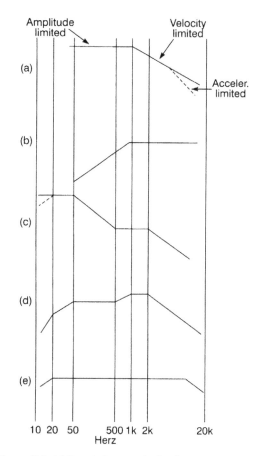

Figure 7.1 (a) Restrictions on the level put onto a vinyl disc. The extra limit of groove curvature—stylus acceleration—is shown dotted. (b) Response of a moving-coil or moving-magnet cartridge to a signal following the maximum contour in Figure 7.1(a). (c) The RIAA replay curve. The IEC amendment is an extra roll-off at low frequency, shown dotted. (d) The combination of (b) and (c). (e) RIAA preamp output limitations. The high-frequency restriction is very common and is often much worse in discrete preamplifier stages with poor load driving capabilities.

is still 36 dB at 100 Hz, but due to the RIAA low-frequency boost this is only 30 mV r.m.s. in absolute terms.

The final complications is that preamplifier output capability almost always varies with frequency. In Preamp '96, the effects have been kept small. The output overload margin voltage—and hence input margin falls to +33 dB at 20 kHz. This is due to the heavy capacitive loading of both the main RIAA feedback path and the pole-correcting *RC* network ($R_{24,25}$ and C_{20}). This could be eliminated by using an op-amp with greater load-driving capabilities, if you can find one with the low noise of a *5534*.

The overload capability of Preamp 96 is also reduced to 31 dB in the bottom octave 10–20 Hz, because the IEC amendment is implemented in the second stage. The 1.f. signal is fully amplified by the first stage, then attenuated by the deliberately slow initial roll-off of the subsonic filter.

Such audio impropriety always carries a penalty in headroom as the signal will clip before it is attenuated. This is the price paid for an accurate IEC amendment set by polyester caps in the second stage, as opposed to the usual method of putting a small electrolytic in the first-stage feedback path, rather than the 220 μF used. Alternative input architectures that put flat amplification before an RIAA stage suffer much more severely from this kind of headroom restriction.[3]

These extra preamp limitations on output level are shown at Figure 7.1(e), and, comparing 1(d), it appears they are almost irrelevant because of the falloff in possible input levels at each end of the audio band.

References

1. Self, D. 'Precision preamplifier '96', *Electronic World*, July/August and September 1996.

2. Holman, T. 'Dynamic range requirements of phonographic preamplifiers', *Audio*, July 1977, p 74.

3. Self, D. 'Precision Preamplifier '96', *Electronics World*, July/August 1996, p 543.

Chapter 8

A balanced view, Part I

(Electronics World, April 1997)

Intended title: *The Technology of Balanced Interconnections Part I*

This is a good example of the less than satisfactory titles that could wind up at the head of an article, imposed by the editor without consultation. This is clearly intended to be a pun of the lamest kind, but just leaves lingering doubt as to whether balanced connections are a bit dubious and require careful weighing of the pros and cons. This is of course untrue; a balanced connection is almost always better than an unbalanced one, and the decision is usually based on economics. There is one qualification to this, and it is significantly missing from the 'Disadvantages of Balancing' list in the article. It is that a balanced link will usually be noisier than an unbalanced one. I am talking about noise internally generated in the link rather than external noise, which of course the balanced input is expected to reject.

If we ignore the noise from the sending circuitry, and assuming its output impedance can be represented as 50Ω to ground, an unbalanced input amplifier (taken to be a 5532/2 unity-gain follower) has a noise output of about −119 dBu. A balanced input performs a subtraction to separate the signal from the ground noise, and that has to be done either with active electronics or a transformer. A 'standard' balanced input amplifier made with a 5532/2 and four 10kΩ resistors has a noise output of about −105 dBu. This is 14 dB worse than the unbalanced link, and hard to explain to people who see a balanced input as a premium input; they expect it to be better in every way. There are ways to make balanced input amplifiers quieter, but all require more electronics [1].

A transformer-balanced input would give much less extra noise (though still some, because of Johnson noise from the winding resistances) and use no power, but it is always a heavy and expensive part, with a doubtful frequency response and dubious low-frequency linearity.

I wrote these two articles during my long tenure as Chief Design Engineer at Soundcraft. Our consoles used unbalanced and balanced inputs, and unbalanced and quasi-floating outputs, plus ground-cancelling outputs in some cases. I don't recall we ever used a straight balanced output. People often got confused when grappling with the various kinds of inputs and outputs; given the many possible combinations this is not surprising. The two articles were an attempt to explain the operation of various kinds of inputs and output combinations, some obvious, some almost unknown, with guidance on using them effectively. This first article covered the basics of balanced interconnections, and then examined the various sorts of output stage; unbalanced, impedance balanced, electronically balanced, ground-cancelling, quasi-floating, and transformer-balanced outputs. The second article examined input stages and the combinations of inputs and outputs.

Balanced or quasi-floating connections have been used for many years in professional circles and are now slowly but apparently steadily increasing in popularity in the more expensive reaches of the hi-fi market. In this area, the outputs are usually balanced by driving the cold line with an inverter rather than going for the full quasi-floating version, and very sensibly so. The input amplifiers are usually the classic one-op-amp differential stage, though now and then some very complicated instrumentation-amp circuits have been used.

For some reason, no one seems to have really picked up on the great merits of ground-cancelling outputs, possibly because of the difficulty of explaining the concept to the paying customer. Not everyone is receptive to the idea of an output which is also partly an input. This is a pity because ground-cancelling technology can give all the benefits of a balanced interconnection at the cost of a couple of resistors, with no extra power consumption and less noise. There is more on ground-cancelling in [2].

References

1. Self, D., *Small-Signal Audio Design,* 2nd edition, Chapter 18, pp. 526–535. Focal Press (Taylor & Francis) 2014. ISBN: 978-0-415-70974-3. Hardback. 978-0-415-70973-6 Paperback. 978-0-315-88537-7 ebook.

2. Ibid., Chapter 19, pp. 539–544.

A BALANCED VIEW, PART I

April 1997

Balanced inputs and outputs have been used for many years in professional audio, but profound misconceptions about their operation and effectiveness still survive. Balanced operation is also making a slow but steady advance into top-end hi-fi, where its unfamiliarity can lead to further misunderstandings. As with most topics in audio technology, the conventional wisdom is often wrong.

A practical balanced interconnect is not always wholly straightforward. Some new variations on input and output stages have emerged relatively recently. For example, a 'ground-cancelling' output is not balanced at all, but actually has one output terminal configured as an input. This can come as a surprise to the unwary.

Despite its non-balanced nature, such a ground-cancelling output can render a ground loop innocuous even when driving an unbalanced input. But even an audio professional could be forgiven for being unsure if it still works when it is driving a balanced input. The answer—which in fact is yes—is explained in a second article on this subject, details of which are given later.

Electronic versus transformer balancing

Electronic balancing has many advantages. These include low cost, low size and weight, superior frequency and transient response, and no problems with low-frequency linearity. While it is sometimes regarded as a second-best, it is more than adequate for hi-fi and most professional applications.

To balance . . .
Balancing offers the following advantages

- Discriminates against noise and crosstalk.
- A balanced interconnect—with a true balanced output—allows 6 dB more signal level on the line.
- Breaks ground-loops, so that people are not tempted to start 'lifting grounds'. This is only acceptable if the equipment has a dedicated ground-lift switch, that leaves the metalwork firmly connected to mains safety earth. In the absence of this facility, the optimistic will remove the mains earth—which is not quite so easy now that moulded plugs are standard— and this practice must be roundly condemned as dangerous.

. . . or not to balance
Balancing also brings with it the following disadvantages.

- Balanced connections are unlikely to provide much protection against r.f. ingress. Both sides of the balanced input would have to demodulate the r.f. with exactly the same effectiveness for common-mode cancellation to occur. This is not very likely.
- There are more possibilities for error when wiring up. For example, it is easy to introduce an unwanted phase inversion by confusing hot and cold in a connector. This can go undiscovered for some time. The same mistake on an unbalanced system interrupts the audio completely.

Transformer balancing has some advantages of its own—particularly for work in very hostile r.f./e.m.c. environments—but many serious drawbacks. The advantages are that transformers are electrically bullet-proof, retain their common-mode rejection ratio performance forever, and consume no power even at high signal levels. Unfortunately they also generate low frequency distortion, and have high frequency response problems due to leakage reactance and distributed capacitance. Transformers are also heavy and expensive.

The first two objections can be surmounted—given enough extra electronic circuitry—but the last two cannot. Transformer balancing is therefore rare, even in professional audio, and is only dealt with briefly here.

Balancing basics

Balanced connections in an audio system are designed to reject both external noise, from power wiring etc., and also internal crosstalk from adjacent signal cables.

The basic principle of balanced interconnection is to get the signal you want by subtraction, using a three-wire connection. In many cases, one signal wire—the hot or in phase conductor—senses the actual output of the sending unit. The other, the cold or phase-inverted, senses the unit's output-socket ground, and the difference between them gives the wanted signal.

Any noise voltages that appear identically on both lines, i.e. commonmode signals, are in theory completely cancelled by the subtraction. In real life, the subtraction falls short of perfection, as the gains via the hot and cold inputs will not be precisely the same. The degree of discrimination actually achieved is called the common-mode rejection ratio (cmrr).

The terms hot and cold for in-phase and out-of-phase respectively, are used throughout this article for brevity.

While two wires carry the signal, the third is the ground wire which has the dual duty of both joining the grounds of the interconnected equipment, and electrostatically screening the two signal wires by being in some way wrapped around them. The 'wrapping around' can mean:

- A lapped screen, with wires laid parallel to the central signal conductor. The screening coverage is not perfect, and can be badly degraded as it tends to open up on the outside of cable bends.
- A braided screen around the central signal wires. This is more expensive, but opens up less when the wire is bent. Screening is not 100%, but certainly better than lapped screen.
- An overlapping foil screen, with the ground wire— called the drain wire in this context for some reason— running down the inside of the foil and in electrical contact with it. This is usually the most effective as the foil cannot open up on the outside of bends, and

should give perfect electrostatic screening. However, the higher resistance of aluminium foil compared with copper braid means that r.f. screening may be worse.

Electrical noise

Noise gets into signal cables in three major ways:

Electrostatic coupling

An interfering signal with significant voltage amplitude couples directly to the inner signal line, through stray capacitance. The situation is shown in Figure 8.1, with C, C representing the stray capacitance between imperfectly-screened conductors; this will be a fraction of a picofarad in most circumstances. This coupling is unlikely to be a problem in hi-fi systems, but can be serious in studio installations with unrelated signals going down the same ducting.

The two main lines of defence against electrostatic coupling are effective screening and low-impedance drive. An overlapping foil screen—such as used on Belden microphone cable—provides complete protection. Driving the line from a low impedance, of the order

of 100 Ω or less, means that the interfering signal, having passed through a very small capacitance, is a very small current and cannot develop much voltage across such a low impedance.

For the best results, the impedance must remain low up to as high as frequency as possible; this can be problem as op-amps invariably have a feedback factor that begins to fall from a low, and possibly sub-audio frequency, and this makes the output impedance rise with frequency.

From the point of view of electrostatic screening alone, the screen does not need to be grounded at both ends, or form part of a circuit.[1] It must of course be grounded at some point.

Electrostatic coupling falls off with the square of distance. Rearranging the cable-run away from the source of interference is more practical and more effective than trying to rely on very good common-mode rejection.

Magnetic coupling

An e.m.f., V_m, is induced in both signal conductors and the screen, Figure 8.2. According to some writers, the screen current must be allowed to flow freely, or its magnetic field will not cancel out the field acting on

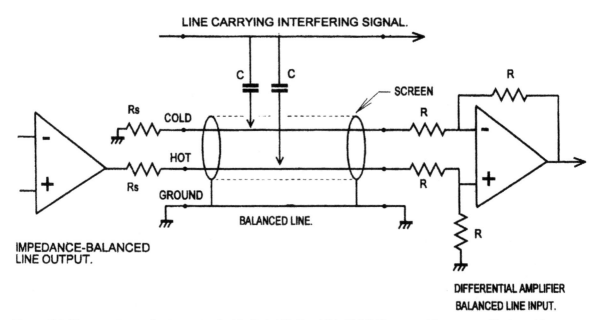

Figure 8.1 Electrostatic coupling into a signal cable, Rs is 100 Ω and R is 10 kΩ. The second Rs to ground in the cold output line makes it an impedance balanced output.

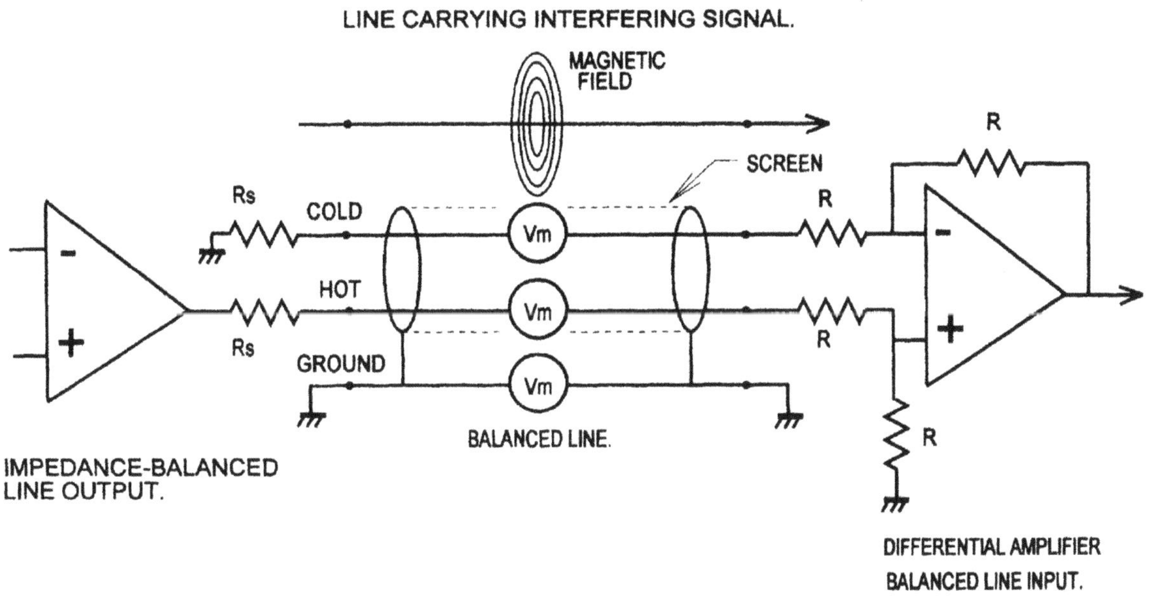

LINE CARRYING INTERFERING SIGNAL.

Figure 8.2 Magnetic coupling into a signal cable, represented by notional voltage-sources Vm.

the signal conductors. Therefore the screen should be grounded at both ends, to form a circuit.[2]

In practice, the field cancellation will be far from perfect. Most reliance is placed on the common-mode rejection of the balanced system, to cancel out the hopefully equal voltages V_m induced in the two signal wires. The need to ground both ends for magnetic rejection is not a restriction, as it will emerge that there are other good reasons why the screens should be grounded at both ends of a cable.

In critical situations, the equality of these voltages is maximised by minimising the loop area between the two signal wires, usually by twisting them tightly together. In practice most audio cables have parallel rather than twisted signal conductors, and this seems adequate most of the time.

Magnetic coupling falls off with the square of distance, so rearranging the cable-run away from the source of magnetic field is usually all that is required. It is unusual for it to present serious difficulties in a domestic environment.

Common-impedance coupling

Ground voltages coupled in through the common ground impedance; often called 'common-impedance coupling' in the literature.[3] This is the root of most

ground loop problems. In Figure 8.3 the equipment safety grounds cause a loop ABCD; the mere existence of a loop in itself does no harm, but it is invariably immersed in a 50 Hz magnetic field that will induce mains-frequency current plus odd harmonics into it. This current produces a voltage drop down the non-negligible ground-wire resistance, and this once again effectively appears as a voltage source in each of the two signal lines. Since the cmrr is finite a proportion of this voltage will appear to be differential signal, and will be reproduced as such.

A common source of ground-loop current is the connection of a system to two different 'grounds' that are not actually at the same a.c. potential. The classic example of this is the addition of a 'technical ground' such as a buried copper rod to a grounding system which is already connected to 'mains ground' at the power distribution board. In most countries this 'mains ground' is actually the neutral conductor, which is only grounded at the remote transformer substation. The voltage-drop down the neutral therefore appears between 'technical ground' and 'mains ground' causing large currents to flow through ground wires.

A similar situation can occur when water-pipes are connected to 'mains ground' except that interference is not usually by a common ground impedance; however the unwanted currents flowing in the pipework generate

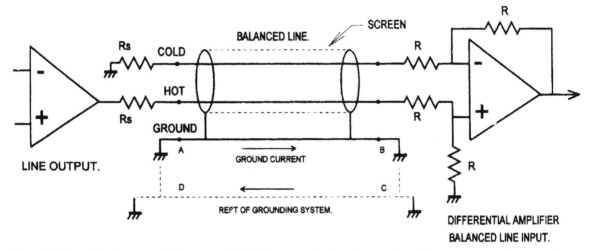

Figure 8.3 Ground-voltages coupling into a signal cable. The ground voltage between A and B is due to ground currents flowing around ABCD.

magnetic fields that may either create ground loops by induction, or interfere directly with equipment such as mixing consoles.

In practice, ground voltages cause a far greater number of noise problems than the other mechanisms, in both hi-fi and professional situations.

Even there is no common-impedance coupling, ground currents may still enter the signal circuit by transformer action. An example of such a situation is where the balanced line is fully floating and not galvanically connected to ground—which is only possible with a transformer-to-transformer connection.

The shield wire or foil acts as a transformer primary while the signal lines act as secondaries; if the magnetic field from the shield wire is not exactly uniform, then a differential noise voltage appears across the signal pair and is amplified as if it were a genuine signal. This effect is often called shield-current-induced-noise, or SCIN, and cables vary in their susceptibility to it according to the details of their construction.[4]

Fortunately the level of this effect is below the noise-floor in most circumstances and with most cables, for once a differential-mode signal has been induced in the signal lines, there is no way to discriminate against it.

From this summary I deduce there are two principle effects to guard against; electrostatic coupling, and the intrusion of unwanted voltages from either magnetic coupling or ground-loop currents.

Electrostatic interference can be represented by notional current sources connected to both signal lines;

these will only be effectively cancelled if the line impedances to ground are the same, as well as the basic cmrr being high. The likely levels of electrostatic interference current in practice are difficult to guess, so the figures I give in the second article are calculated from applying 1 mA to each line; this would be very severe crosstalk, but it does allow convenient relative judgements to be made.

Magnetic and ground-voltage interference can be represented by notional voltage-sources inserted in both signal lines and the ground wire; these are not line-impedance sensitive and their rejection depends only on the basic cmrr, as measured with low-impedance drive to each input. Similarly ground-voltage interference can be represented by a voltage-source in the ground wire only.

Both input and output are voltages so the cmrr can be quoted simply as a ratio in decibels, without specifying any level.

Line outputs

A line output is expected to be able to drive significant loads, partly because of a purely historical requirement to drive 600 Ω, and partly to allow the parallel feed of several destinations. Another requirement is a low source impedance—100 Ω or less—to make the signal robust against capacitive crosstalk, etc.

There are many line output and input arrangements possible, and the results of the various permutations of connection are not always entirely obvious. An examination of the output types in use yields List 1.

Unbalanced output

There are only two physical output terminals—signal and ground, Figure 8.4(a). A third terminal is implied in Figure 8.4(a), emphasising that it is always possible to connect the cold wire in the cable to the ground at the transmitting (output) end.

The output amplifier is almost always buffered from the line shunt-capacitance by a resistor R$_s$ in the range 33 to 100 Ω, to ensure stability. This unbalances the line impedances. If the output resistance is taken as 100 Ω worst-case, and the cold line is simply grounded as in Figure 8.4(a), then the presence of R_s degrades the common-mode rejection ratio to—46 dB, even if the balanced input at the other end of the cable has perfectly matched resistors.

Impedance balanced output

There are now three physical terminals, hot, cold, and ground, Figure 8.4(b). The cold terminal is neither an input nor an output, but a resistive termination with the same resistance R_s as the hot terminal output impedance. This type of output is intended for use with receiving equipment having balanced inputs. The presence of the second R_s terminated to output ground makes the impedance on each signal line almost exactly the same—apart from op-amp output impedance limitations—so that good rejection is achieved for both common-mode ground voltages and electrostatic interference.

If an unbalanced input is being driven, the cold terminal on the transmitting (output) equipment can be either shorted to ground locally or left open-circuit without serious consequences. Either way all the benefits of balancing are lost.

The use of the word 'balanced' is unfortunate as this implies anti-phase outputs, which are not present.

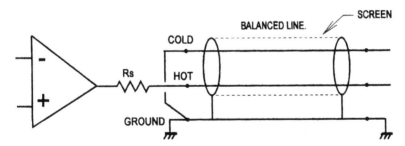

Figure 8.4a An unbalanced line output. The cold output—if it exists at all—is connected directly to ground.

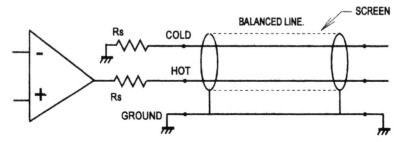

Figure 8.4b An impedance balanced output. The cold output is connected to ground through a second Rs of identical value.

Ground-cancelling output

Also called a ground-compensated output, this arrangement is shown in Figure 8.5(a).

This allows ground voltages to be cancelled out even if the receiving equipment has an unbalanced input. It prevents any possibility of creating a phase error by miswiring. It separates the wanted signal from the unwanted by addition at the output end of the link, rather than by subtraction at the input end.

If the receiving equipment ground differs in voltage from the sending ground, then this difference is added to the output so that the signal reaching the receiving equipment has the same voltage superimposed upon it. Input and ground therefore move together and there is no net input signal, subject to the usual resistor tolerances.

The cold pin of the output socket is now an input, and must have a unitygain path summing into the main signal-output going to the hot output pin. It usually has a very low input impedance equal to the hot terminal output impedance.

It is unfamiliar to most people to have the cold pin of an output socket as a low impedance input, and this can cause problems. Shorting it locally to ground merely converts the output to a standard unbalanced type. If the cold input is left unconnected then there should be only a very small noise degradation due to the very low input impedance of R_s.

Ground-cancelling outputs would appear to be very suitable for hi-fi use, as they are an economical way of making ground-loops innocuous. However, I am not aware that they have ever been used in this field.

Balanced output

The cold terminal is now an active output, producing the same signal as the hot terminal but phase-inverted, Figure 8.5(b). This can be simply done by using an op-amp stage with a gain of minus one to invert the normal in-phase output. Phase spikes are shown on the diagram to emphasise these phase relationships.

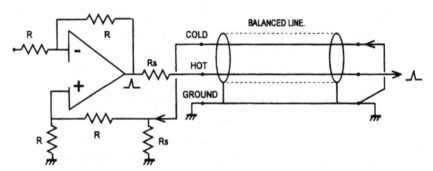

Figure 8.5a A ground-cancelling output, with a unity-gain path from the cold terminal to the hot output. Once more a second Rs balances the line impedances.

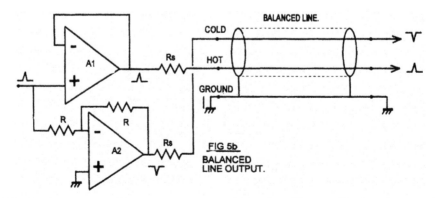

Figure 8.5b A balanced output. A2 is a unity-gain inverter driving the cold output. Line impedances are balanced.

The in-phase signal itself is not degraded by passing through an extra stage and this can be important in quality-critical designs. The inverting output must not be grounded; if not required it can simply be ignored.

Unlike quasi-floating outputs, it is not necessary to ground the cold pin to get the correct gain for unbalanced operation, and it must not be grounded by mistake, because the inverting op-amp will then spend most of its time in current-limiting, probably injecting unpleasant distortion into the preamp grounding system, and possibly suffering unreliability. Both hot and cold outputs must have the same output impedance R_s to keep the line impedances balanced.

A balanced output has the advantage that it is unlikely to crosstalk to other lines, even if they are unbalanced. This is because the current injected via the stray capacitance from each crosstalking line cancels at the receiving end.

Another advantage is that the total signal level on the line is increased by 6 dB, which can be valuable in difficult noise situations. All balanced outputs give the facility of correcting phase errors by deliberately swopping hot and cold outputs. This tactic is however a double-edged sword, because it is probably how the phase became wrong in the first place.

This form of balanced output is the norm in hi-fi balanced interconnection, but is less common in professional audio, where the quasifloating output gives more flexibility.

Quasi-floating output

This kind of output, Figure 8.6, approximately simulates a floating transformer winding; if both hot and cold outputs are driving signal lines, then the outputs are balanced, as if a centre-tapped output transformer were being used.

If, however, the cold output is grounded, the hot output doubles in amplitude so the total level is unchanged. This condition is detected by the current-sensing feedback taken from the outside of the 75 Ω output resistors. Current driven into the shorted cold output is automatically reduced to a low level that will not cause problems.

Similarly, if the hot output is grounded, the cold output doubles in amplitude and remains out of phase; the total hot-cold signal level is once more unchanged. This system has the advantage that it can give the same level into either a balanced or unbalanced input without rewiring connectors. 6 dB of headroom is however lost.

When an unbalanced input is being driven, the quasi-floating output can be wired to work as a ground-cancelling connection, with rejection of ground noise no less effective than the true balanced mode. This

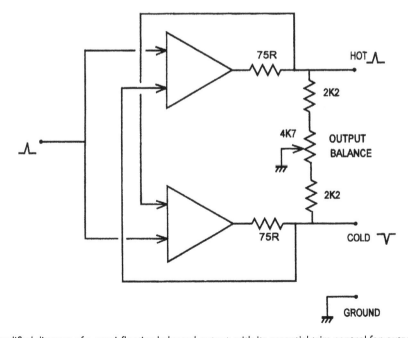

Figure 8.6 Simplified diagram of a quasi-floating balanced output, with its essential trim control for output symmetry.

requires the cold output to be grounded at the remote (input) end of the cable. Under adverse conditions this might cause h.f. instability, but in general the approach is sound. If you are using exceptionally long cable, then it is wise to check that all is well.

If the cold output is grounded locally, i.e. at the sending end of the cable, then it works as a simple unbalanced output, with no noise rejection. When a quasi-floating output is used unbalanced, the cold leg must be grounded, or common-mode noise will degrade the noise floor by at least 10 dB, and there may be other problems. In both of the unbalanced cases the maximum signal possible on the line is reduced by 6 dB.

Quasi-floating outputs use a rather subtle circuit with an intimate mixture of positive and negative feedback of current and voltage. This performs the required function admirably; its only drawback is a tendency to accentuate circuit tolerances, and so a preset resistor is normally required to set the outputs for equal amplitude; the usual arrangement is shown in Figure 8.6.

If the balance preset is not correctly adjusted one side of the output will clip before the other and reduce the total output headroom. After factory setting this preset should not need to be touched unless the resistors in the circuit are replaced; changing the op-amp should make no difference.

The balancing network consists of a loading resistor to ground on each output; in this respect the output characteristics diverge from a true floating output, which would be completely isolated from ground. These loading resistors are lower than the input impedance of typical balanced inputs. So if simple differential amplifiers are used with unequal input impedances, (see the section on line inputs, below) the output balance is not

significantly disturbed and clipping remains symmetrical on the hot and cold outputs.

Quasi-floating outputs are often simply referred to as 'balanced' or 'electronically-balanced', but this risks serious confusion as the true balanced output described earlier must be handled in a completely different way from quasi-floating.

True floating transformer output

This can be implemented with a transformer if galvanic isolation from ground is required. The technique is rarely used.

The second article in this pair looks at line inputs in detail, examines what happens when the different kinds of input and output are connected together, and deals with the philosophy of audio system wiring.

References

1. Williams, T. 'EMC for product designers', Newnes (Butterworth-Heinemann), p 1992, ISBN 0 7506 1264 9, p 176.

2. Williams, T. 'EMC for product designers', Newnes (Butterworth-Heinemann), p 1992, ISBN 0 7506 1264 9, p 173.

3. Muncy, N. 'Noise susceptibility in analog and digital signal processing systems', *JAES*, 3 (6), June 1995, p 447.

4. Muncy, N. 'Noise susceptibility in analog and digital signal processing systems', *JAES*, June 1995, p 441.

Chapter 9

A balanced view, Part II

(Electronics World, May 1997)

Intended title: *The Technology of Balanced Interconnections Part II*

In the second part of the article, I examined the various sorts of input stage and the various combinations of inputs and outputs. While there are an intriguing number of output stage types, for inputs there are only unbalanced, electronically balanced, and transformer-balanced versions to be considered.

Obviously, an article of reasonable length could not cover everything, and two important issues not ventilated were the desirability of a high Common-Mode Rejection Ratio (CMRR) in balanced inputs, and the problem that a basic electronically balanced input stage is much noisier than an unbalanced input stage, typically by some 14 dB when it is built with 10 kΩ resistors.

The CMRR of a balanced input is comprehensively examined in *Small-Signal Audio Design* [1]. At the low frequencies that typically make up hum and buzz, it is purely a matter of resistor accuracy. If you use 1% resistors, the CMRR will be around 46 dB; if you use 0.1% resistors, the CMRR will be about 66 dB. Since 0.1% resistors are ten times as expensive as 1%, and don't forget you need four of them, this is a relatively expensive way to go. If 46 dB is not considered enough then a CMRR trim preset can be used to tweak the value of one of the four resistors. This allows CMRR of better than 85 dB at low frequencies, though adjustment at manufacture is obviously required. At higher frequencies, which can mean above 200 Hz, CMRR worsens steadily due to the limited open-loop bandwidth of the opamp employed. There doesn't seem much to be done about this apart from shelling out for a more expensive opamp, though I can't help wondering if some sort of active feedback configuration would enhance the performance. Balanced inputs typically have small capacitors across the negative feedback resistor to ensure stability, and these need to be matched by a capacitor on the hot input leg; (something I should underlined in the text) mismatches in these capacitors also degrade the high frequency CMRR, but it is not usual to allow trimming with a preset capacitor.

The noise disadvantage suffered by the conventional balanced input is to a large extent due to the high resistor values used; they need to be not less than 10 kΩ to give adequate input impedances. Noise can be reduced by adding unity-gain buffers to each input, so the resistor values around the balanced input can be lower, by using multiple opamps and averaging their outputs to partially cancel the noise [2],or by making use of the recently-appreciated special properties of the instrumentation-amplifier configuration [3].

The initial impetus for this pair of articles was the difficulty of working out what a given interconnection would do, given that there are a minimum of six types of outputs and three types of inputs, giving eighteen combinations. Most of them are relatively easy to work out, but a few are non-obvious. What about ground-cancelling out to electronically balanced in? Quasi-floating out to transformer-balanced in? Some of them had to be simulated to be exactly sure what was happening, and the results were duly listed.

The final part of the article deals with wiring philosophy. In the years since the article was written, the concept of OEO (One End Only) grounding appears to have disappeared completely, and I am strongly persuaded that is no bad thing. The best way is unquestionably to ground every signal lead at both ends, avoid circulating ground currents as much as you possibly can, and then leave the common-mode rejection of balanced inputs or ground-cancelling outputs to suppress whatever ground noise does exist.

References

1. Self, D., *Small-Signal Audio Design,* 2nd edition, Chapter 18, pp. 501–505. Focal Press (Taylor & Francis) 2014. ISBN: 978-0-415-70974-3 Hardback. 978-0-415-70973-6 Paperback.

2. Ibid., Chapter 18, pp. 519–521.

3. Ibid., Chapter 18, pp. 527–535.

A BALANCED VIEW, PART II

May 1997

There are only two kinds of input stage—unbalanced and balanced. For interconnection this is the primary distinction. Apart from balancing requirements, a line-level input, as opposed to a microphone input, is expected to have a reasonably high impedance to allow multiple connections to a single output.

Traditionally, a 'bridging impedance'—i.e. high enough to put negligible loading on historical 600 Ω lines—was 10 kΩ minimum. This is still appropriate for modern low-impedance outputs. However, a higher impedance of 100 kΩ or even more is desirable for interfacing to obsolete valve equipment, to avoid increased distortion and curtailed headroom.

Another common requirement is true variable gain at the balanced input, as putting the gain control further down the signal path means that it is impossible to prevent input amplifier overload. Thus you need a balanced stage that can attenuate as well as amplify, and this is where the circuit design starts to get interesting.

In the following circuitry, small capacitors often shunt the feedback elements to define bandwidth or ensure stability. These are omitted for clarity.

Unbalanced inputs

These are straightforward; variable-gain series-feedback stages are easily configured as in Figure 9.1, providing a minimum gain of unity is acceptable; R_2 sets the gain law in the middle of the pot travel.

It is also simple to make a stage that attenuates as well as amplifies. But this implies a shunt-feedback configuration as in Figure 9.2, with a variable input impedance. The minimum input impedance R_1 cannot

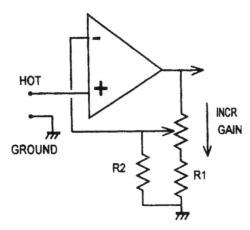

Figure 9.1 Variable-gain series-feedback, unbalanced input stage. Resistor R2 sets mid-position gain.

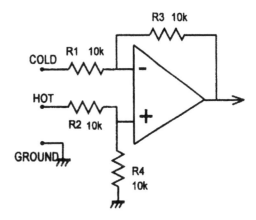

Figure 9.3 Standard one-op-amp differential amplifier, arranged for unity gain.

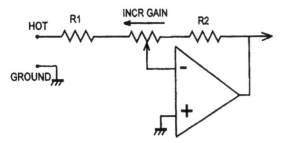

Figure 9.2 Shunt-feedback configuration, with a low and variable input impedance.

be much higher than 10 kΩ or resistor noise becomes excessive.

For a series-feedback stage, the input impedance can be made as high as desired by bootstrapping; an input resistance of 500 kΩ or greater is perfectly possible. This does *not* imply a poorer noise performance, as the noise depends on the source resistance and semi-conductor characteristics.

To ram the point home, my own personal best is 1 GΩ, in a capacitor microphone head amplifier. Although the input impedance is many orders of magnitude greater than the 1 to 2 kΩ of a dynamic microphone preamp, the E_{IN} is −110 dBu, i.e. only 18 dB worse.

Naturally, any unbalanced input can be made balanced or floating by adding a transformer.

Balanced inputs

A standard one-op-amp differential input stage is shown in Figure 9.3. Unlike instrumentation work, a super-high cmrr is normally unnecessary. Ordinary 1% resistors and

no trimming will not give cmrr better than 45 dB; however this is usually adequate for even high-quality audio work.

It is never acceptable to leave either input floating. This causes serious deterioration of noise, hum etc. Grounding the cold input locally to create an unbalanced input is quite alright, though naturally all the balanced noise rejection is lost.

The hot input can be locally grounded instead. In this case, the cold input is driven, to create a phase-inverting input that corrects a phase error elsewhere, but this is not good practice: the right thing to do is to sort out the original phase error.

Balanced input technologies

There are many, many ways to make balanced or differential input amplifiers, and only the most important in audio are considered. These are:

- The standard differential amplifier.
- Switched-gain balanced amp.
- Variable-gain balanced amp.
- The 'Superbal' amp.
- Hi-Z balanced amp.
- Microphone preamp plus attenuator.
- Instrumentation amp.

Standard differential amplifier

The standard one-op-amp differential amplifier is a very familiar circuit block, but its operation often appears somewhat mysterious. The version in Figure 9.3 has a

Table 9.1 Differential amplifier input impedances

Case	Conditions	Hot I/p Z	Cold I/p Z
1	Hot only driven	20 kΩ	Grounded
2	Cold only driven	Grounded	10 kΩ
3	Both driven balanced	20 kΩ	6.7 kΩ
4	Both driven cm, ie together	20 kΩ	20 kΩ
5	Both driven floating	10 kΩ	10 kΩ

gain of R_3/R_1 (= R_4/R_2). It appears to present inherently unequal input impedances to the line; this has often been commented on [1] and some confusion has resulted.

The root of the problem is that a simple differential amplifier has interaction between the two inputs, so that the input impedance on the cold input depends strongly on the signal applied to the hot input. Since the only way to measure input impedance is to apply a signal and see how much current flows into the input, it follows that the apparent input impedance on each leg varies according to the way the inputs are driven. If the amplifier is made with four 10 kΩ resistors, then the input impedances Z are as in Table 9.1.

Some of these impedances are not exactly what you would expect. In Case 3, where the input is driven as from a transformer with its centre-tap grounded, the unequal input impedances are often claimed to 'unbalance the line'. However, since it is common-mode interference we are trying to reject, the common-mode impedance is what counts, and this is the same for both inputs.

The vital point is that the line output amplifier will have output impedances of 100 Ω or less, completely dominating the line impedance. These input impedance imbalances are therefore of little significance in practice; audio connections are not transmission lines (unless they are telephone circuits several miles long) so the input impedances do not have to provide a matched and balanced termination.

As the first thing the signal encounters is a 10 kΩ series resistor, the low impedance of 6.7 kΩ on the cold input sounds impossible. But the crucial point is that the hot input is driven simultaneously. As a result, the inverting op-amp input is moving in the opposite direction to the cold input, due to negative feedback, a sort of anti-bootstrapping that reduces the effective value of the 10 kΩ resistor to 6.7 kΩ.

The input impedances in this mode can be made equal by manipulating resistor values, but this makes the cm impedances (to ground) unequal, which seems more undesirable.

In Case 5, where the input is driven as from a floating transformer with any centre-tap unconnected, the

impedances are nice and equal. They must be, because with a floating winding the same current must flow into each input. However, in this connection the line voltages are *not* equal and opposite: with a true floating transformer winding the hot input has all the signal voltage on it while the cold has none at all, due to the internal coupling of the balanced input amplifier.

This seemed very strange when it emerged from simulation, but a reality-check proved it true. The line has been completely unbalanced as regards talking to other lines, although its own common-mode rejection remains good.

Even if perfectly matched resistors are assumed, the common-mode rejection ratio of this stage is not infinite; with a *TL072* it is about −90 dB, degrading from 100 Hz upwards, due to the limited open-loop gain of the opamp.

Switched-gain balanced amplifier

The need for a balanced input stage with two switched gains crops up frequently. The classic application is a mixing desk to give optimum performance with both semi-professional (−7.8 dBu) and professional (+4 dBu) interface levels.

Since the nominal internal level of a mixer is usually in the range −4 to 0 dBu, the stage must be able to switch between amplifying and attenuating, maintaining good cmrr in both modes.

The obvious way to change gain is to switch both $R_{3,4}$ in Figure 9.3, but a neater technique is shown in Figure 9.4. Perhaps surprisingly, the gain of a differential amplifier can be manipulating by changing the drive to the feedback arm (R_3 etc.) only, without affecting the cmrr. The vital point is to keep the resistance of this arm the same, but drive it from a scaled version of the op-amp output.

Figure 9.4 uses the network $R_{5,6}$, which has the same 2 kΩ output impedance whether R_4 is switched to the output (low gain) or ground (high gain). For low gain, the feedback is not attenuated, but fed through $R_{5,6}$ in parallel.

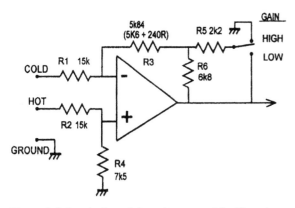

Figure 9.4 Switched-gain balanced input amplifier. The values shown give gains of −6 dB and +6.2 dB, for switching between pro and semi-pro interface levels.

For high gain, $R_{5,6}$ become a potential divider. Resistor R_3 is reduced by 2 kΩ to allow for the $R_{5,6}$ output impedance. The stage can attenuate as well as amplify if R_1 is greater than R_3, as shown here. The nominal output of the stage is assumed to be −2 dBu; the two gains are −6.0 and +6.2 dB.

The differential input impedance is 11.25 kΩ via the cold and 22.5 kΩ via the hot input. Common mode input impedance is 22.5 kΩ for both inputs.

Variable-gain balanced amplifier

A variable-gain balanced input should have its gain control at the very first stage, so overload can always be avoided. Unfortunately, making a variable-gain differential stage is not so easy; dual potentiometers can be used to vary two of the resistances, but this is clumsy and will give shocking cmrr due to pot mismatching. For a stereo input the resulting four-gang potentiometer is unattractive.

The gain-control principle is essentially the same as for the switched-gain amplifier above. To the best of my knowledge, I invented both stages in the late seventies, but so often you eventually find out that you have re-invented instead; any comments welcome.

Feedback arm R_3 is of constant resistance, and is driven by voltage-follower A_2. This eliminates the variations in source impedance at the potentiometer wiper, which would badly degrade cmrr. As in Figure 9.1, R_6 modifies the gain law; however, the centre-detent gain may not be very accurate as it partly depends on the ratio of potentiometer track (often no better than ±10%, and sometimes worse) to 1% fixed resistors.

This stage is very useful as a general line input with an input sensitivity range of −20 to +10 dBu. For a nominal output of 0 dBu, the gain of Figure 9.5 is +20 to −10 dB, with R_6 chosen for 0 dB at the central wiper position.

An op-amp in a feedback path appears a dubious proposition for stability, but here, working as a voltage-follower, its bandwidth is maximised and in practice the circuit is dependably stable.

The 'superbal' amplifier

This configuration [2] gives much better input symmetry than the standard differential amplifier, Figure 9.6. The differential input impedance is exactly 10 kΩ via both hot and cold inputs. Common mode input impedance is 20 kΩ for both inputs. This configuration is less easy to modify for variable gain.

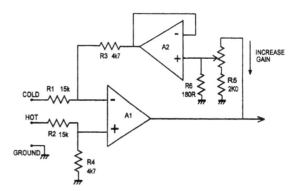

Figure 9.5 Variable-gain balanced input amplifier. Gain range is −10 to +20 dB. Resistor R6 sets the mid-position gain.

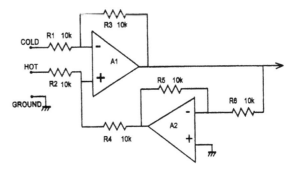

Figure 9.6 The 'Superbal' balanced input stage; input impedance on hot and cold are equal for both differential and common mode.

High-Z balanced amp

High-impedance balanced inputs, above 10 kΩ, are useful for interfacing to valve equipment. Adding output cathode-followers to valve circuitry is expensive, and so the output is often taken directly from a gain-stage anode. Even a light loading of 10 kΩ may seriously compromise distortion and available output swing.

All of the balanced stages dealt with up to now have their input impedances determined by the values of input resistors etc., and these cannot be raised without degrading noise performance. Figure 9.7 shows one answer to this. The op-amp inputs have infinite impedance in audio terms, subject to the need for R5, R6 to bias the non-inverting inputs.[3]

Adding R_g increases gain, but preserves balance. This configuration cannot be set to attenuate.

Microphone preamp with attenuator

It is often convenient to-use a balanced microphone preamp as a line input by using a suitable balanced attenuator, typically 20 to 30 dB. The input impedance of the microphone input stage will be 1 to 2 kΩ for appropriate mic loading, and this constrains the resistor values possible.

Keeping the overall input impedance to at least 10 kΩ means that the divider impedance must be fairly high, with a lot of Johnson noise. As a result, the total noise performance is almost always inferior to a dedicated balanced line-input amplifier. Common-mode rejection ratio is determined by the attenuator tolerances and will probably be much inferior to the basic microphone amp,

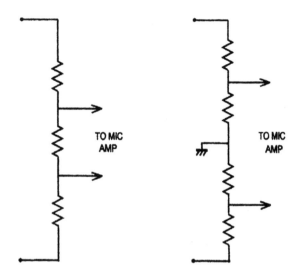

Figure 9.8 At (a), balanced attenuators convert a microphone preamp to line input. Circuit (b) is superior as both differential and commonmode signals are equally attenuated, so common-mode rejection is not degraded more than necessary.

which usually relies on inherent differential action rather than component matching.

Figure 9.8(a) shows a bad way to do it; the differential signal is attenuated, but not the common-mode, so cmrr is degraded even if the resistors are accurate. Figure 9.8(b) attenuates differential and common-mode signals by the same amount, so cmrr is preserved, or at any rate no worse than resistor tolerances make it.

Instrumentation amplifier

All the balanced inputs above depend on resistor matching to set the cmrr. In practice this means better than 45 dB is not obtainable without trimming. If a cmrr higher than this is essential, an IC instrumentation amplifier is a possibility.

Common-mode rejection ratio can be in the range 80 to 110 dB, without trimming or costly precision components. The IC tends to be expensive, due to low production volumes, and the gain is often limited in range and cannot usually be less than unity.

In audio work, cmrr of this order is rarely if ever required. If the interference is that serious, then it will be better to deal with the original source of the noise-rather than its effects.

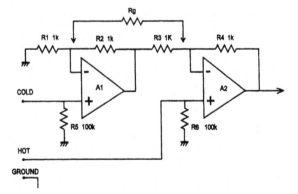

Figure 9.7 High-impedance balanced input stage; R5 and R6 set input impedance, and can be much higher. Add R8 to increase gain.

Input/output combinations

Taking five kinds of output—the rare case of floating output transformers being excluded—and the two kinds of input amplifier, there are ten possible combinations of connection. The discussion below assumes output R_s is 100 Ω, and the differential input amplifier resistors R are all 10 kΩ, as in Figure 9.3.

Unbalanced output to unbalanced input

This is the basic connection. There is no rejection of ground noise (cmrr = unity) or electrostatic crosstalk; in the latter case the 1 mA notional crosstalk signal yields a −20 dBv signal as the impedance to ground is very nearly 100 Ω.

Unbalanced output to balanced input

Assuming the output ground is connected to the cold-line input, then in theory there is complete cancellation of ground voltages. This is true, *unless* the output has a series output resistor to buffer it from cable capacitance,—which is almost always the case—for this will unbalance the line.

If the output resistance is 100 Ω, and the cold line is simply grounded as in Figure 9.8(a), then R_s degrades the cmrr to −46 dB even if the balanced input has exactly matched resistors.

The impedances on each line will be different, but not due to the asymmetrical input impedances of a simple differential amplifier; hot line impedance is dominated by the output resistance R_s on the hot terminal (100 Ω) and the cold line impedance is zero as it is grounded at the output end. The rejection of capacitive crosstalk therefore depends on the unbalanced output impedance. It will be no better than for an unbalanced input, as for the unbalanced output to balanced input case. The main benefit of this connection is ground noise rejection, which solves the most common system problem.

Impedance-balance out to unbalanced in

There is nothing to connect the output cold terminal to at the input end, and so this is the same as the ordinary unbalanced connection for the *unbalanced output to balanced input* configuration.

Impedance-balance out to balanced in

In theory there is complete cancellation of both capacitive crosstalk and common-mode ground voltages, as the line impedances are now exactly equal.

Table 9.2 Impedance-balancing gives better CMRR than the conventional circuit

	Capacitive 1 mA	CMRR (dB)
Conventional	−20 dBv	−46
Impedance-bal 99 Ω	−60 dBv	−101
Impedance-bal 100 R	−∞	−85
Impedance-bal 101 R	−60 dBv	−79

Table 9.2 shows the improvement that impedance-balancing offers over a conventional unbalanced output, when driving a balanced input with exactly matched resistors.

The effect of tolerances in the impedance-balance resistor are also shown; the rejection of capacitive crosstalk degrades as soon as the value moves away from the theoretical 100 Ω, but the cmrr actually has its point of perfect cancellation slightly displaced to about 98.5 Ω, due to second-order effects. This is of no consequence in practice.

Ground-cancelling out to unbalanced in

There is complete cancellation of ground voltages, assuming the ground-cancel output has an accurate unity gain between its cold and hot terminals. This is a matter for the manufacturer.

Ground-cancelling in this way is a very efficient and cost-effective method of interconnection for all levels of equipment, but tends to be more common at the budget end of the market.

Ground-cancelling out to balanced in

This combination needs a little thought. At first there appears to be a danger that the ground-noise voltage might be subtracted twice, which will of course be equivalent to putting it back in in anti-phase, gaining us nothing.

In fact this is not the case, though the cancellation accuracy is compromised compared with the impedance-balanced case; the commonmode rejection will not exceed 46 dB, even with perfect resistor matching throughout. Capacitive crosstalk is no better than for the 'Unbalanced output to balanced input' i.e. approximately −21 dB, which means virtually no rejection. However, this is rarely a problem in practice.

Balanced output to unbalanced input

This is not a balanced interconnection. There is nowhere to connect the balanced cold output to; it must be left open-circuit, its signal unused, so there is a 6 dB loss of headroom in the link. The unbalanced input means the connection is unbalanced, and so there is no noise rejection.

Balanced out to balanced in

A standard balanced system, that should give good rejection of ground noise and electrostatic crosstalk.

Quasi-floating out to unbalanced in

Since the input is unbalanced, it is necessary to ground the cold side of the quasi-floating output. If this is done at the remote (input) end then the ground voltage drop is transferred to the hot output by the quasifloating action, and the ground noise is cancelled in much the same way as a ground-cancelling output.

However, in some cases this ground connection must be local, i.e. at the output end of the cable, if doing it at the remote (input) end cause high-frequency instability in the quasi-floating output stage. This may happen with very long cables. Such local grounding rules out rejection of ground noise because there is no sensing of the ground voltage drop.

Perhaps the major disadvantage of quasi-floating outputs is the confusion they can cause. Even experienced engineers are liable to mistake them for balanced outputs, and so leave the cold terminal unconnected. This is not a good idea. Even if there are no problems with pickup of external interference on the unterminated cold output, this will cause a serious increase in internal noise. I believe it should be standard practice for such outputs to clearly marked as what they are.

Quasi-floating out to balanced in

A standard balanced system, that should give good rejection of ground noise and electrostatic crosstalk.

The hot and cold output impedances are equal, and dominate the line impedance, so even if the line input impedances are unbalanced, there should also be good rejection of electrostatic crosstalk.

Wiring philosophies

It has been assumed above that the ground wire is connected at both ends. This can cause various difficulties due to ground currents flowing through it.

For this reason some sound installations have relied on breaking the ground continuity at one end of each cable. This is called the one-end-only (OEO) rule.[4] It prevents ground currents flowing but usually leaves the system much more susceptible to r.f. demodulation. This is because the cable screen is floating at one end, and is now effectively a long antenna for ambient r.f.

There is also the difficulty that non-standard cables are required. A consistent rule as to which end of the cable has no ground connection must be enforced. The OEO approach may be workable for a fixed installation that is rarely modified, but for touring sound reinforcement applications it is unworkable.

A compromise that has been found acceptable in some fixed installations is the use of 10 nF capacitors to ground the open screen end at r.f. only; however, the other problems remain.

The formal OEO approach must not be confused with 'lifting the ground' to cure a ground loop. Unbalanced equipment sometimes provides a ground-lift switch that separates audio signal ground from chassis safety ground; while this can sometimes be effective, it is not as satisfactory as balanced connections. Lifting the ground must *never* be done by removing the chassis safety earth; this removes all protection against a live conductor contacting the case and so creates a serious hazard. It is also in many cases illegal.

The best approach therefore appears to be grounding at both ends of the cable, and relying on the cmrr of the balanced connection to render ground currents innocuous. Ground currents of 100 mA appear to be fairly common; ground currents measured in amps have however been encountered in systems with serious errors.

A typical example is connecting incoming mains 'Earth'—which is actually 'Neutral' in many cases—to a technical ground such as a buried copper rod. Take a look the section headed 'Electrical Noise' in last month's article for more details.

Ground currents cause the worst problems when they flow not only through cable shields but also the internal signal wiring of equipment. For this reason the preferred practice is to terminate incoming ground wires to the chassis earth of the equipment. This keeps ground currents off pcbs, where the relatively high track resistances would cause bad common-impedance coupling, and preserves r.f. screening integrity.

Grounding is simplified for source equipment that has no other connections, such as double-insulated compact-disc players. These carry a 'square-in-a-square' symbol to denote higher standards of mains insulation, so that external metalwork need not be grounded for safety. Such equipment often has unbalanced outputs, and can usually be connected directly to an unbalanced input with good results, as there is no path for any ground currents to circulate in.

If a balanced input is used, then connecting the hot input to CD signal and the cold to CD 'ground' leaves the CD player ground floating, and this will seriously degrade hum and r.f. rejection. The real ground must be linked to CD player common.

I think this chapter shows that balanced line interconnections are rather more complex than is immediately obvious. Having said that, with a little caution they work very well indeed.

References

1. Winder, S. 'What's the difference?' *Electronics World*, November 1995, p 989.

2. Ballou, G. (ed) 'Handbook For Sound Engineers—The Audio Cyclopaedia' Howard Sams 1987, ISBN 0-672-21983-2 (Superbal).

3. Smith, J.L. 'Modern operational circuit design', Wiley-Interscience, 1971, p 31 and 241.

4. Muncy, N. 'Noise susceptibility in analog and digital signal processing systems', *JAES*, Vol. 3 (6), June 1995, p 440.

Chapter 10

High-quality compressor/limiter

A variable law, low distortion attenuator incorporating
second harmonic cancellation circuitry
(Wireless World, December 1975)

This was the first article I ever wrote for *Wireless World,* as it then was. By the time I was in my
third year at Cambridge, I was deeply involved in audio electronics, and so when the time came
to choose a subject for the design project that was an important part of the course, I went straight
for audio. There were three main considerations: it was important to pick something that was
virtually certain to have a successful outcome, I wanted to have fun doing it, and it should be
suitable for publication in *Wireless World.* The first requirement should not be taken lightly. One of
my fellow students chose for his project the phase-locking of a record turntable to a stable crystal
oscillator. This sounded entirely doable in the time allowed, but I have vivid memories of the sound
of a turntable motor yet again accelerating out of control on the last day of the project period,
accompanied by groans of despair. How that ended, I do not know. Also indelibly marked on my
memory are the components provided by the Cambridge Electronics Lab. These were not exactly
state-of-the-art; in particular the resistors looked very much like old carbon types that had been
salvaged from a 1950s television. They did the job but looked awful. I have suspected ever since
that this was deliberate policy on the part of the Electronics Lab, because nice modern resistors
would have vanished rapidly into the legion of home-made hi-fi amplifiers being constructed by the
student body.

The compressor project was done in the early months of 1973, and took a little to work it through to
the top of the editorial pile at *Wireless World*; in those days, the competition to publish in WW was
fierce.

The technology described in this article is at first sight now somewhat obsolete, though in an area
where directly heated triodes designed just after the First World war are prized, it is a bit difficult
to come up with a working definition of 'obsolete'. I think it will be useful to those looking for
old-school methods of dynamics control. At the time, the junction FET was a great step forward
in voltage-controllable gain; previous approaches included diode bridges, filament-lamp and
photoresistor combinations, and ultrasonic chopping, none of which were very linear or very

satisfactory. The FET VCA was reasonably linear if the signal levels were kept low, and beautifully simple in terms of circuitry, but the Vgs/channel resistance law was (and is) subject to wide production spreads. With a feedback compressor/limiter such as the one described in the article, this was not a great drawback in a single-channel unit, as the sidechain generated whatever voltage was required for the attenuation needed. However, problems arose when two compressor channels were linked together for stereo operation, and a very tedious selection was required to get suitably matched pairs of FETs.

Today most compressor/limiters are built around transconductance-based VCAs, such as those made by THAT Corporation, which have an absolutely predictable control-voltage law that is linear in dB, and are almost distortion-free when compared with FET VCAs. (A little distortion still remains, so the history of VCAs has not ended yet.) These ingenious devices had begun to appear in 1975, but they were initially very expensive, and FET-based compressors remained popular for many years. The widest use of FETs as gain-control elements was in the Dolby cassette noise reduction system.

HIGH-QUALITY COMPRESSOR/LIMITER

December 1975

Compression and limiting play an increasingly important role in the resources of a modern sound studio. The conventional function of signal level control is to avoid overload, but it can be used in the realm of special effects. To date, however, relatively few designs for high-fidelity compressor/limiters have been published.

The main design problem is the voltage-controlled attenuator, v.c.a., which increases attenuation of the input signal in response to a voltage from a control loop as shown in Figure 10.1. In limiting, this circuit block continuously monitors the peak output level from the v.c.a. and acts to maintain an almost constant level if it exceeds a threshold value, or, in compression, allows it to increase more slowly than the v.c.a. input signal. This is illustrated in Figure 10.2, which shows the input-amplitude/output-amplitude characteristic for both compression and limiting. Note that limiting makes use of a much tighter slope to ensure that the output voltage cannot exceed the chosen limit, and that the threshold (point of onset of attenuation) takes place at a higher level than for compression.

Traditionally, studio-quality compressor/limiters (as the two functions are so similar it is logical to produce

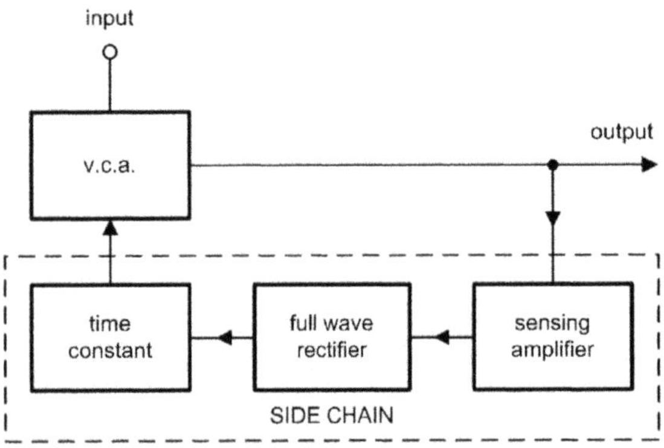

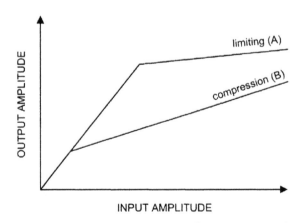

Figure 10.2 Amplitude characteristics for compression and limiting—the last mentioned uses an almost zero slope to prevent the output exceeding a preset level.

Figure 10.1 Voltage-controlled attenuator with d.c. control loop.

a system that can be used for either compression or limiting) used one of two types of v.c.a. Either the audio signal was chopped at an ultrasonic frequency by a variable mark/space square wave—which requires complex circuitry and careful filtering of the audio output to avoid beats with tape-recorder bias frequencies—or it was attenuated by an electronic potential divider one arm of which was a photoresistor, the control signal being applied via a small filament bulb. The last-mentioned has disadvantages because photoresistors are non-linear devices, therefore noticeable distortion is introduced into the audio signal, and the thermal inertia of the bulb filament limits the speed of attenuation onset.

Most modern compression systems use field-effect transistor operated below pinch-off as a voltage-variable resistance in a potential divider. This technique has many advantages; it is a simple, cheap, and fast-acting configuration that can provide an attenuation variable between 0 and 45 dB. The only problem is that an FET is a square-law device, and tends to generate a level of second-harmonic distortion that increases rapidly with signal amplitude. A typical arrangement is shown in Figure 10.3—R_2, R_3 and C_2 allow the source of the FET to be set at a d.c. level above ground, so that a control-voltage that moves positive with respect to ground can be used, to avoid level-shifting problems in the control loop. This d.c. level is isolated from the input and output by C_1.

The distortion introduced by this circuit is at its worst for the 6 dB attenuation condition, because at this point the drain-source resistance equals R_1, and the maximum power level exists in the FET. Table 10.1 shows the level of second-harmonic distortion introduced into a sine-wave signal of 100 mV r.m.s. amplitude, under the 6 dB attenuation condition for three different FET types. Measurements were made with a Marconi TF2330 wave analyser, higher-orders of harmonic distortion proved to be negligible amplitude in all cases. These measurements were made on one sample of each type of FET and, because production spreads are large, the results should be treated with some caution. However, it is clear that these levels of distortion are unacceptable for high-quality applications.

Fortunately, a technique exists for reducing FET distortion to manageable levels, if the control-voltage is applied to the FET gate and summed with a signal consisting of one-half the voltage from drain to source, then the distortion level is dramatically lowered. The configuration in Figure 10.4 shows a simple way of realising this; the signal fraction fed back is not critical and 10% resistors can be used for R_4 and R_5. Surprisingly, this distortion cancellation procedure leaves the attenuation/control-voltage

Table 10.1 Second-harmonic distortion level introduced into a sine-wave of 100 mV r.m.s.

Device	2N3819	2N5457	2N5459
2nd harmonic at −6 dB(%)	13	10	8.9
2nd harmonic with cancellation (%)	0.39	0.12	0.12
attenuation shown (dB)	2	10	2

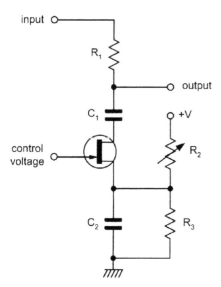

Figure 10.3 Basic v.c.a. circuit providing up to 45 dB of attenuation. This configuration introduces second-harmonic distortion which is greatest at 6 dB of attenuation.

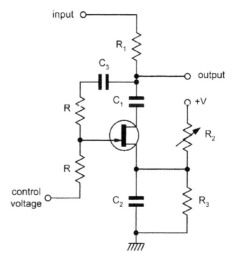

Figure 10.4 Standard circuit technique for reducing FET distortion by summing half of the drain/source voltage with the control voltage.

characteristic almost unchanged. Table 10.1 shows the new maximum distortion values for 100 mV r.m.s. input. (Note that the maximum no longer occurs at 6 dB attenuation, but at a point that varies with the FET type, where cancellation is least effective.) From these results the 2N5457 and 2N5459 are superior, the 2N5459 was used in the final version of the v.c.a.

To determine appropriate signal levels in the v.c.a., measurements were made of maximum distortion generated, i.e. the v.c.a. was set to 2 dB attenuation, against r.m.s. input voltage; results are shown in Table 10.2. The question now arises as to whether this distortion performance is adequate for a high-quality compressor/limiter. There is no general agreement as to the amount of second harmonic distortion that can be introduced into a program signal before it becomes aurally detectable, but 0.1% is a figure that is quoted. This means that the permissible input voltage to the v.c.a. would be restricted to below 100 mV r.m.s. In practice, however, the attenuation level will be constantly changing, and because distortion level peaks fairly sharply with attenuation change, this level of distortion will only be present for a very small percentage of the time. In any case, second harmonic distortion alone has a relatively low 'objectionability factor'. The proof of the pudding is in listening to the compressor output signal; inputs of music around 200 mV r.m.s. produced no trace of audible distortion. (Good class A power amplifiers and headphones were used for monitoring).

The control loop consists of an amplifier which senses the v.c.a. output level. A full-wave rectification system is normal practice because program waveforms have positive and negative peaks that can vary by as much as 8 dB, and an 8 dB uncertainty in the output level is usually unacceptable. A time-constant arrangement is used with the rectification circuit to control the attack and decay rates.

The output sensing amplifier in the system is a non-inverting op-amp which allows a high input impedance because the output impedance of the v.c.a. stage reaches a maximum of about 39 kΩ at zero attenuation.

Table 10.2 Maximum distortion generated by various input voltages at 2 dB attenuation

Input (mV, r.m.s.)	2nd harmonic (%)
20	0.005
50	0.10
100	0.12
200	0.19
500	0.34
1,000	0.56

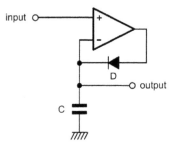

Figure 10.5 Basic precision rectifier circuit where the rectifying element is in the feedback loop of an op-amp.

The full-wave rectification system consists of a transistor phase-splitter driving two op-amp precision-rectifier stages in antiphase. The principle of a precision rectifier is illustrated in Figure 10.5. The rectifying element is placed in the feedback loop of an op-amp, so that the effect of the forward voltage drop on the output voltage is divided by the open-loop gain. During positive half-cycles, if the input voltage exceeds the d.c. level stored on the capacitor C, the op-amp output swings positive and C is charged through diode D until its stored voltage is equal to the input voltage. Thus C takes up a voltage across it equal to that of the positive peak of the input signal. During negative half-cycles, and while the input is less than the voltage on C during positive half-cycles, the op-amp saturates negatively and D remains firmly reverse-biased. Obviously this is only a half-wave rectification circuit, the full-wave version uses two of these driven in antiphase, and charging a common capacitor. A resistance through which the charging currents flow determines the attack time, and another in parallel with C defines the decay time-constant.

The complete circuit is shown in Figure 10.6. The v.c.a. is essentially as described above and the attenuation threshold is set by the variable resistance R_2. As the resistance is increased the level of control voltage required for attenuation to begin is reduced, and the system's input/output characteristic moves smoothly from A to B on Figure 10.2. The threshold decreases and the compression slope becomes less flat as the system turns slowly from a limiter into a compressor by the manipulation of a single control. The output sensing amplifier consists of IC_1 and has a gain of 19 over the audio band. This is rolled off to unity at d.c. by C_5. Transistor Tr_2 and its associated components form a conventional phase-splitter driving IC_2 and IC_3 the precision rectifiers. The rectifier circuitry is more complex than implied above, three modifications have been made to improve the performance. Firstly, IC_2 and IC_3 charge C_9 via current amplifier stages Tr_5 and Tr_6 otherwise the current-limited 741 outputs would be unable to

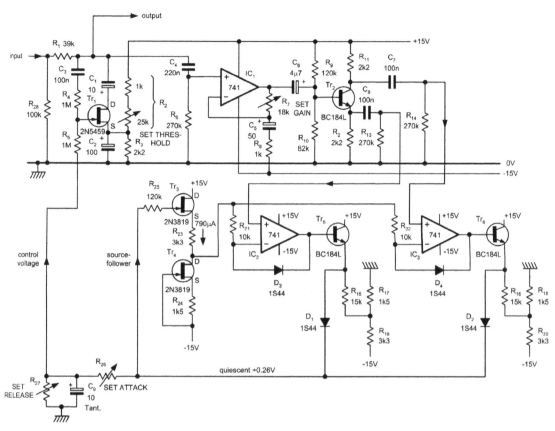

Figure 10.6 Complete circuit where the output is taken directly from the v.c.a.—this may be buffered for loads greater than 100 kΩ.

provide enough current for the faster attack times (less than 1 mS). Secondly, the feedback loop from C_9 to the inverting unputs of IC_2 and IC_3 is completed via a FET source-follower. Without this, C_9 would be loaded by the two 741 inputs, and this would severely limit the maximum decay times available. Incorporating the source-follower allows decay times of several minutes by using large resistance values for R_{27}. The conventional source-follower has a large negative offset voltage and is unusable in this application because due to their the rectifying action IC_2 and IC_3 are unable to provide a voltage on C_9 that is negative of ground. This would be required to allow the source-follower output to be at ground when there is no input to the rectifers. However, if a modified source-follower is used, with a constant-current source and resistance combination in the source circuit, the offset voltage can be varied on either side of zero by manipulation of R_{24} which varies the driving current. The offset voltage is arranged to be plus 0.3 V, to allow a large safety margin for thermal variations, component ageing, etc. This means that under no-signal conditions C_9 takes

up a standing quiescent voltage of plus 0.3 V. The effect of this is taken up in the calibration of R_2.

The third modification is the addition of R_{21}, D_3, and R_{22}, D_4. These two networks prevent IC_2 and IC_3 from saturating negatively, during negative half-cycles of their input voltage, by allowing local negative feedback through D_3 and D_4. This limits the negative excursion of the IC outputs to about 2 V. The prevention of saturation is necessary because the recovery time of the 741s causes the frequency response of the precision rectifier circuit to drop off at about 1 kHz. The addition of the anti-saturation networks provides a frequency response that starts to fall off significantly above about 12 kHz which is ample for our purposes as program signals have very little energy content above this frequency.

The final part of the circuit defines the attenuation time constants. Resistor R_{26} sets the attack time constant and R_{27} the decay time constant; these can range between 0 and 1 MΩ (220 μs and 10 s) for R_{26}, and 1 kΩ and ∞ (10 mS and 20 min) for R_{27}. They can be either switched or variable resistances, depending on the range of variation required.

The circuit in Figure 10.6 shows the compressor output being taken directly from the v.c.a. This is only suitable if the minimum load to the output is greater than 100 kΩ, otherwise the v.c.a. attenuation characteristic will be distorted by excessive loading. If lower resistance loads are to be driven a buffer amplifier stage must be interposed. The IC_1 amplifier stage is suitable for most applications, and its gain is $(R_7 + R_8)/R_8$. For the unity gain case R_8 & C_5 can be eliminated and R_7 replaced by a direct connexion.

The compressor should be driven from a reasonably low impedance output (less than 5 kΩ).

Construction is straightforward; the layout is not critical and the prototype was assembled on 0.1 in matrix Veroboard. To set up the circuit R_{24} is adjusted so that the voltage across C_9 is about $+0.3$ V with no signal input. The value required will vary due to production spreads in the f.e.ts. To calibrate R_2 it is necessary to relate the level of input signal at which attenuation commences, with the voltage across C_2. This can be done with an oscilloscope, or preferably an a.f. millivoltmeter. As a guide the calibration data for the prototype is shown in Table 10.3, along with the values of the compression ratio (number of dBs the input must increase by to increase the output by 1 dB). This data must be regarded as only a guide. It is worth noting that as the controlling factor setting the compression/limiting function is the voltage across C_2 R_2 could be replaced by a 1 kΩ resistor connected to a remote voltage source.

The compressor/limiter is quite straightforward in use, provided a few points are kept in mind. Firstly, if it is being used in the limiting mode to prevent overload of a subsequent device, the fastest possible attack time should be used, to catch fast transients, and a fast decay time (say 100 ms; $R_{27} = 10$ kΩ), to allow the system to recover rapidly when the transient has passed. Secondly, if a noisy programme signal is being compressed a long decay time should be employed, otherwise the noisy background will be faded up during quiet passages, and the familiar compressor 'breathing noises' will be heard. Finally, signals with a large v.l.f. content should

be avoided or filtered, otherwise v.l.f. modulation of the signal will result, if a fast decay time is in use.

If a stereo compressor/limiter is constructed from two of the systems described above it is necessary to gang together R_2, R_{26}, and R_{27} between the two channels. A direct connection between the non-grounded sides of the two C_9s is also needed. It might be necessary to select matched f.e.ts to avoid stereo image shift during compression, due to differing attenuation characteristics in the two v.c.as. A well-smoothed p.s.u. providing ± 15 V should be used to power the compressor/limiter.

Components list

$IC_{1'2'3}$	741	
$Tr_{2'5'6}$	BC184L or equivalent	
Tr_1	2N5459	
$Tr_{3'4}$	2N3819	
$D_{1'2'3'4}$	IS44 or low-leakage equivalent	
R_1	39 k	
R_2	25 k variable, with 1 k in series	
R_3	2.2 k	
$R_{4'5}$	1 M	
R_6	270 k	
R_7	18 k	
R_8	1 k	
R_9	120 k	
R_{10}	82 k	All resistors
$R_{11'12}$	2.2 k	(except R_2)¼ W
$R_{13'14}$	270 k	
$R_{15'16}$	15 k	
$R_{17'18}$	1.5 k	
$R_{19'20}$	3.3 k	
$R_{21'22}$	10 k	
R_{23}	3.3 k	
R_{24}	see text	
R_{25}	120 k	
$R_{26'27}$	see text	
R_{28}	100 k	
C_1	10 µF 25 V electrolytic	
C_2	100 µF 25 V electrolytic	
C_3	100 nF 250 V polyester	
C_4	220 nF 250 V polyester	
C_5	50 µF 40 V electrolytic	
C_6	4.7 µF 40 V electrolytic	
$C_{7'8}$	100 nF 250 V polyester	
C_9	10 µF 16 V tantalum bead	

Table 10.3 Prototype calibration data and compression ratios

VC_2 (V)	Threshold (mV, pk)	Compression ratio
2.9	10	2.3
3.5	20	5.1
5.0	50	10
6.7	100	20
8.5	200	35
9.8	500	50

Chapter 11

Inside mixers

(Electronics World, April 1991)

Intended title: *The Design of Mixing Consoles*

When this was written, I was Chief Design Engineer at Soundcraft Electronics, so mixing console design was my day job, so to speak, and I was doing hi-fi design purely as a spare-time pursuit. Having been responsible for many of the contemporary improvements in mixing console design (such as the padless mic amp and the active panpot), I thought it might be a good plan to publicise these in *Electronics World*. After consulting those set in authority over me, a fine balance was struck between offering solid, accurate technical content and not giving away much to our competitors. The padless microphone amplifier and the active panpot could be described in some detail, as they were fully covered by our patents. Certainly the padless microphone amp patent was very carefully written to give full disclosure, while omitting one or two odd details that would have been most helpful to the proverbial 'person skilled in the art' who was trying to implement the idea. Some other non-patentable bits of technology were glossed over a little, once again with important details, ahem, suppressed. Some of you may think this is not very nice behaviour, but that's business for you.

I was really quite cautious talking about electronic switching as at Soundcraft, we reckoned we had the edge on everybody else, but with the best will in the world, there seemed to be nothing there that was solidly patentable. The whole subject of electronic switching was eventually revealed in two articles in 2004, and they are also collected in this book.

Recording technology has changed greatly since this article was written, and the frequent references to tape-machines seem very dated. Enormous consoles many metres long are also somewhat out of fashion for recording,

Smaller mixing consoles do not present the same formidable challenges as a really big machine. The smaller the mixer, the less the extent of the grounding system, so the less its resistance and the smaller the chance of ground currents causing noise or crosstalk. Shorter mix buses have less capacitance to ground, and less chance of being a significant fraction of an RF wavelength, which makes assuring summing-amp stability rather easier. Currently, large consoles are being cut down to make multiple 'single-bin' consoles composed of eight input modules. Since this provides input modules only, a custom master section has to be added to deal with mix bus summing, PFL, etc.

If I were designing a great big long mixer again (which is pretty unlikely, not least because I have little interest in doing anything of the kind), I would follow the eight-inputs-per-bin format, and each bin would have its own local summing amps, almost certainly balanced. The outputs from these would be combined, once more in a balanced fashion to suppress voltage differences between the bin grounds, and to give the final mix output. This approach, known as devolved summing, distributed summing, or multi-level summing [1], not only gives a summing-noise noise reduction of up to 9 dB, which may be non-intuitive but is very real, but also keeps the mix buses short. This is yet another example of more electronics giving less noise.

Reference

1. Self, D., *Small-Signal Audio Design,* 2nd edition, Chapter 22, pp. 635–639. Focal Press (Taylor & Francis) 2014. ISBN: 978-0-415-70974-3 Hardback. 978-0-415-70973-6 Paperback. 978-0-315-88537-7 ebook.

INSIDE MIXERS

April 1991

A large mixing console arguably represents the most demanding area of audio design. The steady advance of digital media demands that every part of the chain that takes music from performer to consumer must be near-perfect, as the comfortable certainty that everything will be squeezed through the quality bottleneck of either analogue tape or vinyl disc now looks very old-fashioned. This chapter was prompted by the introduction of the Soundcraft 3200 recording console, which is believed to have the highest performance in terms of noise, crosstalk and linearity of any console ever built.

Competition to sell studio time becomes more cut-throat with every passing week, and it is clear that advances in console quality must not harm cost-effectiveness. The only way to reconcile these demands is to innovate and to keep a very clear view as to what is really necessary to meet a demanding specification; in other words the way forward is to use conventional parts in an unconventional way, rather than simply reaching for the most expensive op-amp in the catalogue.

The technical problems that must be over-come in a professional mixing console are many. A large number of signals flow in a small space and they must be kept strictly apart until the operator chooses to mix them; crosstalk must be exceedingly low.

There may be up to 64 input channels, each with many stages, and all having the potential to add distortion and noise to the precious signal. Even summing these signals together, while sounding trivially easy, is in practice a major challenge. In short, requirements are much more demanding than those for the most expensive hi-fi equipment, because degradation introduced at the recording stage can never be retrieved.

Major functions of consoles are largely standardised, although there is much scope for detailed variation. Figure 11.1 shows a typical system diagram for a split (separate groups) mixing console. The technique of multi-track recording is explained in the appendix at the end of this article.

Figure 11.2 shows a typical input channel for a mixing console. The input stage provides switchable balanced mic and line inputs; the mic input has an impedance of 1–2 kΩ, which provides appropriate loading for a 200 Ω mic capsule, while the line input has a bridging impedance of not less than 10 kΩ. This stage gives a wide

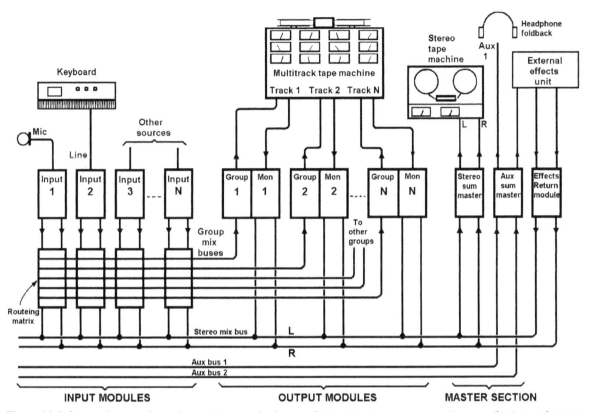

Figure 11.1 System diagram of complete mixing console, showing division into inputs, group monitor contributions and master modules. Routeing matrix determines which group of inputs shall be fed to a given track on the multi-track tape machine. Several channels share one effects device.

range of gain control and is followed immediately by a high-pass filter (usually −3 dB at 100 Hz) to remove low-frequency disturbances.

The tone-control section (universally known in the audio business as 'EQ' or equalisation) typically includes one or more mid-band resonance controls as well as the usual shelving Baxandall-type high and low controls. Channel level is controlled by a linear fader and the panpot sets the stereo positioning, odd group numbers being treated as left, and even as right. The prefade-listen (PFL) switch routes the signal to the master module independently of all other controls; a logic bus signals the master module to switch the studio monitoring speakers from the normal stereo mix bus to the PFL bus, allowing any specific channel to be examined in isolation.

Figure 11.3 shows a typical group module and Figure 11.4 the basics of a master section; a manual source-select switch allows quality checking of the

final stereo recording and two solid-state switches replace the stereo monitor signal with the PFL signal whenever a PFL switch anywhere on the console is pressed.

Microphone inputs

The microphone preamplifier is a serious design challenge. It must provide from 0 to 70 dB of gain to amplify deafening drum-kits or discreet dulcimers, present an accurately balanced input to cancel noise pickup in long cables and generate minimal internal noise. It must also be able to withstand + 48 V DC suddenly applied to the inputs (for phantom-powering internal preamps in capacitor mics) while handling microvolt signals. The Soundcraft approach is to use standard parts, which are proven and cost-effective through quantity production, in new configurations. The latest mic preamplifier design, as used on the

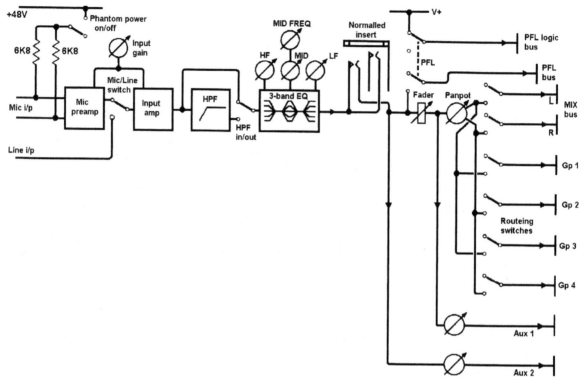

Figure 11.2 One input channel. Gain control is 70 dB and tone control is standard Baxandall shelving type with addition of mid-range lift and cut. Two auxiliary sends are shown.

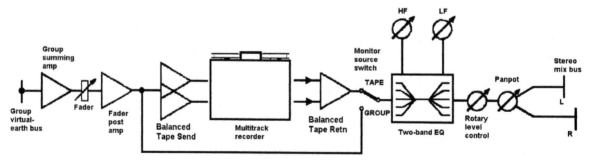

Figure 11.3 Block diagram of typical group module, showing switching between direct output and tape replay for monitoring purposes.

Series 3200, is new enough to be covered by patent protection.

It is now rare to use input transformers to match the low-impedance (150–200 Ω) microphone to the pre-amplifier, since the cost and weight penalty is serious, especially when linearity at low frequencies and high

levels is important. The low-noise requirement rules out the direct use of op-amps, since their design involves compromises that make them at least 10 dB noisier than discrete transistors at low impedance.

This circuit, shown in Figure 11.5, therefore uses a balanced pair of low-noise, low-R_b PNP transistors as

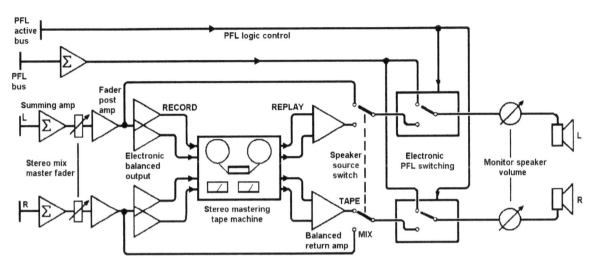

Figure 11.4 Block diagram of master module, with tape send/replay switching and automatic PFL switching.

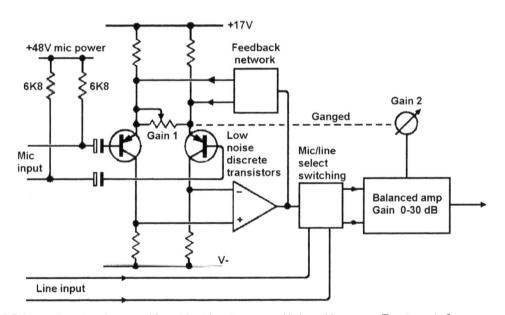

Figure 11.5 Low-noise microphone amplifier with wide gain range and balanced line output. Transistors in first stage avoid noise problem of op-amps.

an input stage, working with two op-amps to provide load-driving capability and raw open-loop gain to linearise signal handling. Preamplifier gain is spread over two stages to give a smooth 0–70 dB gain range with the rotation of a single knob. This eliminates the switched 20 dB attenuator that is normally required to give the lower

gain values, not only saving cost and complication, but also avoiding the noise deterioration and CMRR degradation that switched attenuators impose. The result is an effective input stage that is not only quieter, but also more economical than one using specialised low-noise op-amps.

Equalisation

Since large recording consoles need sophisticated and complex tone-control systems, unavoidably using large numbers of op-amps, there is a danger that the number of active elements required may degrade the noise performance.

A typical mid-band EQ that superimposes a + 15 dB resonance on the flat unity-gain characteristic is shown in Figure 11.6. A signal is tapped from the forward path, put through a state-variable band-pass filter which allows control of centre-frequency and Q, and then added back. To improve noise performance, the signal level at all locations (in all conditions of frequency, Q, and boost/cut) was assessed, and it proved possible to double the signal level in the filter over the usual arrangement, while maintaining full headroom. The signal returned into the forward path is then attenuated to maintain the same boost/cut, and the noise added is thus reduced by about 6 dB.

Auxiliary sends: foldback and effects

The auxiliary sends of a console represent an extra mixing system that works independently of the main groups; the number and configuration of these sends have a large effect in determining the overall versatility of the console. Each send control provides a feed to a console-wide bus; this is centrally summed and then sent out of the console.

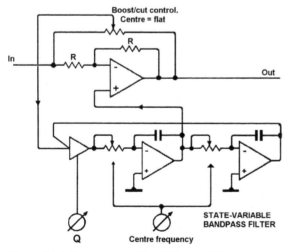

Figure 11.6 Parametric mid-band EQ stage. EQ and centre frequencies are independently variable, being set by the parameters of the state-variable filters.

Sends come essentially in two kinds: prefade sends, which are taken from before the main channel fader, and post-fade sends, which take their feed from after the fader, so that the final level depends on the settings of both. There may be anything from one to twelve sends available, often switchable between pre and post. Traditionally, this means laboriously pressing a switch on every input module, since it is most unlikely that a mixture of pre and post sends on the same bus would be useful; the Series 3200 minimises the effort by setting pre/post selection for each bus from a master switch that controls solid-state pre/post switching in each module.

Prefade sends are normally used for 'foldback'; i.e. sending the artist a headphone feed of what he/she is perpetrating, which is important if electronic manipulation is part of the creative process, and essential if the artist is adding extra material that must be in time with that already recorded. In the latter case, the existing tracks are played back to the artist via the prefade sends on the monitor sections.

Postfade sends are used as effects sends; their source is after the fader, so that the effect will be faded down at the same rate as the untreated signal, maintaining the same ratio. The sum of all feeds to a given bus is sent to an external effects unit and the output of this returned to the console. This allows many channels to share one expensive device (this is particularly applicable to digital reverb) and is often more appropriate than the alternative of patching a processor into the channel insert point.

'Effect returns' may be either modules in their own right or a small subdivision of the master section. The returned effect, which may well now be in stereo, the output of a digital reverb, for example, is usually added to the stereo mix bus via level and pan controls. EQ is also sometimes provided.

Panpot

To give smooth stereo panning without unwanted level changes, the panpot should theoretically have a sine/cosine characteristic; such components exist, but they are prohibitively expensive and so most mixing consoles use a dual linear pot. with its law bent by a pull-up resistor, as shown in Figure 11.7(a).

This not only gives a mediocre approximation of the required law, but also limits the panning range, since the pull-up signal passes through the wiper contact resistance (usually greater than the end-of-track resistance) and limits the attenuation the panpot can provide when set hard left or right. This limitation is removed in the Soundcraft

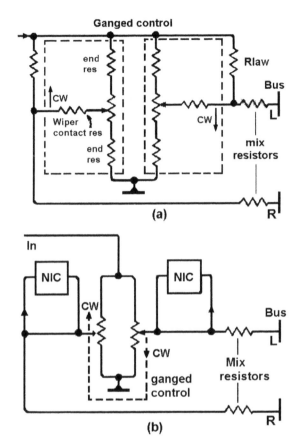

(a)

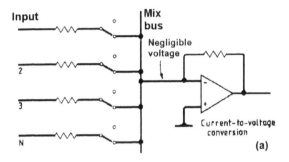

(a)

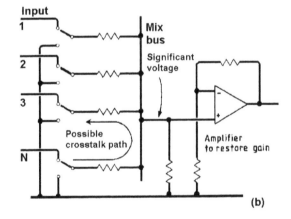

(b)

as in Figure 11.8(a). A summing amplifier with shunt feedback is used to hold a long mixing bus at apparent ground, generating a sort of audio black hole; signals fed into this via mixing resistors apparently vanish, only to reappear at the output of the summing amplifier, as they have been summed in the form of current. The elegance of virtual-earth mixing, as opposed to the voltage-mode

Figure 11.7 Standard panpot circuit at (a) showing how pull-up resistor draws current through wiper contact resistance, which is usually greater than the end resistance of the pot., limiting maximum attenuation. Arrangement at (b) uses NICs to replace pull-up to modulate law with panpot setting. Left/right isolation increased from −65 dB to −90 dB.

active panpot shown in Figure 11.7b by replacing the pull-up with a negative-impedance-converter that modulates the law-bending effect in accordance with the panpot setting, making a close approach to the sine law possible. There is no pull-up at the lower end of the wiper travel, when it is not required, so the left-right isolation using a good-quality pot is improved from approx −65 to −90 dB. This has also been made the subject of patent protection.

Summing

One of the main technical challenges in console design is the actual mixing of signals. This is done almost (but not quite) universally by virtual-earth techniques,

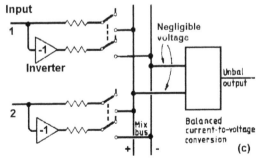

(c)

Figure 11.8 Virtual-earth summer at (a) effectively eliminates cross-talk, since there is almost no signal at the summing point. Voltage-mode circuit at (b) allows crosstalk. Balanced virtual-earth summing circuit at (c) requires a separate inverter for each channel to provide the antiphase signal.

summing technique in Figure 11.8(b), is that signals cannot be fed back out of the bus to unwanted places, as it is effectively grounded, and this can save massive numbers of buffer amplifiers in the inputs.

There is, however, danger in assuming that a virtual earth is perfect; a typical opamp summer loses open-loop gain as frequency increases, making the inverting input null less effective. The 'bus residual' (i.e. the voltage measurable on the summing bus) therefore increases with frequency and can cause inter-bus crosstalk in the classic situation with adjacent buses running down an IDC cable.

Increasing the number of modules feeding the mix bus increases the noise gain; in other words the factor by which the noise of the summing amplifier is multiplied. In a large console, which might have 64 inputs, this can become distinctly problematic. The Soundcraft solution is to again exploit the low noise of discrete transistors coupled to fast opamps, in configurations similar to the mic preamps.

These sum amplifiers have a balanced architecture that inherently rejects supply-rail disturbances, which can otherwise affect LF crosstalk performance.

As a console grows larger, the mix bus system becomes more extensive, and therefore more liable to pick up internal capacitive crosstalk or external AC fields. The 3200 avoids internal crosstalk by the use of a proprietary routeing matrix construction which keeps the unwanted signal on a bus down to a barely measurable 120 dB. This is largely a matter of keeping signal voltages away from the sensitive virtual-earth buses. Further improvement is provided by the use of a relatively low value of summing resistor; this also keeps the noise down, although since it drops as the square-root of the resistor value, at best, there is a clear limit to how far this approach will work before drive power becomes excessive; 4.7 kΩ is a reasonable minimum value.

External magnetic fields, which are poorly screened by the average piece of sheet steel, are rejected by the balanced nature of the Series 3200 mix buses, shown in Figure 11.8c. The operation is much the same as a balanced input; each group has two buses, which run physically as close together as possible and the group reads the difference between the two, effectively rejecting unwanted pickup. The two buses are fed in antiphase from each input, effectively doubling the signal level possible for a given supply voltage. Overall mixing noise is reduced by 3 dB, the signal level is 6 dB up and the noise, being uncorrelated for each bus, only increases by 3 dB.

The obvious method of implementing this is to use two summing amplifiers and then subtract the result. In the 3200, this approach is simplified by using one symmetrical summing amplifier to accept the two antiphase mix buses simultaneously; this reduces the noise level as well as minimising parts cost and power consumption. The configuration is very similar to that of the balanced mic amp, and therefore gives low noise as well as excellent symmetry.

Solid-state switching

There are two main applications for electronic switching in console design. The first is 'hard' switching to reconfigure signal paths, essentially replacing relays with either JFETs (Figure 11.9(a)) or 4016-type analogue gates which, since they are limited to 18 V rails and cannot handle the full voltage swing of an opamp audio path, must be used in current mode, as shown in Figure 11.9(b). Note that when gate 1 is off, gate 2 must be on to ensure that a large voltage does not appear on gate 1 input. Full voltage range gates do exist but are very expensive.

Secondly, there is channel muting; this not a hard switch, since an unacceptable click would be generated

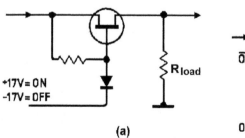
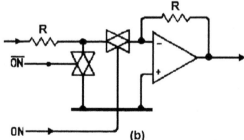

Figure 11.9 Hard switching with JFETs in voltage mode (a) and with analogue gates in the current mode (b), which prevents gate elements from being driven outside their voltage capabilities.

unless the signal happened to be at a zero-crossing at the instant of switching; the odds are against you. The Series 3200 therefore implements muting as a fast-fade that takes about 10 ms; this softens transients into silence while preserving time-precision. It is implemented by a series-shunt JFET circuit, with carefully synchronised ramp voltages applied to the FET gates.

Performance factors

Primary requirements of modern consoles are very low noise and minimal distortion. Since a comprehensive console must pass the audio through a large number of circuit stages (perhaps over 100 from microphone to final mixdown) great attention to detail is essential at each stage to prevent a buildup of noise and distortion; the most important tradeoff is the impedance of the circuitry surrounding the opamp, for if this too high Johnson noise will be increased, while if it is too low an opamp will exhibit nonlinearity in struggling to drive it.

The choice of device is also critical, for cost considerations discourage the global use of expensive chips. In a comprehensive console like the 3200 with many stages of signal processing, this becomes a major concern; nonetheless, after suitable optimisation, the right-through THD remains below 0.004% at 20 dB above the normal operating level. At normal level it is unmeasurable.

Appendix: the technique of multitrack recording

Multitrack recording greatly enhances the flexibility of recording music. The availability of a number of tape tracks (anywhere between 4 and 32 on one reel of tape) that can be recorded and played back separately allows each instrument a dedicated track, the beauty of this being that one mistake does not ruin the whole recording; only a single part need be done again. The multitrack process is in two basic halves; recording individual tracks (or 'tracklaying') and mixdown to stereo.

Recording

Normally only one or two parts are recorded at once, though it quite possible to dedicate five or six tracks to a drum kit. The initial sound, whether captured by a microphone or fed in directly from a synthesiser line output, is usually processed as little as possible before committing it to tape; subsonic filtering and perhaps compression or limiting are used, but most effects are carefully avoided because they are usually impossible to undo later. You can easily add reverberation, for example, but just try removing it.

Recording is performed via the input modules, this being the only place where microphone preamps are fitted. The inputs are mixed together into groups if required; performers doing backing vocals might use four or five microphones, but these would almost certainly be mixed down to a stereo pair of groups at the recording stage, so that only two tape tracks are taken up. A bank of switches on each input module determines which group shall be fed; this is known as the routing matrix. Combined group outputs are then sent to tape; however a 'group' is usually used even if only one signal is being recorded, as this is the part of the console permanently connected to the multitrack.

It is clearly essential that new parts are performed in time with the material already on tape and also that the recording engineer can make up a rough impression of the final mix as recording proceeds. Thus continually replaying already-recorded material is almost as important as recording it in the first place. During recording, the tape tracks already laid down are replayed through 'monitor sections' which are usually much-simplified inputs giving limited control; this keeps the more flexible inputs free for material that is actually being recorded. One of the major features of the Series 3200 is that the monitor sections are unusually capable, having facilities almost identical to the inputs and allowing much more accurate assessment of how the mix is progressing, reducing learning time for operators.

Mixdown

When the tracklaying process is complete, there are 16 or more separate tape tracks that must be mixed down to stereo. Major manipulations of sound are done at this mixdown stage; since the multitrack tape remains unaltered, the resulting stereo being recorded on a separate two-track machine, any number of experiments can be performed without doing anything irrevocable.

Multitrack replay signals now enter the console through the input channels, so that the maximum number of facilities are available. Linear channel faders set the relative levels of the musical parts, while the rotary panpots (panoramic potentiometers) define the placement of instruments in the stereo sound field by setting the proportion of signal going to left and right mix buses. The monitor sections are now redundant, and can therefore be used either as extra inputs to the stereo mix, perhaps for keyboards, or to return effects.

Virtual mixing

The advent of computer-based sequencers has given rise to the term 'virtual mixing'. Keyboard/synthesiser parts of the musical masterwork are not committed to multitrack, but instead stored in the form of MIDI sequencer data. This can be replayed at any time, providing means of synchronising it to the acoustic parts on the multitrack exist; this requires one tape track to be dedicated to some form of timecode.

The advantages are, firstly, that this gives almost any number of extra 'virtual tracks', and secondly, that the synthesiser parts suffer minimal degradation as they avoid one generation of tape storage.

Chapter 12

Electronic analogue switching, Part I: CMOS gates

(Electronics World, January 2004)

In the course of the years spent designing mixing consoles, I came to know a good deal about analogue switching. All but the most basic mixers have a PFL system—the acronym meaning Pre-Fade Listen. In other words, you can press a button on an input channel and have its contribution alone heard through your monitor speakers without disturbing the flow of signals to the main outputs. This requires the feed to the monitors to be switched, and for many years, the only way to do this was a double-pole relay going 'clunk' somewhere in the console. It was a great relief to all concerned when the arrival of the 4016 analogue switch meant that this function could be performed electronically. It was cheap, reliable, and acoustically silent, and if the linearity was not perfect it was quite good enough for monitoring purposes if you used it intelligently. The chapter on mixing console design 'Inside Mixers' in this book gives more details on PFL systems and the like.

The article assumes that the highest possible supply rails will be used to maximise linearity. For the 4000-series devices, this is ±7.5 V; for lower rails, the distortion performance degrades fairly rapidly. In recent times interest in running high-quality audio circuitry from the +5V available on an USB socket has much increased, but 4016s and 4066s give a very poor distortion performance on such a low rail, and specialised parts such as the DG 9424 from Vishay-Siliconix must be used to get low distortion [1]. That part helpfully also has a very low on-resistance.

It is undeniable that this chapter is showing its age more than most of the articles in this book. Nowadays a bank of 4016s and 4066s would be considered a clumsy way to implement any significant amount of audio switching, and there would almost certainly be unkind remarks about the relatively poor linearity.

Take the case of preamplifier input select switching. This would be hard to do economically with 4000-series gates, not least because they cannot handle the maximum input signal which might be experienced, which could easily go up to 10 Vrms. Some sort of initial attenuation would be required, and it would all get very inelegant. Instead you would reach for a modern input-select IC like the Toshiba TC916X series, which contain arrays of analogue gates that can handle full opamp output swings. My experience is with the TC9163, which I designed into the Cambridge Audio 340A integrated amplifier. This is configured so it can be used for both input-select and tape-loop switching for two channels. These devices seem to be reasonably protected against excessive signals and ESD, but they are not totally invulnerable. External clamp diodes on the inputs are not required, but a 1 kΩ series resistor was used to limit the current into the on-chip clamping. All of the series are digitally controlled over a 3-wire link (strobe, clock, and data), so a housekeeping microcontroller is a necessity.

Reference

1. Self, D., *Small-Signal Audio Design,* 2nd edition, Chapter 21, pp. 588–589. Focal Press (Taylor & Francis) 2014. ISBN: 978-0-415-70974-3 Hardback. 978-0-415-70973-6 Paperback. 978-0-315-88537-7 ebook.

ELECTRONIC ANALOGUE SWITCHING, PART I: CMOS GATES

January 2004

Electronic switching

The switching and routing of analogue signals is a fundamental part of signal processing, but not one that is easily implemented if accuracy and precision are required. This chapter focuses on audio applications, but the basic parameters such as isolation and linearity are equally relevant in many fields.

Any electronic switching technique must face comparison with relays, which are still very much with us. Relays give total galvanic isolation between control and signal, zero contact distortion, and in audio terms have virtually unlimited signal-handling capability. They introduce negligible series resistance and shunt leakage to ground is usually not even worth thinking about. Signal offness can be very good, but as with other kinds of switching, this depends on intelligent usage. There will always be capacitance between open contacts, and if signal is allowed to crosstalk through this to nominally off circuitry, the 'offness' will be no better than other kinds of switching. (Throughout this chapter I use the word 'offness' which is not found in any spellchecker but is widely used in the pro audio sector—as the quickest way of referring to the ratio in dB by which an unwanted input is suppressed.)

Obviously relays have their disadvantages. They are big, expensive, and not always as reliable as more than a hundred years of development should have made

them. Their operating power is significant. Some kinds of power relay can introduce disastrous distortion if used for switching audio because the signal passes through the magnetic soft-iron frame; however such problems are likely to be confined to the output circuits of large power amplifiers. For small-signal switching the linearity of relays can normally be regarded as perfect.

Electronic switching is usually implemented with CMOS analogue gates, of which the well-known 4016 is the most common example, and these are examined first. However, there are many special applications where discrete JFETs provide a better solution, so these are dealt in the second part.

Part I: analogue gates

CMOS analogue gates, also known as transmission gates, are quite different from CMOS logic gates, though the underlying process technology is the same. Analog gates are bilateral, which means that either of the in/out leads can be the input or output; this is most emphatically not true for logic gates. The 'analogue' part of the name emphasises that they are not restricted to any fixed logic levels, but pass through whatever signal they are given with low distortion. The 'low' word there requires a bit of qualification, as will be seen later.

There is no 'input' or 'output' marked on these gates, as they are symmetrical. When switched on, the connection between the two pins is a resistance which passes current in each direction as usual, depending on the voltage between the two gate terminals.

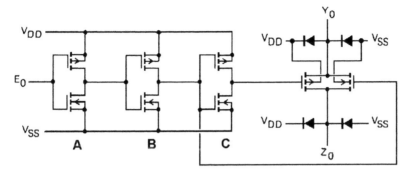

Figure 12.1 The internal circuitry of a 4000-series analogue gate.

Analogue gates have been around for a long time, and are in some ways the obvious method of electronic switching. They do however have significant drawbacks.

Analogue gates such as the 4016 are made up of two MOS FETs of opposite polarity connected back to back. The internal structure of a 4016 analogue gate is shown in Figure 12.1. The two transmission FETs with their protective diodes are shown on the right; on the left is the control circuitry. A and B are standard CMOS inverters whose only function is to sharpen up the rather soggy voltage levels that 4000-series CMOS logic sometimes provides. The output of B directly controls one FET, and inverter C develops the anti-phase control voltage for the FET of opposite polarity, which naturally requires an inverted gate voltage to turn it on or off.

MOS FETS are of the enhancement type, requiring a voltage to be applied to the gate to turn them on; (in contrast JFETs work in depletion mode and require gate voltage to turn them off) so the closer the channel gets to the gate voltage, the more the device turns off. An analogue gate with only one polarity of FET would be of doubtful use because Ron would become very high at one extreme of the voltage range. This is why complementary FETs are used; as one polarity finds its gate voltage decreasing, turning it off, the other polarity has its gate voltage increasing, turning it more on. It would be nice if this process cancelled out so the Ron was constant, but sadly it just doesn't work that way. Figure 12.2 shows how Ron varies with input voltage, and the peaky Ron curve gives a strong hint that something is turning on as something else turns off.

Figure 12.2 also shows that Ron is lower and varies less when the higher supply voltage is used; since these are enhancement FETs the on-resistance decreases as

the available control voltage increases. If you want the best linearity then always use the maximum rated supply voltage.

Since Ron is not very linear, the smaller its value the better. The 4016 Ron is specified as 115 Ω typical, 350 Ω max, over the range of input voltages and with a 15 V supply. The 4066 is a version of the 4016 with lower Ron, 60 Ω typical, 175 Ω max under the same conditions. This option can be very useful both in reducing distortion and improving offness, and in most cases there is no point in using the 4016. The performance figures given below assume the use of the 4066 except where stated.

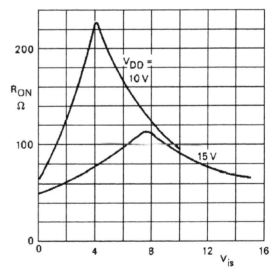

Typical R$_{ON}$ as a function of input voltage.

Figure 12.2 Typical variation of the gate series resistance Ron.

CMOS gates in voltage mode

Figure 12.3 shows the simplest and most obvious way of switching audio on and off with CMOS analog gates. This series configuration is in a sense the 'official' way of using them; the snag is that by itself it doesn't work very well.

Figure 12.4 shows the measured distortion performance of the simple series gate using the 4016 type. The distortion performance is a long way from brilliant, exceeding 0.1% just above 2 V r.m.s. These tests, like most in this section, display the results for a single sample of the semiconductor in question. Care has been taken to make these representative, but there will inevitably be some small variation in parameters like Ron. This may be greater when comparing the theoretically identical products of different manufacturers.

Replacing the 4016 gate with a 4066 gives a reliable improvement due to the lower Ron. THD at 2 V r.m.s. (10K load) has dropped to a third of its previous level. There seems to be no downside to using 4066 gates instead of the more common and better-known 4016, and they are used exclusively from this point on.

The distortion is fairly pure second harmonic, except at the highest signal levels where higher-order harmonics begin to intrude. This is shown in Figures 12.5 and 12.6 by the straight line plots beginning to bend upwards above 2 V r.m.s.

Analogue gate distortion is flat with frequency from 10 Hz up to 30 kHz at least, and so no plots of THD versus frequency are shown; they would merely be a rather uninteresting set of horizontal lines.

This circuit (Figure 12.3) gives poor offness when off, and poor distortion when on. The offness is limited by the stray capacitance in the package feeding through into the relatively high load impedance. If this is 10 K the offness is only −48 dB at 20 kHz, which would be quite inadequate for many applications. The load impedance

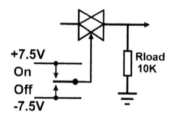

Figure 12.3 Voltage-mode series switching circuit using analogue gate.

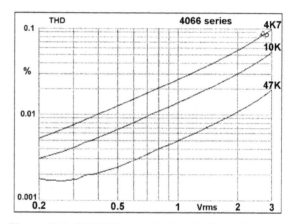

Figure 12.5 4066 THD versus level, with different load resistances.

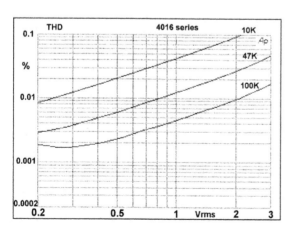

Figure 12.4 4016 series-gate THD versus level, with different load resistances.

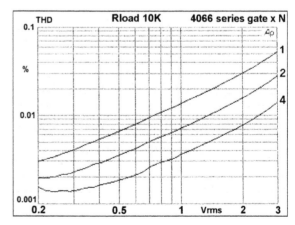

Figure 12.6 THD versus level, for different numbers of series 4066 gates.

could be reduced below 10 K to improve offness—for example, 4K7 offers about a 7 dB improvement—but this degrades the distortion, which is already poor at 0.055% for 3 V r.m.s., to 0.10%. Using 4066 gates instead of 4016 does not improve offness in this configuration. The internal capacitance that allows signals to leak past the gate seems to be the same for both types.

The maximum signal level that can be passed through (or stopped) is limited by the CMOS supply rails and conduction of the protection diodes. While it would in some cases be possible to contrive a bootstrapped supply to remove this limitation, it is probably not a good route to head down.

Figure 12.8 above shows a CMOS three-way switch. When analogue gates are used as a multi-way switch, the offness problem is much reduced, because capacitive feedthrough of the unwanted inputs is attenuated by the low Ron looking back into the (hopefully) low impedance of the active input, such as an opamp output. If this is not the case then the crosstalk from nominally

off inputs can be serious. In this circuit the basic poor linearity is unchanged, but since the crosstalk problem is much less, there is often scope for increasing the load impedance to improve linearity. This makes Ron a smaller proportion of the total resistance. The control voltages must be managed so that only one gate is on at a time, if there is a possibility of connecting two opamp outputs together.

It may appear that if you are implementing a true changeover switch, which always has one input on, the resistor to ground is redundant, and just a cause of distortion. Omitting it is however very risky, because if all CMOS gates are off together even for an instant, there is no DC path to the opamp input and it will register its displeasure by snapping its output to one of the rails. This does not sound nice.

Figure 12.9 shows the offness of a changeover system, for two types of FET-input opamps. The offness is much improved to −87 dB at 20 kHz, an improvement of 40 dB over the simple series switch; at the high-frequency end however it still degrades at the same rate of 6 dB/octave. It is well-known that the output impedance of an op-amp with negative feedback increases with frequency at this rate, as the amount of internal gain falls, and this effect is an immediate suspect. However, there is actually no detectable signal on the opamp output, (as shown by the lowest trace in Figure 12.9) and is also not very likely that two completely different opamps would have exactly the same output impedance. I was prepared for a subtle effect, but the true explanation is that the falling offness is simply due to feedthrough via the internal capacitance of the analogue gate.

It now remains to explain why the OPA2134 apparently gives better offness in the flat low-frequency region.

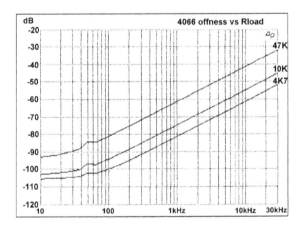

Figure 12.7 Offness versus load resistance. −48 dB at 20 kHz with a 10 K load.

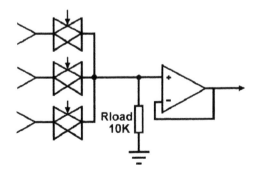

Figure 12.8 A one-pole, three way switch made from analogue gates.

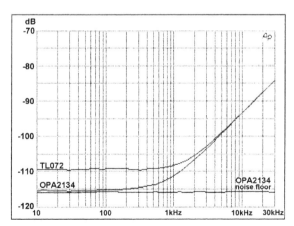

Figure 12.9 Voltage-mode changeover circuit offness for TL072 and OPA2134. Rload = 10 K.

In fact it does not; the flat parts of the trace represent the noise floor for that particular opamp. The OPA2134 is a more sophisticated and quieter device than the TL072, and this is reflected in the lower noise floor.

There are two linearity problems. Firstly, the on-resistance itself is not totally linear. Second, and more serious, the on-resistance is modulated when the gates move up and down with respect to their fixed gate voltages.

It will by now probably have occurred to most readers that an on/off switch with good offness can be made by making a changeover switch with one input grounded. This is quite true, but since much better distortion performance can be obtained by using the same approach in current mode, as explained below, I am not considering it further here.

Figure 12.10 above shows a shunt muting circuit. This gives no distortion in the 'ON' state because the signal is no longer going through the Ron of a gate. However the offness is limited by the Ron, forming a potential divider with the series resistor R; the latter cannot be very high in value or the circuit noise will be degraded. There is however the advantage that the offness plot is completely flat with frequency. Note that the ON and OFF states of the control voltage are now inverted.

Table 12.1 below gives the measured results for the circuit, using the 4066. The offness can be improved by putting two or more of these gates in parallel, but since doubling the number N only gives 6 dB improvement, it is rarely useful to press this approach beyond four gates.

CMOS gates in current mode

Using these gates in current mode—usually by defining the current through the gate with an input resistor and dropping it into the virtual-earth input of a shunt-feedback amplifier—gives much superior linearity. It removes the modulation of channel resistance as the gate goes up and down with respect to its supply rails, and, in its more sophisticated forms, can also remove the signal voltage limit and improve offness.

Figure 12.11 shows the simplest version of a current-mode on/off switch, and it had better be said at once that it is a bit too simple to be very useful as it stands. An important design decision is the value of Rin and Rnfb, which are often equal to give unity gain. Too low a value increases the effect of the non-linear Ron, while too high a value degrades offness, as it makes the gate stray capacitance more significant, and also increases Johnson noise from the resistors. In most cases 22 K is a good compromise.

Table 12.2 gives the distortion for dBu (7.75 V r.m.s.) in/out, and shows that it is now very low compared with voltage-mode switchers working at much lower signal levels; compare with Figures 12.5 and 12.6. The increase in THD at high frequencies is due to a contribution from the opamp. However, the offness is pretty poor, and would not be acceptable for most applications. The problem is that with the gate off, the full signal voltage appears at the gate input and crosstalks to the summing node through the package's internal capacitance. In practical double-sided PCB layouts the inter-track capacitance

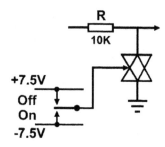

Figure 12.10 Voltage-mode shunt CMOS circuit.

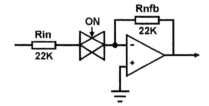

Figure 12.11 The simplest version of a current-mode on/off switch.

Table 12.1 Offness versus number of shunt gates

N gates	Offness (dB)
1	−37
2	−43
4	−49

Table 12.2 Distortion and offness for *4016* series current-mode switching

	I kHz	10 kHz	20 kHz
THD via 4016, dBu (%)	0.0025	0.0039	0.0048
THD: 4016 shorted, dBu (%)	0.0020	0.0036	0.0047
Offness (dB)	−68	−48	−42

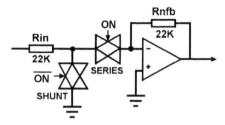

Figure 12.12 Current-mode switch circuit with breakthrough prevention resistor Rin2.

can usually be kept very low by suitable layout, but the internal capacitance is inescapable.

In Figures 12.11 and 12.12, the CMOS gate is powered from a maximum of +/–7.5 V. This means that in Figure 12.11, signal breakthrough begins at an input of 5.1 V r.m.s. This is much too low for opamps running off their normal rail voltages, and several dB of headroom is lost. Figure 12.12 shows a partial cure for this. Resistor Rin2 is added to attenuate the input signal when the CMOS gate is off, preventing breakthrough. There is no effect on gain when the gate is on, but the presence of Rin2 does increase the noise gain of the stage.

Series-shunt current mode

We now extravagantly use two 4016 CMOS gates, as shown in Figure 12.13.

When the switch is on, the series gate passes the signal through as before; the shunt gate is off and has no effect. When the switch is off the series gate is off and the shunt gate is on, sending almost all the signal at A to ground so that the remaining voltage is very small. The exact value depends on the 4016 specimen and its Ron value, but is about 42 dB below the input voltage. This deals with the offness (by greatly reducing the signal that can crosstalk through the internal capacitance) and also increases the headroom by several dB, as there is now effectively no voltage signal to breakthrough when it exceeds the rails of the series gate.

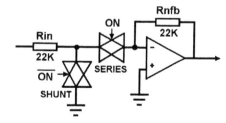

Figure 12.13 A series-shunt current-mode switch.

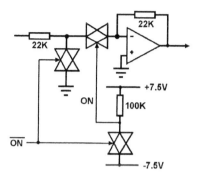

Figure 12.14 Generating the control signals with a spare analogue gate.

Two antiphase control signals are now required. An excellent way to generate the inverted control signal is to use a spare analogue gate as an inverter, as shown in Figure 12.14.

The distortion generated by this circuit can be usefully reduced by using two gates in parallel for the series switching, as in Table 12.3 below; this gate-doubling reduces the ratio of the variable Ron to the fixed series resistor and so improves the linearity. Using two in parallel is sufficient to render the distortion negligible. (The higher distortion figures at 10 and 20 kHz are due to distortion generated by the TL072 opamp used in the measurements.)

As before the input and output levels are +20 dBu, well above the nominal signal levels expected in opamp circuitry; measurements taken at more realistic levels would show only noise.

Discrete FETs have lower Ron than analogue gates. If a J111 JFET is used as the shunt switching element the residual signal at A is further reduced, to about 60 dB below the input level, with a consequent improvement in offness, demonstrated by the final entry in Table 12.3. This could also be accomplished by using

Table 12.3 Distortion and offness for 4016 series-shunt current-mode switching

	1 kHz	10 kHz	20 kHz
THD via 4016 × 1, +20 dBu	0.0016%	0.0026%	0.0035%
THD via 4016 × 2, +20 dBu	0.0013%	0.0021%	0.0034%
THD 4016 shorted, dBu	0.0013%	0.0021%	0.0034%
Offness 4016 × 1 (dB)	−109	−91	−86
Offness 4016 × 1, J111 (dB)	Less than −116	−108	−102

two or more CMOS gates in parallel for the shunt switching.

There is more on discrete FETs in Part Two of this article (Chapter 13).

Control voltage feedthrough in CMOS gates

When an analogue gate changes state, some energy from the control voltage passes into the audio path via the gate-channel capacitance of the switching FETs, through internal package capacitances, and through any stray capacitance designed into the PCB. Since the control voltages of analogue gates move snappily, due the internal inverters, this typically puts a click rather than a thump into the audio. Attempts to slow down the control voltage going into the chip with RC networks are not likely to be successful for this reason. In any case, slowing down the control voltage change simply converts a click to a thump; the FET gates are moving through the same voltage range, and the feedthrough capacitance has not altered, so the same amount of electric charge has been transferred to the audio path—it just gets there more slowly.

The only certain way to reduce the effect of transient feedthrough is to soak it up in a lower value of load resistor. The same electric charge is applied to a lower resistor value (the feedthrough capacitance is tiny, and controls the circuit impedance) so a lower voltage appears. Unfortunately reducing the load tends to increase the distortion, as we have already seen; the question is if this is acceptable in the intended application.

Chapter 13

Electronic analogue switching, Part II: discrete FETs

(Electronics World, February 2004)

The signal handling capabilities of 4000-series analogue gates is limited to about half of what is needed in an opamp-based signal path. You will also have seen from the previous article that good linearity and the ability to handle a full op-amp output swing with these ICs requires the use of current-mode, with a virtual-earth (shunt-feedback) amplifier. This lands with you with an extra op-amp for each switch, which needs power and board space, and gives added noise and distortion in return. It also, crucially, introduces a phase inversion which usually needs to be immediately undone with another op-amp, because the preservation of correct phase in a mixing console is an absolute necessity.

There was a powerful incentive in the early 1980s to develop something less clumsy. This was because mixing consoles grew more complex and sophisticated in the 1970s and 1980s, and came to incorporate much more electronic switching. The 'in-line' mixer format, in which input channel and output group lived in the same module and shared some facilities, became popular as analogue tape machines grew from 8- to 16-, and to 24-track, because it gave a much more compact console. The downside of this was that there was only one equaliser (tone-control) section, which needed to be used by the input channel while recording and the output group at mixdown time. It therefore had to be switchable from one signal path to the other, preferably by a single global control rather than pressing forty-eight or more individual switches. This demanded electronic switching—it takes six switches to move an equalisation section from one path to another and replace it with a direct connection in the path where it no longer lives—as the cost, weight, and power consumption of the number of relays required made an electromechanical solution very unpalatable. There was also a clear marketing benefit to had from claiming 'electronic switching!' Before surface-mount components became commonplace, doing it with 4000-series gates in DIL packages took up a lot of PCB area, and to handle the full opamp level, you needed either extra opamps and unwelcome phase inversions, or more sophisticated and much more expensive analogue gate types like the DG308.

The aim was to use discrete JFET switching, which promised to handle the full signal range, with low distortion, while using a TO-92 packaged device that took up minimal board area. The low on-resistance meant you could implement a phase-preserving, very linear switch with one JFET, a diode and a resistor, and that economy is why this technology was developed, by me amongst others, in the early 1980s. The only downside is that an extra −23 or −24V supply rail was required to make sure that the JFETS stayed firmly off on negative signal peaks when required to do so, because the gate voltage had to be taken at least 6V negative of the source to accomplish this. The current drawn from this rail was very low, and the economic penalty was very small.

I had the technology described here fully in place by the end of 1983, and it was first used in the Soundcraft TS24 recording console, the first in-line console they produced. It made extensive use of J111 JFETs for signal switching and J112s for soft muting.

ELECTRONIC ANALOGUE SWITCHING, PART II: DISCRETE FETs

February 2002

Discrete FET switching

Having looked in detail at analogue switching using CMOS gates, and having seen how well they can be made to work, you might be puzzled as to why anyone should wish to perform the same function with discrete FETs. There are at least two advantages in particular applications. Firstly, JFETs can handle the full output range of opamps working from maximum supply rails, so higher signal levels can often be switched directly without requiring opamps to convert between current and voltage mode. Secondly, the direct access to the device gate allows relatively slow changes in attenuation (though still measured in milliseconds, for reasons that will emerge) rather than the rapid on–off action which CMOS gates give as a result of their internal control-voltage circuitry. This is vital in creating mute circuits that essentially implement a fast fade rather than a sharp cut, and so do not generate clicks and thumps by abruptly interrupting the signal.

The downside is that they require carefully-tailored voltages to drive the gates, and these cannot always be conveniently derived from the usual opamp supply rails.

Discrete FETs in voltage mode: the series JFET switch

The basic JFET series switching circuit is shown in Figure 13.1. With the switch open there is no other connection to the gate other than the bootstrap resistor, Vgs is zero, and so the FET is on. When the switch is closed, the gate is pulled down to a sufficiently negative voltage

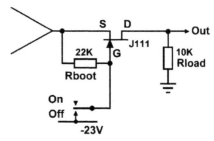

Figure 13.1 The basic JFET switching circuit, with gate bootstrap resistor.

to ensure that the FET is biased off even when the input signal is at its negative limit.

The JFET types J111 and J112 are specially designed for analogue switching and are pre-eminent for this application. The channel on-resistances are low and relatively linear. This is a depletion-mode FET, which requires a negative gate voltage to actively turn it off. The J111 requires a more negative Vgs to ensure it is off, but in return gives a lower Rds(on) which means lower distortion.

The J111, J112 (and J113) are members of the same family—in fact they are same the device, selected for gate/channel characteristics, unless I am much mistaken. Table 13.1 shows how the J111 may need 10 V to turn it off, but gives a 30 Ω on-resistance or Rds(on) with zero gate voltage. In contrast the J112 needs only 5.0 V at most to turn it off, but has a higher Rds(on) of 50 Ohms. The trade-off is between ease of generating the gate control voltages, and linearity. The higher the Rds(on), the higher the distortion, as this is a non-linear resistance.

FET tolerances are notoriously wide, and nothing varies more than the Vgs characteristic. It is essential to take the full range into account when designing the control circuitry.

Table 13.1 Characteristic of the J111 FET series

	J111	J112	J113
Vgs(off) min	−3.0	−1.0	−0.5 V
Vgs(off) max	−10	−5.0	−3.0 V
Rds(on)	30	50	100

Both the J111 and J112 are widely used for audio switching. The J111 has the advantage of the lowest distortion, but the J112 can be driven directly from 4000 series logic running from 7.5 V rails, which is often convenient. The J113 appears to have no advantage to set against its high Rds(on) and is rarely used—I have never even seen one.

The circuits below use either J111 or J112, as appropriate. The typical version used is shown, along with typical values for associated components.

Figure 13.1 has Source and Drain marked on the JFET. In fact these devices appear to be perfectly symmetrical, and it seems to make no difference which way round they are connected, so further diagrams omit this. As JFETs, in practical use they are not particularly static-sensitive.

The off voltage must be sufficiently negative to ensure that Vgs never becomes low enough to turn the JFET on. Since a J111 may require a Vgs of −10 V to turn it off, the off voltage must be 10 V below the negative saturation point of the driving opamp—hence the −23 V rail. This is not exactly a convenient voltage, but the rail does not need to supply much current and the extra cost in something like a mixing console is relatively small.

To turn a JFET on, the Vgs must be held at 0 V. That sounds simple enough, but it is actually the more difficult of the two states. Since the source is moving up and down with the signal, the gate must move up and down in exactly the same way to keep Vgs at zero. This is done by bootstrap resistor Rboot in Figure 13.1. When the JFET is off, d.c. flows through this resistor from the source; it is therefore essential that this path be d.c.-coupled and fed from a low impedance such as an opamp output, as shown in these diagrams. The relatively small d.c. current drawn from the opamp causes no problems.

Figure 13.2 is a more practical circuit using a driver transistor to control the JFET. (If you had a switch contact available, you would presumably use it to control the audio directly.) The pull-up resistor Rc keeps diode D reverse-biased when the JFET is on; this is its sole function, so the value is not critical. It is usually high to reduce power consumption. I have used anything between 47 K and 680 K with success.

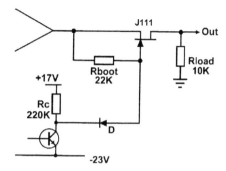

Figure 13.2 Using a transistor and diode for gate control.

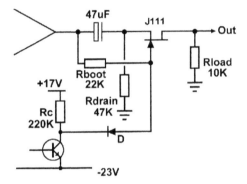

Figure 13.3 The JFET switching circuit with a d.c. blocking capacitor.

Sometimes d.c.-blocking is necessary if the opamp output is not at a d.c. level of 0 V. In this case the circuit of Figure 13.3 is very useful; the audio path is d.c.-blocked but not the bootstrap resistor, which must always have a d.c. path to the opamp output. Rdrain keeps the capacitor voltage at zero when the JFET is held off.

Figure 13.4 shows the distortion performance with a load of 10 K. The lower curve is the distortion from the opamp alone; the low THD level should tell you immediately it was a 5532. The signal level was 7.75 V r.m.s. (+20 dBu).

Figure 13.5 shows the distortion performance with heavier loading, from 10 K down to 1 K. As is usual in the world of electronics, heavier loading makes things worse. In this case, it is because the non-linear Ron becomes a more significant part of the total circuit resistance. The signal level was 7.75 V r.m.s. (+20 dBu).

Figure 13.6 shows the distortion performance with different values of bootstrap resistor. The lower the value, the more accurately the drain follows the source

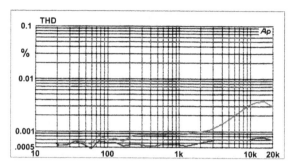

Figure 13.4 The JFET distortion performance with a load of 10 K.

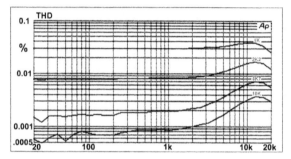

Figure 13.5 The JFET distortion performance versus loading.

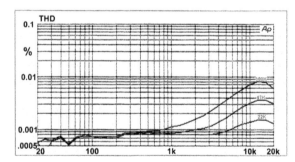

Figure 13.6 The distortion performance with different values of bootstrap resistor.

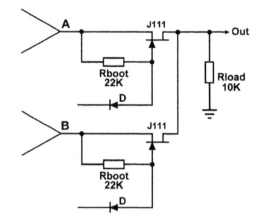

Figure 13.7 A JFET changeover switch.

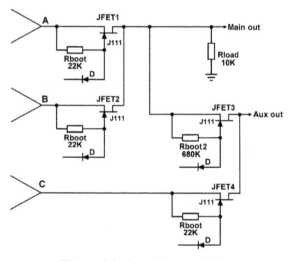

Figure 13.8 Cascaded FET switches.

at high audio frequencies, and so the lower the distortion. The signal level was 7.75 V r.m.s. (+20 dBu) once again. There appears to be no disadvantage to using a bootstrap resistor of 22 K or so, except in in special circumstances, as explained below.

Two series JFET switches can be simply combined to make a changeover switch, as shown in Figure 13.7. The valid states are A on, B on, or both off. Both on is not a good option because the two opamps will then be driving each other's outputs through the JFETs.

It is possible to cascade FET switches, as in Figure 13.8, which is taken from a real application. Here the main output is switched between A and B as before, but a second auxiliary output is switched between this selection and another input C by JFET3 and JFET 4. Cascading FET switches in this way removes the need for a buffer opamp between JFET1 and JFET3. The current drawn by the second bootstrap resistor Rboot2 must flow through the Rds(on) of the first FET, and will thus generate a small click. Rboot2 is therefore made as high as possible to minimise this effect, accepting that the distortion performance of the JFET3 switch will be compromised at HF; this was acceptable in the application as the second output was not a major signal path. The bootstrap resistor of JFET4 can be the desirable lower value of 22 K as this path is driven direct from an opamp.

The offness of this type of series FET switch is not usually an issue because it is almost always used in the changeover format, where capacitative crosstalk from the off-JFET is made negligible by the low resistance of the on-JFET. If you simply want to turn a signal off, there are better ways to do it; see below.

The shunt JFET switch

The basic JFET shunt switching circuit is shown in Figure 13.9. Like the shunt analogue gate mute, it gives poor offness but good linearity in the ON state, so long as its gate voltage is controlled so it never allows the JFET to begin conducting. Its great advantage is that the depletion JFET will be in its low-resistance before and during circuit power-up, and can be used to mute switch-on transients. Switch-off transients can also be effectively muted if the drive circuitry is configured to turn on the shunt FETs as soon as the mains disappears, and keep them on until the various supply rails have completely collapsed.

The circuit of Figure 13.9 was used to mute the turn-on and turn-off transients of a hifi preamplifier. Since this is an output that is likely to drive a reasonable length of cable, with its attendant capacitance, it is important to keep R1 as low as possible, to minimise the possibility of a drooping treble response. This means that the Rds(on) of the JFET puts a limit on the offness possible. The output series resistor R1 is normally in the range 47–100 Ω, when it has as its only job the isolation of the output opamp from cable capacitance. Here it has

a value of 1 K, which is a distinct compromise. The muting obtained with 1 K was not quite enough so two J111s were used in parallel, giving a further −6 dB of attenuation, and yielding in total −33 dB across the audio band, which was sufficient to render the transients inaudible. The offness is not frequency dependent as the impedances are all low and so stray capacitance is irrelevant.

Discrete FETs in current mode

JFETs can be used in the current mode, just as for analogue gates. Figure 13.10 shows the basic muting circuit, with series FET switching only. Ring prevents breakthrough. The stage as shown has less than unity gain; this can be corrected by increasing Rnfb.

Soft JFET muting: crosstalk/linearity trade-off

When switching audio signals, a instantaneous cut of the signal is sometimes not what is required. When a non-zero audio signal is abruptly interrupted there is bound to be a click. Perhaps surprisingly, clever schemes for making the instant of switching coincide with a zero-crossing give little improvement. There may no

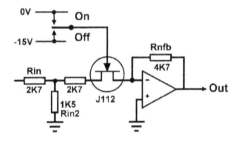

Figure 13.10 Circuit of series-only JFET mute bloc. Note phase-inversion. For performance see Figures 13.11 and 13.12.

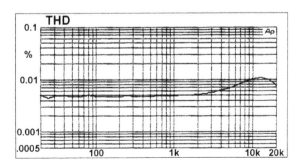

Figure 13.11 THD of the single-FET circuit in Figure 13.10 +20 dBu.

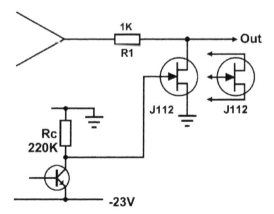

Figure 13.9 The basic JFET shunt switching circuit. Adding more JFETs in parallel increases the offness, but each −6 dB requires doubling the number.

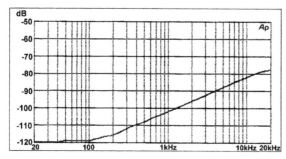

Figure 13.12 Offness of the single-FET circuit in Figure 13.10.

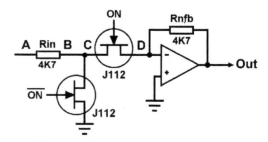

Figure 13.13 Series-shunt mode mute bloc circuit.

longer be a step-change in level, but there is still a step-change in slope and the ear once more interprets this discontinuity as a click.

What is really required is a fast-fade over about 10 ms. This is long enough to prevent clicks, without being so slow that the timing of the event becomes sloppy. This is normally only an issue in mixing consoles, where it is necessary for things to happen in real time. Such fast-fade circuits are often called 'mute blocks' to emphasise that they are more than just simple on–off switches. Analogue gates cannot be slowly turned on and off due to their internal circuitry for control-voltage generation. Therefore discrete JFETs must be used. Custom chips to perform this function have been produced, but the ones I have evaluated have been expensive, single-source, and give less than startling results for linearity and offness. This situation is of course subject to change.

In designing a mute bloc, we want low distortion and good offness at the same time, so the series-shunt configuration, which proved highly effective with CMOS analogue gates, is the obvious choice. The basic circuit is shown in Figure 13.13. A small capacitor C is usually required to ensure HF stability, (Figure 13.14) due to the FET capacitances hanging on the virtual-earth node at D.

The control voltages to the series and shunt JFETs are complementary as before, but now they can be slowed down by RC networks to make the operation gradual, as shown in Figure 13.14. The exact way in which the control voltages overlap is easy to control, but the Vgs/resistance law of the FET is not (and is about the most variable FET parameter there is) and so the overlap of FET conduction is rather variable. However, I should say at once that this system does work, and works well enough to go in top-notch mixing consoles. The distortion performance is shown in Figure 13.15. As you go into the muted condition the series JFET turns off and

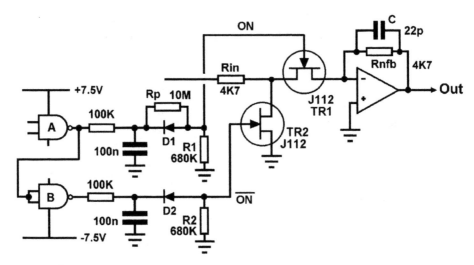

Figure 13.14 Circuitry to generate drive voltages for series-shunt JFET mute bloc.

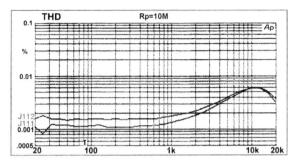

Figure 13.15 The THD of the mute bloc in Figure 13.13. The increase in FET distortion caused by using the J112 rather than J111 is shown +20 dBu.

Table 13.2 Performance of the series-shunt JFET mute circuit in Figure 13.14

	1 kHz	10 kHz	20 kHz
THD +20 dBu (%)	0.0023	0.0027	0.0039
Offness (dB)	−114	−109	−105

the shunt JFET turns on, and if the overlap gets to be too much in error, the following bad things can happen:

1. If the shunt FET turns on too early, while the series JFET is still mostly on, a low-resistance path is established from the opamp virtual-earth point to ground, causing a large but brief rise in stage noise gain. This produces a 'chuff' of noise at the output as muting occurs.

2. If the shunt FET turns on too late, so the series JFET is mostly off, the large signal voltage presented to the series FET causes visibly serious distortion. I say 'visibly' because it is well-known that even quite severe distortion is not obtrusive if it occurs only briefly. The transition here is usually fast enough for this to be the case; it would however not be a practical way to generate a slow fade. The circuit of Figure 13.14 generates no audible distortion and only a very small chuff.

The drive circuitry

The mute bloc requires two complementary drive voltages, and these are easily generated from 4000-series CMOS running from ±7.5 V rails. NAND gates are shown here as they are convenient for interfacing with other bits of control logic, but any standard CMOS output can be used. It is vital that the JFET gates get as close to 0 V as possible, ensuring that the series gate is fully on and gives minimum distortion, so the best technique is to and use diodes to clamp the gates to 0 V. Thus, in Figure 13.14, when the mute bloc is passing signal, the signal from NAND gate A is high, so D1 is reverse-biased and the series JFET TR1 gate is held at 0 V by R1, keeping it on. (The role of Rp will be explained in a moment) Meanwhile, D2 is conducting as the NAND-gate output

driving it is low, so the shunt JFET TR2 gate is at about −7 V and it is firmly switched off. This voltage is more than enough to turn off a J112, but cannot be guaranteed to turn off a J111, which may require −10 V (See Table 13.2). This is one reason why the J112 is more often used in this application—it is simpler to generate the control voltages. When the mute bloc is off, the conditions are reversed, with the output of A low, turning off TR1, and the output of B high, turning on TR2.

Reducing THD by on-biasing

The distortion generated by this circuit bloc is of considerable importance, because if the rest of the audio path is made up of 5532 opamps—which is likely in professional equipment—then this stage can generate more distortion than the rest of the signal path combined, and dominate this aspect of the performance. It is therefore worth examining any way of increasing the linearity.

We have already noted that to minimise distortion, the series JFET should be turned on as fully as possible to minimise the value of the non-linear Rds(on). When a JFET has a zero gate-source voltage, it is normally considered fully on. It is, however, possible to turn it even more on than this.

The technique is to put a small positive voltage on the gate, say about 200–300 mV. This further reduces the Rds(on) in a smoothly continuous manner, without forward biasing the JFET gate junction and injecting d.c. into the signal path. This is accomplished in Figure 13.14 by the simple addition of Rp, which allows a small positive voltage to be set up across the 680 K resistor R1. The value of Rp is usually in the 10–22 MΩ range, for the circuit values shown here.

Care is needed with this technique, because if temperatures rise the JFET gate diode may begin to conduct after all, and d.c. will leak into the signal path, causing thumps and bangs. In my experience 300 mV is about the upper safe limit for equipment that gets reasonably warm internally, i.e. about 50 °C. Caution is the watchword here, for unwanted transients are much less tolerable than slightly increased distortion.

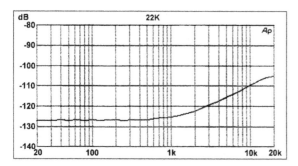

Figure 13.16 Offness of mute bloc in Figure 13.13 with Rin = Rnfb = 22 K.

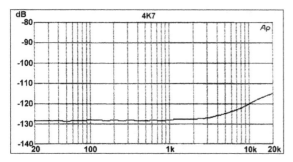

Figure 13.17 Offness of mute bloc in Figure 13.13 with Rin = Rnfb = 4K7. Offness is 10 dB better at 10 kHz and the noise floor (the flat section below 2 kHz) has been lowered by about 2 dB.

As with analogue CMOS gates, the choice of the resistors Rin and Rnfb that define the magnitude of the signal currents is an important matter. Figures 13.16 and 13.17 show the performance of the circuit is affected by using values of 4K7 and 22 K. Usually 47 would be the preferred value; choosing 22 K as the value makes the noise floor higher, as well as the signal leakage. Values below 4K7 are not usual as distortion will increase, as the JFET Rds(on) becomes a larger part of the total resistance in the circuit. The loading effect of Rin on the previous stage must also be considered.

Layout and offness

The offness of this circuit is extremely good, providing certain precautions are taken in the physical layout. In Figure 13.18, there are two possible crosstalk paths that can damage the offness. The path C–D, through the internal capacitances of the series JFET,

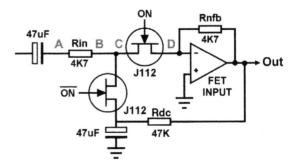

Figure 13.18 Circuit of JFET mute showing stray capacitances and d.c. handling.

is rendered innocuous as C is connected firmly to ground by the shunt JFET. However, point A is still alive with full amplitude signal, and it is the stray capacitance from A to D that defines the offness at high frequencies.

Given the finite size of Rin, it is often necessary to extend the PCB track B–C to get A far enough from D. This is no problem if done with caution. Remember that the track B–C is at virtual earth when the mute bloc is on, and so vulnerable to capacitative crosstalk from other signals straying into the area.

Dealing with the d.c.

The circuits shown so far have been stripped down to their bare essentials to get the basic principles across. In reality, things are (surprise) a little more complicated. Opamps have non-zero offset and bias voltages and currents, and if not handled properly these will lead to thumps and bangs. There are several issues:

1. If there is any d.c. voltage at all passed on from the previous stage, this will be interrupted along with the signal, causing a click or thump. The foolproof answer is of course a d.c. blocking capacitor, but if you are aiming to remove all capacitors from the signal path, you may have a problem. d.c. servos can partly make up the lack, but since they are based on opamp integrators they are no more accurate than the opamp, while d.c. blocking is totally effective.

2. The offset voltage of the opamp. If the noise gain is changed when the mute operates (which it is) the changing amplification of this offset will change the d.c. level at the output. The answer is shown

in Figure 13.18. The shunt FET is connected to ground via a blocking capacitor to prevent DC gain changes. This capacitor does not count as 'being in the signal path' as audio only goes through it when the circuit is muted. Feedback of the opamp offset voltage through Rd_c to this capacitor renders it innocuous.

3. The input bias and offset currents of the opamp. These are potentially much more of a problem and are best dealt with by using JFET opamps such as the OPA2134, where the bias and offset currents are negligible at normal equipment temperatures.

Soft changeover circuit

This circuit (Figure 13.19) is designed to give a soft changeover between two inputs—in effect a fast cross-fade. It is the same mute block but with two separate inputs, either or both of which can be switched on. The performance at +20 dBu in/out is summarised in Table 13.2.

The THD increase at 20 kHz is due to the use of a TL072 as the opamp. J112 JFETs are used in all positions.

This circuit is intended for soft-switching applications where the transition between states is fast enough for a burst of increased distortion to go unnoticed. It is not suitable for generating slow crossfades in applications like disco mixers, as the exact crossfade law is not very predictable.

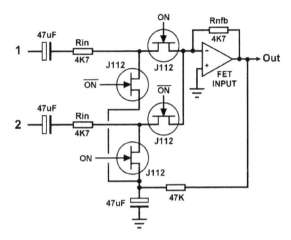

Figure 13.19 Circuit of JFET soft changeover switching.

Control voltage feedthrough in JFETs

All discrete FETs have a small capacitance between the gate and the device channel, so changes in the gate voltage will therefore cause a charge to be transferred to the audio path, just as for CMOS analogue gates. As before, slowing down the control voltage change tends to give a thump rather than a click to a thump; the same amount of electric charge has been transferred to the audio path, but more slowly. Lowering the circuit impedance reduces the effects of feedthrough, but can only be taken so far before distortion begins to increase as the non-linear Rds(on) of the JFET becomes a greater part of the total circuit resistance.

Chapter **14**

Preamplifier 2012

Part 1: introduction and line-in/tone/volume board
(Elektor, April, 2012)

This was the first preamplifier I had designed for some time, the previous one being the Precision Preamplifier '96, published in the July/August 1996 and September 1996 editions of *Wireless World*. That's a sixteen-year gap. Preamp 2012 was actually designed in 2011, but not published until 2012.

The project was, amongst other things, something of a study in low-impedance design [1]. The idea was to find out how far the concept could be pushed before things got awkward. The need to source dual-gang pots, preferably with a centre-detent for the balance and tone controls, put a lower limit on their track value of 1 kΩ. Anything lower implies that rotary switches and resistor ladders will need to be used, as indeed they have been in some later designs. Driving low impedances is often most effectively done by using multiple opamps to share the current required, and this gives an extra benefit because multiple amplifiers with their outputs averaged, which can be done exceedingly simply by connecting them together with 10 Ω resistors, have lower noise due to its partial cancellation.

Taking 1 kΩ as the pot value defines the impedances of most of the circuitry, and as a result various techniques are required to drive these low impedances. For example, the low-noise balanced input also implements the active balance control. The low-value resistors around the actual differential amplifier IC2A are driven by buffers IC1A, B so they do not load the source equipment. The negative feedback resistors around IC2A are split into R8 and R9, and each is driven by its own unity-gain buffer IC3A, B. The Baxandall tone-control also has its associated impedances defined at low values by the 1 kΩ pots, and so both the input and feedback connection of the Baxandall network need special measures to drive them. The bass and treble sections are driven separately, in what I call a split-drive tone-control. The treble section input is driven directly from the input/balance stage, while the bass section input is driven by unity-gain buffer opamp IC2B, which does nothing except provide greater drive capability. The treble section feedback is driven directly by the tone-control opamp IC4A, but the bass section feedback is driven by another unity-gain buffer IC4B. The treble section rather than the bass is driven directly because the extra phase-shift in IC4B would be likely to cause instability if the high-frequency feedback went through it.

Similar techniques are used in the active volume-control stage. Four inverting amplifier are used to reduce noise, and as a side-benefit they provide more than enough power to drive the feedback end of the volume pot P4A. There are also four buffers, one driving each inverter, but they are there to reduce noise rather than increase drive-power. Driving the input end of the volume pot requires an extra buffer IC9B, which takes over half of the load which would otherwise fall entirely on IC4A. I call this configuration (which strangely seems almost unknown) Mother's Little Helper [2].

Several changes were made to the preamplifier design by the *Elektor* staff (without consulting me, but that's how they work). First, it was laid out as several separate PCBs, and partly as a consequence, all the switching was converted to relays. A tone-cancel facility was added, which simply switched out the tone stage; this was a little unfortunate as it means a phase reversal when tone is cancelled. All my other recent preamplifiers have preserved absolute phase.

This preamplifier is MRP12 in my numbering system.

References

1. Self, D., *Small-Signal Audio Design,* 2nd edition, Preface, p. xxii. Focal Press (Taylor & Francis) 2014. ISBN: 978-0-415-70974-3 Hardback. 978-0-415-70973-6 Paperback. 978-0-315-88537-7 ebook.

2. Ibid., Chapter 1, p. 33.

PREAMPLIFIER 2012

Audio lovers, sit up. Besides presenting a truly high end audio control preamplifier for home construction, this article series also aims to show how low-impedance design and multiple-amplifier techniques can be used to significantly reduce the noise levels in analog circuitry. The result of the design effort is a top-notch preamp that's brilliant not just sonically but also in terms of cost/performance ratio.

It is now some time since I published a preamplifier design—the Precision Preamp in 1996 [1]. Inevitably technology has moved on. In that design the recording output level was as low as 150 mV$_{rms}$ to get a good disc input overload margin, the amplitude being well within the reach of the average tape recorder input level control. Nowadays most audio line inputs will be from digital sources, typically at 1 V$_{rms}$ unbalanced or 2 V$_{rms}$ balanced, and recognition of that requires a completely new design, especially in the MM/MC phono section.

The preamplifier described here demonstrates how very low noise levels can be achieved in analog circuitry without using exotic parts. It was originally conceived as being entirely made up of 5532 opamps, like the earlier Elektor *5532 OpAmplifier* [2] but during the design

process it became clear that adding a few LM4562s (which are now a good deal cheaper than they used to be) would avoid some awkward compromises on distortion performance, because of their superior load-driving ability.

In addition, the preamplifier has a very versatile MC/MM phono input stage with gain-switching, which I believe can optimally handle any cartridge on the market; this is guided by an innovative level indicator that provides more information than just on/off from one LED and does away with the need for bar-graph metering.

A block diagram of the proposed preamplifier is shown in Figure 14.1. In practice, the project is divided into several circuit boards, each of which will be discussed separately starting this month with the circuitry comprising the line amplifier, tone control, volume control, and output stage.

Noise of three kinds

There are three main causes of electronic noise in analog circuitry: Johnson noise from resistors, and current and voltage noise from semiconductors.

All resistances (including those forming an integral part of other components, such as the base spreading

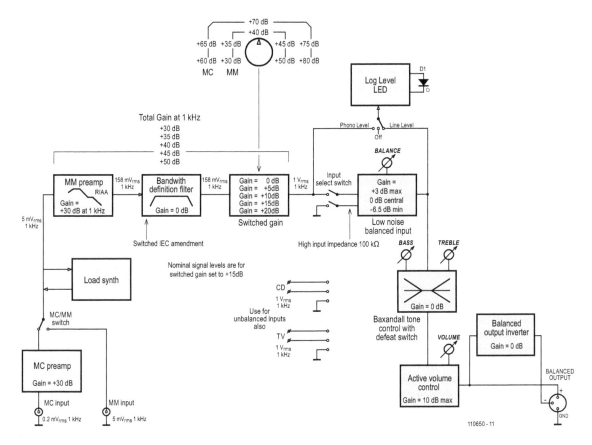

Figure 14.1 Architecture of the Preamplifier 2012. Although functions appear as individual blocks here, several of these are comprised together on circuit boards, for example, the four orange-colored blocks discussed in this article, i.e. from line input, through tone control, volume control up to balanced output.

resistance of a bipolar transistor) generate **Johnson noise** at a level that depends on the resistance and the absolute temperature. There is usually little you can do about the ambient temperature, but the resistance is under your control. Johnson noise can therefore be minimized by Low Impedance Design, in other words keeping the circuit impedances as low as possible.

In the early 1970s audio circuitry commonly used 25 kΩ or 50 kΩ potentiometers and associated components of proportionally high impedance, mainly because the discrete transistor circuitry of early opamps used had poor load-driving abilities. When the (NE)5532 appeared, and equally importantly, came down to a reasonable price, it was possible to reduce impedance levels and use 10 kΩ pots. This may not seem very ambitious when you consider that a 5532 can drive about 800 Ω while keeping its good distortion performance, but the pot value does not always give a good idea of, for example, the input

impedance of the circuit block in which it is used. It is not widely known that a standard Baxandall tone control constructed with two 10 kΩ pots has an input impedance that can easily fall to less than 1 kΩ. There is a cunning way round this problem; the Bass and Treble tone control networks can be driven separately; more on this below.

Fixed resistors are available in almost any value, but the pot values available are much more restricted, and the lowest practical value in that two-gang pots are available for stereo, is 1 kΩ.

Volume/balance/tone control board performance

Test conditions: supply voltage ±17.6 V; all measurements symmetrical; tone control defeat disabled. Test equipment: Audio Precision Two Cascade Plus 2722 Dual Domain (@Elektor Labs)

THD+N (200 mV in, 1 V out)	0.0015% (1 kHz, B = 22 Hz–22 kHz)	Max. output voltage	
	0.0028% (20 kHz, B = 22 Hz–80 kHz)	(200 mV in)	1.3 V
THD+N (2 V in, 1 V out)	0.0003% (1 kHz, B = 22 Hz–22 kHz)	Balance	+3.6 dB to −6.3 dB
	0.0009% (20 kHz, B = 22 Hz–80 kHz)		
		Tone control	±8 dB (100 Hz)
S/N (200 mV in)	96 dB (B = 22 Hz–22 kHz)		±8.5 dB (10 kHz)
	98.7 dBA		
		Crosstalk R to L	−98 dB (1 kHz)
Bandwidth	0.2 Hz–300 kHz		−74 dB (20 kHz)
		Crosstalk L to R	−102 dB (1 kHz)
			−80 dB (20 kHz)

Current noise is associated with opamp inputs. It only turns into voltage noise when it flows through an impedance, so it can be reduced by Low Impedance Design. The use of low value pots also means that opamp bias currents create less voltage drop in them and so there are less likely to be intrusive noises when the wiper is moved.

Voltage noise is the third type of noise. It is already in voltage format, being equivalent to a voltage noise generator in series with the opamp inputs, and so cannot be reduced by Low-Impedance Design. At first it seems that the only thing that can be done to minimize it to use the quietest opamps available. Opamps exist that are quieter than the 5532 or the LM4562, but they are expensive and have high current noise. A typical example is the AD797, which has the further drawback of only being available in a single package, putting the cost up more.

A better path is the powerful technique of using of multiple cheap amplifiers with their outputs summed— or to be more accurate, averaged. With two amplifiers connected in parallel, the signal gain is unchanged. Their outputs are averaged simply by connecting a low value resistance to each amplifier and taking the output from their junction. The two noise sources are uncorrelated, as they come from physically different components, so they partially cancel and the noise level drops by 3 dB (√2). The amplifier outputs are very nearly identical, so little current flows from one opamp to another and distortion is not compromised. The combining resistor values are so low (typically 10 Ω) that their Johnson noise is negligible. This strategy can be repeated by using four amplifiers, giving a signal-to-noise improvement of 6 dB, eight amplifiers give 9 dB, and so on. Obviously there are limits on how far you can take this

Peter Baxandall

I have tried hard in this project to design the highest performance preamplifier possible, and it is significant that in two out of the three stages, ideas put forward by Peter Baxandall have proved to be the best solution. He was a great man.

Editor's Note. Besides reprints of selected articles from *Wireless World / Electronics World,* Jan Didden's book "Baxandall and Self on Audio Power" presents a previously unpublished exchange of letters between Peter Baxandall and Douglas Self, both correspondents actively tracing the various causes of distortion in high end audio power amplifiers.

The book is published by Linear Audio, www. linearaudio.net.

sort of thing. Multiple opamps in parallel are also very useful for driving the low impedances of Low Impedance Design, so the two techniques work beautifully together. The Elektor *5532 OpAmplifier* [2] pretty much took this to its logical conclusion.

The multiple-amplifier approach gets unwieldy when the feedback components around each amplifier are expensive or over-numerous.

To use two amplifiers in a standard Baxandall tone control you would have to use quad-gang rather than dual pots for stereo, and also duplicate all the resistors and capacitors. While this would reduce the effect of all the noise mechanisms in the circuit by √2, giving a solid 3 dB improvement, it is not an appealing scenario.

If instead you simply halved the impedance of the Baxandall network by halving the pot and resistance values and doubling the capacitor values, the situation is not equivalent. In the second case you have halved the effect of the opamp current noise flowing in the circuit impedances, and reduced the Johnson noise by root-two, but the opamp voltage noise remains unaffected and will often dominate.

Line input and balance control stage

This is a balanced input stage with gain variable over a limited range to implement the balance control function. Maximum gain is +3.7 dB and minimum gain −6.1 dB, which is more than enough for effective balance control. Gain with balance central is +0.2 dB. Looking at the circuit diagram in Figure 14.2 and discussing the left (L) channel only, IC2A is the basic balanced amplifier; it is an LM4562 for low voltage noise and good driving ability. The resistances around it are low to reduce noise so it is fed by unity-gain buffers IC1A/B which give a high input impedance of 50 kΩ that improves the CMRR. Note the EMC filters R1-C1 and R2-C2 and the very start of the circuitry. The stage gain is set by 1 kΩ pot P1A, the negative feedback to IC2A being applied through two parallel unity-gain buffers IC3A/B so the variation in output impedance of the gain-control network will not degrade the CMRR. The dual buffers reduce noise and give also more drive capability. In this stage combining the buffer outputs is simple because the feedback resistance can be split into two halves; R8 and R9. This requires R11 and R12 to be paralleled to get exactly the right resistance value and so preserve the CMRR. The noise output of this stage is very low: −109 dBu with the balance control central; −106 dBu

at maximum gain and −116 dBu at minimum gain. (all 22 Hz–22 kHz, rms)

The tone control stage

It is not obvious but this is (mostly) a conventional Baxandall tone-control. Once more 1 kΩ pots are used, requiring large capacitors to set the turnover frequencies, C7 at 1 µF sets the bass frequency and C8, C9 at 100 nF set the treble frequency. The cut/boost is ±10 dB maximum for both. The stage has a low input impedance, especially when set to boost; to deal with this the Bass and Treble parts of the tone-control network are driven separately. The Treble network C9-P3B-C8 is driven directly by IC2A in the previous stage, while the Bass network R15-C7-P2B-R14 is driven separately by the extra unity-gain buffer IC2B, the other half of the LM4562 in the line input stage. I call this a split-drive Baxandall circuit.

The Treble network is the two-capacitor version rather than the one-capacitor types; this has the advantage that the treble pot is uncoupled from the circuit at low frequencies and reduces the loading.

The main tone-control opamp is IC4A, which drives the Treble feedback path directly, while unity-gain buffer IC4B gives separate drive to the Bass feedback path. Polypropylene capacitors are strongly recommended as they are free from distortion while polyester types show significant non-linearity. Unfortunately they are also physically larger and more expensive, but well worth it in my view.

The noise output of the tone-control stage alone is only −113 dBu with controls set to flat.

Relays RE1 and RE2 implement a tone control defeat function by enabling the active volume control stage to be driven directly from IC2A. To prevent clicks and other noises when the relays switch over, R18 and R58 have been added, effectively keeping up the bias to IC9B and IC18A. To keep crosstalk down to a minimum each channel has its own relay. As a bonus, two contacts can be connected in parallel, preventing any risk of signal degradation and at the same time increasing life span.

Active volume control stage

The volume control is of the active Baxandall type which gives low noise at low volume settings and also synthesizes a quasi-logarithmic control law from a linear pot, giving much superior channel balance. Maximum gain is +16 dB, with 0 dB obtained with the control

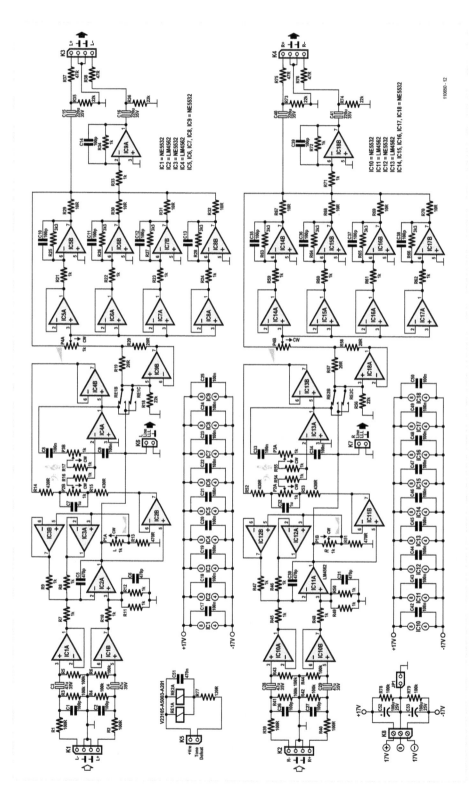

Figure 14.2 The circuit diagram of the line/tone/volume/output board is largely dominated by NE5532 and LM4562 opamps for good reasons. Note the unusually low value of the potentiometers used.

central. The input impedance of the volume control stage, implemented with another 1 kΩ pot P4A, falls to low values at high volume settings. It is therefore driven by the 'load-splitting arrangement' where buffer IC9B provides half of the drive from the tone-control stage. Resistors R19, R20 ensure that IC4A and IC9B share the load between them.

The conventional Baxandall active volume control as in [1] uses a single buffer and one inverting amplifier, such as IC5A and IC5B. Here four of these circuits are used in parallel to reduce noise by partial cancellation of the uncorrelated noise from the four amplifier paths, and to give sufficient drive capability to drive the back end of the 1 kΩ volume pot. The use of four paths reduces the noise by 6 dB. The multiple shunt amplifiers have no common-mode voltage on their inputs and so no CM distortion, and the associated buffers handle less than a third of the output voltage so stage distortion is very low. The enhanced drive capability means that it is not necessary to resort to LM4562s which still get a little expensive if you are using a lot of them. Four 10 Ω resistors R29-R32 are used to average the four outputs.

Component list

Resistors

(1% tolerance; metal film; 0.25W)
R1,R2,R39,R40 = 100Ω
R3-R6,R41-R44,R78,R79 = 100kΩ
R7-R12,R16,R17,R21-R24,R33,R34,R45-R50,R54,R55,
R59-R62,R71,R72 = 1kΩ
R13,R51 = 470Ω
R14,R15,R52,R53 = 430Ω
R18,R35,R36,R56,R73,R74 = 22kΩ
R19,R20,R57,R58 = 20Ω
R25-R28,R63-R66 = 3.3kΩ
R29-R32,R67-R70 = 10Ω
R37,R38,R75,R76 = 47Ω
R77 = 120Ω
P1,P2,P3,P4 = 1kΩ, 10%, 1W, stereo potentiometer,
linear law, e.g. Vishay Spectrol cermet type
14920F0GJSX13102KA. Alternatively, Vishay Spectrol
conductive plastic type 148DXG56S102SP (RS Components p/n 484–9146).

Capacitors

C1,C2,C10-C14,C26,C27,C35-C39 = 100pF 630V,
1%, polystyrene, axial
C3,C4,C28,C29 = 47µF 35V, 20%, bipolar, diam.
8mm, lead spacing 3.5mm, e.g. Multicomp p/n

NP35V476M8X11.5
C5,C6,C30,C31 = 470pF 630V, 1%, polystyrene, axial
C7,C32 = 1µF 250V, 5%, polypropylene, lead spacing
15mm
C8,C9,C33,C34 = 100nF 250V, 5%, polypropylene,
lead spacing 10mm
C15,C16,C40,C41 = 220µF 35V, 20%, bipolar,
diam. 13mm, lead spacing 5mm, e.g. Multicomp p/n
NP35V227M13X20
C17-C25,C42-C50 = 100nF 100V, 10%, lead spacing
7.5mm
C51 = 470nF 100V, 10%, lead spacing 7.5mm
C52,C53 = 100µF 25V, 20%, diam. 6.3mm, lead spacing
2.5mm

Semiconductors

IC1,IC3,IC5-IC10,IC12,IC14-IC18 = NE5532, e.g. ON
Semiconductor type NE5532ANG
IC2,IC4,IC11,IC13 = LM4562, e.g. National Semiconductor type LM4562NA/NOPB

Miscellaneous

K1-K4 = 4-pin straight pinheader, pitch 0.1" (2.54mm),
with mating sockets
K5,K6,K7 = 2-pin pinheader, pitch 0.1" (2.54mm), with
mating sockets
JP1 = 2-pin pinheader, pitch 0.1" (2.54mm), with
jumper
K8 = 3-way PCB screw terminal block, pitch 5mm
RE1,RE2 = relay, 12V/960Ω, 230VAC/3A, DPDT, TE
Connectivity/Axicom type V23105-A5003-A201
PCB # 110650–1

Note: all parts available from Farnell (but not exclusively), except PCB 110650–1

At sustained maximum sinewave output (about 10 V_{rms}) the volume pot gets perceptibly warm, as a consequence of the Low-Impedance Design approach. This may appear alarming but the heating is well within the specification of the hotpot. This does not occur with music signals.

The noise output of the active volume stage alone is −101 dBu at maximum gain and −109 dBu for 0 dB gain. For low gains around −20 dB, those most used in practice, noise output is about −115 dBu. Rather quiet.

Here I have quoted the noise performance for each stage separately, to demonstrate the noise-reduction techniques. In a complete preamplifier the noise levels

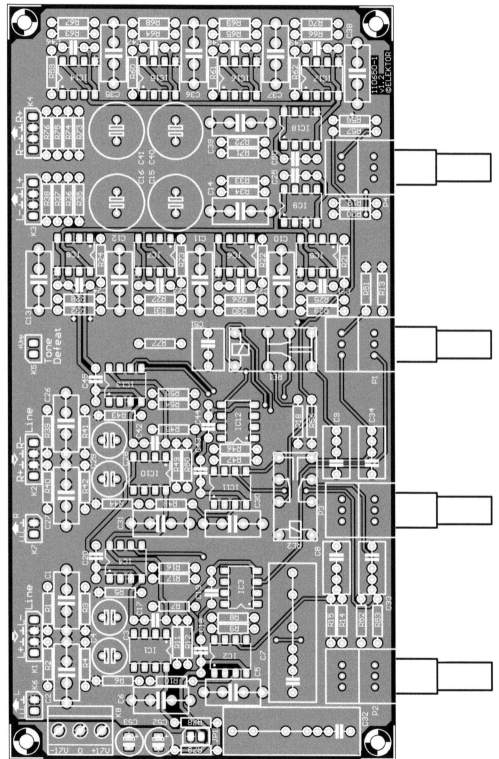

Figure 14.3 Component overlay of the circuit board designed by Elektor Labs for the line / tone / volume / output section of the Preamplifier 2012. Ready made boards may be ordered at www.elektorPCBservice.com.

add up as a signal goes through the system, though how this happens depends very much on the control settings.

Balanced output stage

The balanced output consists simply of a unity-gain inverter IC9A which generates the cold (phase-inverted) output. The balanced output is therefore at twice the level of the unbalanced output, as in normal hi-fi practice.

Construction notes

The project employs standard leaded components throughout. A high quality circuit board designed by Elektor Labs for the project is available through www.elektor-PCBservice.com. The component overlay appears in Figure 14.3.

It is recommended to use a flip over type PCB jig. Start with low-profile components and finish with the taller ones.

The finished board pictured in Figure 14.4 should be taken as an example to work from—success is guaranteed if you strive to achieve this level of perfection in construction.

Do not forget to fit JP1 for the ground through connection, a similar jumper is present on the MC/MM board. This allows you to determine empirically which ground connection works best. When the wiring is complete one jumper has to be fitted, or both.

Selected performance graphs

Line / tone / volume / output board # 110650–1 only.
Test equipment: Audio Precision Two Cascade Plus 2722 Dual Domain (@Elektor Labs)
All measurements symmetrical.

Figure 14.4 The prototype of the line / tone / volume / output board impeccably built by Elektor Labs.

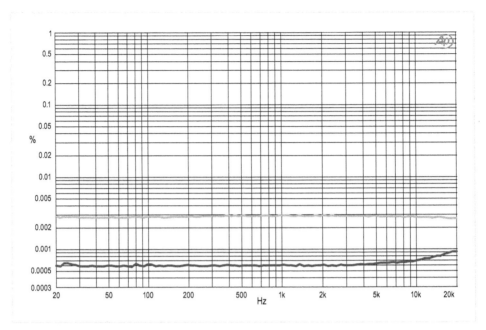

THD+N vs. Frequency. Measurements at 200 mV in and 2 V in (lower curve) and 80 kHz bandwidth. 1 V out. At 200 mV in there is noise only (approx. 96 dB measured at 22 kHz BW), and even lower levels at 2 V. Distortion not evident until above a few kHz.

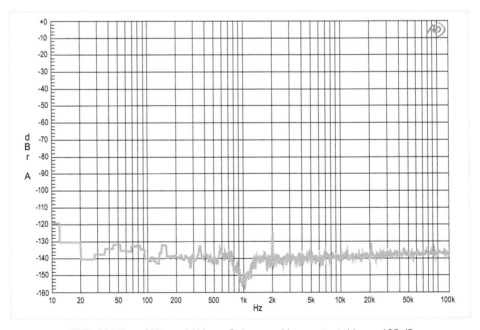

FFT of 1 kHz at 2 V in and 1 V out. Only second harmonic visible at −125 dB.

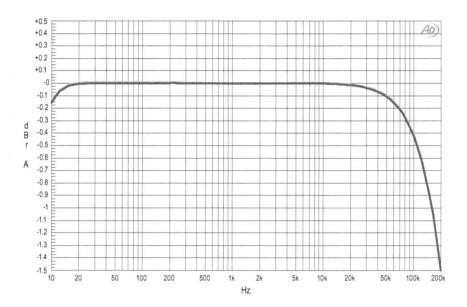

Amplitude vs. Frequency. Note the resolution on amplitude scale is just 0.1 dB. Tone defeat enabled. However, identical response obtained with tone control enabled and controls at mid position.

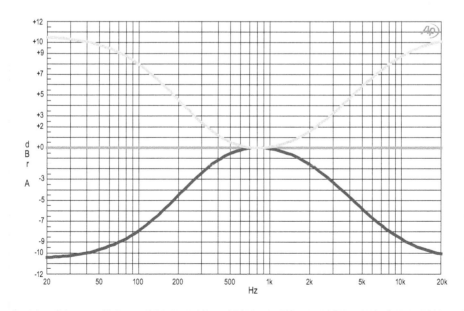

Response at max. min. and center positions of tone control (3 curves). Bass and treble set to identical extreme positions.

Axial polystyrene caps require care in determining where to bend the legs. Their size is not standardized and subject to various tolerances compared to other parts.

Finally, on the potentiometers, plastic types from Vishay Spectrol may be used instead of cermet. At the time of writing, supply of the cermet versions was subject to a lead time of up to 93 working days at Farnell's. Next month's installment will discuss the very high quality MC/MM preamplifier board.

(110650)

References

1. Precision Preamplifier 96, by Douglas Self. Wireless World July/August & September 1996.

2. The 5532 OpAmplifier, by Douglas Self. Elektor October & November 2010, www.elektor.com/100124 and www.elektor.com/100549.

Chapter 15

Preamplifier 2012

Part 2: moving-coil/moving-magnet (MC/MM) board (Elektor, May 2012)

The 2012 preamplifier has a flat gain stage after the RIAA phono section that can be switched from 0 to +20 dB in 5 dB steps. This was mainly aimed at accommodating the very wide range of MC cartridge sensitivities that exists. It is equally applicable to optimising the gain for MM cartridges, but these vary much less in output.

Analysis of gains and signal levels showed that if the MC head amplifier had a fixed gain of +30 dB, followed by an MM RIAA stage with a fixed gain of +30 dB, overload could never occur in either of those stages, no matter what combination of cartridge and technically possible recorded velocities was used. This is true even though the MC stage actually works at a gain of +45 dB, with its output tapped down the feedback path so that it appears to have a gain of +30 dB. The signals here are very small and so this apparently daft configuration really does not cause loss of headroom. It was therefore safe to put the switched-gain stage at the end of the phono signal chain, so it could optimise the gain for both MC and MM inputs. It was not necessary to build gain-switching into the MC stage, as was done in the design described in Chapter 4, and this was a very welcome simplification.

The switched-gain stage has to be set correctly to get the best results; this can be done either by calculation from the cartridge sensitivity, with guesses at recorded velocities, or simply and more scientifically by some means of level indication. Since nobody except hi-fi reviewers change their cartridges on a regular basis, elaborate and expensive means of level indication are not appropriate. A single LED usually only gives an indication that the signal has exceeded a given threshold, but it can be driven by more sophisticated circuitry rather than a single comparator, so the mark/space ratio of its illumination gives more information about the signal level than mere on/off. Such an indicator, the LLLL, was described in detail in the third article of the series.

The MC head amplifier proved very hard to improve upon. The last time I looked at this subject was in designing the Precision Preamp '96, where noise was reduced significantly by replacing the original three paralleled 2N4403's with the same number of 2SB737's. The 2SB737 is now inexplicably obsolete, so simply adding more of them for lower noise was hardly an option. After much experimentation, four paralleled 2SA1085 transistors were used, which gave a reliable noise

reduction of 1.0 dB over the previous version. However, their availability is now also in question. The vital parameter of base spreading resistance (rbb) for this device is not quoted on the Hitachi data sheet, but they work very well in this application.

The other area in which I attempted some improvement was the rate at which the MC stage settled into steady operating conditions after switch-on. I felt this could be a bit quicker, but after much labour, I had made little progress. However, given that this design has been around for more than 25 years, and no one has ever complained about it, the issue cannot be too serious.

The MM and RIAA equalisation stage is in Configuration A, which later proved to make the least efficient use of its expensive precision capacitors compared with other configurations; see Chapter 19 on RIAA network optimisation.

This preamplifier is MRP12 in my numbering system.

PREAMPLIFIER 2012

Just in case you didn't know, vinyl records are making a comeback and there are even under-25 musicians releasing new material on CD cheerfully along with vinyl, preferably of the 180-gram variety. Also, high-end turntables are available at extragalactic prices but none of this makes any sense if you do not have a preamplifier to match your MC or MD cartridge optimally and that's exactly what the present design does—rather successfully.

Referring back to the block diagram of the Preamplifier 2012 shown in part 1 of the article [1], this month we discuss the blocks identified as 'MC preamp', 'Load synth', MM preamp', Bandwidth definition filter' and 'Switched gain'. Note that the switch drawn with Switched gain' block is actually an on-PCB jumper block. All units are comprised on a single circuit board, the second of a total of seven that make up our very high end audio control amplifier. Let's see how it all works by taking a tour of the circuit diagram in Figure 15.1.

Moving-coil (MC) stage

This stage built around transistors T1-T4 and opamps IC1A and IC2A gives very low noise with the low impedances of moving-coil cartridges. It provides a fixed gain to its output of +30 dB. Gain switching to cope with the very wide range of MC cartridge sensitivities is done later in the switched-gain stage. There are no compromises on noise or headroom with this architecture, and no necessity to switch the gain of the MC stage, which simplifies things considerably.

The total gain of the stage is actually +45 dB, to allow a sensibly high value of feedback resistance defined by R8 and R9. Only part of this gain is used, tapped off via C7. The extra 15 dB of gain causes no headroom problems as the following MM stage will always clip long before the MC stage.

The DC conditions for the 2SA1085 input transistors are set by R3 and R4. The DC conditions for the opamp IC1A are set independently by the DC integrator servo IC2A, which enforces exactly 0 V at the output.

This MC stage design gives a 1 dB improvement in noise performance (for 3.3 Ω and 10 Ω source resistances) compared with earlier versions of this circuit. This results from using four paralleled 2SA1085 pnp transistors, which should be easier to obtain than the obsolete 2SB737; the latter can however be used if you have them.

Component positions R1 and C1 are provided so the cartridge loading can be modified. This has only a marginal effect on MC cartridge response in most cases because the cartridge impedance is so low. However, if you want to experiment then the appropriate range for R1 is 10 Ω- 1 kΩ, and for C1 0–10 nF.

Moving-magnet (MM) stage

This is a relatively conventional stage, except that it uses multiple polystyrene capacitors to obtain the required value(polyester capacitors have worse tolerance and introduce non-linear distortion) and to improve RIAA accuracy (because random errors in the capacitor values tend to cancel). Multiple RIAA resistors R22-R23 and R24-R25 are used to improve accuracy in the same way.

The value of C12 is large as the IEC amendment is not implemented in this stage.

Performance graph

MC/MM board # 110650–2 only.

Test equipment: Audio Precision Two Cascade Plus 2722 Dual Domain (@Elektor Labs).

Here we have the AP-2 supplying an amplitude corrected signal according to RIAA pre-equalisation curve. This allows the deviation from the ideal RIAA curve (amplitude error) to be visualised conveniently. The curve with the higher roll-off point was plotted with the IEC Amendment relay energised. The error at 20 kHz is less than 0.06 dB, measured on the left-channel MC input. Measurements on the right-channel MD input gave practically identical results, the curves matching extremely closely.

In conclusion it is safe to say that the investment in a large number of relatively costly polystyrene capacitors in this section of the Preamplifier 2012 is justified.

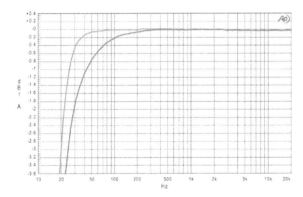

The HF RIAA characteristic is corrected for the relatively low gain of the stage by R26, R27, and C22. Once again two resistors are used to improve accuracy, and C22 is polystyrene.

Note that an NE5534A is used here for IC3 as it is quieter than half an NE5532, and considerably quieter than an LM4562 with its higher current noise. The high inductance of an MM cartridge makes low current noise important. Cartridge loading, and capacitance in particular, has a much greater effect on MM cartridges. Component positions R13 and C8 are provided so it can be modified. The appropriate range for C8 is 0–330 pF. Adding extra loading resistance is rarely advocated; if used here it will partly undo the noise reduction given by the load synthesiser. The lowest recommended value for R13 is 220 kΩ.

The load synthesiser

A load-synthesis circuit around IC4 is used to make an electronic version of the required 47 kΩ loading resistor from the 1 MΩ resistor R16. The Johnson noise of the resistor is however not emulated and so noise due to the rising impedance of the MM cartridge inductance is eliminated. R16 is made to appear as 47 kΩ by driving its bottom end in anti-phase to the signal at the top. IC4B shows a high impedance to the MM input while IC4A is an inverting stage. Multiple resistors R19-R20 and R17-R18 are used to improve gain accuracy and therefore the accuracy of the synthesised impedance.

MM/MC board performance

Test conditions: supply voltage ±17.6 V, B = 80 kHz; measured at Volume/Balance/Tone control board output (# 110650–1); volume set to 1 V out.

Test equipment: Audio Precision Two Cascade Plus 2722 Dual Domain (@Elektor Labs)

MD: 5 mV in, 1 kHz, JP1/2 = 15 dB (source 750 Ω)	THD+N	0.008%
	S/N	82 dB
	S/N	86 dBA
	S/N (input shorted)	88 dBA
MC: 0.2 mV in, 1 kHz, JP1/2 = 15 dB (source 1 Ω)	THD+N	0.016%
	S/N	76 dB
	S/N	79.5 dBA
MC stage gain		29.8 dB

Low roll-off (−3 dB)	19.8 Hz (L)
	20 Hz (R)
	23.3 Hz (L, IEC Amendment on)
	24.8 Hz (R, IEC Amendment on)
Deviation from straight line:	−0.06 dB (100 Hz to 20 kHz)

Gain definitions on JP1/JP2 (dB)	L	R
0	0	0
5	5.22	5.23
10	10.95	10.97
15	14.71	14.72
20	19.52	19.51

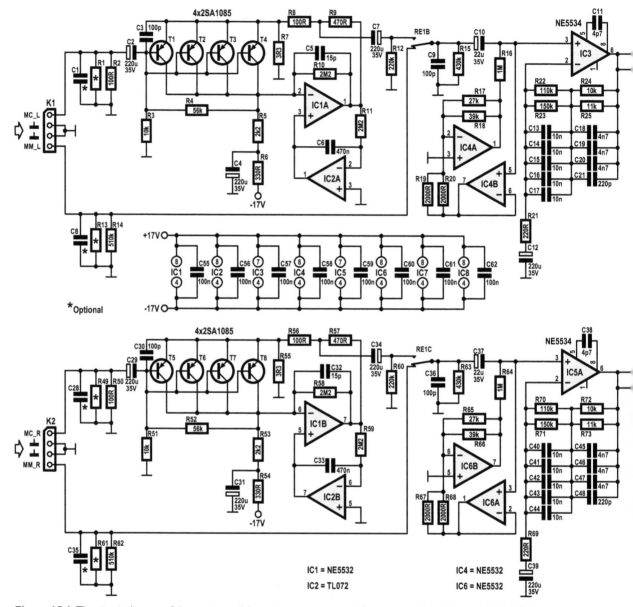

Figure 15.1 The circuit diagram of the moving coil / moving magnet preamplifier section of our Preamplifier 2012. Everything is designed with low noise in mind, as well as perfect adaptability to a wide variety of MC or MD cartridges out there. Check the figures in the Performance inset to see if we've been anywhere near successful.

Subsonic filter

This is a two-stage 3rd-order Butterworth highpass filter that is −3 dB at 20 Hz. Multiple resistors R28-R29 and R30-R31 are again used to improve accuracy. My previous preamp designs have used a single-stage version of this, but I have found the two-stage configuration is preferred when seeking the best possible distortion

performance [2]. An LM4562 is used here (IC7A) as it significantly reduces distortion.

Switchable IEC amendment

The IEC amendment is an extra LF roll-off that was added to the RIAA spec at a later date. Most people

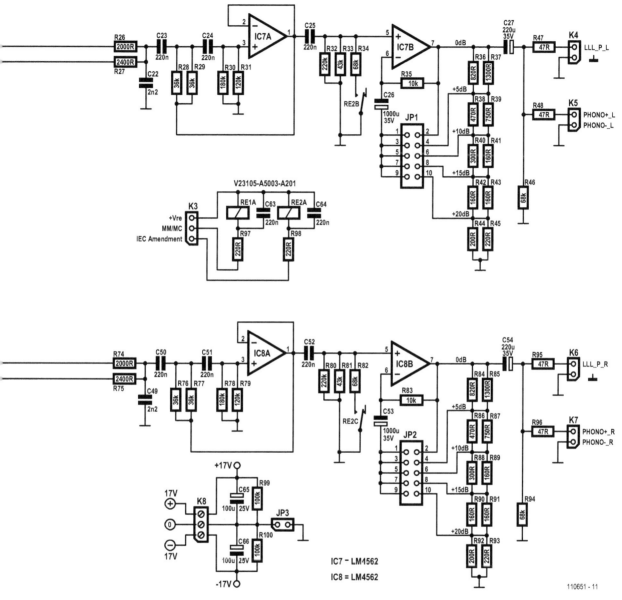

Figure 15.1 (Continued).

regard it as unwelcome, so it is often omitted. Here it can be switched in by placing an extra resistance R34 across the subsonic filter resistances R32-R33. This is something of an approximation, but saves an opamp stage and is accurate to ±0.1 dB down to 29 Hz. Below this the subsonic filter roll-off begins and the accuracy is irrelevant.

The switched-gain stage

This stage around IC7B allows every individual MC and MM cartridge on the market to receive the amount of gain required for optimal noise and headroom. The gain is varied in 5 dB steps by a jumper on jumper block JP1 selecting the desired tap on the negative-feedback divider R36–R45.

Each divider step is made with two paralleled resistors to get the exact value required, and improve accuracy. R35 provides continuity of DC feedback when the switch is altered.

The drive signal to the Log-Law Level LED stage (LLLL) is tapped off via R47 and appears on connector K4. The LLLL circuit and circuit board will be discussed next month.

Construction

The circuit is constructed on double-sided through-plated printed circuit board # 110650–2 (note number) of which the silkscreen (component overlay) is shown in Figure 15.2. As with the board we discussed in the previous instalment, assembly is largely a routine matter since only through-hole parts and conventional soldering are involved. For assembly we again recommend the use of a grill or the even better a flip-over type of PCB assembly jig. Assuming you have positively identified each and every part using the components list, the flip-over jig enables the parts leads to be inserted first. Next, the parts are held securely in place at the top side of the board by a thick layer of packaging foam and a clamp-on panel. The board then gets flipped over allowing the wires to be soldered one by one without the parts (now at the underside) dropping or dislocating. Experienced users do the low-profile parts first for obvious reasons.

The end result should be a board that's as thoughtfully built as the circuit was designed—check your personal effort against our prototype pictured in Figure 15.3.

References

1. Preamplifier 2012 part 1, Elektor March 2012; www. elektor.com/110650.

2. Peter Billam 'Harmonic Distortion in a Class of Linear Active Filter Networks', *Journal of the Audio Engineering Society* June 1978 Volume 26, No. 6, p 426.

Component list

Resistors

(1% tolerance, metal film, 0.25W)
R1,R13,R49,R61 = optional, see text
R2,R8,R50,R56 = 100Ω
R3,R24,R35,R51,R72,R83 = 10kΩ
R4,R52 = 56kΩ
R5,R53 = 2.2kΩ
R6,R54 = 330Ω
R7,R55 = 3.3Ω
R9,R38,R57,R86 = 470Ω
R10,R11,R58,R59 = 2.2MΩ

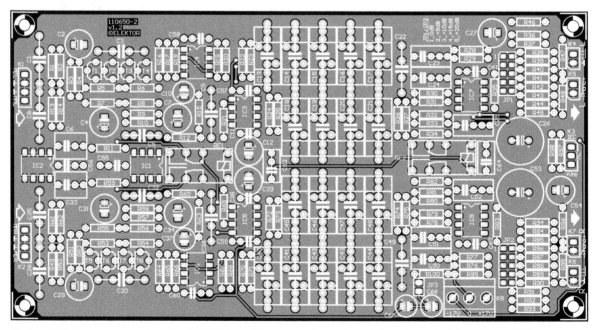

Figure 15.2 Component overlay of the MM/MC board. The high quality ready-made board is available from ElektorPCBservice.com.

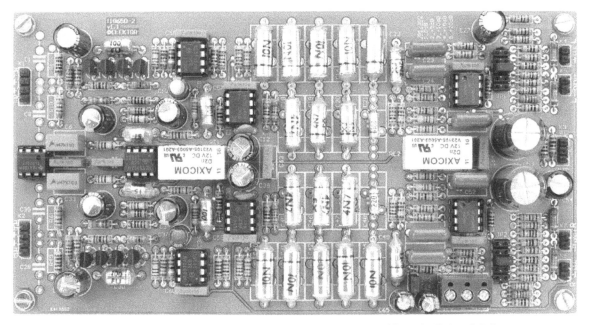

Figure 15.3 Fully assembled and tested MM/MD board "escaped from the Elektor Labs".

R12,R32,R60,R80 = 220kΩ
R14,R62 = 510kΩ
R15,R63 = 430kΩ
R16,R64 = 1MΩ
R17,R65 = 27kΩ
R18,R66 = 39kΩ
R19,R20,R26,R67,R68,R74 = 2.00kΩ
R21,R45,R69,R93,R97,R98 = 220Ω
R22,R70 = 110kΩ
R23,R71 = 150kΩ
R25,R73 = 11kΩ
R27,R75 = 2.4kΩ
R28,R29,R76,R77 = 36kΩ
R30,R78 = 180kΩ
R31,R79 = 120kΩ
R33,R81 = 43kΩ
R34,R46,R82,R94 = 68kΩ
R36,R84 = 820Ω
R37,R85 = 1.3kΩ
R39,R87 = 750Ω
R40,R88 = 300Ω
R41,R42,R43,R89,R90,R91 = 160Ω
R44,R92 = 200Ω
R47,R48,R95,R96 = 47Ω
R99,R100 = 100kΩ

Capacitors

C1,C8,C28,C35 = optional, see text
C2,C4,C7,C12,C27,C29,C31,C34,C39,C54 = 220µF 35V, 20%, diam. 8mm, lead spacing 3.5mm
C3,C9,C30,C36 = 100pF 630V, 1%, polystyrene, axial
C5,C32 = 15pF ±1pF 160V, polystyrene, axial
C6,C33 = 470nF 100V, 10%
C10,C37 = 22µF 35V, 20%, diam. 6.3mm, lead spacing 2.5mm
C11,C38 = 4.7pF ±0.25pF 100V, lead spacing 5mm
C13-C17,C40-C44 = 10nF 63V, 1%, polystyrene, axial
C18,C19,C20,C45,C46,C47 = 4.7nF 160V, 1%, polystyrene, axial
C21,C48 = 220pF 630V, 1%, polystyrene, axial
C22,C49 = 2.2nF 160V, 1%, polystyrene, axial
C23,C24,C25,C50,C51,C52 = 220nF 250V, 5%, polypropylene, lead spacing 10mm
C26,C53 = 1000µF 35V, 20%, diam. 13mm, lead spacing 5mm
C55-C62 = 100nF 100V, 10%, lead spacing 7.5mm
C63,C64 = 220nF 100V, 10 %, lead spacing 7.5 mm
C65,C66 = 100µF 25V, 20%, diam. 6.3mm, lead spacing 2.5mm

Semiconductors

T1-T8 = 2SA1085, Hitachi, e.g. Reichelt.de # SA 1085;
RS Components # 197–9834
IC1,IC4,IC6 = NE5532, e.g. ON Semiconductor type
NE5532ANG
IC2 = TL072
IC3,IC5 = NE5534, e.g. ON Semiconductor type
NE5534ANG
IC7,IC8 = LM4562, e.g. National Semiconductor type
LM4562NA/NOPB

Miscellaneous

K1,K2 = 4-pin straight pinheader, pitch 0.1″ (2.54mm)
Socket headers for K1,K2

K3 = 3-pin straight pinheader, pitch 0.1″ (2.54mm)
Socket header for K3
K4-K7,JP3 = 2-pin straight pinheader, pitch 0.1″
(2.54mm)
Socket header for K4-K7
Jumper for JP1,JP2,JP3
JP1,JP2 = 10-pin (2×5) pinheader, pitch 0.1″ (2.54mm)
K8 = 3-pin screw terminal block, lead pitch 5mm
RE1,RE2 = relay, DPDT, 12V/960Ω, 230V/3A, PCB
mount, TE Connectivity/Axicom type V23105-
A5003-A201
PCB # 110650–2 (www.elektorpcbservice.com)
Note: parts available from Farnell (but not exclusively),
except T1–T8 and PCB 110650–2.

Chapter 16

Preamplifier 2012

Part 3: level indicator, source selector, power supply
(Elektor, June 2012)

This is the third and final article on Preamplifier 2012, covering the level indication circuitry, the input selection system, and the power supply.

The Log Law Level LED, or LLLL, was thought up to help with the correct setup of phono input with switched gain options. It is, to the best of my knowledge, a new idea. A single LED level driven by a comparator typically goes from fully off to fully on for less than a 2 dB change in level when measuring actual music rather than steady sinewaves. The LLLL also uses a comparator to drive the indicating LED hard on or hard off, but before that is a simple log-converter so that the level range from LED always-off to always-on is much increased, to about 10 dB with music. The on-off ratio indicates where the level lies in that range. It is straightforward to set it up so that the level is correct when the LED is on about 50% of the time. This gives a much better indication than simple comparator methods.

In this application, a stereo version of the LLLL is required; it would be possible to use separate LLLL's for each channel, but this hardly seems necessary. The version chosen uses a single LED to indicate the greater of the two input signals, and there are two rectifier and log-converter sections, with the output of comparator IC5A (Left) wire-OR'ed with the output of IC5C (Right). Looking at this approach with the famed clarity of hindsight, a more elegant and economical way to make a stereo version would be to combine the outputs of the two peak rectifiers IC1A, IC1B and IC3A, IC3B to charge a single capacitor C2. This would save a good number of components as there is now only one log-converter and one comparator. I must confess that I have not yet got around to actually trying it out, but it looks solid.

The Preamplifier 2012 was originally designed with a rotary switch for input selection. *Elektor* replaced this with a relay system. Relays are also used to select between MM and MC inputs, to switch the IEC Amendment in and out, and to switch the tone-control in and out. The switched-gain stage is not controlled by relays; the gain option is selected by a jumper.

The power supply is completely conventional, using LM317 and LM337 IC regulators. A well-designed opamp preamplifier with a competent grounding system does not require very low levels of supply-rail noise and this power supply is more than adequate for the task. *Elektor* added the 7815 regulator IC3 to power the relays.

Elektor produced a suite of PCBs for the Preamplifier 2012. The circuitry was split into the phono board, the balanced-input/tone-control/volume-control board, the LLLL board, the PSU, a board holding rotary switches to control the relay system, and two input/output connector boards.

Elektor are unusual in that when they publish a project, they measure it exhaustively in their own laboratories. I am glad to report that the results were entirely in line with my own measurements. Some of the *Elektor* measurements were published at the end of the first article.

PREAMPLIFIER 2012

This month we close off the discussions of the individual boards that make up the preamplifier, both in terms of theory of operation and construction.

The Log-Law Level LED (LLLL) stage

The Log Law Level LED, or LLLL, is to the best of my knowledge, a new idea. A normal single-LED level indicator is driven by a comparator, and typically goes from fully-off to fully-on with less than a 2 dB change in level when fed with music (not sine waves). It therefore does not give much information. The LLLL, on the other hand, incorporates a simple logarithmic converter so the range from LED-always-off to LED-always-on is extended to 10 dB. The preamplifier internal level is correct when the LED is on about 50% of the time. This gives a much better indication as you can judge the level approximately from the on/off ratio.

Referring to the circuit diagram in Figure 16.1, IC1A with R1, R2, D1, D2 is a precision half-wave rectifier circuit. Its output when combined with the feed through R3 provides a full-wave rectified signal to IC1B; this is another precision rectifier that establishes the peak level of the signal on C2. This is buffered by IC2A to prevent loading by the next stage, and the peak voltage is applied to the very simple log-law converter IC2B. For low level signals the gain is set to unity by R6 and R7. As the input voltage increases, first D5 begins to conduct, reducing the gain of the stage by increased

negative feedback through R8. At a higher voltage set by divider R9 and R10, D6 also conducts, reducing the gain further by negative feedback through R9. This simple circuit clearly gives only a very approximate version of a log law but it transforms the operation of the indicator.

The processed voltage is applied to comparator IC5A, and if the threshold set by R21, R22 is exceeded the open-collector output of IC5A goes Low and the output of IC5B, which simply acts as a logical inverter, goes High, removing the short across the LED connected to pinheader K6 and allowing it to be powered by current source T1. There are two channels for stereo operation and the outputs of comparators IC5A and IC5C are wire-OR'd together to pull-up resistor R23 so that a signal on either channel will make the LED illuminate. IC5D is not used. The LLLL can be switched to read either the phono input directly (so the switched-gain stage can be set up correctly) or the signal after the input select relays and balanced line input stage, where it can be used to check the incoming level of any input. This switching is done by relay contacts RE1B, RE1C. Series relay contacts RE2B, RE2C allow the LLLL to be disabled when its job is done and the flickering LED may become irritating.

The LLLL is built on circuit board no. 110650–6 shown in Figure 16.2. It contains through-hole components only, hence nothing should hinder you from producing a working board straight off if you are careful about placing and soldering the parts. The same applies to the other three (four) boards discussed below. The finished LLLL board is pictured in Figure 16.3.

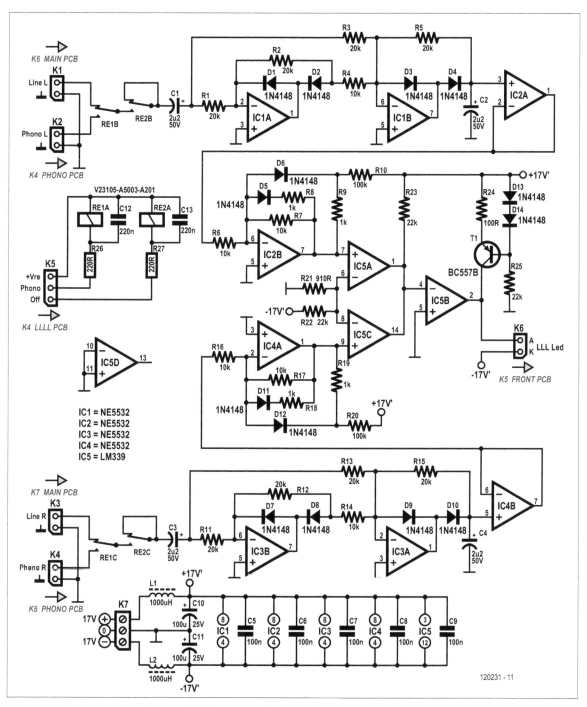

Figure 16.1 Schematic of the Log-Law Level LED (LLLL) board.

Component list

LLLL Board (# 110650–6)

Resistors

(0.25W, 1%)
R1,R2,R3,R5,R11,R12,R13,R15 = 20kΩ
R4,R6,R7,R14,R16,R17 = 10kΩ
R8,R9,R18,R19 = 1kΩ
R10,R20 = 100kΩ
R21 = 910Ω
R22,R23,R25 = 22kΩ
R24 = 100Ω
R26,R27 = 220Ω

Capacitors

C1-C4 = 2.2µF 20% 100V, diam. 6.3mm, 2.5mm pitch
C5-C9 = 100nF 10% 100V, 7.5mm pitch
C10,C11 = 100µF 20% 25V, diam. 6.3mm, 2.5mm pitch
C12,C13 = 220nF 10%, 100V, 7.5mm pitch

Inductors

L1,L2 = 1mH 3.6Ω 370mA, axial

Semiconductors

D1-D14 = 1N4148
T1 = BC557B
IC1-IC4 = NE5532
IC5 = LM339

Miscellaneous

K1-K4,K6 = 2-pin pinheader, 0.1" pitch (2.54mm)
Socket headers for K1-K4,K6
K5 = 3-pin pinheader, 0.1" pitch (2.54mm)
Socket header for K5
K7 = 3-way PCB terminal block, 5mm pitch
RE1,RE2 = V23105-A5003-A201, 12V/960Ω, 230V/3A, DPDT
PCB # 110650–6 (www.elektorpcbservice.com)

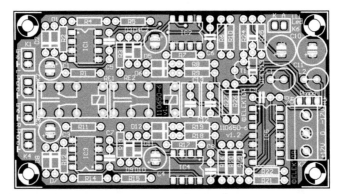

Figure 16.2 The combined metering and driver circuitry of the Log-Law Level LED indicator is built on this board. Board available through www.elektorpcbservice.com.

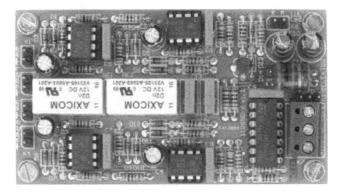

Figure 16.3 Prototype of the LLLL board built by Elektor Labs.

The input PCBs

The input circuit boards (one per channel) carry the input and output connectors plus the relays that select which input is in use. Their circuit diagram appears in Figure 16.4. The four line-level inputs can work in either unbalanced or balanced mode, though of course balanced mode is strongly recommended because it rejects electrical noise on the ground paths. Bear in mind this is true *even if the source is unbalanced.* In this case the cold (phase-inverted) balanced input must be connected to ground at the *source* end of the cable—a 3-way cable is still required. There is therefore no technical reason to provide RCA connectors but they are included for convenience. If they are to be used the jumpers JP1–JP4 must be installed.

For Input 1, relay contacts RE1B, RE1C connect to the preamplifier input when relay coil RE1A is energized, and the signal goes to the main preamplifier board via connector K15.

Facilities are also provided for using the XLR inputs in unbalanced mode; in this case the jumper S1 is set so that RE1B connects to ground rather than pin 3 of the XLR. The same applies to Input 2 and Input 3. However, if you have spent the money for an XLR you really should make proper use of it by employing balanced mode.

Input 4 is the phono input if the phono facility is included in the preamplifier. Input 4 is used for either MM or MC mode, as this selection is made on the phono PCB discussed in part 2. In this case K7 and K8 are not

fitted. If the phono PCB is not used Input 4 can be used as another Line input, with K7 and K8 fitted. In this case there is no connection to K13.

There is no facility for using the MC and MM cartridge inputs in balanced mode so only RCA connectors are provided for these inputs.

This circuit is constructed on circuit board no. 110650–3 pictured in Figure 16.5; the assembled board appears in Figure 16.6.

The front panel PCB

The front panel board of which the schematic is shown in Figure 16.7, carries the input select switch, the LLLL source select switch, the three mode switches, and the power and LLL LEDs.

S5A is the input select switch, controlling the relays on the input boards Only the first four positions are used, the other two being blocked off by a mechanical stop. S5B is not used. The switching between MM/MC and Line input is done by a separate switch S2 on the front panel board.

S4A and S4B control the operation of the LLLL indicator. In the first position the LLLL is fed from after the input select and balanced input stage and can be used to check any input. In the second position it is fed from the phono stage only, whichever input is selected. The third position removes the signal fed to the LLLL and

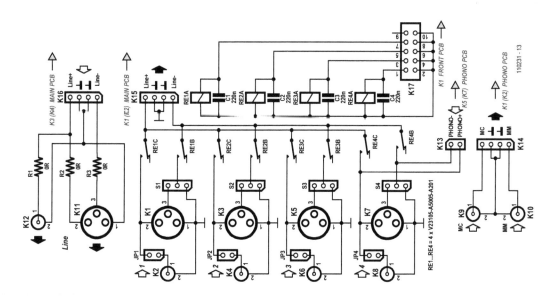

Figure 16.4 Circuit diagram of the input section of the preamplifier.

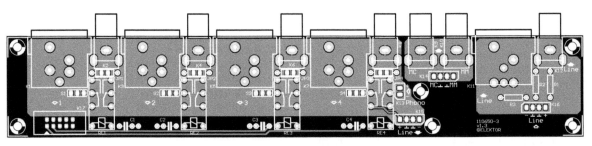

Figure 16.5 The preamplifier's input board (here at 75% of its real size). Two of these are required for stereo operation. Board available through www.elektorpcbservice.com.

Figure 16.6 Fully assembled Input board.

Component list

Input Board (#110650–3)

Resistors

R1, R2 = 0Ω

Capacitors

C1-C4 = 220nF 10% 100V, 7.5mm lead pitch

Miscellaneous

K1, K3, K5, K7 = XLR socket, PCB mount, horizontal, 3-way
K11 = XLR plug, PCB mount, horizontal, 3-way
K2, K4, K6, K8-K10, K12 = RCA/Phono socket, PCB mount, Pro Signal type PSG01545 (red) (R) or PSG01546 (white) (L)
K13, JP1-JP4 = 2-pin pinheader, 0.1" pitch (2.54mm)
Socket header for K13
K14, K15, K16 = 4-pin pinheader, 0.1" pitch (2.54mm)
Socket headers for K14, K15, K16
K17 = 10-way boxheader, straight, 0.1" pitch (2.54mm)
S1-S4 = 3-pin pinheader, 0.1" pitch (2.54mm)
Jumpers for S1-S4, JP1-JP4
RE1-RE4 = V23105-A5003-A201, 12V/960Ω, 230V/3A, DPDT
PCB # 110650–3 (www.elektorpcbservice.com); 2 pcs required

so disables it. Only the first three positions of the switch are used (Line—Phono—Off) and the other three are blocked off by a mechanical stop. Switch S1 controls the tone-control defeat relay. When S1 is closed the tone-control is bypassed completely.

Switch S2 selects between moving-magnet (MM) and moving-coil (MC) operation, using relay contacts RE1B, RE1C on the phono PCB. When S2 is closed the MC input is active.

S3 selects the IEC amendment, which is a largely unwanted extra roll-off at 20 Hz that was later added to the RIAA spec. When S3 is closed the IEC amendment is applied. The small board no. 110650–4 for securing to the front panel of your preamplifier case is pictured in Figure 16.8, along with the Component List. Check Figure 16.9 for the assembled board.

The power supply PCB

The power supply is fairly conventional as the preamplifier does not require ultra-low noise on the supply rails to give its best performance. Referring to Figure 16.10, a center-tapped 18–0–18 V power transformer, a bridge rectifier D3–D6 and reservoir capacitors C8, C10 produce the unregulated supply voltages. Snubbing capacitors C14–C17 prevent RF being generated by the rectifier diodes. LEDs D7 and D8 are mounted on the PCB and indicate

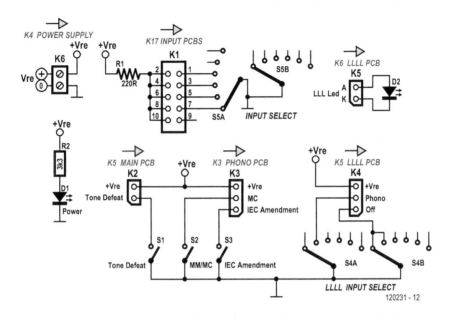

Figure 16.7 This sub circuit of switches, LEDs, resistors and connectors forms a separate board intended for fitting to the inside of the preamplifier front panel.

Component list

Front Panel Board (#110650–4)

Resistors

(0.25W, 1%)
R1 = 220Ω
R2 = 3.3kΩ

Semiconductors

D1 = LED, green, 3mm, through hole
D2 = LED, red, 3mm, through hole

Miscellaneous

K1 = 10-way boxheader, straight, 0.1″ pitch (2.54mm)
K2,K5,D1,S1-S3 = 2-pin pinheader, 0.1″ pitch (2.54mm)
Socket headers for K2,K5,D1,S1-S3
K3,K4 = 3-pin pinheader, 0.1″ pitch (2.54mm)
Socket headers for K3,K4
K6 = 2-way PCB terminal block, 5mm pitch
S1,S2,S3 = switch, SPDT (On-On)
S4,S5 = 2-pole, 6-position rotary switch, PCB mount, e.g. Lorlin type CK1050
PCB # 110650–4 (www.elektorpcbservice.com)

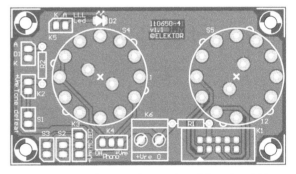

Figure 16.8 The front panel circuit board. Board available through www.elektorpcbservice.com.

Figure 16.9 The Front Panel board, ready for use.

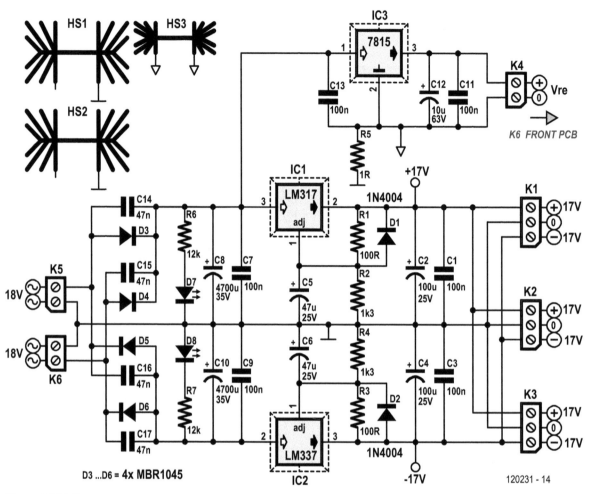

Figure 16.10 The circuit diagram of the power supply holds common-or-garden components hence few surprises. Note that the relays have their own 15 V voltage regulator.

that the unit is powered and both unregulated supplies are present. The ±17 V supplies for the opamps are generated by regulators IC1 and IC2. The networks R1, R2 and R3, R4 set the voltage required while C5 and C6 improve the ripple performance of the regulators. D1 and D2 prevent the stored charge on C5 and C6 damaging the regulators if the output is short-circuited. Capacitors C1-C4 reduce the output impedance of the supply at high frequencies.

A separate +15 V regulator IC3 is used to power the relays. This is powered from the unregulated positive supply that also feeds the +17 V regulator. Note that the audio ground and the relay ground must be connected together at one point only, right back at the power

supply; otherwise relay currents flowing in the audio ground will cause unpleasant clicking noises in the pre-amp output.

Number 110650–5 is the last board you'll be building to construct the Preamplifier 2012; it's shown in Figures 16.11 (PCB only) and 16.12 (assembled board).

Wiring

Comprising a total of seven boards, the Preamplifier 2012 is quite an intricate job to wire up. For your convenience the connectors ('Kx') in this month's schematics have been given what/where/to labels.

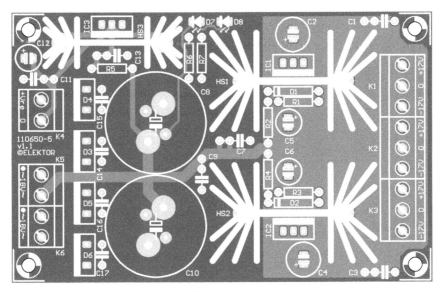

Figure 16.11 The power supply board. Board available through www.elektorpcbservice.com.

Figure 16.12 Preamplifier PSU board assembled by Elektor Labs.

Component list

Power Supply Board (# 110650–5)

Resistors

(0.25W, 1%)
R1,R3 = 100Ω
R2,R4 = 1.3kΩ
R5 = 1Ω
R6,R7 = 12kΩ

Capacitors

C1,C3,C7,C9,C11,C13 = 100nF 10% 100V, 7.5mm pitch
C2,C4 = 100µF 20% 25V, 6.3mm diam., 2.5 mm pitch
C5,C6, = 47µF 20%, 25V, 6.3 mm diam., 2.5 mm pitch
C8,C10 = 4700µF 20% 35V, 22mm diam., 10mm pitch
C12 = 10µF 20% 63V, 6.3mm diam., 2.5mm pitch
C14-C17 = 47nF 50V, ceramic, 5mm pitch

Semiconductors

D1,D2 = 1N4004
D3-D6 = MBR1045
D7 = LED, red, 3mm, through hole
D8 = LED, green, 3mm, through hole
IC1 = LM317
IC2 = LM337
IC3 = 7815

Miscellaneous

K1,K2,K3 = 3-way PCB terminal block, 5mm pitch
K4,K5,K6 = 2-way PCB terminal block, 5mm pitch
HS1,HS2 = heatsink, 50.8 mm, Fischer Elektronik type SK 129 50,8 STS
HS3 = heatsink, 50.8 mm, Fischer Elektronik type SK 104 50,8 STS
PCB # 110650–5 (www.elektorpcbservice.com)
Not on PCB:
Tr1 = AC power transformer, 115/230V primary,

2×18V/50VA secondary, e.g. Multicomp # MCTA050/18, Farnell (Newark) # 9530380
F1 (230VAC) = fuse, 0.315A antisurge, 5×20mm, panel mount
F1 (115VAC) = fuse, 0.63A antisurge, 5×20mm, panel mount
S1 = 115/230V AC line voltage selector, e.g. Arcolectric Switches type T22205BAAC
S2 = AC power switch

Most of the board interconnections consist of short runs of 2, 3 or 4 light duty, flexible wires of different colors (agree on black for ground though) twisted together and their ends soldered to the socket for plugging on the pinheader. Look at the symbol with the mating pinheader on the PCB to identify the bevelled edge, this will help you get the pin-to-pin connections right.

The 10-way connection between the front panel PCB and the input PCB is the only one requiring a length of flatcable (a.k.a. ribbon cable) with IDC sockets pressed on at the ends.

All wiring and components 'ahead of' and including power transformer Tr1 is at AC powerline potential hence should comply with safety regulations and recommendations in place for electrical safety. When in doubt, ask for help or advice from a qualified electrician.

At the time of writing the Preamplifier 2012 is fully wired up on the audio test bench at Elektor House, but not yet fitted in a case. The two input boards are mounted like a sandwich using four PCB pillars. Depending on your response to the project we may do a final article later this year on installing the boards in a high quality enclosure. For now we leave that part of the project to you, as well as enjoying the superb sound quality of the Preamplifier 2012.

Chapter 17

Self-improvement for capacitors

The linearisation of polyester capacitors
(Linear Audio, Volume 1, April 2011)

This article described a phenomenon I stumbled across when checking for capacitor distortion. An ideal capacitor is a perfectly linear component, but some real types of capacitor create easily measurable distortion when they have a significant signal voltage across them. For electrolytics, this is as low as 80 mV rms, but some non-electrolytic capacitors (there really should be a better name for them than that), such as those with a polyester dielectric, require voltages of 9 Vrms to generate about 0.0015 % THD, so the problem is on another scale entirely. This is just as well, as while electrolytic capacitor distortion can usually be eliminated just by making the capacitor (typically a DC-blocking or coupling component) big enough, non-electrolytics with a signal voltage across them are usually implementing a vital time-constant, so you can't change the value. Very often, an electrolytic coupling capacitor has only opamp offset voltage across it, and its function is simply to stop switches clicking and pots rustling; in this case a voltage rating as low as 6V3 works perfectly, and is not more susceptible to distortion than higher-voltage capacitors.

I recall being most disconcerted when I first encountered these effects in polyester caps—somehow I felt someone was cheating somewhere; capacitors are supposed to be linear components, no? This was long before anything was published on the subject, to the best of my knowledge. Fortunately polystyrene-dielectric capacitors are free of this effect, which solves the problem up to 10 nF. For bigger values, polypropylene-dielectric capacitors are also free from distortion but are expensive and bulky compared with their polyester equivalents. In a some circumstances, you can cut down the cost and size by using capacitance-multiplier techniques to make a small capacitor appear to be a large one [1], but unless it has one side connected to ground, which makes it easy, the circuitry gets get complicated and requires multiple extra opamps. These are likely to impact the noise and distortion performance themselves.

Through an intermediary I asked what someone who had spent much time studying and measuring capacitor defects what he thought of the article. I expected that he would be intrigued. However, he clearly thought I was treading on his turf because his only comment, so far as I am aware, was to pronounce that the article was wholly worthless because I failed to provide the exact provenance and complete history of the components measured. I found that a tad ungenerous.

I have to admit that since this article was published, the world has shown no great interest in the phenomenon. It is hard to argue that it has great practical use; pre-conditioning polyester capacitors by subjecting them to high signal voltages only reduces their distortion; it does not eliminate it, and it also partially returns when the applied signal is removed. If capacitor distortion is anything of an issue, using polystyrene or polypropylene parts will be a much better bet.

From my point of view I am still interested in the questions left unanswered at the end of the article. Will a 10 kHz test signal linearise capacitors ten times as fast? Would exercising the capacitors with a sawtooth waveform be equally effective? They remain unanswered at present. Do feel free to have a go.

Reference

1. Self, D., *The Design of Active Crossovers*, pp. 333–335. Focal Press 2011. ISBN 978-0-240-81738-5.

SELF-IMPROVEMENT FOR CAPACITORS

We are all resignedly familiar with components that deteriorate over time, such as electrolytic capacitors drying out and contacts that tarnish. It is therefore a relief to be able to reveal a case where components actually improve, like the maturing of a fine wine.

Some capacitor types introduce distortion when they have a significant signal voltage across them. Electrolytic capacitors are notorious for this [1] but they present few problems, as when they are used for coupling and DC-blocking you just have to make the capacitance large enough so that the signal voltage is suitably small at the lowest frequency of interest. In my experience 'suitably small' means less than 80 mVrms. This usually requires much more capacitance than is needed just to avoid a significant low-end rolloff, resulting in a component that looks over-sized for the job, but since the voltage rating can be kept very low, for example 6.3V, the CV product and hence the volume and cost of the capacitor is only increased slightly, if at all.

A rather more serious problem is presented by the use of non-electrolytic capacitors to define a time-constant, as in RIAA equalisation or filters. The capacitor has by definition a large signal voltage across it at some frequencies, and some sorts of non-electrolytic capacitor, notably those with a polyester dielectric, can generate quite large amounts of distortion. I will say at once that this problem can be eliminated by using either polystyrene capacitors (for 10 nF or less) or polypropylene

capacitors; (for values above 10 nF) as described in my Small Signal Audio Design book.[2] The problem is that large polypropylene capacitors are both big and expensive compared with polyester types.

I therefore decided to try to establish more closely just how bad polyester capacitors were. The test circuit shown in Figure 17.1 is a simple RC low-pass filter which is-3 dB at 723 Hz. With a 9 Vrms input and a 220nF 100V microbox polyester capacitor this gives obvious third-harmonic distortion almost reaching 0.002%, as seen in Figure 17.2; interestingly the maximum THD levels occur well below the-3 dB point at 300 to 400 Hz, corresponding with 0.7 to 1.2 dB of attenuation. The amount of distortion generated by nominally identical capacitors varies over a range of at least two to one. At all frequencies the distortion is visually pure third harmonic, which is what we would expect from a

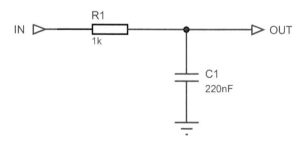

Figure 17.1 Simple RC lowpass test circuit for capacitor distortion; -3 dB at 723 Hz.

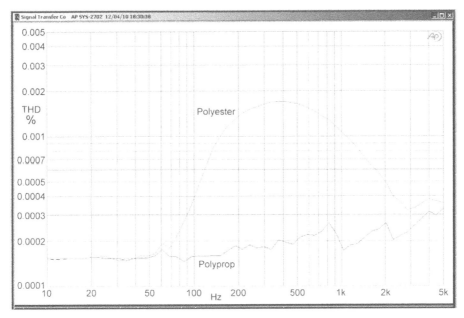

Figure 17.2 The distortion (all third-harmonic) from a 220nF 100V polyester capacitor in the low-pass test circuit, compared with a distortion-free 220nF polypropylene. Sample cap #14. Input level 9 Vrms.

symmetrical component. Most of the capacitors tested have been in my possession for a long time, and I am afraid I have no idea who the manufacturer was, but they appear to be standard ivory-coloured microbox polyester 100V capacitors. Brand-new parts behave in the same way.

You may be thinking that 0.002% is not a lot of THD; but it is well above the amount produced by opamps in well-designed circuitry- including that using the veteran 5532. This is also just one capacitor, and RIAA equalisers and most tone-controls will have at least two. We could well live without this problem.

Distortion Versus Time

As the tests proceeded, there emerged a strange inconsistency in the results. Whenever a capacitor sample was re-measured, its distortion was lower than the time before. Plotting distortion against time revealed that under the test conditions the distortion dropped, at first quickly and then at a slower rate. Figure 17.3 shows a typical test on a 'virgin' unused capacitor; the initial low distortion is with the resistor only present. When the capacitor is plugged in the reading jumps up to 0.002%, falls to 0.0016% in the first 20 seconds, and then continues a very slow reduction to 0.0014 % over 5 minutes,

with no sign of levelling out. Note that the test frequency is 1 kHz just for convenience; this is not the frequency of maximum distortion.

Over a longer time-scale, the reduction in non-linearity is more dramatic, as shown for Capacitor #5 in Figure 17.4. On an overnight run of 40,000 seconds (11.1 hours) the THD reading halved from 0.0008% to 0.0004%, and on starting readings again the next morning, (the test signal was present all the time) there was a further reduction from 0.00035% to 0.00030%. The capacitor linearity has improved by nearly three times, and that is a remarkable result. But is it permanent?

Ninety days of rest later, Capacitor #5 was tested again. It did not come as a complete surprise that it had partly relapsed to a high-distortion state. When exercised again the distortion dropped rapidly over about 200 seconds to 0.00030% which is as good as it was before; see Figure 17.5. This is interesting, because originally it took some 18 hours for the THD to drop from 0.0008% to 0.0003%, but now it recovered in some 400 seconds to be as good as it was. After that a very slow improvement continued. And another 90 days after that . . . when Capacitor #5 was again tested, after an initial rapid fall there was still more slow improvement over 300 seconds from 0.00026% to 0.00025% (Figure 17.5)

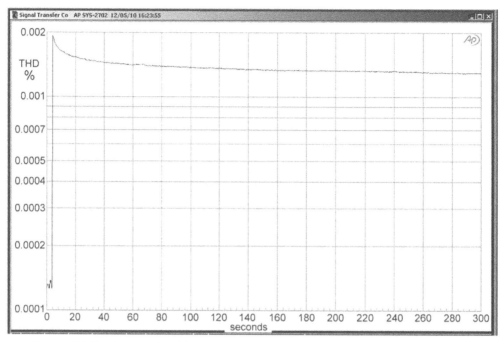

Figure 17.3 The distortion level falls rapidly at first, then more slowly. Sample cap #16, 9 Vrms 1 kHz.

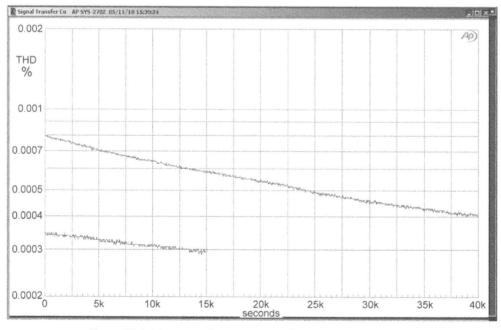

Figure 17.4 A long-term distortion test on Capacitor #5. 9 Vrms 1 kHz.

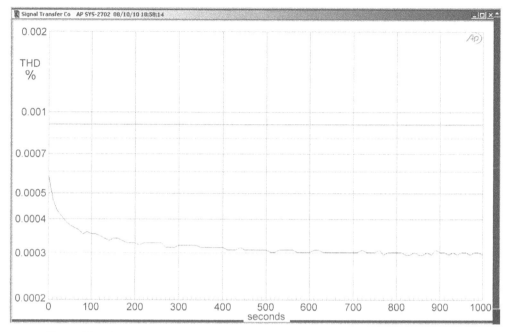

Figure 17.5 Capacitor #5 bounces back to low-distortion after a 90-day lay-off. 9 Vrms 1 kHz.

To the best of my knowledge, this is the first time this long-term linearisation effect has ever been reported, though Cyril Bateman [3] noted some short-term effects. Many different types of polyester capacitor were tested, including some quite old minibox types, and all gave similar results. A difficulty with the experimentation is that each capacitor is a one-off, and it is impossible to repeat a test because of the permanent changes that occur in the component.

Conclusions

Firstly, what's going on here? Obviously some sort of long-term change in the capacitor dielectric, but that hardly answers the question. The way that the distortion partly- but only partly- returns over time implies that there might be two linearisation mechanisms- one giving a permanent improvement, while the other decays over time.

Secondly, is it useful? Should we envisage a ball-room floored with prototype board so thousands of polyester capacitors can be linearised at once? It is doubtful if this is a good idea. Linearised polyester capacitors will never be as sweetly distortion-free as polypropylene, and the linearising effect is only semi-permanent.

Finally, have I stumbled on an effect that explains why some people insist on the need to burn-in so-called high-end hifi for days before it sounds right? Could it be that lurking in that high-end equipment there might be some polyester capacitors that need straightening out? Well, let's see. The effect only applies to capacitors acting to define time-constants, so it would be restricted to RIAA equalisation networks, tone-controls if you have them, and just possibly sub-sonic filters. It would also only work if the 'burn-in' was accomplished by having signals of 9 Vrms or so continuously present in the circuitry. Unless your preamplifier has a very funny gain structure indeed, this is not going to happen, though you could put some extra amplification between the cartridge and the preamplifier input. You would also need to change the vinyl continuously to keep the signal coming in. That is not, as far as I know, what even the most devoted audiophiles do. Looks like the hypothesis is untenable.

A lot of questions remain unanswered. Will a 10 kHz test signal linearise capacitors ten times as fast? Would exercising the capacitors with a sawtooth waveform be equally effective? Presumably it would-it's not as if the capacitor is being taught what a sine-wave looks like. However, given the availability of polypropylene capacitors that never distort at all, it's questionable how much time should be devoted to this interesting effect.

I would like to thank Gareth Connor for providing sample capacitors for these experiments.

References

1. Douglas Self, *Small Signal Audio Design* Focal Press 2010, Chapter 2, p58, ISBN 978–0–240–52177–0

2. Douglas Self, *Small Signal Audio Design* Focal Press 2010, Chapter 2, p53 ISBN 978–0–240–52177–0

3. Cyril Bateman, *Capacitor Sound?* Parts 1–6, Electronics World July 2002—Mar 2003

A low-noise preamplifier with variable-frequency tone controls

(Linear Audio, Volume 5, April 2013)

One day in 2012, I was on the diyAudio forum, and mentioned in passing that I had designed an improved version of the variable-frequency tone control used in the Precision Preamplifier '96, with lower noise, lower distortion, and improved control laws. There was quite a clamour on for me to write it up properly and publish it. Jan Didden was enthusiastic for its publication in *Linear Audio,* so I set to work. Since a stand-alone tone-control is of limited utility, I added a line input, balance and volume control, and *voila*, the *Linear Audio* Low-Noise Preamplifier was born.

I didn't have anything very new to say about phono inputs at the time, so it is a line-only preamplifier, the first I have published. Those of you who have followed my writings know I have a rather low opinion of vinyl as a music-delivery medium, but the replay electronics present some interesting technical challenges, and so research on RIAA stages continued after this design. The latest fruit of this is Chapter 19, originally published in March 2014, which shows how to make accurate RIAA networks more economically. The phono stage from Preamplifier 2012, with its ability to match the sensitivity of any cartridge on the market, could be added to this design without any difficulties. Likewise, it would be straightforward to add the balanced output of the Preamplifier 2012; this just requires a unity-gain inverting stage and a male XLR output connector.

In the balanced input stage, we are seeking a resistor value of 310.7436 Ω to get optimal CMRR. Three parallel resistors R11 = 680 Ω, R12 = 680 Ω, and R13 = 3k6 are used to get a nominal combined value of 310.6598 Ω, which is only 0.0270% low, much less than any likely resistor tolerance. If 1% tolerance resistors are used, then the overall tolerance of the parallel combination is 0.65%. The nominal value can be made more accurate with other combinations; for example R11 = 560 Ω, R12 = 820 Ω, and R13 = 4k7 give a nominal value of 310.7528 Ω, which is only 0.0029% low; however the overall tolerance of the parallel combination has worsened to 0.67 % because the resistor values are further away from equality. In this case, the first combination is likely to give better CMRR results because the nominal value was already very much more accurate than the tolerance.

Looking at the design again, with an eye to development and further improvements, one possibility is the addition of CMRR trimming to the balanced input stage, by adding a multi-turn trimpot to the R11-R13 network. This would allow the low-frequency CMRR to be improved from to about −50 to −85 dB.

This preamplifier is MRP13 in my numbering system. At the time of writing, this is the most recent design I have published.

However, preamplifier design has not stopped there. In 2013, I designed a preamplifier for a client which I think moved things on a bit in terms of sophistication. There is a balanced line input stage which also implements an active balance control with a truncated panpot law [1]; in other words the maximum attenuation of either channel is only 10 dB. This is enough to correct even quite eccentric loudspeaker placements, and the spread-out control law makes adjustment easier. In some of the articles here, I have called this a vernier balance control. The design uses the Self Input configuration [2], so there are no errors from balance pot tolerances. After this comes a distributed volume control using a Baxandall active stage followed by a passive attenuator to give zero noise at zero volume. There is no tone-control. Prototypes have been fully built and tested, and if I may say so, the noise performance and volume matching are superb, and if they're not world-beating, I shall be extremely surprised. In my numbering system, this is the MRP14.

The practice of having an active volume-control stage followed by a passive attenuator may seem to be foolproof and require little thought. Not so. There is an issue with how the active and the passive gain laws interact. If the active stage still has a gain of greater than one when the passive control is attenuating, a headroom bottle-neck is created, as the signal is amplified then attenuated and so will clip in the active stage before anywhere else. I have always called this loss of headroom the 'overlap penalty' as it results from the way the active and passive gain laws overlap; I appreciate it sounds like it might have something to do with the offside rule in football, but it really does not. Ideally, as volume control is turned back from maximum, the active stage gain would decrease, but the passive gain would be held at unity. When the active stage gain fell to unity, the passive control would begin attenuating. Implementing this with pots would not be easy (though perhaps something could be done with centre-tapped pots), but it is quite straightforward with a switched volume control.

This leads us to the MRP15. This a development of the MRP14 with increased headroom and reduced noise, achieved by eliminating the above-mentioned overlap penalty in the active-passive gain control. So far, it has not reached the PCB stage. Preamplifier development continues . . . On, on, ever on.

References

1. Self, D., *Small-Signal Audio Design*, 2nd edition, Chapter 14, pp. 389–391. Focal Press (Taylor & Francis) 2014. ISBN: 978-0-415-70974-3 Hardback. 978-0-415-70973–6 Paperback. 978-0-315-88537-7 ebook.
2. Ibid., Chapter 18, pp. 513–514.

A LOW-NOISE PREAMPLIFIER WITH VARIABLE-FREQUENCY TONE CONTROLS

Preamble

In 1996 I published a design for what is commonly known as the Precision Preamplifier in Electronics World; one of its main features was a treble/bass tone control with the frequencies variable over a 10:1 range. When I mentioned on the DIYaudio board in May 2012 [1] that I had developed an improved version of this tone-control, I was surprised at the enthusiasm for its immediate publication.

I could of course have just written up the tone-control, but it occurred to me that since it inherently phase-inverts it needs another inversion before or after it to preserve absolute phase. A stage that merely inverts to correct phase has never appeared to me an efficient use of components, so I thought that it would be better to make use of it as a Baxandall active volume-control; I have therefore designed one intermediate in complexity and performance between the original 1996 one-path volume control and the 4-path volume control used in my recent Elektor preamp.[2] This 2-path volume control is placed after the tone-control because in normal use it is likely to be set for a gain less than unity and so noise from the earlier stage will be attenuated.

You can see where this is heading- with a comprehensive tone-control and an active volume-control we almost have a line-only preamplifier. We just need to add an input select switch. Actually things are slightly more complicated than that (surprise) as the tone-control can have quite a low input impedance and some sort of input buffer is needed. It could be a simple voltage-follower, but a balanced input amplifier will cost very little more and give the great advantage of rejecting noise from ground currents. Job done?

Well, maybe, but a conventional unity-gain balanced input amplifier made with four 10 kΩ resistors and a 5532/2 is a relatively noisy thing, giving some -104 dBu noise out. This would make it the noisiest block in the

signal path by quite some margin, which is not helpful for a design intended to showcase a low-noise tone-control. The noise can be reduced significantly by adding a little more complexity in the form of two unity-gain buffers to drive a balanced input amplifier built from much lower value resistors. With 820 Ω resistors and using LM4562 sections in all three positions we get -113.9 dBu; almost a 10 dB improvement and now significantly quieter than the tone-control.

However, our work here is not yet done. I think it is difficult to claim that a preamplifier is usable without a balance control; we don't all have symmetrical listening spaces. On the other hand, it seems inelegant to add an extra stage just for the simple balance function, so there is a strong incentive to try to build it into one of the three existing stages. The tone-control is relatively complex and I don't relish the task of trying to add variable gain (non-interactive with all the tone functions, of course) to it. Trying to bodge it onto the active gain control without losing its supreme property- that volume depends on angular control setting and nothing else does not appear promising. That leaves us with the balanced input stage. If this is configured so that it can either attenuate or add gain, it makes a very effective vernier balance control; 'vernier' in that it has more than enough range to move the stereo scene fully from right to left, but is not capable of fading one channel to zero. That is quite unnecessary. I have used such vernier balance stages many times, but I think I have moved the technology on another step, as you will see below.

We wind up in Figure 18.1 with something much simpler than the Elektor preamp; not with the same remarkable noise performance, but still better than almost anything else around. And it has the unique feature of the variable-frequency tone-control. A 4-pole input select switch is placed before the input amplifier.

Until recently my design policy has been to use a careful mix of cheap 5532s and expensive LM4562s, the latter only being used when they made a significant improvement to the performance. The price of LM4562s has continued to slowly decrease, so for this design I decided to use them throughout.

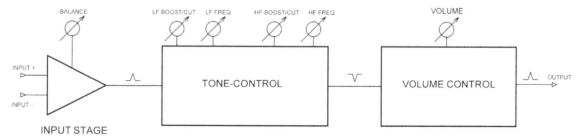

Figure 18.1 Preamplifier block diagram. Spikes show the phase of the signal.

Balanced input amplifier

The variable-gain balanced input stage for the Elektor preamp was the state of the art- or at any rate the state of my art- when it was designed in 2011. Having thought more on the subject since, I think it can be improved.

As is now well-known, the great merit of the Baxandall active volume-control is that its control law depends only on the setting of a linear pot and not on the value of the pot track or the ratio of that track to fixed resistors. Since pot tracks are usually specified at ±20 %, dependence on their value leads to significant errors. With the Baxandall control excellent inter-channel balance is obtained with inexpensive pots. You can take it from me that customers really do complain if there is image shift when they adjust the volume control.

But what about the other preamplifier controls? The tone-control uses the same principle up to a point; the boost/cut controls are linear pots and their effects is independent of their track resistance, but the track values of the HF and LF frequency controls do have some effect on the frequency. It is, principle at least, possible to design out that track-resistance-dependence, but to do it without compromising noise and distortion might be quite a challenge. However, we can eliminate track-resistance-dependence in the balance control . . .

Figure 18.2a is the standard balanced input. Figure 18.2b is the low-noise variant, in which the addition of the two input buffers allows much lower resistances to be used around the differential amplifier, greatly reducing the effects of Johnson noise and opamp current noise. With the values shown, and using all LM4562s the improvement is 9.1 dB over a conventional 10 kΩ & 5532 balanced input as in Figure 18.2a. Making the gain of a balanced input stage variable is not just a matter of making two of the four resistors variable; the four-gang pot is never a welcome component, and in this application it would need to have quite impossible mechanical and electrical accuracy to give a good Common-Mode-Rejection-Ratio. (CMRR)

Figure 18.2c shows the way to do it. Since the feedback arm is driven from a constant and very low impedance, the CMRR does not vary at all when the gain control is altered. This is in almost every way a very satisfactory configuration and I have used it times without number in mixing consoles. If resistors R2 and R4 are made less than R1 and R3 then the stage can attenuate as well as amplify,

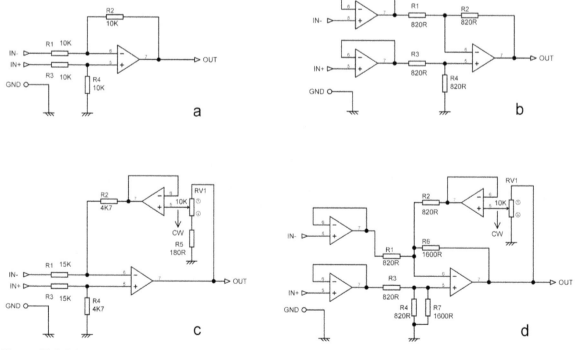

Figure 18.2 Balanced input stages: a) Conventional; b) with buffers for low-noise; c) Variable gain with CMRR preserved d) Low-noise, variable gain, and no pot-dependence. Gain increases as control is moved clockwise.

which is ideally what we want for a balance control. The values in Figure 18.2c are optimised for a general purpose line input gain control rather than a balance control, the maximum gain being +25 dB and minimum gain -10 dB. This means a reverse log pot is required to get middle gain at middle control setting. For the more restricted gain range of a vernier balance control a linear pot is satisfactory so the uncertainties of dual slope controls are avoided. The resistors have relatively high values as input buffers are not used. However, input buffers can be added just as in Figure 18.2b, allowing noise reduction.

A snag remains. You will note that the maximum gain of Figure 18.2c is set by the end-stop resistor R5 at the bottom of the pot. The pot track resistance will probably be specified at ± 20 %, while the resistor will be a much more precise 1%. The variation of pot track resistance can therefore cause quite significant differences between the two channels. While this is less important than the volume control matching, in that it only causes a fixed balance error rather than varying image-shift, it would be nice to eliminate it. If we could lose the end-stop resistor then the pot would be working as a pure potentiometer with its division ratio controlled by the angular position of the wiper alone.

Obviously if we simply do away with the end-stop resistor the gain is going to be uncontrollably large with the pot wiper at the bottom its track. We need a way to limit the maximum gain that keeps other resistors away from the balance pot.

I was just going to bed at dawn the other day (the exigencies of the audio service, ma'am) when it occurred to me that the answer is to add a resistor that gives a separate feedback path around the differential opamp only. R6 in Figure 18.2d provides negative feedback independently of the gain control and so limits the maximum gain. It is not intuitively obvious (to me, at any rate) but the CMRR is still preserved when the gain is altered, just as for Figure 18.2c. You will note in Figure 18.2d that two resistors R4, R7 are paralleled to get a value exactly equal to R2 in parallel with R6. Figure 18.2d shows resistor values that give a minimum gain of -3.6 dB and a maximum gain of +5.8 dB, and low value resistors are used with input buffers once more to reduce noise. The gain with the pot wiper central is -0.1 dB which I would suggest is close enough to unity for anyone. This concept eliminates the effect of pot track resistance on gain, and I propose to call it the Self Input. Alright?

In Figure 18.2d two resistors R4, R7 are paralleled to equal R2 in parallel with R6. The nominal values must be exactly equal for the best CMRR; in practice we are at the mercy of resistor tolerances, but we should at least start out with an exact nominal value. The first prototype gave a CMRR of -56.5 dB, (flat 20–20 kHz) which is rather better than you would expect from 1% resistors; they are commonly more accurate than their official tolerance. If it came out as, say, -30 dB you could be sure that something was wrong with either the design or construction.

Figure 18.3 shows the final schematic for the balanced input amplifier, and incorporates not only the extra components required for practical use, such as

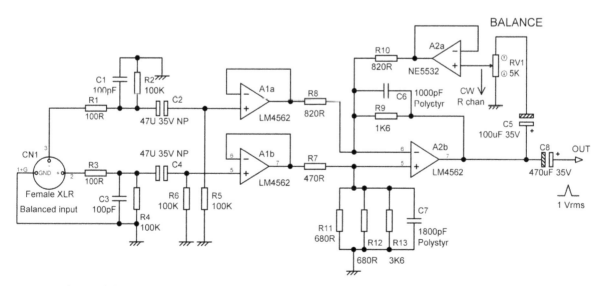

Figure 18.3 Final arrangement of the variable-gain balanced input amplifier. The balance pot is a linear type.

EMC filters, but also some extra minor improvements. In Figure 18.3, making R7 equal to R8, R10 equal to R11, and R9 equal to R12 is the simplest way to get the nominal resistance values equal for good CMRR. However, so long as the ratio between the feedback and non-inverting arms is correct the resistors need not have the same value. If we scale down all the resistors in the non-inverting arm by the same ratio we reduce the effective resistance at the non-inverting input of A2b by that factor and so reduce both Johnson noise and the effect of the current noise of A2b at its non-inverting input. Dividing them all by ten would be simple and effective as regards noise but places far too much loading on the buffer A1b, so we have to use a smaller factor. We have to divide the resistors by this smaller factor, and here some minor troubles begin. Assuming we stick with E24 resistor values, we are not going to find a combination that gives the exact ratio we want, even though there are two in parallel which gives us some extra freedom.

We will here divide the resistor values by roughly a factor of two, so that R is changed from 820 Ω to 470 Ω. This is a conservative approach to avoid too heavy a loading on A1b. Applying the ratio 470/820 to the combination of R11 and R12 in parallel we get a target value for their combined resistance of 310.7436 Ω; you will see the need for such precision in a moment. The closest approach we can make with two E24 values is R = 620 Ω and R = 620 Ω, giving a nominal combined value of 310 Ω. The measured CMRR was now only -43.6 dB. Since different resistor specimens were used this doesn't actually prove anything, but it is rather worrying. 310 Ω is only 0.239% low compared with 310.7436 Ω, but that represents a quarter of the tolerance of a 1% resistor.

We can get much closer to the desired value if we permit ourselves three parallel resistors. R11 = 680 Ω, R12 = 680 Ω, and R13 = 3k6 gives us a nominal combined value of 310.6598 Ω, which is only 0.0270% low, much less than the resistor tolerance. Using this network gave a measured CMRR of -49.3 db, which is rather more satisfactory. The nominal value is now so close to that required that the individual tolerances of the resistors will completely determine the CMRR behaviour.

As I have written elsewhere,[3] a resistance made up of two equal resistors, either in series or parallel, will on average be more accurate by a factor of √2 than the resistor tolerance because random errors in the values tend to partially cancel. Three equal resistors are more accurate by √3, four equal resistors by a factor of √4 = 2, and so on. Because 3k6 is much larger than the two 680 Ω resistors it has little effect on the final value so the accuracy improvement here is only better than √2 by a tiny amount. Still, it all helps.

Table 18.1 Balanced input stage output noise at 3 gain settings. R7 = 470 Ω. Corrected for AP noise at -119.2 dBu; measurement bandwidth 22 Hz–22 kHz, rms sensing, unweighted.

		Min gain	Mid gain	Max gain
Stage gain dB		−3.6	−0.1	+5.8
Noise out dBu		−116.6	−112.4	−107.3
EIN	dBu	−113	−112.3	−113.1

So, after a somewhat effortful reduction of resistance, how much lower is the noise? Frankly, the results are not stunning. Noise is only reduced by 0.5 dB at minimum gain and 0.3 dB at higher gains. There is some scope for scaling down R7 etc further without overloading A1b, but very little is likely to be gained. Nonetheless the improvement, although small, should be wholly robust and costs only one extra resistor. A bit of work now and slightly lower noise forever.

Table 18.1 gives the noise performance after scaling down the resistor values. I think you will agree it is fairly quiet.

It is however slightly regrettable that the EIN is about 0.7 dB worse at the central balance setting, where it is most likely to be used. This is because the pot wiper sees the maximum impedance when in the middle of the track, so this is the worst case both for Johnson noise from the pot, and the effect of A2a current noise flowing in that impedance. Their calculated contributions are of the same order. There doesn't seem to be much that can be done about this except to use a 1 kΩ pot as in the Elektor preamp, with a suitable increase in the value of C5. I prefer to keep all the pots the same value.

The final step is to add the capacitors C6, C7 to define the HF bandwidth at the first practical opportunity. The values shown (which are larger than usual) give -1.0 dB at 100 kHz. The value of C7 has to be scaled up in the same ratio as the resistors R7 etc were scaled down; fortuitously, 1800 pF is close enough to the exact value to maintain good CMRR up to 20 kHz and beyond. Polystyrene capacitors should be used, preferably with a 1% tolerance. Be aware that the circuit may show signs of parasitic oscillation if these capacitors are not installed. It will be stable without the capacitors if A2 is a 5532, but the noise performance is then 3 dB worse.

The distortion of the balanced input amplifier (with balanced input signal) is shown in Figure 18.4. The output level of 9 Vrms is at least 6 dB above the highest level the stage is likely to handle.

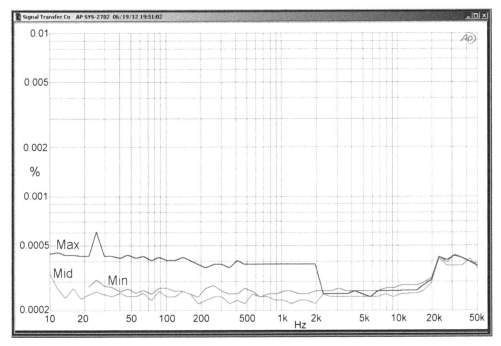

Figure 18.4 Balanced input amplifier distortion at minimum gain, mid gain, and maximum gain. 9 Vrms out for each case. The step in the upper trace is solely due to internal gain-switching in the testgear.

The variable-frequency tone control

The tone-control stage has two bands- bass (LF) and treble, (HF) and in that respect resembles a conventional Baxandall tone control. However it is a great deal more flexible; in each band the frequency at which control begins is variable over a ten to one range, making it much more useful for correcting speaker deficiencies etc. Very few manufacturers have offered this facility. The only one that comes immediately to mind is the Yamaha C6 preamp (1980–81) which had LF and HF frequency variable over wide ranges (they could overlap in the middle) and Q controls for each band as well. Some improvements have been made to the tone-control compared with the 1996 design; they are:

- All opamps changed from 5532 to LM4562 to reduce noise and distortion
- Circuit redesigned to use 5k linear pots throughout instead of 10k linear, again to reduce noise
- Frequency control laws improved so the middle of the frequency range corresponds with the middle position of the control (centre-detent).
- Number of expensive 220 nF polypropylene capacitors reduced

- DC blocking added to LF frequency control to stop DC flowing through it
- Noise in tone-cancel mode reduced

The boost/cut range is +/−10 dB, the LF frequency range is 100 Hz–1 kHz, and the HF frequency range is 1 kHz–10 kHz. The response curves do not level out at their boosted or cut level, but smoothly return to unity gain outside the audio band. This is sometimes called an RTF (Return-To-Flat) characteristic; in the mixing console world it would be known as a 'peaking' EQ. It is a valuable feature because boosting 10 kHz is one thing, but boosting 200 kHz quite another, and can lead to stability or EMC problems. Likewise a touch of boost at 40 Hz may improve a loudspeaker response, but boosting 5 Hz by any amount is likely to be a bad idea. The RTF time-constants are fixed so the boost/cut ranges are necessarily less than +/−10 dB toward the frequency extremes, where the RTF effect starts to overlap the variable boost/cut frequencies. The measured responses for maximum boost and maximum cut at minimum, middle, and maximum frequency settings are shown in Figures 18.5 and 18.6.

The schematic of the tone-control is shown in Figure 18.7.

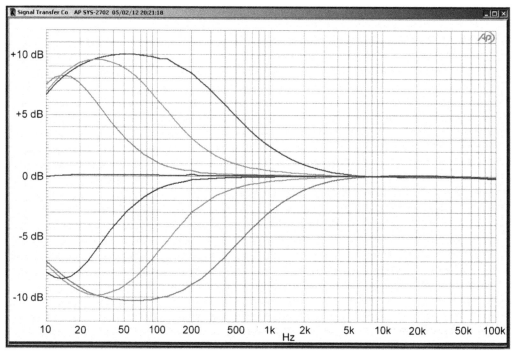

Figure 18.5 LF tone-control frequency response, max cut/boost, at minimum, middle, and maximum frequencies.

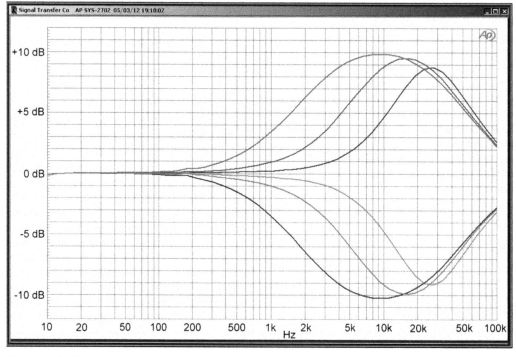

Figure 18.6 HF tone-control frequency response, max cut/boost, at minimum, middle, and maximum frequencies.

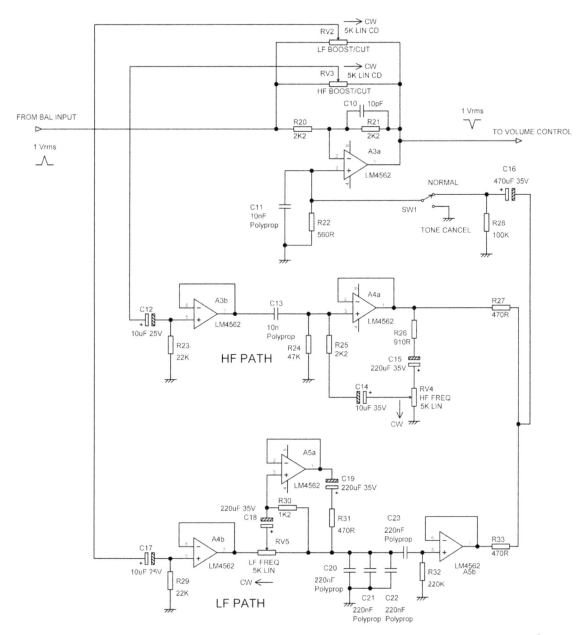

Figure 18.7 The tone-control schematic, showing the two separate paths for HF and LF control. All pots are 5 kΩ linear.

The tone-control stage consists of an inverting amplifier A3a, and two separate frequency-selective paths that provide LF and HF control. The LF path is essentially a variable-frequency first-order low-pass filter, and the associated LF cut/boost control can only act on the frequencies this lets through. Similarly the HF path is essentially a variable-frequency high-pass filter. The filtered signals from the LF and HF paths are summed and returned to amplifier A3a via its non-inverting input. Returning these signals at unity gain would give excessive levels of maximum cut and boost, so an attenuation of 9 dB is introduced into the return path. This limits the maximum cut and boost to +/– 10 dB. This attenuation is introduced by the loading of R22 on the combination of R27 and R33. The gain of in the two paths is unity before this attenuation is applied, and so there are no problems with them clipping before the output of A3a. It is therefore possible to put the 9 dB loss after the two paths rather than before, so it very handily attenuates the noise from them. The loss attenuator is made up of the lowest value resistors that can be driven without distortion, to minimise both the Johnson noise thereof and the effect of the current noise of the non-inverting input of op-amp A3a.

The Tone Cancel switch SW1 disconnects the return signal from any contribution to the main path, preventing 5 out of 6 opamps from contributing noise. R22 is also shorted to ground, removing its Johnson noise and the effect of the current noise from A3a. Tone-cancel usefully reduces the stage output noise by about 4 dB, depending on the HF freq setting. It leaves only A3a in circuit; if this was switched out its phase-inversion would be lost and (assuming absolute phase in the whole signal chain is preserved with the tone-control active) a phase error introduced. This tone-cancel configuration also has the advantage that the signal does not briefly disappear as the cancel switch moves between its two contacts. This minimises transients due to suddenly chopping the waveform and makes valid tone in/out comparisons much easier. Such interruptions are bound to occur when an entire stage is by-passed.

The LF path begins with buffer A4b which prevents loading on the cut/boost control RV2 and so avoids boost/frequency interaction. The low-pass RC time-constant capacitance is the parallel combination of C20, C21 and C22; the use of three capacitors increases the accuracy of the total value by $\sqrt{3}$. The associated time-constant resistance is a combination of RV5, R30, and R31. The LF frequency law is made approximately logarithmic by A5a; at minimum frequency RV5 is fully anti-clockwise, so the input of buffer A5a is the same as at the C20 end of R31, which is thus bootstrapped and has no effect. When RV5 is fully clockwise R30 and R31 are effectively in parallel with RV5 and the turnover frequency is at a maximum. The presence of R30 gives a roughly logarithmic law; its value is carefully chosen so that the centre of the frequency range is at the centre of the control travel. Sadly, there is some pot-dependence here.

The original design had DC excluded from all the pots except the LF frequency control, which had about 2 mV across its track due to the offset voltage of A5a; the control noise from this was audible at very high gain and so it had to be dealt with. Putting a blocking capacitor after A4b does not work as this leaves A5a without any DC reference; adding a bias resistor to ground immediately after the blocking cap also did not work too well as

a tendency for the DC conditions to wander around a bit remained. Putting C19 in series with the output of A5a instead worked nicely. There is absolutely no effect on the LF response.

The HF path begins with buffer A3b to prevent loading on cut/boost control RV3. C13 with R24 and R25 make up the basic high-pass time-constant. The effective value of the time-constant resistance can be altered over a 10:1 range by varying the amount of drive that R25 receives from A4a, the potential divider effect and the rise in source resistance of RV4 in the centre combining to give a good approximation to a logarithmic frequency/rotation law R26 is the frequency endstop resistor that limits the maximum effective value of R24; its value is carefully chosen so that the centre of the frequency range is at the centre of the control travel. C11 is the HF RTF capacitor- at frequencies above the audio band it shunts the sidechain signal to ground, so the HF cut/boost control no longer has any further effect. The HF frequency setting has a strong effect on the noise performance; this is true even when the HF boost /cut control is set to flat, as the HF path is still connected to A3a.

Compared with the HF path, the LF path contributes very little extra noise to the tone stage because most of its circuitry is before the low-pass filter, which almost eliminates its noise contribution; only the noise from buffer A5b returns unfiltered. This can be seen in Table 18.2. The RTF time-constant for the LF path is set by C23 and R32, which block very low frequencies and so limit the lower extent of LF control action. C23 was originally 1 uF, which would be very expensive in a no-distortion polypropylene type, and so it was reduced to 220 nF and resistor R32 increased accordingly to maintain the same roll-off frequency. The impedance at this point is therefore quite high at low frequencies and care should be taken that C23 does not pick up electrostatic hum.

A complete evaluation of the performance of the tone-control is a lengthy business because of the large number of permutations of the controls. If we just look at extreme and middle positions for each control, we have maximum boost, flat, and maximum cut for both HF and LF, and maximum, middle and minimum for the two frequency controls, yielding $3 \times 2 \times 3 \times 2 = 36$ permutations. It is a pain to measure every possibility so the measurements here are restricted to those that put the greatest demands on the circuitry. (eg HF freq at minimum) Table 18.2 shows the noise output of the tone control at these settings:

The final improvement planned for the tone-control was to reduce the noise in the tone-cancel position by reducing the resistors R20 and R21 from 4k7 to 2k2. The output noise with tone-cancel engaged is −110.2

Table 18.2 The noise output of the tone control at various settings. Not corrected for AP noise at −119.2 dBu. (Difference negligible). Measurement bandwidth 22 Hz–22 kHz, rms sensing, unweighted.

HF level	HF freq	LF level	LF freq	Noise out dBu
Flat	Min	Flat	Max	−105,1
Flat	Mid	Flat	Max	−106,2
Flat	Max	Flat	Max	−107,2
Flat	Max	Flat	Mid	−106,8
Flat	Max	Flat	Min	−107,1
Flat	Min	Flat	Min	−105,6
Max boost	Min	Flat	Min	−100,5
Max cut	Min	Flat	Min	−107,4
Max cut	Min	Flat	Max	−107,6
Flat	Min	Max boost	Max	−103,8
Flat	Min	Max cut	Max	−105,9
Flat	Min	Max cut	Min	−105,6
Flat	Min	Max boost	Min	−105,0
Tone-cancel				−110,2

Table 18.3 Input impedance of the tone control at various settings.

HF level	HF freq	LF level	LF freq	Input impedance Ω
Flat	Min	Flat	Mid	987
Flat	Mid	Max boost	Mid	481
Flat	Mid	Max cut	Mid	1390
Max boost	Mid	Flat	Mid	480
Max cut	Mid	Flat	Mid	1389

dBu with R20, R21 = 4k7, and with R20, R21 = 2k2 it drops handily to −112.6 dBu. Since purists will no doubt do most of their listening with tone-cancel engaged this seems a useful modification. But . . .

It is essential to bear in mind that circuits like this tone-control can show unexpected input impedance variations. A standard Baxandall tone-control made with 10 kΩ pots can have an input impedance that falls to 1 kΩ or less at high frequencies where the capacitors have a low impedance. It is not obvious but the alternative tone-control configuration used here also has serious input impedance variations.

Looking at the circuit with the original 4k7 resistors, it would be easy to assume that because the input terminal connects only to a 4k7 resistor and two 5 kΩ pots the input impedance cannot fall below their parallel combination ie 4k7 || 5k || 5k = 1.63 kΩ. It would also be quite wrong, because while the other end of the 4k7 resistor is connected to virtual ground, the two 5 kΩ pots are

connected to the stage output. When the controls are set flat, or tone-cancel is engaged, this carries an inverted version of the input signal. The effective value of the pots is therefore halved, with zero voltage occurring halfway along the pot tracks. The true input impedance when flat is therefore 4k7 || 2.5k || 2.5k = 987 Ω, which is confirmed by simulation.

When the tone-control is not set flat, but to boost, then at those frequencies the inverted signal at the output is larger than the input. This makes the input impedance lower than for the flat case; when the circuit is simulated it can be seen that the input impedance varies with frequency inversely to the output amplitude. Conversely, when the tone-control is set to cut, the inverted signal at the output is reduced and the input impedance is higher than in the flat case. This is summarised in Table 18.3, with R20, R21 = 4k7.

As you can see from the table, the input impedance falls to the worryingly low figure of 480 Ω at maximum

HF and LF boost. It is true that in this case the gain is +10 dB, and so the input voltage into this impedance cannot exceed 3 Vrms without the output clipping, limiting the current required of the preceding stage, but it is not clear that driving a low voltage (3 Vrms) into an impedance that would cause distortion when driven with a high voltage (9 Vrms) gives suitably low distortion. An investigation of this appears to be work that needs doing. The planned reduction of R20, R21 from 4k7 to 2k2 to reduce noise with tone-cancel engaged brings the minimum input impedance down to 388 Ω, and this now looks like rather less of a good idea unless you are planning to use the tone-control after a stage with plenty of drive capability.

The electrolytic capacitors in the tone-control are used solely for DC blocking, and play no part at all in determining the frequency response. This would be highly undesirable, both because of the wide tolerance on value, and the distortion generated by electrolytics when they have significant signal voltage across them. The only design criterion for these capacitors was that they should be large enough to introduce no distortion at 10 Hz and the original values were chosen using the precision algorithm "looks about right" though I hasten to add this was followed up by THD measurements to confirm all was well (See also 'Electrolytic capacitor distortion' at the end of this article).

The distortion of the tone-control was measured for various other permutations of maximum boost/cut and maximum/minimum frequency. There were no surprises there so I will not use up valuable space displaying the results; if there is interest I will put them on my website at douglas-self.com. Suffice it to say that THD at 9 Vrms in/out never gets above 0.001 % except now and then it just barely exceeds it at 20 kHz. General THD levels are in the range 0.0003–0.0007 %.

Figure 18.8 shows the distortion performance when the controls are set flat, for active and with tone-cancel. The slightly higher level with tone active is due solely to extra noise from the LF and HF paths. I have spent some time on the design issues of the tone-control because it is the original raison d'etre of this article. There are many things to consider, and no doubt the design could be further improved with some more work. In terms of mixing console design this is a simple EQ circuit; a big console would have four separate control bands instead of two, and each band would have variable Q as well as frequency. The top and bottom bands would also have a peak/shelving

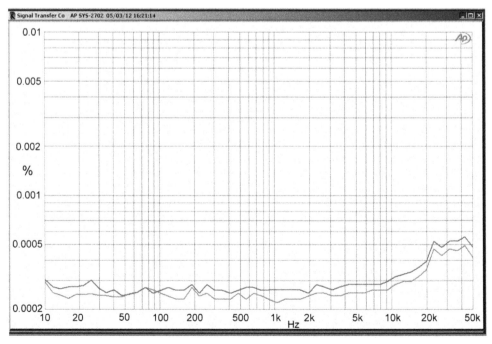

Figure 18.8 Tone-control distortion at 9 Vrms output. Lower trace is for tone-cancel mode, upper trace is for tone-control active but set flat.

switch allowing defeat of the RTF feature. The complexity is considerable.

The active volume control

As previously mentioned, the great value of the Baxandall active volume-control is that its gain depends only on the angular setting of the control, and not the ratio of the track resistance (with its ±20% tolerance) to other resistors, or the dubieties of two-slope "log" pots. This is because the maximum gain is set not by end-stop resistors but by an amplifier stage with its gain set by two fixed resistors. As a result even quite ordinary dual linear pots can give very good channel matching; the one problem that the Baxandall configuration cannot solve is channel imbalance due to mechanical deviation between the wiper positions. A disadvantage is that the gain/rotation law is determined solely by the maximum gain, and it is not possible to bend it about by adding resistors without losing the freedom from pot-value-dependence.

Figure 18.9 shows how this works. Pot rotation is described here as Marks from Mk 0 to Mk 10 for full rotation. (No provision is made for rock bands seeking

controls going up to Mk 11 [5]). Changing the maximum gain has a much smaller effect on the gain at the middle setting (Mk 5). In this preamplifier design a maximum gain of +10 dB is used, giving −4 dB with the volume control central.

Figure 18.10 shows the volume stage. Noise is reduced compared with the 1996 design by the use of a 5 kΩ rather than a 10 kΩ pot, and by doubling the shunt-feedback gain stages so their noise partially cancels and gives a 3 dB advantage. A6a is a unity-gain buffer that prevents the gain stage loading the pot, and A6b, A7a are the shunt-feedback stages whose gain sets the maximum gain of the complete control. Their outputs are averaged by the 10 Ω resistors R43, R46. At a first glance the Baxandall volume configuration looks like a conventional shunt feedback gain control, but the vital difference is that the maximum gain is set by R41, R42 and R44, R45. The stage inherently gives a phase inversion which here neatly cancels that of the inverting tone-control stage, so preserving absolute phase. It requires a low-impedance drive such as an opamp output if it is to give the designed gain range and freedom from law-dependence on the pot value.

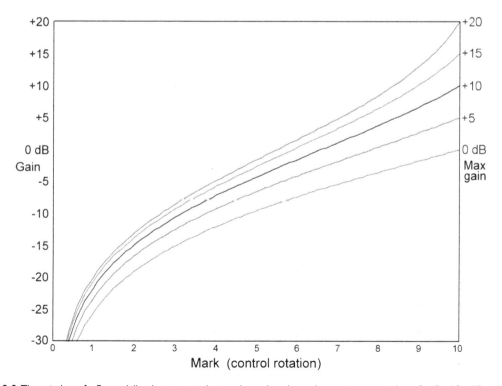

Figure 18.9 The gain law of a Baxandall volume control stage depends only on the maximum gain; here 0, +5, +10, +15 and +20 dB.

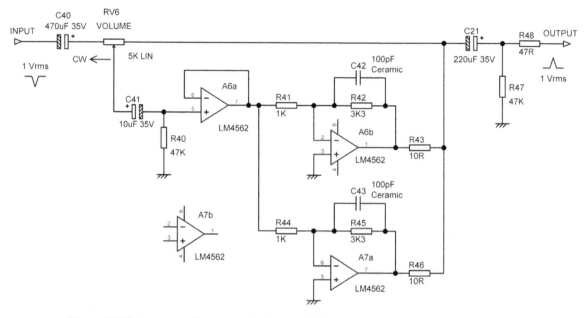

Figure 18.10 Baxandall volume-control with single buffer, dual gain stages. Maximum gain is +10 dB.

With circuits like this that are not wholly obvious in their operation, it pays to keep a wary eye on all the loading conditions. There are three loading conditions to consider:

Firstly, the input impedance of the stage. This varies from the pot track resistance at Mk 0, to a fraction of this set by the maximum gain of the stage. It falls proportionally with control rotation as the volume setting is increased. Figure 18.11 shows how the minimum input impedance becomes a smaller proportion of the track resistance as the maximum gain increases. With a maximum gain of +10 dB the minimum input impedance is 0.23 times the track resistance, which for a 5 kΩ pot gives 1.162 kΩ. Since the preceding stage is based on an LM4562 it will have no trouble at all in driving this.

Secondly, the loading on the buffer stage A6a. A consequence of the gain of the 2-path A6b, A7a stage is that the signals handled by the unity-gain buffer A6a are never very large; less than 3 Vrms if output clipping is avoided. This means that R41, R42 and R44, R45 can all be kept low in value to reduce noise without placing an excessive load on unity-gain buffer A6a.

Thirdly, the loading on the gain stages A6b, A7a. At Mk 10 the loading is a substantial fraction of the value of the pot, but it gets heavier as volume is reduced, as demonstrated in the rightmost column of Table 18.4. We

note thankfully that the loading stays at a reasonable level over the mid volume settings. Only when we get down to a setting of Mk 1 does the load get down to a slightly worrying 383 Ω; however at this setting the attenuation is −21.6 dB, so even a maximum input of 10 Vrms would only give an output of 830 mV. We also have two opamps in parallel to drive the load, so the opamp output currents are actually quite small. Note that the loading considered here is only that of the pot on the gain stages. The gain stage opamps also have to drive their own feedback resistors R42 and R45, and whatever load is connected to the preamp output sockets, so this should be taken into account. I have heard doubts expressed about a possible rise in distortion at low volume settings, because the gain stages A6b and A7a see a very low impedance. This may be so, but the current to be absorbed by the gain stages is very limited because almost the whole of the pot track is in series with the input at low settings. To prove there is not a problem here, I set the volume to Mk 1 and pumped 20 Vrms in, getting 1.6 Vrms out. The THD residual was indistinguishable from the GenMon output of the AP SYS-2702. In use the input cannot exceed 10 Vrms as it comes from an opamp.

To push things further, I set the volume to Mk 0.2, (ie only 2% off the endstop) and shoved 20 Vrms in to get only 300 mVrms out. The THD+N residual was

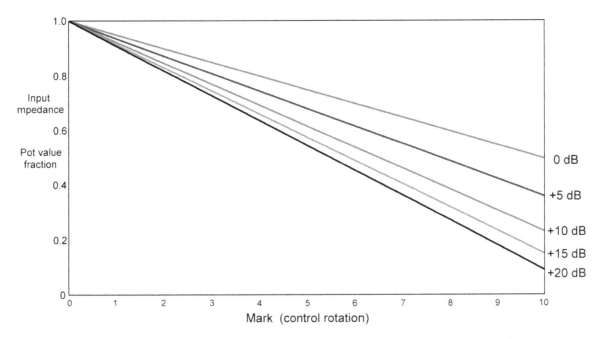

Figure 18.11 The input impedance of the volume control stage as a proportion of the pot track resistance falls more rapidly when the stage is configured for higher maximum gains.

Table 18.4 Volume stage gain, noise, input impedance, and gain stage loading versus control setting. Corrected for AP noise at −119.2 dBu. Measurement bandwidth 22 Hz–22 kHz, rms sensing, unweighted.

Control position (Mk)	Gain dB	Noise output dBu	Input impedance Ω	Opamp load Ω
10	10,37	−109,0	1162	3900
9	6,98	−110,3	1547	3456
8	4,03	−111,4	1929	3069
7	1,29	−112,7		2687
6	−1,38	−113,7		2305
5	−4,11	−114,9	3083	1918
4	−7,07	−115,7		1534
3	−10,48	−117,1		1151
2	−14,83	−118,4		767
1	−21,61	−119,8		383
0	−infinity	−121,3	5000	

0.0007%, composed entirely of noise with no trace of distortion. I then replaced the 4562s in the A6b, A7a positions with Texas 5532s (often considered the worst sort for distortion) and the results were just the same except the noise level was higher giving a THD+N of 0.0008%. As noted, Table 18.4 shows the gain and the noise output at the various control settings.

I haven't bothered to fill in all the entries for input impedance, as it simply changes proportionally with control setting as seen in Figure 18.12, ranging from the minimum of 1162 Ω to 5 kΩ, the resistance of the pot track.

Figure 18.12 shows the distortion performance at maximum gain (Mk 10) and 9 Vrms out. A small amount

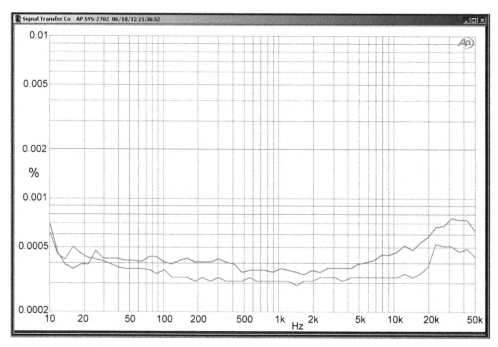

Figure 18.12 Volume-control distortion at max gain. Upper trace is volume-control output; lower trace is testgear reference output. (GenMon) 9 Vrms output.

of third-harmonic is discernable in the noise of the THD residual. Results at Mk 9 and below are indistinguishable from the testgear output apart from a small amount of added noise.

Overall performance

Now to look at the noise and distortion performance when the input/balance, tone control, and volume control blocks are plugged together. The noise will accumulate down the signal path in an orderly rms-sum fashion, but distortion products may reinforce or cancel making the outcome unpredictable.

The noise from the balanced input amplifier with the balance control central is -112.4 dBu. The noise from the tone-control depends on its settings, but if set for a flat response is -107.2 dBu with the HF frequency control at maximum and -105.6 dBu with the HF frequency control at minimum. If we add the balanced input noise to those figures we get -106.0 dBu (HF freq max) and -104.8 dBu (HF freq min). The difference is now only 1.2 dB; we will take the worst case of -104.8 dBu and see what happens when we send that to the volume control.

Figure 18.13 shows how the noise from the volume-control stage increases with gain. When added to the noise coming from upstream scaled by the stage gain, we get the "total noise" line. The noise output with the control central (it will probably spend most of its time around there) is a satisfyingly low -108 dBu. The dotted line shows the upstream noise scaled by the volume gain but with the volume stage noise not added; in other words the result with an ideal noiseless volume control. The point of this sort of noise analysis is to find out if any stage is dominating the noise situation, and if so, is it worthwhile spending time and money making it quieter? The volume stage noise contribution is negligible at higher settings, but becomes more significant as the volume is turned down, until at a setting around Mk 3, the contributions are equal. The gain here is 25 dB below the +10 dB maximum, and the total noise out only -116 dBu. There does not seem to be a pressing need to make the volume stage quieter, which is a pity as it would be relatively straightforward. If we want to reduce noise then the tone-control stage is the obvious candidate for development.

It would certainly be possible to do this. For lower noise in Tone-Cancel mode, the 5 kΩ boost/cut pots could be scaled down to 1 kΩ, as in the Elektor pre-amplifier, and the extra opamps required to drive the

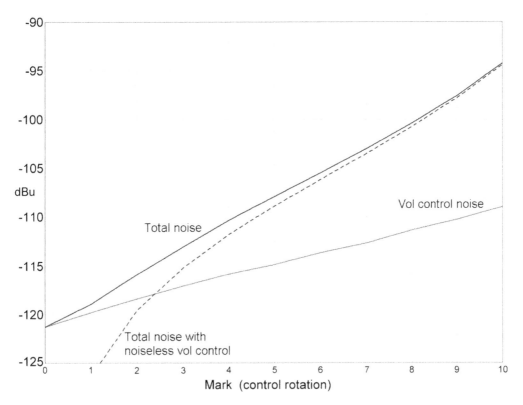

Figure 18.13 Total noise output and volume control stage noise versus volume control setting. The dotted line shows the noise output with a noiseless volume control.

resulting low impedances would give partial cancellation of their voltage noise. For lower noise with the tone-control active, which is more useful for most of us, the impedances in the HF and LF paths could also be scaled down by 5:1; a bigger ratio is difficult due to the problems of sourcing dual-gang pots with a value of less than 1 kΩ. The HF path is the priority as it generates much more noise than the LF path. Such a modified tone-control would however be a formidable piece of electronics, with some dauntingly large polypropylene capacitors, and it would probably be hard to reduce the noise by more than 6 dB before the number of opamps required got out of hand.

A little-known disadvantage of the Baxandall configuration, and indeed of any active-gain control, is that the output noise does not go to zero when the volume control is set to Mk 0. It is very low here, at -121.3 dBu, and much quieter than a fixed-gain amplifier following a passive volume control, but it is not zero. The shunt amplifiers in the active-gain stage work at

a noise gain of one when volume is set to Mk 0, so the voltage noise of the opamp(s) will always appear at the output, even though the signal gain is zero. The only obvious way to solve this is to combine an active-gain stage with a final passive attenuator, to form a distributed volume control system. With the passive attenuator at zero there is a short-circuit across the output and no noise, unless you want to get pedantic about the Johnson noise of the internal wiring (of the order of -155 dBu, I would estimate). This does of course require 4-gang pots for stereo. Distributed volume techniques using two passive controls with a fixed gain stage in between have been used in high-end Japanese preamplifiers in the past.

The overall distortion performance of the complete preamplifier is summarised in Figure 18.14. It keeps below 0.001 % from 10–20 kHz at 9 Vrms, which is well above normal operating levels. The volume was set to Mk 6 (gain -2 dB) to keep a high level throughout the signal chain.

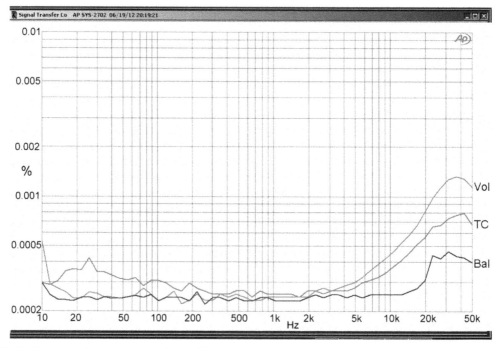

Figure 18.14 Distortion of complete preamplifier at 9 Vrms, measured at the outputs of the balanced input, (Bal) the tone-control, (TC) and the volume-control. (Vol) Balance mid, tone-control in but flat, volume = Mk 6.

Variations

You will note that in Figure 18.11 there is a spare LM4562 section left floating helplessly in space. This could be used as an inverter to provide a fully-balanced output, which raises the signal level over the interconnect by 6 dB and so gives more immunity from ground noise when used with a suitable input [6]. Alternatively, the section could be added to the volume control in parallel with the existing shunt-feedback gain stages to further reduce noise.

A CMRR trim preset could be added to the input/balance stage, with one of R11, R12 or R13 in Figure 18.3 replaced by a preset in series with a fixed resistor. This allows the CMRR to be trimmed to better than-80 dB up to at 500 Hz, and 70 dB at 2 kHz; it falls off at higher frequencies with opamp open-loop gain. Sophisticated testgear is not required as you are simply tuning for minimal signal; any sort of oscilloscope should be adequate.

It would be possible to build the preamplifier using 5532s rather than LM4562s, without changing any other components. The noise will be greater, and distortion will inevitably be somewhat higher, but I think that the loading conditions are light enough to allow this to work reasonably well.

In conclusion

What started as a simple showcase for the improved variable-frequency tone–control has evolved into a complete pre-amplifier design, with new material on both balanced line inputs and active volume-controls. Hope you like it.

References

1. diyaudio.com http://www.diyaudio.com/forums/solid-state/209004-new-doug-self-preamp-design-22.html

2. Douglas Self *Preamplifier 2012,* Elektor April, May, June 2012

3. Douglas Self *Active Crossover Design,* Focal Press 2011

4. Douglas Self *Small Signal Audio Design,* Focal Press 2010, p60 ISBN 978–0–240–52177–0

5. Rob Reiner, Director: *This is Spinal Tap* (Film, 1984)

6. Douglas Self *Small Signal Audio Design,* Focal Press 2010, p386 ISBN 978–0–240–52177–0

Electrolytic Capacitor Distortion

While writing this article, I decided to take a more penetrating look at those electrolytic capacitor values. It might be that some of them were grossly over-sized and could be smaller and cheaper types. Electrolytic capacitor distortion is examined in Small Signal Audio Design, [4] where I reached the tentative conclusion that a criterion of less than 80 mV rms signal voltage of across the capacitor would make the THD unmeasurably low. The circuit of Figure 18.7 was simulated with the four combinations of max/min LF frequency, and maximum cut/boost, and a 1 Vrms input. The voltage across each of the capacitors was then evaluated.

As expected, C14 and C15 in the HF path never see more than a few hundred microvolts across them, and it appears that C15 in particular could be substantially reduced in value. In the end I decided not to spend the time working out how much as the cost saving is trivial, and it helps with purchasing to keep as many components as possible the same value.

In the LF path, the worst case signal voltage across C8 is 2.6 mV at 55 Hz (max LF freq, max boost) and the worst case for C11 is 6.7 mV at 57 Hz. (max LF freq, max boost) If we scale these up for a maximal 9 Vrms output at 50 Hz, (which means a 3 Vrms input) the worst case overall is 20 mV across C11, which is comfortably less than our 80 mV criterion.

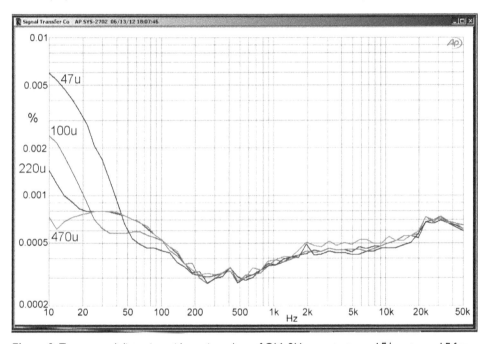

Figure A Tone-control distortion with varying values of C16. 9 Vrms output; max LF boost, max LF freq.

In the original 1996 design, the worst offender was in neither the HF or LF path; it was the 220 uF return capacitor C16, which had 75 mV across it at 10 Hz (max LF freq, max boost). This is a result of the relatively high turnover frequency of R22 and C16 which is 1.29 Hz. With a maximal 3 Vrms input the voltage across C16 scales up to 222 mV at 10 Hz, (it falls from that value as frequency increases) and we can expect added distortion at low frequencies as shown in Figure A. When we do the actual measurements we find that the distortion increase is very modest with a 220 uF capacitor, but the situation is definitely improved by increasing it to 470 uF. The results for 47 uF and 100 uF are also given as a Dreadful Warning against skimping on coupling capacitor size.

Linear Audio

Chapter 19

Optimising RIAA realisation

(Linear Audio, Volume 7, Mar 2014)

This article is the sort of thing that happens when I sit on the sofa and let my mind wander. For no special reason, I began to think of the four different ways to make an RIAA network that were dealt with in Stanley Lipshitz's famous paper [1]. An accurate RIAA network requires both precision resistors (which are relatively cheap) and precision capacitors (which are not). Given that it is hard to improve on the 5534 for noise performance with a Moving-Magnet (MM) cartridge, and it is not costly, it soon becomes clear that those precision capacitors will dominate the cost of an MM RIAA preamp stage. So the question was, for those four RIAA configurations, were there differences in how efficiently they used their capacitors? In other words, if all four were working at the same impedance, would some need more capacitance (and hence more money) to get the correct response? The answer was a resounding yes, my elation only slightly diminished by the realisation that I had been using the least efficient configuration for years. We live, and hopefully, learn.

Any accurate RIAA network will contain non-standard resistor and capacitor values. The resistors can be made up by using parallel pairs, and this not only gives a much more accurate nominal value than the nearest E24 resistor, but improves the accuracy of the combined value because random errors partially cancel. This is described in detail in the new edition of *Small Signal Audio Design* [2]. The capacitors are another story because the values available are usually much sparser than E24, often being E6.

For the resistors, I used these rules:

1. The nominal value of the combination shall not differ from the desired value by more than half the component tolerance. For 1% parts, this means within ±0.5%.

2. The resistors are to be as near-equal as possible, to get the maximum improvement possible over the tolerance of a single component. Thus two equal resistors, whether in series or in parallel, have the effective tolerance reduced by a factor of $\sqrt{2}$, while three have it reduced by $\sqrt{3}$, and so on.

3. The E24 series of preferred values will be used.

The article reproduced here gives the subject a fairly thorough examination, and you may think the topic is pretty much exhausted. Not a bit of it. The examples given in the article use two resistors in

parallel, and the relatively small number of combinations available means that the nominal value is not always as accurate as we would like; for example the 0.44% error in Table 19.3, which only just meets Rule 1 above.

This can be addressed by using three resistors in parallel. Given the cheapness of resistors, the economic penalties of using three rather than two to approach the desired value very closely are small, and the extra PCB area required is modest. However, the design process is significantly harder. Writing a bit of Javascript that explores all the feasible combinations of two-resistor in parallel is straightforward; you start with a resistor that is just above twice the target value, and put resistors from the E24 series in parallel until you have bracketed it with one result too high and the other too low. If neither answer is close enough to the target, increase the first resistor by one E24 step, then rinse and repeat. This process was used to generate all the resistor pairs in the article, and inevitably some are closer to the target than others. It does not look promising for dealing with three resistors because of the large number of combinations available.

To solve this problem, I made use of a table created by Gert Willmann, which he very kindly supplied to me. It lists all the three-resistor parallel combinations and their combined value. It covers only one decade but is naturally still a very long list, running to 30,600 entries. I applied it first to Figure 19.9 in the article. (+30dB gain, C1 set to exactly 33nF). The two-resistor values are shown in Table 19.3.

I started with R0, which has a desired value of 211.74Ω. The Willmann table was read into a text editor, and using the search function to find '211.74' takes us straight to an entry at line 9763 for 211.7439674Ω, made up of 270Ω, 1100R, and 9100Ω in parallel. This is more than accurate enough—in fact it might be wise to increase the precision of the input value of 211.74Ω. However, since the resistor values are a long way from equal, there will be little improvement in accuracy; the effective tolerance calculates as 0.808%, which is not much improvement over 1%.

Looking up and down the Willman table by hand, so to speak, some better combinations were spotted that were more equal than others. For example. 390Ω 560Ω 2700Ω at line 9774 has a nominal value only 0.012% in error, while the tolerance is improved to 0.667%, and this is clearly a better answer. Scanning the table by eye is a bit of a tedious business, but it can be speeded up. The best result for improving the tolerance would be three equal resistors, but 620Ω 620Ω 620Ω has a nominal value 2.4 % too low, and 680Ω 680Ω 680Ω has a nominal value more than 7% too high, so obviously they are no good. But this does suggest that the first resistor should be either 560Ω or 620Ω, and armed with this clue searching by eye is faster. I found the best result for R0 is 560Ω 680Ω 680Ω at line 9754, which has a nominal value only −0.09% in error and an effective tolerance of 0.580%, which is very close to the best possible 0.577% (1/√3).

In contrast, the original two-resistor solution for R0 has a nominal value −0.33% in error and an effective tolerance of 0.718%.

Repeating the process for R1 (66.18kΩ desired) and searching for '661.8' takes us to 1000Ω 2000Ω 91000Ω. Obviously with resistors outside the range 100Ω–999Ω, we need to scale by

9750	211.47068247	<E12>220	10000	12000
9751	211.50208097	<E24>430	430	13000
9752	211.52542373	<E24>300	1300	1600
9753	211.53846154	<E24>220	11000	11000
9754	**211.55555556**	**<E12>560**	**680**	**680**
9755	211.57127346	<E24>330	620	12000
9756	211.59062885	<E24>240	3300	3900
9757	211.63593539	<E24>390	470	30000
9758	211.64995936	<E24>360	560	6200
9759	211.6741501	<E24>220	7500	22000
9760	211.6838488	<E12>220	5600	-
9761	211.68437026	<E24>360	620	3000
9762	211.73156673	<E12>220	8200	18000
9763	**211.74396741**	**<E24>270**	**1100**	**9100**
9764	211.7577373	<E24>220	10000	13000
9765	211.76470588	<E24>240	1800	-
9766	211.76470588	<E24>240	2000	18000
9767	211.76470588	<E24>240	3600	3600
9768	211.76470588	<E24>300	750	18000
9769	211.76470588	<E24>300	1200	1800
9770	211.77174906	<E12>390	470	33000
9771	211.77416398	<E24>220	9100	15000
9772	211.81001284	<E24>240	2200	10000
9773	211.82108626	<E24>390	510	5100
9774	**211.85600345**	**<E12>390**	**560**	**2700**
9775	211.85860163	<E24>330	620	13000

Preface Figure 19.1 Part of the three-resistor Willmann table. This is the area around 211.7Ω

factors of 10; in this case by 100 times. This result has a very accurate nominal value but once more the tolerance is not brilliant at 0.74%. We need to cast our net wider. 660Ω x 3 is 1980Ω, so we look for 1800Ω as the first resistor, and we get 1800Ω 2000Ω 2200Ω. Scaling that up gives us 180kΩ 200kΩ 220kΩ. The nominal value is only 0.061% low, and the tolerance is 0.579%. It is hard to see how the latter value in particular could be much bettered.

Repeating again for R2 (9432Ω desired), I got 22kΩ 33kΩ 33kΩ. The nominal value is only 0.036% low, and the tolerance is 0.589%.

The resistor R3 in the HF correction pole is less critical than the other components, as it only gives a minor tweak at the top of the audio band, but I applied the three-resistor process to it anyway. The desired value is 1089.3Ω. There are no suitable three-resistor combinations starting with 3000Ω, and the best I found was 2700Ω 2700Ω 5600Ω which has a nominal value error of −0.13% and a tolerance of 0.602%.

The final result is shown in Preface Table 19.1, which is Table 19.3 from the article rewritten with paralleled resistor triples; the errors in the nominal value column are now much smaller by factors between 12 and 1.2. It is an interesting question as to what the average improvement factor over a large number of two-resistor to three-resistor changes would be; I suspect (but am completely incapable of proving mathematically) that it's going to be in the area of 3 to 4. The effective tolerances have been added as an extra column at the right, and you can see that all of them are

Preface Table 19.1 Table 19.3 in the article redone using paralleled resistor triples.

Component	Desired Value	Actual Value	Parallel Part A	Parallel Part B	Parallel Part C	Nominal Error	Tolerance
R0	211.74 Ω	211.03 Ω	560 Ω	680 Ω	680 Ω	−0.087 %	0.580%
R1	66.18 kΩ	65.982 kΩ	180 kΩ	200 kΩ	220 kΩ	+0.062 %	0.579%
C1	33 nF	33 nF	33nF	–	–	0 %	
R2	9.432 kΩ	9.474 kΩ	22 kΩ	33 kΩ	33 kΩ	−0.036 %	0.589%
C2	11.612 nF	11.60 nF	4n7	4n7	2n2	−0.10 %	
R3	1089.2 Ω	1090.9Ω	2.7 kΩ	2.7 kΩ	5.6 kΩ	−0.13 %	0.602%

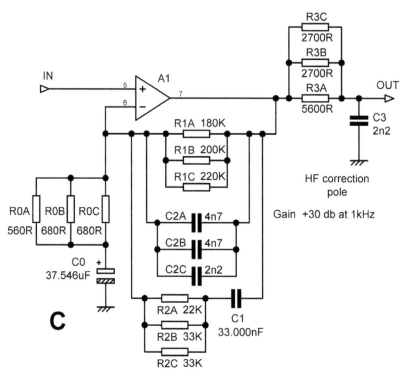

Preface Figure 19.2 The RIAA preamplifier in Preface Figure 19.9 of the article, with C1 = 33nF, redesigned with three parallel resistors to approach each non-standard value more closely, and have a smaller tolerance

quite close to the best possible value of 0.577% (1/√3). Preface Figure 19.2 shows the resulting schematic when Figure 19.9 in the article is redrawn with the new values. There is only one E24 resistor out of twelve, and all the others are E12. This is purely happenstance; no effort whatever was made to avoid E24 values. There is very little point in doing so unless you are using exotic parts that only exist as E12.

I think it's pretty clear that using three resistors instead of two gives much more accurate nominal values, and at the same time, a usefully smaller tolerance that almost halves the tolerance errors

compared with a single resistor. Frequently, there are several suitable combinations, and you can choose between a more accurate nominal value or a smaller tolerance percentage.

The obvious question (to me, anyway) is: would four resistors be better? Not really. There is no point in having super-accurate nominal values if you are starting off with 1% parts. The tolerance is now halved, at best, but the improvement depends on the square-root of the number of resistors so we are heading into diminishing returns. If you want a better tolerance than three 1% resistors can give, the obvious step is to go to 0.1% resistors which are freely available, though they cost ten times as much as the 1% parts. Anything more accurate than that would be a specialised and very expensive item, and not obtainable from the usual component distributors.

I have applied the same process to the 3 × 10 nF design in Figure 19.12 of the article, and the result is shown in Preface Figure 19.3. There are now four E24 values out of twelve. The total value of C2 has been made more accurate (now −0.066%) by correcting the value of C2D; this has nothing to do with the three-resistor process.

I also applied the process to the 4 × 10 nF design in Figure 19.14 of the article, and the result is shown in Preface Figure 19.4. There are now three E24 values out of twelve; I am starting

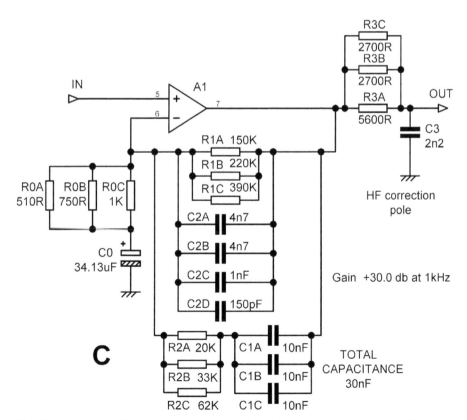

Preface Figure 19.3 The RIAA preamplifier in Figure 19.12 of the article, with C1 = 3 × 10nF, redesigned with three parallel resistors

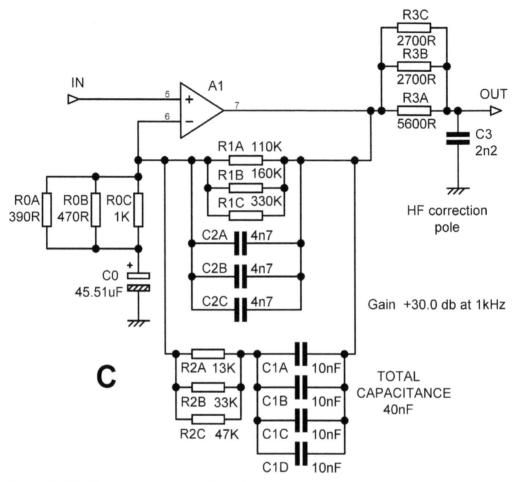

Preface Figure 19.4 The RIAA preamplifier in Figure 19.14 of the article, with C1 = 4 × 10nF, redesigned with three parallel resistors

to wonder if there is some mathematical property that means that E24 values are always in the minority. It seems unlikely, but if anyone with mathematical skills would like to tackle the question, the answer might be enlightening.

Errata

This article is the only one which has not been reproduced here exactly as it was published, because unfortunately, a few numerical errors crept in that could cause confusion; they have been corrected here. My thanks to Gert Willmann for pointing these out. In the original article the errors were:

- In Figure 19.1 the correct value of T2 is 7950us, (not 7960us as in the original article) the correct value of f5 is 2122 Hz (not 2112 Hz as in the original), and the correct value of T6 is 1.35us (not 0.075us as in the original).

- In Table 19.4, the desired value for R1 should be 72.583kΩ, and the actual value 72.639kΩ. C2 should be 10.52nF instead of 11.60nF. The +0.34% error given in the final column is correct.

- In Figure 19.11, R1 should be 72.639kΩ.

- In Table 19.5 the error for R1 should be +0.19%. Also the desired value for R2 should be 7.782kΩ, as shown correctly in Figure 19.13; the actual value and the % error in Table 19.5 are also correct.

References

1. Lipshitz, S.P., On RIAA Equalisation Networks, *J. Audio Eng Soc,* June 1979, p. 458 onwards.

2. Self, D, *Small-Signal Audio Design,* 2nd edition, Chapter 2, pp. 48–56. Focal Press (Taylor & Francis) 2014. ISBN: 978-0-415-70974-3 Hardback. 978-0-415-70973-6 Paperback. 978-0-315-88537-7 ebook.

OPTIMISING RIAA REALISATION

RIAA equalisation is associated with moving-magnet (MM) pickups, but is no less necessary for moving-coil (MC) pickups. A very common arrangement is to have an MM input stage, with a suitable 47 kΩ input impedance, which can be switched between a directly-connected MM cartridge, and an MC cartridge followed by a flat low-noise gain stage. The noise conditions for the MM stage are very different in the two cases, but in either RIAA equalisation is required. This can be done in many ways, but some approaches can be discarded at once. Shunt-feedback RIAA stages for MM pickups are some 14 dB noisier than an equivalent series-feedback stage [1]; this is inherent because of the high inductance of MM cartridges, so we can lose that approach at once. Other methods split the various RIAA time-constant across separate stages, in various mixtures of passive and active circuits; this makes the calculations easy but uses more hardware and invariably involves compromises in headroom and noise performance.

If that sort of thing is the "brawn" hardware-heavy option, the "brains" option is to do all the RIAA equalisation in one stage, using series feedback. The "brains" part comes in because the interaction of the RIAA time-constants in a single feedback path makes the calculations for accurate RIAA much more difficult. Most of the hard work was done for us by Stanley Lipshitz, who in 1979 published comprehensive design equations for a large number of RIAA network configurations [2]. He deserves the thanks of us all. Nevertheless, these equations are pretty long and complicated, and adapting them to spreadsheet or MathCAD use is not a trivial business, so it is wise to assess which configurations are worth the effort before you undertake it.

At this point I am going to firmly state that the best way to implement RIAA equalisation is the conventional one-stage series-feedback method, using a gain of between +30 and +35 dB (at 1 kHz). Without going into all the detail of recorded levels and cartridge sensitivities, those gains will give more than adequate maximum input levels of 316 mVrms and 178 mVrms respectively. The output levels will be 158 and 281 mVrms respectively with a 5 mVrms input (1 kHz). The input stage will normally be followed by some form of controlled gain to get the signals up to the desired final level.

In this article RIAA accuracy is often worked out to a hundredth of a dB. That does not of course mean that such accuracy is essential for good listening, which is just as well as normal component tolerances would make it quite impossible. Instead the aim is to get the *nominal* results spot-on, so that whatever tolerances may do to us, we at least know we are aiming at the right values, and the practical results will be centred on them.

Equalisation and its discontents: The RIAA characteristic

The RIAA characteristic is probably the most complex standard equalisation curve in common use. Figure 19.1 shows the response asymptotes of its main features. The most important three corners in its response curve are f3 at 50.05 Hz, f4 at 500.5 Hz, and f5 at 2.122 kHz, which

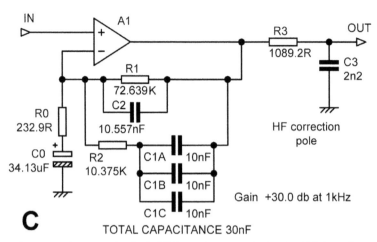

Figure 19.1 The practical response for series-feedback RIAA equalisation, including the IEC Amendment which gives an extra roll-off at f2 (20.02 Hz).

are set by three time-constants of 3180 μ sec, 318 μ sec, and 75 μsec. To put it concisely, the f3—f4 region allows for the constant-amplitude recording used over that range, while the plateau at f4—f5 allows for the high-frequency recording pre-emphasis which reduces noise by de-emphasis at playback.

In Figure 19.1 there is a further response corner at f2 (20.02 Hz), corresponding to a time-constant of 7950 μs. This LF roll-off is called the "IEC Amendment" because it was added to the RIAA standard in 1976. It was intended to reduce the subsonic output, but its introduction is slightly mysterious. It was certainly not requested or wanted by either equipment manufacturers or their customers. Some manufacturers refused to implement it or just ignored it. It still attracts trenchant criticism today. The likeliest explanation for its introduction seems to be that various noise reduction systems (such as dbx), were then being promoted for use with vinyl and their operation was interfered with by subsonic disturbances (none of the systems caught on). In some preamps the IEC Amendment is switchable in/out.

The response in Figure 19.1 flattens out at f6, when the gain demanded by the RIAA network falls to unity, whereas to achieve the RIAA standard it ought to continue to drop at 6 dB/octave. This unwanted response corner is inherent in the series-feedback configuration, and is its only drawback. The gain in Figure 19.1 is +35 dB at 1 kHz, which puts f6 at 118 kHz, and this causes the gain to rise above the RIAA characteristic around 10–20 kHz; here it gives an excess gain of only 0.10 dB at 20 kHz, but the effect is more severe at lower gains because f6 occurs at lower frequencies. If the gain is

reduced to +30 dB at 1 kHz to get more overload margin, the final zero f6 is now at only 66.4 kHz, introducing an excess gain at 20 kHz of 0.38 dB. This problem is completely cured by adding a lowpass time-constant after the stage; this is usually called an "HF correction pole" which means precisely the same thing. If the correct time-constant is used the gain and phase are corrected exactly.

The frequency f1 simply shows where the gain flattens out at the low-frequency end as once again it cannot fall below unity; it is usually of little or no importance.

A basic RIAA equalisation stage

Figure 19.2 shows a single-stage series-feedback MM preamp with a gain of +30 dB (1 kHz). It gives 158 mV out for 5 mV in (1 kHz), with an input overload level of 316 mV rms. This means an excellent overload margin of 36 dB, though more (variable) gain will be required later in most cases.

There are several different ways to arrange the resistors and capacitors in an RIAA network, all of which give identically exact equalisation when the correct component values are used. Of the four series RIAA configurations examined by Lipshitz, this is configuration A. It has the advantage that it is the simplest case when the Lipshitz equations are put into practice. Details that are essential for practical use, like input DC-blocking capacitors, DC drain resistors, and EMC/cartridge-loading capacitors have been omitted for clarity; in later diagrams the 47 kΩ input loading resistor is also omitted. The overall

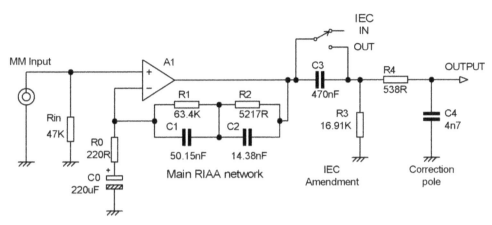

Figure 19.2 Series-feedback RIAA equalisation in configuration A, designed for 30.0 dB gain (1 kHz) which allows a maximum input of 316 mV rms (1 kHz). The switchable IEC Amendment is implemented by C3, R3. HF correction pole R4, C4 is added to keep RIAA accuracy within ± 0.1 dB 20 Hz to 20 kHz.

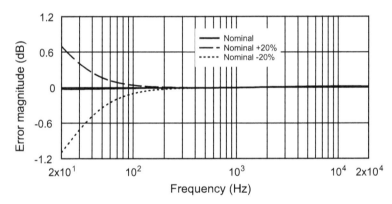

Figure 19.3 The effect of a ±20% tolerance for C0 when it is used to implement the IEC amendment in an amplifier; R0 = 220 Ω, gain +30 dB (1 kHz).

impedance of the RIAA network has been kept reasonably low by making R0 220 Ω; further reduction has virtually no effect on the noise performance and results in large and expensive RIAA capacitors. No attempt is made at this point to deal with the awkward values of R1 and R2. The notation R0, C0, R1, C1, R2, C2 is as used by Lipshitz; C1 is always the larger of the two.

The unloved and unwanted IEC Amendment was no doubt intended to be implemented by restricting the value of C0; in this case the correct value of C0 would be 36.18 uF. This however leaves us wide open to significant errors due to the tolerance of an electrolytic capacitor. Figure 19.3 shows that the gain will be +0.7 dB up at 20 Hz for a +20% C0, and -1.1 dB down at 20 Hz for a -20% C0. The possible effect of

C0 is less than ±0.1 dB above 100 Hz, but this is obviously not a great way to make dependably accurate RIAA networks. It also requires non-standard values for either R0 or C0.

Instead C0 is increased to 220 uF so that its -3 dB roll-off does not occur until 3.29 Hz. Even this wide frequency spacing introduces an unwanted 0.128 dB loss at 20 Hz (assuming no IEC amendment is used), and perfectionists will want to use 470 uF here, which reduces the error to 0.06 dB. The IEC amendment is now implemented passively after the amplifier stage, by non-electrolytic capacitor C3 (which can have a tolerance of ±1 %) in conjunction with R3, giving the required -3 dB roll-off at f2 (20.02 Hz). It is easier to make it switchable in/out. In Figure 19.2 we use a standard E3 capacitor value for C3; 470 nF was chosen, and as usual an unhelpful resistor value results; in this case R3 = 16.91 kΩ. Using E24 resistors, this can be implemented exactly as 16 kΩ + 910 Ω, though it has to be said near-equal series or parallel resistor pairs would give more accuracy for a given tolerance; more on that later. The switch as shown may not be entirely click-free because of the offset voltage at A1 output, but that is not important as it will probably only be operated a few times in the life of the preamp, if indeed ever.

We made C0 220 uF, which will be a handily compact component. Because it is not infinitely large there is, as we saw, an error of-0.128 dB at 20 Hz. It is possible to compensate for this by tweaking the IEC amendment frequency f2. If R3 is changed from 16.91 kΩ to 17.4 kΩ the overall response is made accurate to ±0.005 dB. The compensation is not mathematically exact- there is a + 0.005 dB hump around 20 Hz- but I suggest it is good enough for most of us. This process gets the nominal response right but does not of course do anything to reduce the effects of the tolerance of C0; these are however small, around + 0.05 dB.

Since the gain is +30 dB (1 kHz) f6 is at 66.4 kHz and an HF correction pole is essential. This is implemented by R4, C4 in Figure 19.2. There is no significant interaction with the IEC amendment, so we have complete freedom in choosing C4 and use a standard E3 value and then get the pole frequency exactly right by using two resistors in series for R4-470 Ω and 68 Ω. Since these components are only doing a little fine tuning at the top of the frequency range, the tolerance requirements are relaxed compared with the main RIAA network. The design considerations are a) that the resistive section R4 should be as low as possible in value to minimise Johnson noise, and on the other hand b) that the shunt capacitor C4 should not so large as to load the opamp output excessively at 20 kHz. It is assumed that an opamp such as the 5534 with good load-handling capability is used.

The IEC network is placed before the HF correction pole, as in Figure 19.2, so that R4 is not loaded by R3, which would cause a 0.3 dB loss, compromising noise or headroom slightly. As it is C3 is loaded by C4, but the loss is much smaller at 0.09 dB. It is assumed there is no significant external loading on the output at C4. Often the RIAA stage will be feeding the high-impedance input of a non-inverting gain stage, but if not buffering will be required so the two passive output networks behave as designed.

There is our basic RIAA stage. It works very well in both practice and theory, but is it the best solution possible? Here we set out on our voyage of optimisation.

Series-feedback RIAA-network configurations

Four possible configurations are described by Lipshitz in his classic paper.[2] These, using his component values, are shown in Figure 19.4; I have used the same identifying letters as Lipshitz. They are all accurate to within ±0.1 dB when implemented with a 5534 opamp, but in the case of configuration A the error is getting close to-0.1 dB at 20 Hz due to the relatively high closed-loop gain (+46.4 dB at 1 kHz) and the finite open-loop gain of the 5534. All have their RIAA networks at a relatively high impedance, and have relatively high gain and therefore a low maximum input. In each case the IEC amendment is implemented by the value of C0.

Burkhard Vogel in his weighty book *The Sound of Silence* describes configuration A as a Type-Eub network, configuration B a Fub-B network, and configuration C a Fub-A network [3]. He does not deal with configuration D.

Long ago I wrote a spreadsheet to implement the Lipshitz equations for configuration A simply because it was the easiest one to do. Repeating that for the remaining three configurations would be a fair amount of work, so the question arises, do any of the other three configurations have advantages over A that would make that work worthwhile?

Four things to examine suggested themselves:

1. Each configuration in Figure 19.4 contains two capacitors, a large C1 and a small C2 that set the RIAA response. If they are close-tolerance (to get accurate RIAA) and non-polyester (to prevent capacitor distortion) then they will be expensive, so if there is a configuration that makes the large capacitor smaller, even if it is at the expense of making the small capacitor bigger, it is well worth pursuing. That large capacitor is almost certainly the most expensive component in the RIAA MM amplifier by a long way.

2. I started by assuming that the signal voltages across each capacitor would be different. If polyester capacitors must be used for cost reasons, then if there is a configuration that puts less voltage across a capacitor then that capacitor will generate less distortion. Capacitor distortion at least triples, and may quadruple, as the voltage across it doubles, so choosing the configuration that minimises the voltage might be worthwhile. However, a good deal of simulation tells us that there is actually little difference between the four configurations as regards the magnitude of the signal voltage across the capacitors. Not a helpful result or an interesting result, but sometimes things just are that way.

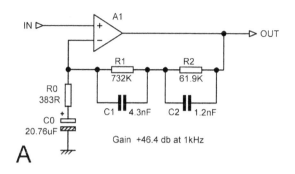

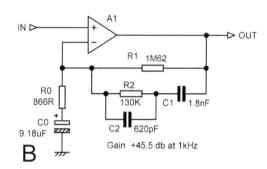

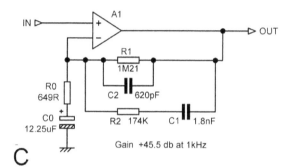

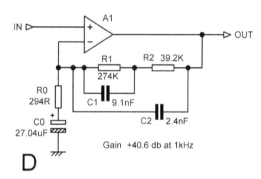

Figure 19.4 The four RIAA feedback configurations in the Lipshitz paper.

3. If we assume a certain amount of non-linearity in one or both of the RIAA capacitors, are the configurations different in their sensitivity to that non-linearity? In other words, how much distortion will appear at the output? This could be resolved in SPICE simulation, by using non-linear capacitor models constructed with Analog Behavioural Models, but it would be a lot of work, and since the emphasis of this article is on high quality, where we can presumably afford a polypropylene capacitor or two, I have put that one on the back-burner, though I hope it will not fall completely off the back of the cooker.

4. My implementation of the Lipshitz equations starts off with the value of R0. Combined with the desired gain, it sets the impedance level of the whole RIAA network. A low impedance level reduces noise, but increases the size and cost of the precision capacitors C1 and C2. So just how low should R0 be for the best performance? This issue is examined at the end of this article.

That leaves us with 1) and 4), and in these cases we *do* get some useful results. Now read on . . .

The RIAA configurations compared for capacitor cost

We need to put the four configurations A, B, C, and D into a common form where they can be directly compared. In Figure 19.4 they are working at different impedance levels, as shown by the differing values of R0, and at different gains. To make the impedance levels the same we scale all the RIAA component values to make R0 exactly 200 Ω, as in Figure 19.5. C0 is then 39.75 uF in each case, and implements the IEC amendment. The impedance scaling does not affect the gain or the RIAA accuracy; this was checked by simulation for each configuration. The new values for C1 and C2 are shown in Table 19.1.

Since the gains of configurations A, B, and C are nearly the same, we can roughly compare the values for C1, and it's already starting to look as if A might in fact be the worst case for capacitor size. To be quite sure about this we can alter the gain of A to be exactly the same as B and C at +45.5 dB, which is easy as we already have the spreadsheet for it. We don't have the tool for D. Configuration D has significantly bigger

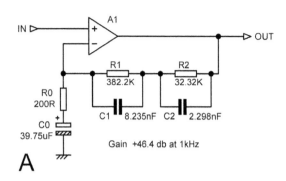

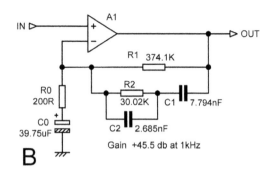

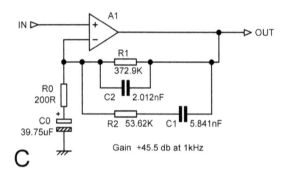

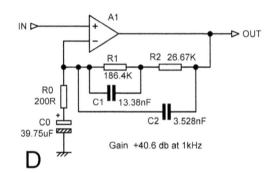

Figure 19.5 The four RIAA feedback configurations, with component values scaled so that R0 = 200 Ω in each case.

Table 19.1 The values of C1 and C2 in Figure 19.5, with networks scaled so R0 = 200 Ω in each case.

Configuration	Gain at 1 kHz	Large cap C1	Small cap C2	C1/C2 ratio
A	+46.4 dB	8.235 nF	2.298 nF	3.583
B	+45.5 dB	7.794 nF	2.685 nF	2.903
C	+45.5 dB	5.841 nF	2.012 nF	2.903
D	+40.6 dB	13.38 nF	3.528 nF	3.791

capacitors than A, B, and C because it has about half the gain but the same value of R0.

We can change the gain of D by simply by scaling the RIAA network, with R0 kept constant. This will not be very accurate but should be good enough for us to compare the configurations. We have to increase the gain of D by 4.948 dB, or 1.767 times, so we divide capacitors C1 and C2 by this factor, and multiply resistors R1 and R2 by it. This gives the values shown in Figure 19.6 and Table 19.2. The C1/C2 ratios are naturally unchanged.

After this gain alteration, configuration D is definitely less accurate than before, but is still almost within a ±0.1 dB error band, only exceeding this (and not by much) over small frequency ranges. This is good enough to allow proper assessment of each configuration.

From Figure 19.6 and Table 19.2, we can see that C1 in configuration C is only 63% the size of C1 in configuration A. I expected this, but was afraid it might be accompanied by an increase in C2 in configuration C, but this is also smaller at 78% of C2 in A. I have to ruefully admit that configuration A (which I have been using for years) makes the least efficient use of its capacitors, since they are essentially in series, reducing the effective value of both of them. Configurations B and D have intermediate values for C1, but of these two D has a significantly smaller C2.

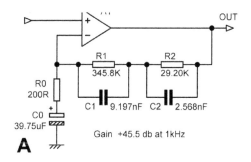

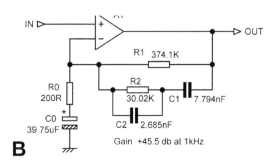

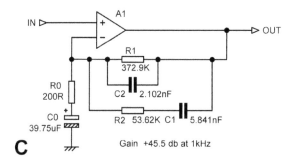

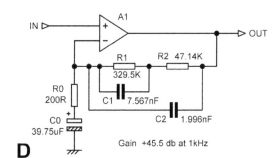

Figure 19.6 The RIAA feedback configurations, with component values scaled so that R0 = 200 Ω and the gain is +45.5 dB at 1 kHz in each case. Note that the RIAA response of D is here not completely accurate. The high gain means that the maximum input in each case is only 53 mV rms (1 kHz) which is not in general adequate.

Table 19.2 The values of C1 and C2 in Figure 19.6, after scaling so R0 = 200 Ω and gain =+45.5 dB (1 kHz) for all four configurations

Configuration	Gain at 1 kHz	Large cap C1	Small cap C2	C1/C2 ratio
A	+45.5 dB	9.197 nF	2.568 nF	3.581
B	+45.5 dB	7.794 nF	2.685 nF	2.903
C	+45.5 dB	5.841 nF	2.012 nF	2.903
D	+45.5 dB	7.567 nF	1.996 nF	3.791

Therefore configuration C would appear to be the optimal solution in terms of capacitor size, and hence cost and PCB area used. B and D appear to have no special advantages. It was now time to put together a software tool for the Lipshitz equations of C, so it could be designed accurately for gains other than +45.5 dB (1 kHz). As anticipated it was somewhat more difficult than it had been for configuration A, but no big deal.

We have already noted that a gain as high as +45.5 dB (1 kHz) gives an unacceptably poor overload margin. It has only used here so far because it was the gain adopted

in the Lipshitz paper. If we assume an opamp can provide 10 Vrms out, then the maximum input at 1 kHz is only 53 mV rms. The gain of an MM input stage should not, in my opinion, much exceed 30 dB (1 kHz) See the earlier example in Figure 19.2.

My Precision Preamplifier design [4] has an MM stage gain of +29 dB (1 kHz), permitting a maximum input of 354 mV rms (1 kHz). The more recent Elektor Preamplifier 2012 [5] has an MM stage gain of +30 dB (1 kHz), allowing a maximum input of 316 mV rms; it is followed by a flat switched-gain stage which allows for

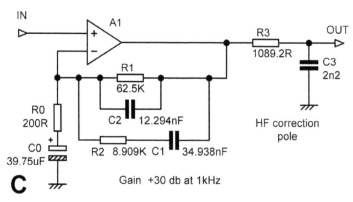

Figure 19.7 Configuration C with values calculated from the Lipshitz equations to give +30.0 dB gain at 1 kHz, and an accurate RIAA response within ±0.01 dB; the lower gain now requires HF correction pole R3, C3 to maintain accuracy at the top of the audio band.

the large range in MC cartridge sensitivity. Both designs use configuration A.

A preamp with a gain of exactly +30 dB (1 kHz) using configuration C was therefore designed with the new software tool, and there it is in Figure 19.7. It has an RIAA response, including the IEC amendment that is accurate to within ±0.01 dB from 20 Hz to 20 kHz (assuming C0 is exactly correct in value). The relatively low gain means that an HF correction pole is required to maintain accuracy at the top of the audio band, and this is implemented by R3 and C3. If it was omitted the response would be 0.1 dB high at 10 kHz, and 0.37 dB high at 20 kHz. We use the E6 value of 2n2 for capacitor C3, so R3 is a non-preferred value.

In Figure 19.7, and in the examples that follow, I have implemented the IEC amendment by using the appropriate value for C0, rather than by adding an extra time-constant after the amplifier as in Figure 19.2. We have already seen that using C0 to do this is not the best method, but I have stuck with it here as it is instructive how the correct value of C0 changes as other alterations are made to the RIAA network.

Making C1 a single E6 capacitor

So far we have optimised our RIAA network by switching to configuration C. A further stage of refinement is possible after minimising C1 and C2. There's nothing special about the value of R0 at 200 Ω (apart from the bare fact that it's an E24 value), it just needs to be suitably low for a good noise performance. We can therefore scale the RIAA components again, not to make R0 a specific value, but to make at least one of the capacitors a preferred value, the larger one being the

obvious candidate. Compared with the possible savings on expensive capacitors here, the cost of making up a non-preferred value for R0 is negligible.

In Figure 19.7, C1, at 34.9 nF, is suggestively close to 33nF. If we twiddle the new software tool for configuration C so that C1 is exactly 33 nF, we get the circuit in Figure 19.8. R0 has only increased by 6% to 212 Ω, and so the effect on noise performance will be negligible; I calculate it as 0.02 dB worse (with MM cartridge input load). All the other values in the RIAA feedback network have likewise altered by about 6%, including C0, but the HF correction pole is unchanged; that would only need to be altered if we altered the gain of the stage. Like Figure 19.7, the RIAA accuracy of this circuit is well within ±0.01 dB from 20 Hz to 20 kHz when a 5534 is used.

The circuit of Figure 19.8 now has two preferred-value capacitors, C1 and C3, but that is the best we can do. All the other values are, as expected, thoroughly awkward. In my book *Active Crossover Design* [6] I describe how to make up arbitrary resistor values by paralleling two or more resistors, and how the tolerance errors partly cancel. This also applies to capacitors. The optimal way to do this is with components of as near-equal values as you can manage. If the values are exactly equal, the combined accuracy is √2 time better than the individual components; the further away from equality, the less the improvement. Nothing more esoteric than the E24 series is assumed for the resistors. The optimal parallel pairs were selected in a jiffy using a specially written software tool. The parallel combination for C2, using E6 values, was done by manual bodging, though as you can see in Table 19.3, we have been rather lucky with how the values work out, with three parallel capacitors getting us very close to the desired value. Figure 19.9 shows the resulting circuit.

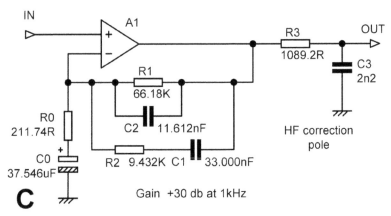

Figure 19.8 Configuration C from Figure 19.7 with Ro tweaked to make C1 exactly the E6 preferred value of 33.000 nF. Gain is still 30.0 dB at 1 kHz, and RIAA accuracy within ±0.01 dB. The HF correction pole R3, C3 is unchanged.

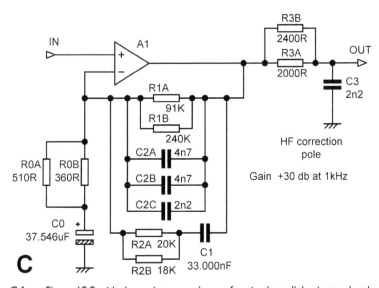

Figure 19.9 Configuration C from Figure 19.8 with the resistors made up of optimal parallel pairs to closely approach the correct value. C2 is made up of three parts. Gain +30.05 dB at 1 kHz, RIAA accuracy is worsened but still within ±0.048 dB.

Table 19.3 Approximating the exact values of Figure 19.8 with paralleled components, giving Figure 19.9.

Component	Desired value	Actual value	Parallel part A	Parallel part B	Parallel part C	Error
R0	211.74 Ω	211.03 Ω	360 Ω	510 Ω	–	−0.33%
R1	66.18 kΩ	65.982 kΩ	91 k Ω	240 k Ω	–	−0.30%
C1	33 nF	33 nF	33nF	–	–	0%
R2	9.432 kΩ	9.474 kΩ	18 kΩ	20 kΩ	–	0.44%
C2	11.612 nF	11.60 nF	4n7	4n7	2n2	−0.10%
R3	1089.2 Ω	1090.9Ω	2 kΩ	2.4 kΩ	–	0.16%

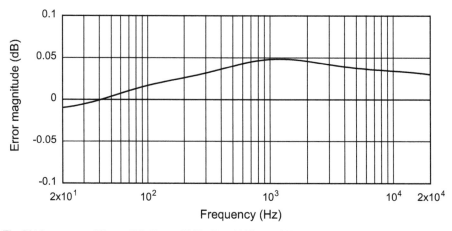

Figure 19.10 The RIAA accuracy of Figure 19.9. Gain is 30.05 dB at 1 kHz, and RIAA error reaches a maximum of +0.048 dB midband.

When selecting the parallel resistor pairs, the criterion I used was that the error in the nominal value should be less than half of the component tolerance, assumed to be ±1 %. R2 only just squeaks in at +0.44 %, but its near-equal values give almost all of the √2 improvement. Don't forget that in Table 19.3 we are dealing with *nominal* values, and the % error in the nominal value shown in the rightmost column has nothing to do with the component tolerances.

Naturally the small errors in nominal value seen in the rightmost column of Table 19.3 have their effect. Figure 19.10 shows that the RIAA response now has a gentle peak of +0.048 dB at 1 kHz, which is by coincidence the worst-case frequency. Above that, the error slowly declines to +0.031 dB at 20 kHz. Most of this deviation is caused by the +0.44 % error in the nominal value of R2 (see Table 19.3).

C1 made up of multiple 10nF capacitors

We have just scaled the RIAA network so that the major capacitor C1 is a single preferred value; at 33nF it will probably be a rather expensive 1% polypropylene. RIAA optimisation can however be tackled in another way, depending on component availability. In many polystyrene capacitor ranges 10 nF is the highest value that can be obtained with a tolerance of 1%. Paralleling several of these is a good deal more cost-effective than a single 1% polypropylene part. To exploit this we need to redesign the circuit of Figure 19.8 so that C1 is either exactly 30 nF or exactly 40 nF. Figure 19.11 shows the result for C1 = 30 nF. Ro has now increased to 233 Ω, which I calculate will degrade the noise by only 0.06 dB compared with R0 = 200 Ω (with MM cartridge input load).

The 40 nF version costs a bit more but gives a total capacitance that is twice as accurate as one capacitor (as √4 = 2), while the 30 nF version only improves accuracy by √3 (= 1.73) times. Figure 19.13 shows the result for C1 = 40 nF. R0 has now dropped to 175 Ω, which gives a calculated noise improvement of 0.04 dB. Not to belabour the point, but R0 has only a very minor effect on the noise-we will come back to that later. C0 is now larger at 45.51 uF.

Since in each case the gain is unchanged the values for the HF correction pole R3, C3 are also unchanged.

As for the previous example with C1 = 33nF, the awkward resistor values are made up with optimally-selected parallel pairs. The results of this process for C1 = 30nF are shown in Table 19.4 and Figure 19.12. In this case we have been unlucky with the value of C2, which needs to be trimmed with a 120 pF capacitor to meet the criterion that the error in the nominal value will not exceed half the component tolerance.

The same process can be applied to the 4 × 10 nF version in Figure 19.13, giving the component values in Table 19.5 and in Figure 19.14.

This time we are much luckier with the value of C2; three 4n7 capacitors in parallel give almost exactly the required value. On the other hand we are very unlucky with R0, where 180 Ω in parallel with 6.2 kΩ is the most "equal" solution falling within the error criterion.

Both my Precision Preamplifier '96 [4] and the more recent Elektor Preamplifier 2012 [5] have MM stage gains close to +30 dB (1 kHz) like the examples above, but both use configuration A, and five paralleled 10 nF capacitors are required for C1. A third existing design using configuration A was modified to configuration C as in Figure 19.12, and thoroughly tested with an AP SYS-2702. The RIAA was exactly correct within the

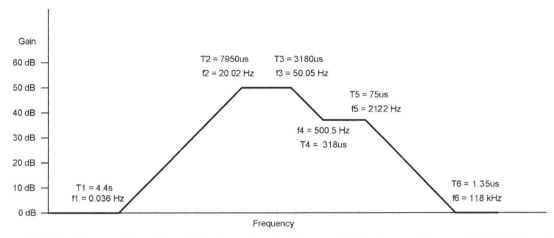

Figure 19.11 Configuration C from Figure 19.8 redesigned so that C1 is now 30 nF, made up with three paralleled 10 nF capacitors. Gain +30.0 dB at 1 kHz, RIAA accuracy is within ±0.01 dB.

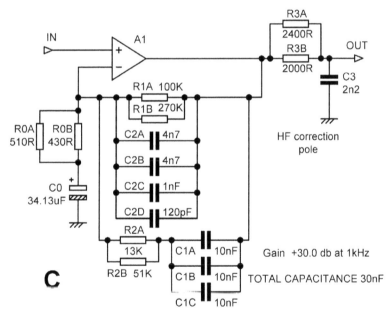

Figure 19.12 Configuration C from Figure 19.11 with resistors made up of optimal parallel pairs, and C2 made up of four parts. Gain +30.0 dB at 1 kHz, RIAA accuracy is within ±0.01 dB.

Table 19.4 Approximation to the exact values in Figure 19.11 by using parallel components, giving Figure 19.12.

Component	Desired value	Actual value	Parallel part A	Parallel part B	Parallel part C	Error
R0	232.9 Ω	233.3 Ω	430 Ω	510 Ω	–	0.17%
R1	72.80 kΩ	72.97 kΩ	100 k Ω	270 k Ω	–	0.24%
C1	30 nF	30 nF	10 nF	10 nF	10 nF	0%
R2	10.375kΩ	10.359 kΩ	13 kΩ	51 kΩ	–	−0.15%
C2	10.557 nF	11.60 nF	4n7	4n7	1nF + 120 pF	0.34%
R3	1089.2 Ω	1090.9Ω	2 kΩ	2.4 kΩ	–	0.16%

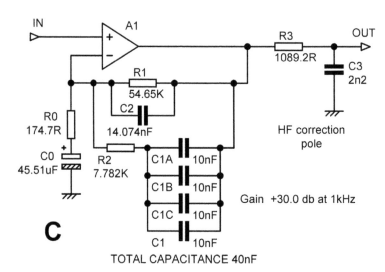

Figure 19.13 Configuration C from Figure 19.8 redesigned so that C1 is 40 nF, made up with four paralleled 10 nF capacitors. Gain +30.0 dB at 1 kHz, RIAA accuracy is within ±0.01 dB.

Table 19.5 Approximation to the exact values in Figure 19.13 by using parallel components, giving Figure 19.14.

Component	Desired value	Actual value	Parallel part A	Parallel part B	Parallel part C	Parallel part D	Error
R0	174.7 Ω	174.9 Ω	180 Ω	6.2 kΩ	–	–	0.10%
R1	54.44 kΩ	54.54 kΩ	100 kΩ	120 kΩ	–	–	0.41%
C1	40 nF	40 nF	10 nF	10 nF	10 nF	10 nF	0%
R2	7.821kΩ	7.765 kΩ	12 kΩ	22 kΩ	–	–	−0.22%
C2	14.074 nF	14.1 nF	4n7	4n7	4n7	–	0.18%
R3	1089.2 Ω	1090.9Ω	2 kΩ	2.4 kΩ	–	–	0.16%

limits of measurement and the noise performance was unchanged. It gave a hefty parts-cost saving of some £2 on the product concerned.

Noise & the value of R0

We have seen already that minor changes in the value of R0 have a negligible effect on the noise performance. The noise calculations were performed with MAG-NOISE2, an upgraded version of my MAGNOISE tool. This gives results within 0.2 dB of measurements and I believe it trustworthy.

In my published designs, R0 has gone from 470 Ω in 1979 [7] to 330 Ω in 1983 [8] and then 220 Ω in 1996 [4] as part of a general drive towards low noise by low-impedance design. I considered 220 Ω about as low

as could be achieved while using a reasonable number (five) of paralleled 10 nF polystyrene capacitors for C1.

In MAGNOISE2, any source of noise can be switched off so its contribution can be assessed. R0 produces noise in two ways; its own Johnson noise, and the voltage produced by the opamp current noise flowing through it. At these impedances, the latter is negligible. Table 19.6 shows the results for 30 dB gain (1 kHz) using a 5534AN opamp and cartridge parameters of R = 610 Ω, L = 470 mH, for the values of R0 used in the discussions above, plus some more. The noise results are shown to two decimal places simply so that the small changes are visible; in practice they will be swamped by variations in opamp noise. R0 has only a minor effect; most of the noise comes from the Johnson noise of the cartridge resistance and the 47 kΩ input load, the opamp

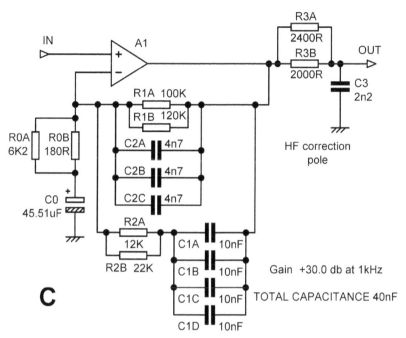

Figure 19.14 Configuration C from Figure 19.13 with resistors made up of optimal parallel pairs. C2 is made up of three parts. Gain +30.0 dB at 1 kHz, RIAA accuracy is within ±0.01 dB.

voltage noise, and the opamp current noise at the non-inverting input.

The first thing we see is that with R0 at my usual value of 220 Ω, we are only 0.39 dB noisier than some impossible circuit in which R0 did not exist at all. That is not an audible difference, and suggests that if we can accept just a little more noise we would be able to increase the impedance of the RIAA network and further reduce the amount of expensive capacitance required.

Looking at configuration C in Figure 19.11, with R0 = 233 Ω, we can redesign it, using the new software tool, so that C1 = 2 × 10 nF. Running the numbers, we get R0 = 349 Ω, as in Table 19.7, which Table 19.6 shows will be noisier than Figure 19.11 by only 0.18 dB. In many cases this will be an acceptable cost/performance trade-off. If we were happy with a 1.0 dB noise increase, R0 can go as high as 611 Ω.

For comparison Table 19.7 also shows the component values for configuration A, with C1 constrained to be 20 nF. Because of its inefficient use of capacitance, this requires R0 to be a good deal higher at 552 Ω, and the noise penalty referred to R0 = 233 Ω is 0.50 dB rather than 0.19 dB. This is rather less acceptable and confirms that configuration C is the best. Table 19.7 was checked by simulation and the RIAA accuracy is within ±0.01 dB from 20 Hz to 20 kHz.

Let's push this a little further and assume we want C1 to be a single 10 nF 1% capacitor for real economy. What is the noise penalty? The software tools give us Table 19.8 for configurations C & A, and the noise penalties are inevitably larger, though perhaps not by as much as you would expect for fairly radical changes in the impedance of the RIAA network. Table 19.8 was checked by simulation and the RIAA accuracy is within ±0.01 dB from 20 Hz to 20 kHz.

All the noise calculations above are for a preamplifier driven by an MM cartridge with the parameters given above. As we noted at the start, the MM RIAA stage may also be driven from an MC head-amp. The noise conditions for the opamp are quite different, as it is now fed from a very low impedance, probably dominated by a series resistor in series with the MC amp output to give stability against stray capacitances. My current MC head-amp design has an EIN of -141.5 dBu with a 3.3 Ω input source resistance [5]. Its gain is +30 dB, so the noise at its output is -111.5 dBu. The series output resistor is 47 Ω and its Johnson noise at -135 dBu is negligible, and so is the effect of the RIAA stage input noise flowing in it. The noise at the output of the RIAA stage is then -85.7 dBu, which is considerably higher than any of the figures in Table 19.6. In this application the value of R0 is uncritical.

Table 19.6 Calculated noise output for different values of R0 (22–22 kHz bandwidth, with RIAA), unweighted.

Value of R0	Noise out	Noise ref R0 = 0 Ω	Where used
0 Ω	−92.90 dBu	0.00 dB	
100 Ω	−92.72 dBu	+0.18 dB	
175 Ω	−92.59 dBu	+0.31 dB	Figure 19.13
200 Ω	−92.55 dBu	+0.35 dB	Figure 19.7
212 Ω	−92.53 dBu	+0.37 dB	Figure 19.8
220 Ω	−92.51 dBu	+0.39 dB	Ref [4]
233 Ω	−92.49 dBu	+0.41 dB	Figure 19.11
330 Ω	−92.33 dBu	+0.57 dB	Ref [8]
349 Ω	−92.30 dBu	+0.60 dB	Table 19.7 C
470 Ω	−92.11 dBu	+0.79 dB	Ref [7]
552 Ω	−91.99 dBu	+0.91 dB	Table 19.7 A
611 Ω	−91.90 dBu	+1.00 dB	
699 Ω	−91.77 dBu	+1.13 dB	Table 8 C
1000 Ω	−91.36 dBu	+1.54 dB	
1104 Ω	−91.22 dBu	+1.68 dB	Table 8 A
2000 Ω	−90.20 dBu	+2.70 dB	
2310 Ω	−89.90 dBu	+3.00 dB	

Table 19.7 Calculated RIAA components with C1 constrained to be 20 nF. IEC amendment implemented by C0.

Configuration	R0	C0	C1	C2	R1	R2
C	349.36 Ω	22.76 uF	20 nF	7.038 nF	108.87 kΩ	15.562 kΩ
A	552.24 Ω	14.39 uF	20 nF	5.727 nF	158.99 kΩ	13.094 kΩ

Table 19.8 Calculated RIAA components with C1 constrained to be 10 nF; IEC amendment implemented by C0.

Configuration	R0	C0	C1	C2	R1	R2	Noise ref R0 = 233 Ω
C	698.7 Ω	11.37 uF	10 nF	3.519 nF	217.74 kΩ	31.126 kΩ	+0.72 dB
A	1104.4 Ω	7.198 uF	10 nF	2.864 nF	317.99 kΩ	26.189 kΩ	+1.27 dB

Conclusions

Our investigations have shown that there are very real differences in how efficiently the various RIAA networks use their capacitors, and it looks clear that using configuration C rather than configuration A will cut the cost of the expensive capacitors C1 and C2 in an MM stage by 36% and 19% respectively, which I suggest is both a new result and well worth having. From there we went on to find that different constraints on capacitor availability lead to different optimal solutions for configuration C.

We have also demonstrated that accepting very small amounts of extra noise allows us to increase the impedance of the RIAA network substantially, further reducing the capacitance required and saving some more of our hard-earned money.

I hope you will forgive me for not making public the software tools mentioned in this article. They are part of

my stock-in-trade as a consultant engineer, and I have invested significant time in their development.

References

1. Self, Douglas, "Small Signal Audio Design" Focal Press 2010, p171; ISBN 978-0-240-52177-0

2. Lipshitz, S P, "On RIAA Equalisation Networks" *J. Audio Eng Soc,* June 1979, p458 onwards.

3. Vogel, Burkhard, *The Sound of Silence,* Springer 2nd edition 2011, p523; ISBN 978-3-642-19773-d

4. Self, Douglas, "Precision Preamplifier 96" *Electronics World* July/August and September 1996

5. Self, Douglas, "Elektor Preamplifier 2012" *Elektor* April, May, June 2012

6. Self, Douglas, "Active Crossover Design" Focal Press, 2011, pp342–350, ISBN 978-0-240-81738-5

7. Self, Douglas, "High-Performance Preamplifier" *Wireless World*, Feb 1979, p40

8. Self, Douglas, "Precision Preamplifier" *Wireless World*, Oct 1983, p31

Power amplifiers

Chapter 20

Sound mosfet design

(Electronics World, September 1990)

Intended title: *A Hybrid BJT/FET Power Amplifier*

This was the first article on power amplifiers that I published, though I had been designing them for manufacture since 1975. This investigation into the concept of combining power FET output devices with bipolar drivers was done some years before I undertook my major investigation into the root causes of power amplifier distortion (see 'Distortion in Power Amplifiers', Parts 1–8, later in this book), which led in turn to the book *Audio Power Amplifier Design* [1]. In the throes of the design process, I realised with greater force than hitherto that the distortions in the small-signal part of a power amplifier were (a) far from negligible and (b) susceptible to analysis by a mixture of SPICE simulation and a few well-chosen experiments. I also determined that SPICE could be extremely useful in the analysis of output stages, if used to create an incremental gain plot.

Delving back into my history of power amplifier design, I note that at this stage I wasn't all that enthusiastic about DC-coupled power amplifiers. The 'kamikaze krowbars' mentioned were the sort of untrustworthy and unworkable compromise that was sometimes adopted before everyone realised that a DC-coupled amplifier really did have to be fitted with proper offset-detection and a reliable relay capable of breaking the fault current. The crowbar was a triac (so it could conduct both ways) connected across the amplifier output, and triggered via a simple RC lowpass filter. It usually only triggered once because to protect the loudspeaker, the entire charge of the reservoir capacitors had to be shorted to ground, exploding the triac and vaporising its associated PCB tracking. If it triggered when it should not, it would probably destroy perfectly sound output devices as well, by putting a dead short across the output that would severely test the short-circuit protection. Essentially it was a matter of destroying the amplifier to save the speaker, much as an expensive power transistor will sacrifice itself to save a fuse. It was a rotten idea but it was perpetrated by at least one much respected company that should have known better. Compared with this sort of thing, an output capacitor seemed delightfully simple, though in those days I did not know much about the higher distortion that can occur when you put large signal currents through electrolytic capacitors [2]. Putting the output capacitor at least partly inside the global negative feedback loop should help, but it's not so easy to do.

It might be of interest to give some details of the last AC-coupled power amplifier I designed, and if not you're going to get them anyway. It was a 100W/8Ω design developed around 1979 for multiple applications, including driving big intercom rings in large venues. Reliability and low cost rather than high-end hi-fi was the prime directive, and so it had a quasi-complementary output stage using two 2N3773s in TO-3 packages. These parts had a reputation for being tough high-voltage devices, and were used in the famous Crown DC300 amplifiers. They were supplied to us by Jermyn. Nowadays the 2N3773 is showing its age, with an unimpressive minimum beta of 15 at an Ic of 8 Amps, falling to a feeble 5 at the maximum Ic of 16 Amps according to the manufacturer's data sheet. The output coupling capacitor was 4700uF/100V, which was twice the size normally used, and in those days a pretty hefty component. The schematic is shown in Preface Figure 20.1.

There is a single input transistor Q1 which performs the feedback subtraction; in an AC-coupled amplifier the DC precision of a differential input pair is not necessary. The second-harmonic generated by a single transistor was acceptable for this application and the parts count was minimised. Q1 is a transconductance stage in the same way as a differential pair, and it passes its output current into the base of VAS Q2. A VAS buffer Q3 prevents the output stage loading the VAS and is included in the loop of dominant-pole capacitor C4. This arrangement was inspired by the Equin amplifier published in one of the first issues of *Elektor* in 1976 [3], though no other Equin features were adopted apart from the use of capacitor output coupling. I have no recollection as to why C5 was included; it looks like a nod to the 'input-inclusive compensation' often advocated by John Linsley-Hood. All I can say at this point is that the amplifier was reliably stable. The collector of the VAS and the emitter of the VAS buffer are bootstrapped by C8, my thinking being that it would be preferable to using two current sources because it would be more reliable, with no transistors to fail, and more economical because resistors were then significantly cheaper than transistors; also there was no need for biasing components for current-sources. Q4 acts as a current-limiter for the VAS buffer Q3, to prevent it passing excessive current when in negative clipping or SOAR protection, as it attempts to obey the negative feedback and pull the output down; looking back at the old schematics, I was surprised that I was using this technique so early. It is usually very necessary if an amplifier is survive prolonged clipping, but does not appear in all designs by any means. Standard single-slope SOAR limiting was fitted; this is not shown in the schematic for clarity.

The output stage is a standard quasi-complementary configuration with a Baxandall diode (D3). Looking back, I wonder if the value of R23 should have been adjusted to match that of R25. The output emitter resistors R21 and R22 are 0.33 Ω; nowadays I would make them 0.1 Ω, which would give better linearity at the cost of increased quiescent current.

The output inductor L1 followed contemporary practice, being a very small component as it was a VHF suppression choke wound on a small ferrite core; what this core might have done to the linearity I do not know. I thought up a beautifully simple RC network for de-thumping at switch-on and no output relay was required. When amplifiers with this sort of input stage are switched on, they usually make an almighty bang. This is despite the large value of filter capacitor C2, which makes the bias voltage come up slowly. The problem is that as the bias comes up, Q1 remains firmly off because of the extra delay in charging C3 via feedback resistor R9. As a result the VAS collector, and therefore the amplifier output, shoots up at the same rate as the supply rail rises, rather than rising slowly with the voltage on C2. This is effectively prevented by adding C7, R10, and D2;

the VAS collector can now only rise slowly because it has to charge C7 with a limited amount of current. C7 is also charged by R10, and eventually reaches rail voltage, and D2 is reverse-biased even at full output swing. With the means of measurement available to me, the presence of a reverse-biased D2 seemed to have no ill-effects on linearity. D1 is intended to speed-up the discharge of C7 on powering down.

The single supply rail reduced PSU costs somewhat compared with dual-rail; there was a single 4700uF/100V reservoir capacitor for two channels. The output capacitor C11 makes short-circuit protection easier because however low the load impedance, only one capacitor's worth of electrical charge can be transferred per cycle. I don't recall ever hearing that one of these amplifiers had failed.

The construction was, naturally, somewhat old-school. The amplifier PCBs were single-sided, with no solder-resist (the company concerned soldered everything manually—no wave-soldering then) and no component ident, once again to keep costs down. That may sound silly but with the technology of the day, adding a silk-screen component ident added significant cost; it also had to be tediously laid out by hand on plastic film with Letraset rather than generated automatically by a CAD package. I succeeded in laying out the PCB without a single link being necessary; back in the day this saved assembly time and got definite kudos. As usual this required some rather convoluted routing of non-critical tracks. Now that PTH PCBS are so economical, this sort of thing is probably becoming a lost art.

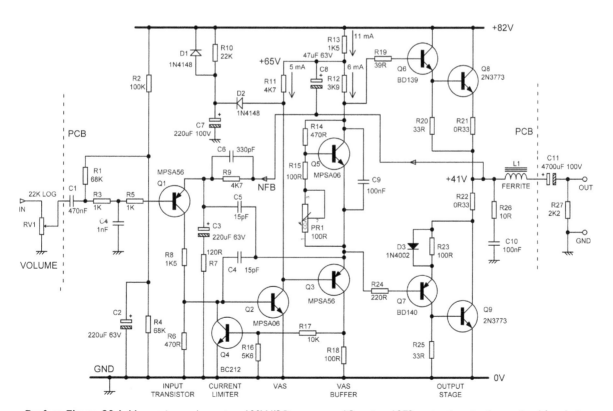

Preface Figure 20.1 My quasi-complementary 100W/8Ω power amplifier, circa 1979: protection circuitry omitted for clarity

Preface Figure 20.2 Two 100W/8Ω quasi-complementary power amplifier modules make up a stereo amplifier, circa 1979.

A stereo amplifier using two of the amplifier PCBs is shown in Preface Figure 20.2 The output transistors are mounted on the underside of the extrusion at the top; in its centre can be seen a thermal cutout switch. The PSU reservoir capacitor is on the left, and the output capacitors are on the right. Note the vintage E&I transformer and the ancient DIN socket input.

Going back to the article, the motivation for my FET investigations had its roots many years earlier, when I was trying to design a new sort of ultra-linear output stage. This ambitious project was beset by many difficulties, but I decided that some of them could be overcome by using faster output devices than the BJT power transistor pairs available at the time. I knew from my experiments with simple FET power amplifiers in 1980 that FETs were non-linear compared with BJTs, and so I developed the plan of combining BJT drivers with faster FET power devices, (I was sure they were faster because everybody said so) to combine the linearity of a local feedback loop with a speed that would enable HF stability to be assured. The first stage was to build and measure a BJT-FET power amplifier, and that resulted in the article you see here. Amongst other things, I discovered that the speed of power FETs is mixed blessing; parasitic oscillation that would pass almost un-noticed with a BJT amplifier (until you looked at the horrible effects on the distortion residual) would cause prompt disassembly of an output stage. In other words the FETs would explode, often violently. The design here is safe, but getting to it was a bit of a trial, and that did little to make me feel happier about FET amplifiers. This problem meant that no progress was made with the ultra-linear output stage.

The article may have suffered a bit as, it being my first article on power amplifiers, I tried to cram in as much information as I could. I was surprised to be reminded that I was using model amplifiers at the this early date. Model amplifiers are power amplifiers with the output stage replaced by a very linear Class-A emitter-follower stage that only needs to be capable of driving the feedback network; this allows the distortion of the amplifier small-signal stages to be measured in isolation. Model amplifiers proved extremely useful in the great assault on the North Face of Amplifier Distortion in 1993.

I have to point out that the final circuit falls somewhat short of the Blameless performance standards set by the *Distortion in Power Amplifiers* series. The input stage may look symmetrical, but in fact is grotesquely unbalanced, with 16 uA flowing through the left transistor of the pair, and 580 uA through the right. This stage must have generated far more second-harmonic distortion than necessary, and with the benefit of hindsight I am not at all proud of it. This imbalance also means that the input stage transconductance is much reduced compared with the balanced condition, which precludes any emitter degeneration of the input pair, and this probably also explains why a dominant-pole capacitor as small as 15 pF is enough for stability. My only excuse is ignorance, and in engineering that is no excuse at all. The voltage-amplifier stage could be greatly improved in linearity by adding another transistor within the Cdom local feedback loop.

I was also a bit unlearned in those days as to how important it was to avoid inductive distortion, caused by the half-wave output stage currents getting into the signal path, and this may well have affected the results. With the clarity of hindsight, I should have spotted that a while back. I once worked for a company that made an integrated amplifier; it was not one of my designs. The screened-cable signal connection between the preamplifier and power amplifier sections wound its way in an arcane fashion across the top of the single PCB, its route carefully adjusted for minimum distortion. Looking back, this process can only have been the optimising of the cancellation of inductive distortion.

References

1. Self, D., *Audio Power Amplifier Design*, 6th edition. Focal Press 2013. ISBN 978-0-240-52613-3.

2. Ibid., p. 107.

3. *Elektor*, Equin. Part 1 April 1976. Part 2 May 1976.

SOUND MOSFET DESIGN

September 1990

Mosfet amplifiers undoubtedly present a tantalising prospect for simple circuitry. Unfortunately, practical application requires careful consideration of their many foibles. But my aim here is to stimulate thinking about possible improvements to FET power amps, and describe two new avenues of development. Each produces a practical result, though neither should be regarded as a foolproof recipe for success.

Power-amplifier design is as prone as any other branch of audio to folklore and confusion. One of the less extreme myths holds that high levels of negative feedback are 'A Bad Thing' because they require heavier compensation for HF stability, leading to low slew rates and generally indolent and sluggish behaviour.

As far as it goes this is true. But only a poor designer would lose all sight of slew rate while adjusting amplifier compensation. Despite much study of TIM, DIM, SID, 'internal overload', 'delayed feedback' and the rest, everything comes back to slew rate.

If an amplifier can reproduce a 20 kHz sine wave at full amplitude without excessive distortion (say under 0.1%) it can be regarded as blameless in respect of speed. Apply as much feedback as is decent, but always keep an eye on stability and slew rate.

Design fundamentals

Ground rules for this design study require DC output coupling, one preset only (for quiescent current) and as simple a circuit as possible.

In the current audio market, almost any technological approach appears to be acceptable (an idiosyncratic hybrid with valves driving power fets is one recent design, [1]) with the possible exception of capacitor output coupling.

While problems can include capacitor distortion at LF (of the real and measurable kind [2]) and perhaps grounding difficulties, the overwhelming simplicity of this method still has its attractions; the almost unnatural reliability of the capacitor-coupled Quad 303 should be mentioned at this point.

But a designer must still prove that he knows what a differential pair is, and d.c. coupling has therefore been adopted. However, remember that proper offset protection is not a trivial problem, as the cost of a reliable output relay and d.c. detection circuitry can add 30% to amplifier electronics costs. A relay seems unavoidable, as my own experiences with kamikaze crowbars have been distinctly unhappy, though I would not claim this to be a definitive judgment.

So a simple unregulated power supply looks to be the best. I realise this brings me into head-to-head confrontation with Mr Linsley Hood, [3] so I shall quickly explain my preference.

Putting expensive power semiconductors in a high-current dual supply can easily double an amplifier's electronic-component cost and there is much more to go wrong.

Ensuring PSU HF stability can be difficult, and the PSU compensation required, threatening a steadily rising output impedance versus frequency, can lead to some awkward amplifier stability problems.

Finally, the unregulated supply can deliver more power on a transient basis—which is exactly what is required for audio.

The price to be paid for unregulated simplicity is the attention to be paid to the amplifier's supply-rail rejection. But since it is physically impossible for the voltage on large reservoir capacitors to change very quickly, this rejection need only be extremely good at low frequencies.

The excellent supply-rail rejection of IC op-amps—and a power amplifier is, after all, only a big op-amp—shows the problem to be distinctly soluble, although I admit that op-amp PSRR often differs markedly between the two rails, and usually declines above 1 kHz. This sort of difficulty can be simply solved in a power amplifier with a little RC decoupling.

Harmonic distortion should be kept as low as possible, but without spending significant money specifically on its reduction. THD in commercial equipment varies more widely than any other performance parameter, ranging from 0.003% to 1% at similar powers [4] with the most expensive units often giving the worst performance.

In marketing circles, there are clearly two routes to take: make the THD vanishingly small to show you know what you're doing; or make it poor and imply that this very practical parameter has been sacrificed in favour of some intangible and unmeasurable sonic benefit.

I have always gone for the former and so have concentrated on linearity as being the prime determinant of amplifier topology.

Distortion performance is not easy to specify completely, ideally requiring a spectrum analysis of every combination of level, frequency and load impedance. But this is not practical and so I have summarised it as THD plotted against frequency, into 8 Ω, at different levels where appropriate.

Pros and cons of mosfet output devices have been thoroughly ventilated, [5, 6] but one point needs qualification. They have been praised for having a large crossover region between the two halves of a Class B stage, but my experiments show this to be a very dubious advantage.

Mosfet outputs, with or without the augmentations described below, may be uncritical of quiescent current setting, but this really means that nothing is exactly right. Bipolar stages, with sharper crossover regions, do at least make it obvious where to set the quiescent current, providing it is set by observing the distortion residual—and it certainly should be.

Mosfet distortion residuals typically present a rather gnarled appearance, with plenty of harmonics at least to the seventh. Though it runs counter to conventional wisdom, in my experience complementary-pair bipolar residuals tend to be smoother.

Another vital point is that mosfet complementary-pairs are rare, not particularly complementary, and definitely more expensive than the profusion of strictly N-channel devices intended for switching.

Determining performance

Mosfet power amps have suffered more than most branches of technology from 'application-note cloning' though some original designs have been published, ranging from the complex [7, 8] to the very complex. [9] These are well worth ferreting out, though perhaps unattractive commercially.

The 'standard' mosfet amplifier circuit (Figure 20.1) differs from the equally standard bipolar-style circuit (Figure 20.2) mainly in possessing a sort of push–pull/ current-mirror configuration in the voltage amplifier stage, probably intended to provide better charge/discharge of the mosfet input capacitances. This stage is sometimes called the pre-driver, but it is less confusing to call the first full-voltage-swing stage 'the voltage amplifier stage', or VAS.

The first question to be asked is what improvement is made by this push–pull arrangement (Figure 20.3). The

linearity is not very different and the benefit at HF is not startling. As always in science, it pays to be skeptical.

Given that the simpler bipolar configuration of Figure 20.2 is workable, there are established ways to improve its overall linearity and some of these are shown in Figure 20.4.

Linearity can be enhanced simply by increasing open-loop gain (Figure 20.4a and b) or by a cascode arrangement (Figure 20.4c) which attempts to linearise the VAS by eliminating Early effect. Cascode arrangements are relatively ineffective at reducing distortion in power amplifiers, since the Early effect seems to be dominated by non-linear loading of the high impedance at the interface with the output stage (points A and B).

The added emitter-follower and current-mirror enhancements (Figure 20.4a and b) work because the input pair act as a transconductance amplifier (voltage-difference in, current out) feeding a VAS that is basically a Miller integrator, thanks to the dominant-pole capacitance C_{dom}.

Emitter-follower Tr_3 increases open-loop gain by enhancing the current-gain of the VAS. The current

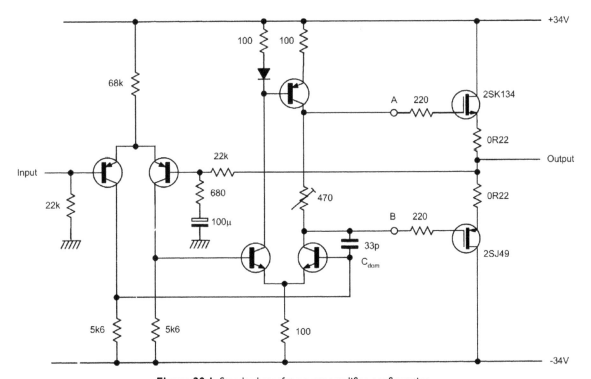

Figure 20.1 Standard mosfet power amplifier configuration.

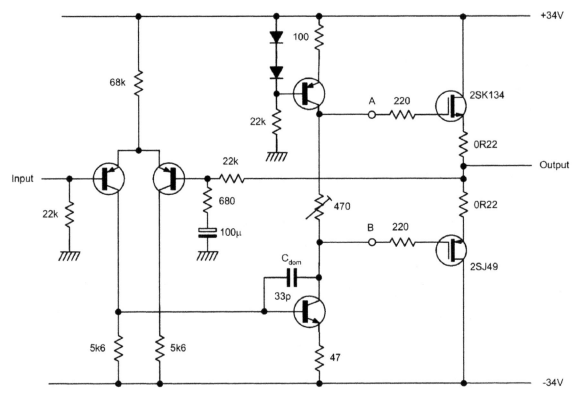

Figure 20.2 Standard bipolar power amplifier configuration, using mosfets.

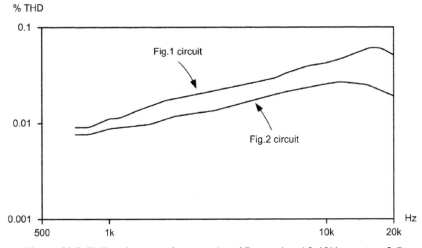

Figure 20.3 THD vs frequency for examples of Figures 1 and 2. 10V r.m.s. into 8 Ω.

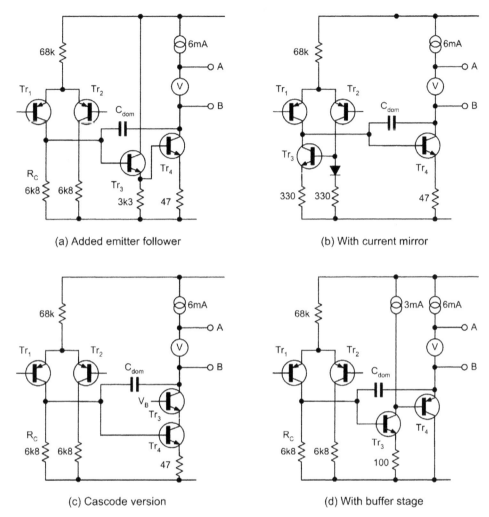

(a) Added emitter follower

(b) With current mirror

(c) Cascode version

(d) With buffer stage

Figure 20.4 Enhancements to the basic amplifier configuration.

mirror does the same thing by doubling input-pair transconductance; also slew-rate is greater and symmetrical (30 V/μs was obtained) as the input stage can now sink current as effectively as it sources it. Furthermore, input collector currents are kept balanced, valuable if the d.c. offset at the amplifier output is to be kept within acceptable bounds (say ±50 mV) without adjustment. The input collector currents, and hence the base currents drawn through input and feedback resistors, must be roughly equal.

Total input pair current is a critical parameter, since it affects input-stage transconductance and hence open-loop gain, and also defines the maximum slew rate,

setting the maximum current that can flow in and out of the dominant-pole compensation capacitor C_{dom}. Though not obvious, the input-pair current is an important influence on the HF stability of the amplifier. Power amplifier design has always been impeded by the fact that the crucial VAS, with its high-impedance collector, has to drive an output stage with markedly non-linear input impedance. The resulting interaction means that, when distortion occurs, it is not clear whether it arises in the output stage itself, or at the VAS due to non-linear loading.

Distortion not caused by the output can be studied by replacing the output stage with a very linear

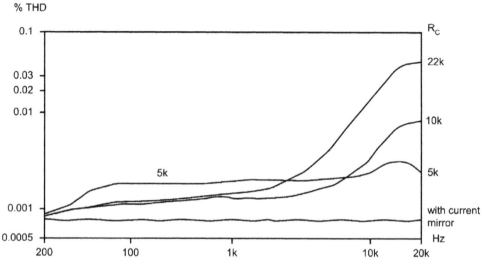

Figure 20.5 Effect of Rc and current-mirror, input and VAS only.

Class-A voltage-follower. Resistor R_c has a strong effect on HF distortion (Figure 20.5) with a clear improvement when R_c is replaced by a simple diode/ transistor current mirror. This underlines the crucial nature of the interface between the input pair and the VAS.

Similarly, the output stage can be tested in isolation by driving it directly from a low-impedance oscillator [6] and this technique was used to study the two output arrangements described below.

Neither method, however, allows study of the VAS/ output interaction directly. One strategy for side-stepping this problem is to use a unitygain buffer between the VAS and the nonlinear load of the output stage (e.g. emitter-follower Tr_4, Figure 20.4d).

This approach has not been pursued with mosfets, but I have used it in a commercial design for a bipolar quasi-complementary amplifier. It proved effective at reducing distortion and is an approach which seems unfairly neglected.

Reducing costs

In the search for a better amplifier, two routes were examined. The first is to reduce the cost of a mosfet

power amp by using two N-channel devices in a form of quasi-complementary output stage.

The second is to increase linearity and improve quiescent stability by using bipolar drivers with local feedback around each output FET.

The quasi-complementary approach is directly analogous to that used in the early days of transistor amplifiers; a P-N-P transistor is combined with an N-channel FET to emulate a P-channel FET, and this works very well.

The output devices are now the same (promising bulk-buying economy) and can be chosen without reference to complements. That used was the IRF530, offering 100 V, 14 A, and 75 W in a TO-220 package, a pair costing £5.00 against £9.50 for the 2SK134/2SJ49 pair. Open-loop distortion of the output stage alone driven from a low impedance was shown (Figure 20.6) to be 1.1%.

Closed-loop performance (Figure 20.7), (yielded by the practical circuit shown in Figure 20.8) demonstrated that, predictably, output symmetry is not wonderful, and crossover effects on the residual were clear.

But this performance is acceptable from such a simple and economical circuit, and could almost certainly be improved at minor cost by adopting added emitter-follower and/or current mirror.

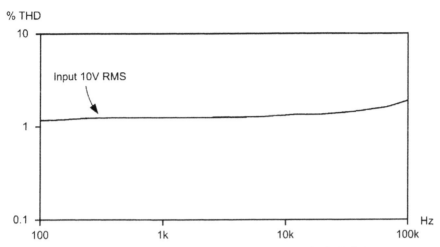

Figure 20.6 Quasi-complementary output stage only (open-loop) showing almost flat THD. 8 Ω resistive, measurement bandwidth 22–7500 Hz.

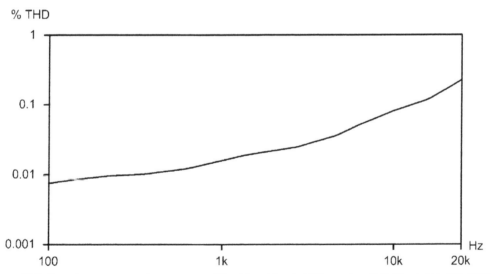

Figure 20.7 Complete quasi-complementary amplifier (closed-loop). 10 V r.m.s. into 8 Ω. (Bandwidth 100–80 kHz.)

Tests showed that the value of R_c in Figure 20.8 is non-critical, so there seems no difficulty in driving the bottom mosfet gate capacitance.

Quiescent setting is by transistor bias-generator; this is purely a regulator, and is *not* thermally coupled to the output devices. In an attempt to improve output symmetry, a Baxandall diode [10] was inserted at point D in the driver emitter. Sadly, THD was unchanged, despite the dependable improvement that this modification gives in bipolar quasi-complementary designs.

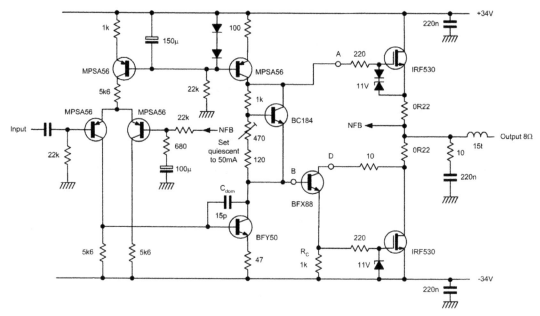

Figure 20.8 Practical circuit for quasi-complementary amplifier.

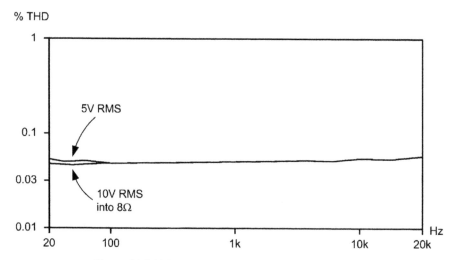

Figure 20.9 Hybrid output stage only, operated open-loop.

Reducing distortion

If performance outweighs economy, a true complementary output pair is retained, with local feedback linearising each mosfet.[6] An important variable here is the value of emitter-resistor R_e in Figure 20.11. A high value increases output distortion, since it reduces the feedback factor within each hybrid bipolar-mosfet loop, while a low value makes quiescent-current setting unduly critical. As a compromise, 10–22 Ω works well.

Figure 20.9 shows the output stage alone in open loop giving 0.05%, and Figure 20.10 shows closed-loop performance as given by the practical design in Figure 20.11, plus the effect of varying C_{dom}.

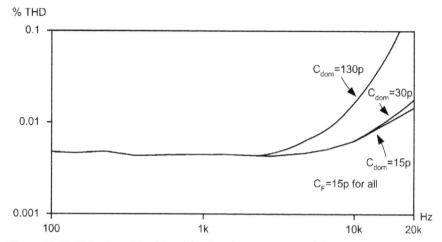

Figure 20.10 Hybrid amplifier (closed-loop) with varying values of Cdom. 16 V r.m.s. into 8 Ω.

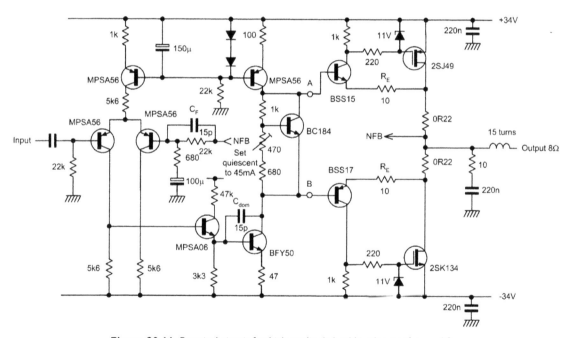

Figure 20.11 Practical circuit for high-quality hybrid bipolar-mosfet amplifier.

An emitter-follower is added to increase the feedback factor, and C_f is now needed for HF stability with $C_{dom} = 15$ pF. The compensation is not necessarily optimised for all possible real-life loads.

Distortion across the band is now very low. While it is desirable to define the closed-loop bandwidth of any audio device, do not put a simple RC filter at the input, as this makes the bandwidth dependent on the output impedance of the upstream equipment.

The temptation to implement it by increasing C_f should also be resisted, because if it is large enough to provide a suitable roll-off, there is a danger that it will induce mysterious VHF instability in Tr_2_a common effect. It is only revealed by distortion that vanishes at the touch of a cautious finger.

I have not tried to give the final word on this subject, though I hope it illuminates a few new directions. The use of current mirror topology in particular looks promising.

References

1. Kessler, K. 'Valfet Amplifier', *HiFi News and Record Review*, June 1989, p 65.

2. Self, D. 'Electrolytics & Distortion', *EW + WW*, February 1989, pp 43–44.

3. Linsley Hood, J. 'Class AB Mosfet Power Amp', *EW + WW*, March 1989, p 264.

4. Self, D. 'Science vs Subjectivism', *EW & WW*, July 1988, p 695.

5. Brown, I. 'Feedback and FETs in Audio Power Amps', *EW + WW*, February 1989, pp 123–124.

6. Self, D. 'Mosfet Audio Output', Letter, *EW + WW*, May 1989, p 514.

7. Borbely, E. 'High Power Amplifier', *Wireless World*, March 1983, pp 69–75.

8. Tilbrook, D. '150W Mosfet Amplifier', *ETI*, June 1982, pp 48–55.

9. Cordell, R. 'Mosfet Power Amplifier with Error-Correction', *JAES*, January/February 1984, pp 6–17.

10. Baxandall, P. 'Symmetry in Class B', Letter, *Wireless World*, September 1969, p 416.

Chapter 21

FETs versus BJTs

The linearity competition
(Electronics World, May 1995)

Intended title: *Power Fets and Bipolars: The Linearity Competition*

This short article was published when debate about the linearity of power FETs was raging, or at any rate smouldering, in the letters columns of *Electronics World.* Many contributors were content to point out that FETs must be more linear than bipolar transistors because everyone says so. This sort of argument has never had much appeal for me, and I carried out several investigations to see if there was any way in which FETs could be claimed to produce less distortion. One of the problems with this is making a meaningful comparison between two rather different kinds of active device.

Here I tried to level the playing field by making the transconductances the same; the BJT wins heavily on linearity when degenerated to have the same low transconductance as a FET. In a complete power amplifier, the situation is naturally rather more complex; BJT output devices need BJT drivers (I suppose you could use FET drivers, but I think it a most unpromising route to head down) which introduce more distortion than you might expect, while many FET power output stages can dispense with drivers altogether, so long as the VAS is capable of charging and discharging those rather large gate-capacitances. Nevertheless, I am confident that in a fair contest, a BJT amplifier will always have lower distortion than its FET equivalent. In particular, it will have smaller and less nasty crossover artefacts.

Not everyone felt that this contribution settled the matter for good—the SPICE models used to simulate the FETs were the focus of particular attention. Fortunately, since then a lot more has been published on SPICE FET models, and it appears my conclusions were generally correct. While there may be debate (and there has been a great deal) about how well the various models represent the crossover region in Class-B output stages, there is no argument about the basic transconductance; bipolar devices have much more of it, and it can be put to work in many useful ways. The great predictability of bipolars, at least as regards their fundamental operation as represented by the transistor equation, is also a great advantage. In contrast, FETs have significant Vgs variations.

Since the results of SPICE simulation depend on the accuracy of the FET models used, people have put much effort put into these models. Two well-known examples are Ian Hegglun's two articles published in *Electronics World* in 1999 [1], and Cyril Batemans's monumental 28-page study in *Electronics World* in 2004–5 [2].

MOSFET power amplifier designs were also published in *Electronics World* by John Linsley-Hood in 1982 [3] and 1989 [4], by Peter Wilson in 1982 [5] (though I have to say the THD of 0.08% at 20W/8Ω did not impress me much) and David White in 2001 [6]. All of these used FETs driven directly from the VAS, rather than the BJT-FET combination I used in my design here. Several MOSFET amplifier designs also appeared in magazines like *Electronics Today International*.

References

1. Hegglun, I, Hotter Spice (Power MOSFET models), *Electronics World May,* July 1999 (2 parts).

2. Bateman, C, Simulating Power MOSFETs, *Electronics World,* Oct., Nov., Dec. 2004, Jan. 2005 (4 parts).

3. Linsley-Hood, J, JLH 80–100W MOSFET Audio Amplifier, *Electronics World,* June, July, Aug. 1982 (3 parts).

4. Linsley-Hood, J, Class A/AB MOSFET Power Amplifier, *Electronics World,* Mar. 1989, pp. 261–264.

5. Wilson, P, Simple Power Amplifier, *Electronics World,* April 1982, p. 56.

6. White, D, MOSFET power—pure and simple, *Electronics World,* Aug. 2001, pp. 578–583.

FETs VERSUS BJTs

May 1995

There has been much debate recently as to whether power FETs or bipolar junction transistors (BJTs) are superior in power amplifier output stages. Reference 1 is a good example. It has often been asserted that power FETs are more linear than BJTs, usually in tones that suggest that only the truly benighted are unaware of this.

In audio electronics it is a good rule of thumb that if an apparent fact is repeated times without number, but also without any supporting data, it needs to be looked at very carefully indeed. I therefore present my own view of the situation here, in the hope that the resulting heat may generate some light.

I suggest that it is now well-established that power FETs, when used in conventional Class-B output stages, are a good deal less linear than BJTs.[2] Gain deviations around the crossover region are far more severe for FETs than the relatively modest wobbles of correctly biased BJTs, and the shape of the FET gain-plot is inherently jagged, due to the way in which two square-law devices overlap.

The incremental gain range of a simple FET output stage is 0.84 to 0.79, range 0.05, and this is actually much greater than for the bipolar stages in Ref. 2; the emitter-follower stage gives 0.965 to 0.972 into 8 Ω, with a range of 0.007, and the complementary feedback pair gives 0.967 to 0.970 with a range of 0.003. The smaller ranges of gain-variation are reflected in the much lower THD figures when PSpice data is subjected to Fourier analysis.

However, the most important difference may be that the bipolar gain variations are gentle wobbles, while all FET plots seem to have abrupt changes. These are much harder to linearise with negative feedback that must decline with rising frequency. The basically exponential I_c/V_{be} characteristics of two BJTs approach much more closely the ideal of conjugate mathematical functions—i.e. always adding up to 1. This is the root cause of the much lower crossover distortion.

Close-up examination of the way in which the two types of device begin conducting as their input voltages increase shows that FETs move abruptly into the square-law part of their characteristic, while the exponential behaviour of bipolar devices actually gives a much slower and smoother start to conduction (see Figures 21.4 and 21.5).

Similarly, recent work shows that less conventional approaches, such as the common-collector/common-emitter configuration of Bengt Olsson, also suffer from the non-conjugate nature of FETs. They also show sharp changes in gain. Gevel [3] shows that this holds for both versions of the stage proposed by Olsson, using both N and P-channel drivers. There are always sharp gain-changes.

Class A stage

It occurred to me that the idea that FETs are more linear was based not on Class-B power-amplifier applications, but on the behaviour of a single device in Class-A. You might argue that the roughly square-law nature of afet's I_d/V_{gs} law is intuitively more 'linear' than the exponential I_c/V_{be} law of a BJT, but it is difficult to know quite how to define 'linear' in this context. Certainly a square-law

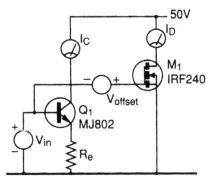

Figure 21.1 Linearity test circuit. Voltage Voffset adds 3 V to the d.c. level applied to the FET gate, purely to keep the current curves helpfully adjacent on a graph.

device will generate predominantly low-order harmonics, but this says nothing about the relative amounts produced.

In truth the BJT/FET contest is a comparison between apples and aardvarks, the main problem being that the raw transconductance (g_m) of a BJT is far higher than for any power FET. Figure 21.1 illustrates the conceptual test circuit; both a TO3 BJT *MJ802* and an *IRF240* power FET have an increasing d.c. voltage, V_{in}, applied to their base/gate, and the resulting collector and drain currents from PSpice simulation are plotted in Figure 21.2.

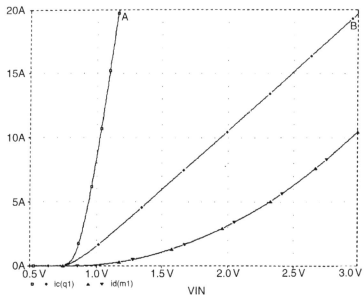

Figure 21.2 Graph of Ic and Id for the BJT and the FET. Curve A shows Ic for the BJT alone, while Curve B is the result for Re = 100 mΩ. The curved line is the Id result for a power FET without any degeneration.

Voltage V_{offset} is used to increase the voltage applied to FET M_1 by 3.0 V because nothing much happens below a V_{gs} of 4 V, and it is helpful to have the curves on roughly the same axis. Curve A, for the bjt, goes almost vertically skywards, as a result of its far higher g_m. To make the comparison meaningful, a small amount of local negative feedback is added to Q_1 by R_e. As this emitter degeneration is increased from 0.01 to 0.1 Ω, the I_c curves become closer in slope to the I_d curve.

Because of the curved nature of the FET I_d plot, it is not possible to pick an R_e value that allows very close g_m equivalence; a value of 0.1 Ω was chosen for R_e, this being a reasonable approximation; see Curve B. However, the important point is that I think no-one could argue that the FET I_d characteristic is more linear than Curve B.

This is made clearer by Figure 21.3, which directly plots transconductance against input voltage. There is no question that FET transconductance increases in a beautifully linear manner-but this 'linearity' is what results in a square-law I_d increase. The near-constant g_m lines for the BJT are a much more promising basis for the design of a linear amplifier.

To forestall any objections that this comparison is nonsense because a BJT is a current-operated device, I add here a small reminder that this is untrue. The BJT is a voltage operated device, and the base current that flows is merely an inconvenient side-effect of the collector current induced by said base voltage. This is why beta varies more than most BJT parameters; the base current is an unavoidable error rather than the basis of transistor operation.

The PSpice simulation shown was checked against manufacturers' curves for the devices, and the agreement was very good—almost unnervingly so. It therefore seems reasonable to rely on simulator output for these kind of studies; it is certainly infinitely quicker than doing the real measurements. In addition, the comprehensive power-FET component libraries that are part of PSpice allow the testing to be generalised over a huge number of component types without you needing to buy them.

To conclude, I think it is probably irrelevant to simply compare a naked BJT with a naked FET. Perhaps the vital point is that a bipolar device has much more raw transconductance gain to begin with, and this can be

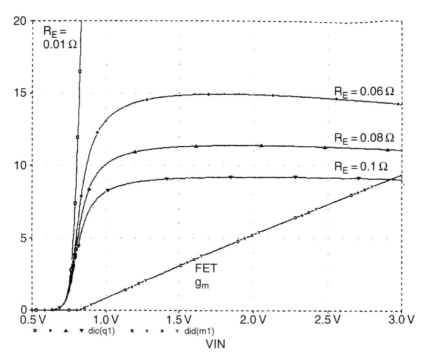

Figure 21.3 Graph of transconductance versus input voltage for BJT and FET. The near-horizontal lines are BJT gm for various RE values.

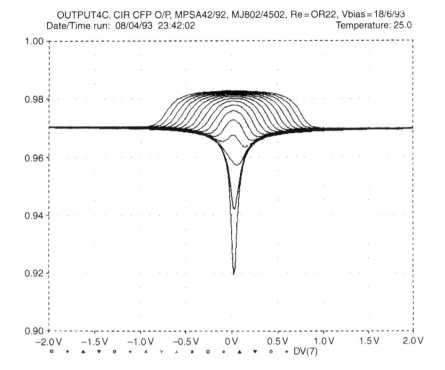

OUTPUT4C. CIR CFP O/P, MPSA42/92, MJ802/4502, Re = OR22, Vbias = 18/6/93
Date/Time run: 08/04/93 23:42:02　　　　　　　　　　Temperature: 25.0

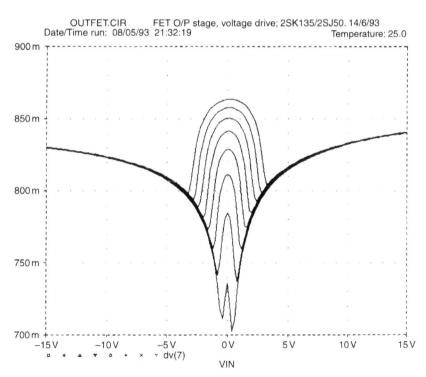

OUTFET.CIR　　FET O/P stage, voltage drive; 2SK135/2SJ50. 14/6/93
Date/Time run: 08/05/93 21:32:19　　　　　　　　　　Temperature: 25.0

Figures 21.4 and 21.5 Top are curves for a bipolar complementary feedback pair, crossover region ±2 V, Vbias as a parameter. Fourth curve up provides good optimal setting—compare with curves below, for a FET source follower crossover region with ±15 V range.

handily converted into better linearity by local feedback, i.e. adding a little emitter degeneration.

If the transconductance is thus brought down roughly to FET levels, the bipolar has far superior large-signal linearity. I must admit to a sneaking feeling that if practical power BJTs had come along after FETs, they would have been seized upon with glee as a major step forward in power amplification.

References

1. Hawtin, V. Letters, *EW +WW* December 1994, p 1037.

2. Self, D. 'Distortion in power amplifiers', Part 4, *EW + WW*, November 1993, pp 932–934.

3. Gevel, M. Private Communication, January 1995.

Chapter 22

Distortion in power amplifiers

Part I: the sources of distortion
(Electronics World, August 1993)

For many years I felt that the output stages of power amplifiers presented very great possibilities for creative design, and I actually got round to exploring some of them. The hybrid bipolar-FET output stage, also collected in this volume, was one of them. One of the first difficulties I met was the problem of determining how much of the total distortion was due to the output stage, and how much was being produced in the small-signal sections. This latter contribution turned out to be larger and more important than I expected. Very little reliable information appeared to exist on amplifier distortion, and I found myself embarked on a major effort to track down all the sources of distortion in the typical solid-state power amplifier. The initial investigations were not very illuminating, until I realised that changing a single component value in the typical power amplifier circuit very often alters two or more distortion mechanisms simultaneously, making the results hard to interpret. I was sitting in my armchair late one spring night when the full force of this struck home, and thereafter I devised ways to simulate or measure the distortion mechanisms in isolation. This approach is well-illustrated in Parts 2 and 3 of the series, dealing with the input pair and the Voltage-Amplifier Stage, respectively.

Things then began to fall into place, and one day I put together all the various minor and, apparently insignificant, improvements I had made to the utterly conventional amplifier circuit I had started with. The distortion was exhilaratingly low, stability was good, and the circuit was generally predictable and docile. I called it a Blameless amplifier to emphasise that the good distortion performance was due to avoiding a series of mistakes (for example, an unbalanced input pair) rather than breathlessly exciting new circuitry. The design of such a linear amplifier was not actually my original aim, which was simply to characterise the distortions of the small-signal section so I could experiment with output stages in an informed way. It was in fact rather disconcerting that it worked as well as it did with a conventional Class-B output stage, because the need for me to invent brilliant new output configurations (if, indeed, I were capable of that) seemed a lot less pressing. I think I can say without fear of successful contradiction that it really was a major step forward. In fact, since then I have only made public one new output stage concept: the Crossover Displacement (XD) principle. But I can assure you that much more has gone on behind the scenes.

After further experimentation, I felt I could write a pretty comprehensive guide to distortion and power amplifier design. *Electronics World* were good enough to give me all the space I asked for, and I believe I succeeded; here, in eight chapters, is the result.

The first chapter introduced the surprisingly simple equations that determine the open-loop gain of the conventional three-stage amplifier, pointed out that the input stage gives current-drive to the VAS and is not a voltage amplifier (a vital fact that a surprising number of people appeared unaware of) and gave a concise summary of the known distortion mechanisms as they stood at that time. There were only seven then, but now there are thirteen at the last count. I also described the extremely important technique of model amplifiers. These are amplifiers where the input and VAS stages are standard, but the Class-B output stage is replaced by a very linear Class-A output stage which only needs to be able to drive the feedback path without adding significant distortion. This allows the behaviour of the small-signal section to be studied in isolation, without the crossover distortion of the output stage, or its loading effects on the VAS. The supply rails are usually 'full-size' as some distortion mechanisms are dependent on the signal voltage, but useful lessons can still be learned when running from opamp supply rails; for example input stage distortion is unaffected.

DISTORTION IN POWER AMPLIFIERS

August 1993

It seems surprising that in a world which can build the Space Shuttle and detect the echoes of the birth of the universe, we still have to tolerate distortion in power amplifiers. Leafing through recent reviews and specifications shows claims for full-power total harmonic distortion ranging more than three orders of magnitude between individual designs, a wider range than any other parameter.

Admittedly the higher end of this range is represented by subjectivist equipment that displays dire linearity, presumably with the intention of implying that other nameless audio properties have been given priority over the mundane business of getting the signal from input to output without bending it.

Given the juggernaut rate of progress in most branches of electronics this seems to me anomalous, and especially notable in view of the many advanced analogue techniques used in op-amp design; after all power amps are only op-amps with boots on. One conclusion seems inescapable: a lot of power amplifiers generate much more distortion than they need to.

This series attempts to show exactly why amplifiers distort, and how to stop them doing it, culminating in a practical design for an ultra-linear amplifier. It should perhaps be said at the outset that none of this depends on excessively high levels of negative feedback. Many of the techniques described here are also entirely applicable to discrete op-amps, headphone drivers, and similar circuit blocks. Since we are almost in the twenty-first century I have ignored valve amplifiers.

Since mis-statements and confusions are endemic to audio, I have based these articles almost entirely on my own experimental work backed up with spice circuit simulation; much of the material relates specifically to bipolar transistor output stages, though a good deal is also relevant to mosfet amplifiers. Some of the statements made may seem controversial, but I believe they are all correct. If you think not, please tell me, but only if you have some real evidence to offer.

The fundamental reason why amplifier distortion persists is, of course, because it is a difficult technical problem to solve. A Science proverbially becomes an Art when there are more than seven variables, and since it will emerge that there are seven major distortion mechanisms to the average amplifier, we would seem to be nicely balanced on the boundary of the two cultures. Given so many significant sources of unwanted harmonics, overlaid and sometimes partially cancelling, sorting them out is a nontrivial task.

Make your amplifier as linear as possible before applying NFB has long been a cliche, (one that conveniently ignores the difficulty of running a high gain amp without any feedback) but virtually no dependable advice on how to perform this desirable linearisation has been published. The two factors are the basic linearity of the forward path, and the amount of negative feedback applied to further straighten it out. The latter cannot be increased beyond certain limits or high-frequency stability is put in peril, whereas there seems no reason why open-loop linearity could not, in principle, be improved without limit, leading us to the Holy Grail of the distortionless amplifier. This series therefore takes as its prime aim the understanding and improvement of open-loop linearity. As it proceeds we will accrete circuit blocks to culminate in two practical amplifier designs that exploit the techniques presented here.

How an amplifier (really) works

Figure 22.1 shows the usual right trusty and well-beloved power amp circuit drawn as standard is possible. Much has been written about this configuration, though its subtlety and quiet effectiveness are usually overlooked, and the explanation below therefore touches on several aspects that seem to be almost unknown. It has the merit of being docile enough to be made into a workable amplifier by someone who has only the sketchiest of notions as to how it works.

The input differential pair implements one of the few forms of distortion cancellation that can be relied upon to keep working in all weathers. This is because the transconductance of the input pair is determined by the physics of transistor action rather than matching of variable parameters such as beta; the logarithmic relation between I_c and V_{be} is proverbially accurate over some eight or nine decades of current variation.

The voltage signal at the voltage amplifier stage (hereafter VAS) transistor base is typically a couple of millivolts, looking rather like a distorted triangle wave. Fortunately the voltage here is of little more than academic interest, as the circuit topology essentially consists of a transconductance amp (voltage-difference input to current output) driving into a transresistance (current-to-voltage converter) stage. In the first case the exponential V_{be}/I_c law is straightened out by the differential-pair action, and in the second the global (overall) feedback factor at LF is sufficient to linearise the VAS, while at HF shunt negative feedback (hereafter NFB) through C_{dom} conveniently takes over VAS-linearisation while the overall feedback factor is falling. The behaviour of Miller dominant-pole compensation in this stage is exceedingly elegant, and not at all just a case of finding the most vulnerable transistor and slugging it. As frequency rises and C_{dom} begins to take effect, negative feedback is no longer applied globally around the whole amplifier, which would include the higher poles, but instead is seamlessly transferred to a purely local role in linearising the VAS. Since this stage effectively contains a single gain transistor, any amount of NFB can be applied to it without stability problems.

The amplifier operates in two regions; the LF, where open-loop gain is substantially constant, and HF, above the dominant-pole breakpoint, where the gain is decreasing steadily at 6 dB/octave. Assuming the output stage is unity-gain, three simple relationships define the gain in these two regions:

$$\text{LF gain} = g_m \times \text{beta} \times R_c \qquad (1)$$

At least one of the factors that set this (beta) is not well-controlled and so the LF gain of the amplifier is to a certain extent a matter of potluck; fortunately this doesn't matter as long as it is high enough to give a suitable level of NFB to eliminate LF distortion. The use of the word 'eliminate' is deliberate, as will be seen later. Usually the LF gain, or HF local feedback-factor, is made high by increasing the effective value of the VAS collector impedance R_c, either by the use of a currentsource collector-load, or by some form of bootstrapping.

The other important relations are:

$$\text{HF gain} = g_m / (\omega \times C_{dom}) \qquad (2)$$

$$\text{Dominant pole freq } P1 = \frac{1}{(\omega \cdot C_{dom} \cdot \beta \cdot R_c)} \qquad (3)$$

where $\omega = 2\pi f$.

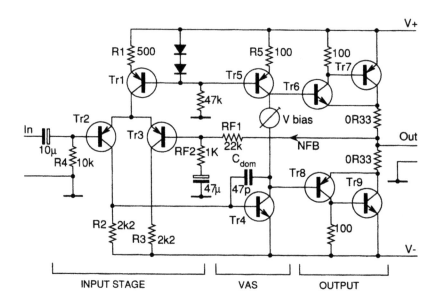

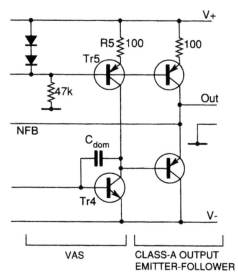

Figure 22.1 Figure 22.1(a) Conventional class-B power amp circuit. The apparent simplicity of circuitry conceals a series of sophisticated operating mechanisms. The lower drawing (Figure 22.1(b)) shows an adaptation of the output stage for small signal modelling.

In the HF region, things are distinctly more difficult as regards distortion, for while the VAS is locally linearised, the global feedback-factor available to linearise the input and output stages is falling steadily at 6 dB/octave. For the time being we will assume that it is possible to define an HF gain (say N dB at 20 kHz) which will assure stability with practical loads and component

variations. Note that the HF gain, and therefore both HF distortion and stability margin, are set by the simple combination of the input stage transconductance and one capacitor, and most components have no effect on it at all.

It is often said that the use of a high VAS collector impedance provides a current drive to the output devices, often with the implication that this somehow allows the stage to skip quickly and lightly over the dreaded crossover region. This is a misconception—the collector impedance falls to a few kΩ at HF, due to increasing local feedback through C_{dom}. In any case it is very doubtful if true current drive would be a good thing since calculation shows that a low-impedance voltage drive minimises distortion due to beta-unmatched output halves,[1] and it certainly eliminates distortion mechanism four described later.

The seven distortions

In the typical amplifier THD is often thought to be simply due to the Class-B nature of the output stage, which is linearised less effectively as the feedback factor falls with increasing frequency. However the true situation is much more complex as the small-signal stages can generate significant distortion in their own right in at least two different ways. This can easily exceed the output stage distortion at high frequencies. It seems inept to allow this to occur given the freedom of design possible in the small-signal section.

Include all the ills that a class-B stage is prone to and then there are seven major distortion mechanisms.

Distortion in power amplifiers arises from:

1. *Non-linearity in the input stage.* If this is a carefully-balanced differential pair then distortion is typically only measurable at HF, rises at 18 dB/octave, and is almost pure third harmonic.

   If the input pair is unbalanced (which from published circuitry it usually is) then the HF distortion emerges from the noise floor earlier. As frequency increases, it rises at 12 dB/octave as it is mostly second harmonic.

2. *Non-linearity in the voltage amplifier stage* surprisingly does not always figure in the total distortion. If it does, it remains constant until the dominant-pole frequency P1 is reached, and then rises at 6 dB/octave. With the configurations discussed here, it is always second harmonic.

   Usually the level is very low due to linearising negative feedback through the dominant-pole capacitor. Hence if you crank up the *local* VAS open-loop gain, for example by cascoding or putting more current-gain into the local VAS/C_{dom} loop, and attend to mechanism four below, you can usually ignore VAS distortion.

3. *Non-linearity in the output stage,* which is naturally the obvious source. This, in a Class-B amplifier, will be

a complex mix of large-signal distortion and crossover effects, the latter generating a spray of high-order harmonics, and in general rising at 6 dB/octave as the amount of negative feedback decreases. Large-signal THD worsens with 4 Ω loads and worsens again at 2 Ω. The picture is complicated by dilatory switch-off in the relatively slow output devices, ominously signalled by supply current increasing in the top audio octaves.

4. *Loading of the VAS by the non-linear input impedance of the output stage.* When all other distortion sources have been attended to, this is the limiting distortion factor at LF (say below 2 kHz). It is simply cured by buffering the VAS from the output stage. Magnitude is essentially constant with frequency, though overall effect in a complete amplifier becomes less as frequency rises and feedback through C_{dom} starts to linearise the VAS.

5. *Non-linearity caused by large rail-decoupling capacitors feeding the distorted signals on the supply lines into the signal ground.* This seems to be the reason many amplifiers have rising THD at low frequencies. Examining one commercial amplifier kit, I found that rerouting the decoupler ground-return reduced THD at 20 Hz by a factor of three.

6. *Non-linearity caused by induction of Class-B supply currents into the output, ground, or negative-feedback lines.* This was highlighted by Cherry [3] but seems to remain largely unknown; it is an insidious distortion that is hard to remove, though when you know what to look for on the THD residual, it is fairly easy to identify. I suspect that a large number of commercial amplifiers suffer from this to some extent.

7. *Non-linearity resulting from taking the NFB feed from slightly the wrong place near where the power-transistor Class-B currents sum to form the output.* This may well be another common defect.

Having set down what Mao might have called The Seven Great Distortions—Figure 22.2 shows the location of these mechanisms diagrammatically—we may pause to put to flight a few Paper Tigers. The first is common-mode distortion in the input stage, a spectre that tends to haunt the correspondence columns. Since it is fairly easy to make an amplifier with less than < 0.00065% THD (1 kHz) without paying any special attention to this, it cannot be too serious a problem. A more severe test is to apply the full output voltage as a common-mode signal, by running the amplifier as a unity-gain voltage-follower. If this is

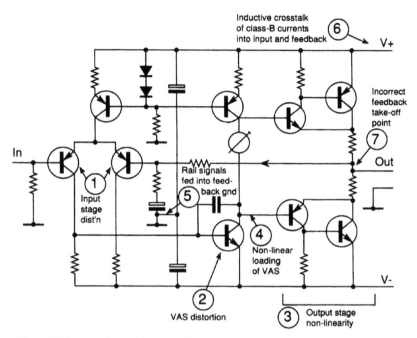

Figure 22.2 A topology of distortion: the location of the seven distortion mechanisms.

done using a model (see below for explanation) small-signal version of Figure 22.1, with suitable attention to compensation, then it yields less than 0.001% at 8 V r.m.s. across the audio band. It therefore appears that the only real precaution required against common-mode distortion is to use a tail current-source for the input pair.

The second distortion conspicuous by its absence in the list is the injection of distorted supply-rail signals directly into the amplifier circuitry. Although this putative mechanism has received a lot of attention, [4] dealing with Distortion five above by proper grounding seems to be all that is required . . . Once again, if triple-zero THD can be attained using simple unregulated supplies and without specifically addressing power supply rejection ratio, (which it reliably can be) then much of the work done on regulated supplies may be of doubtful utility. However, PSRR does need some attention if the hum/noise performance is to be of the first order.

A third mechanism of doubtful validity is thermal distortion, allegedly induced by parameter changes in semiconductor devices whose instantaneous power dissipation varies over a cycle. This would presumably manifest itself as a distortion increase at very low frequencies, but it simply does not seem to happen.

The major effects would be expected in Class-B output stages where dissipation can vary wildly over a cycle. However drivers and output devices have relatively large junctions with high thermal inertia. Low frequencies are of course also where the NFB factor is at its maximum.

The advantages of being conventional

The input pair not only provides the simplest way of making a d.c. coupled amp with a dependably small output offset voltage, but can also (given half a chance) completely cancel the second-harmonic distortion which would be generated by a single-transistor input stage. One vital condition must be met; the pair must be accurately balanced by choosing the associated components so that the two collector currents are equal. (The 'typical' component values shown in Figure 22.1 *do not* bring about this most desirable state of affairs.)

The input devices work at a constant and near-equal V_{ce}, giving good thermal balance.

The input pair has virtually no voltage gain so no low-frequency pole can be generated by Miller effect in the Tr_2 collector-base capacitance. All the voltage gain is provided by the VAS stage, which makes for easy compensation. Feedback through C_{dom} lowers VAS input and output impedances, minimising the effect of input-stage and output stage capacitance. This is often known as pole-splitting; [2] the pole of the VAS is moved downwards in frequency to become the dominant pole, while the input-stage pole is pushed up in frequency.

The VAS Miller compensation capacitance smoothly transfers NFB from a global loop which may be unstable, to the VAS local loop that cannot be. It is quite wrong to state that *all* the benefits of feedback are lost as the frequency increases above the dominant pole, as the VAS is still being linearised. This position of C_{dom} also swamps the rather variable C_{cb} of the VAS transistor.

To return to our list of the unmagnificent seven, note that only Distortion three is *directly* due to O/P stage non-linearity, though numbers 4–7 all result from the Class-B nature of the typical output stage.

The performance

The THD curve for the standard amplifier is shown in Figure 22.3. As usual the distortion increases with frequency and, as we shall see later, would give grounds for suspicion if it did not. The flat part of the curve below 500 Hz represents non-frequency-sensitive distortion rather than the noise floor, which for this case is at about the 0.0005% level. Above 500 Hz the distortion rises at an increasing rate, rather than a constant

number of dB/octave, due to the combination of Distortions 1, 2, 3 and 4. (In this case Distortions 5, 6 and 7 have been carefully eliminated to keep things simple. This is why the distortion performance looks good already, and the significance of this should not be overlooked.) It is often written that having distortion constant across the audio band is a good thing. This is a most unhappy conclusion as the only practical way to achieve this with a Class-B amplifier is to *increase* the distortion at LF, for example by allowing the VAS to distort significantly.

It should now be clear why it can be hard to wring linearity out of a snake-pit of contending distortions. A circuit-value change is likely to alter at least two of the distortion mechanisms, and probably change the open-loop gain as well. In the coming articles I shall demonstrate how each of these mechanisms can be measured and manipulated separately.

Determining open-loop linearity

Improving something demands its measurement, and so it is essential to examine the open-loop linearity of typical power-amp circuits. This cannot in general be done directly, so it is necessary to measure the NFB factor and calculate open-loop distortion from the usual closed-loop data. It is assumed that the closed-loop gain is fixed by operational requirements.

Finding the feedback-factor is at first sight difficult, as it means determining the open-loop gain. The

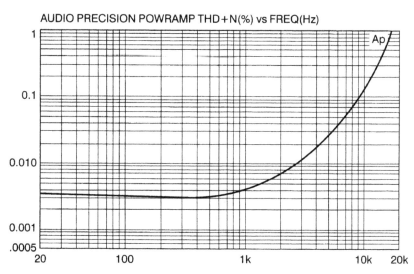

Figure 22.3 The distortion performance of the class-B amplifier shown in Figure 22.1(a).

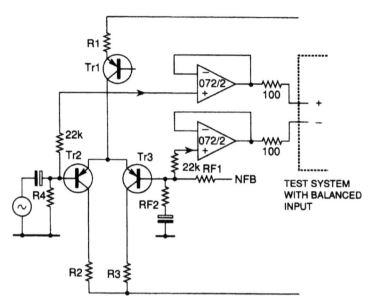

Figure 22.4 Test circuit for measuring open-loop gain directly. The measurement accuracy depends on the test gear CMRR.

standard methods for measuring op-amp open-loop gain involve breaking feedback-loops and manipulating closed-loop gains, procedures that are likely to send the average power-amplifier into fits. However, the need to measure this parameter is inescapable, as a typical circuit modification—e.g. changing the value of R_2—will change the open-loop gain as well as the linearity, and to prevent total confusion it is necessary to keep a very clear idea of whether the observed change is due to an improvement in open-loop linearity or merely because the open-loop gain has risen. It is wise to keep a running check on the feedback-factor as work proceeds, and so the direct method of open-loop gain measurement shown in Figure 22.4 was evolved.

Direct open-loop gain measurement

Since the amplifier shown in Figure 22.1 is a differential amplifier, its open-loop gain is simply the output divided by the voltage difference between the inputs. If the output voltage is kept effectively constant by providing a swept-frequency constant voltage at the +ve input, then a plot of open-loop gain versus frequency is obtained by measuring the error-voltage between the inputs, and referring this to the output level. This yields an upside-down plot that rises at HF rather than falling, as the differential amplifier requires more input for

the same output as frequency increases, but the method is so quick and convenient that this can be lived with. Gain is plotted in dB with respect to the chosen output level (+16 dBu in this case) and the actual gain at any frequency can be read off simply by dropping the minus sign. Figure 22.5 shows the plot for the amplifier in Figure 22.1.

The HF-region gain slope is always 6 dB/octave unless you are using something special in the way of compensation and, by the Nyquist rules, must continue at this slope until it intersects the horizontal line representing the feedback factor provided that the amplifier is stable. In other words, the slope is not being accelerated by other poles until the loop gain has fallen to unity, and this provides a simple way of putting a lower bound on the next pole P2; the important P2 frequency (which is usually somewhat mysterious) must be above the intersection frequency if the amplifier is seen to be stable.

Given test gear with a sufficiently high common-mode-rejection-ratio balanced input, the method of Figure 22.4 is simple; just buffer the differential inputs from the cable capacitance with *TL072* buffers, placing negligible loading on the circuit if normal component values are used. Be particularly wary of adding stray capacitance to ground to the—ve input, as this directly imperils amplifier stability by adding an extra feedback pole. Short wires from power amplifier to buffer IC

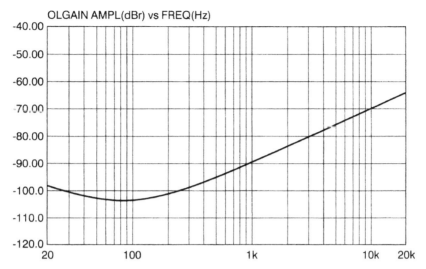

OLGAIN AMPL(dBr) vs FREQ(Hz)

Figure 22.5 Open-loop gain versus frequency plot for Figure 22.1. Note that the curve rises as gain falls, because the amplifier error is the actual quantity measured.

can usually be unscreened as they are driven from low impedances.

The test gear input CMRR defines the maximum open-loop gain measurable; I used an Audio Precision System-1 without any special alignment of CMRR. A calibration plot can be produced by feeding the two buffer inputs from the same signal; this will probably be found to rise at 6 dB/octave, being set by the inevitable input asymmetry. This must be low enough for amplifier error signals to be above it by at least 10 dB for reasonable accuracy. The calibration plot will flatten out at low frequencies, and may even show an LF rise due to imbalance of the test gear input-blocking capacitors; this can make determination of the lowest pole P1 difficult, but this is not usually a vital parameter in itself.

Model amplifiers

The first two distortions on the list can dominate amplifier performance and need to be studied without the complications introduced by a Class-B output stage. This can be done by reducing the circuit to a model amplifier that consists of the small-signal stages alone, with a very linear Class A emitter-follower attached to the output to allow driving the feedback network. Here 'small-signal' refers to current rather than voltage, as the model amplifier should be capable of giving a full power-amp voltage swing, given sufficiently high rail voltages. From

Figure 22.2 it is clear that this will allow study of Distortions 1 and 2 in isolation, and using this approach it will prove relatively easy to design a small-signal amplifier with negligible distortion across the audio band. This is the only sure foundation on which to build a good power amplifier.

A typical plot combining Distortions 1 and 2 from a model amp is shown in Figure 22.6, where it can be seen that the distortion rises with an accelerating slope, as the initial rise at 6 dB/octave from the VAS is contributed to and then dominated by the 12 dB/octave rise in distortion from an unbalanced input stage.

The model can be powered from a regulated current-limited PSU to cut down the number of variables, and a standard output level chosen for comparison of different amplifier configurations. The rails and output level used for the results in these articles was ±15 V and +16 dBu. The rail voltages can be made comfortably lower than the average amplifier HT rail, so that radical bits of circuitry can be tried out without the creation of a silicon cemetery around your feet. It must be remembered that some phenomena such as input-pair distortion depend on absolute output level, rather than the proportion of the rail voltage used in the output swing, and will be worse by a mathematically predictable amount when the real voltage swings are used. The use of such model amplifiers requires some caution, and cannot be applied to bipolar output stages whose behaviour is heavily influenced by the sloth and low current gain of the power devices. As

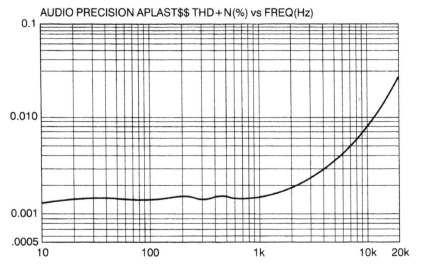

Figure 22.6 The distortion from a model amplifier, produced by the input pair and the voltage amplifier stage. Note increasing slope as input pair distortion begins to add to VAS distortion.

another general rule, if it is not possible to lash on a real output stage quickly and get a stable and workable power amplifier; the model may be dangerously unrealistic.

Glossary

Several abbreviations will be used throughout this series to keep its length under control.

l.f.   Relating to amplifier action below the dominant pole, where the open-loop gain is assumed to be essentially flat with frequency.

h.f.   Amplifier behaviour above the dominant pole frequency, where the open-loop gain is usually falling at 6 dB/octave.

i/p   Input.

p1   The first open-loop response pole, and its frequency in Hz.

nfb   Negative feedback.

References

1. Oliver *Distortion in Complementary-Pair Class-B Amplifiers*, Hewlett-Packard Journal, February 1971, p 11.

2. Feucht *Handbook of Analog Circuit Design*, Academic Press 1990, p 256 (Pole-splitting).

3. Cherry 'A new distortion mechanism in class-B amplifiers', *JAES*, May 1981, p 327.

4. Duncan 'PSU regulation boosts audio performance', *EW + WW*, October 1992, p 818.

Chapter 23

Distortion in power amplifiers

Part II: the input stage
(Electronics World, September 1993)

The input stage of a power amplifier only has to handle small signals compared with the output signal, and so it might be thought that its contribution to the distortion of a complete power amplifier would be negligible. This is not so.

The vast majority of amplifiers use Miller dominant-pole compensation, in which a small capacitor essentially turns the VAS stage into an integrator. This gives very dependable stability, and has many other advantages, but the downside is that as the frequency increases, the amount of current that has to be pumped in and out of the Miller capacitor increases proportionally. Therefore the error voltage across the two inputs of the differential pair, which drives this current, also increases proportionally due to the global negative feedback. The signal levels here at say 20 kHz are surprisingly large, and make it essential to consider the linearity of the input stage carefully. In practical terms this means using emitter degeneration resistors of 100Ω, and a current-mirror to phase-sum the collector currents.

SPICE simulation of the input stage was found to be very helpful, demonstrating clearly the effectiveness of emitter degeneration in linearising the voltage-current relationship. It is simple to simulate an input stage in isolation, so long as you take the precaution of providing a subsequent virtual-earth stage so the output current can be absorbed while keeping a constant voltage on the input stage output node.

The beauty of the differential pair is that it is one of the few places where the much-mentioned "cancellation of second-harmonic distortion" really does work reliably and without adjustment, thanks to the great predictability of the Vbe-Ic relationship of the bipolar transistor. To make this work, it is necessary to keep the collector currents of the two devices accurately equal, and this can be done very elegantly by a current-mirror in the collectors. This circuit element also doubles the open-loop gain of the overall amplifier, and doubles its slew-rate, so there is no doubt it is earning its keep.

The article is by no means a fully comprehensive guide to input stages. Other important issues were investigated later. These include common-mode distortion, which might be expected to cause difficulties [1] as the error signal between the two inputs is usually much smaller than the common-mode voltage on both of them. However, CM distortion only becomes a measurable problem with very low closed-loop gains (say two times) which are not customary in power amplifiers.

Another issue is the increased distortion and hum that occurs when the amplifier input is driven from a significant source impedance, say more than 200Ω. This occurs because the base currents drawn by the input transistor pair are not linear even if the amplifier output is completely distortion free [2]. This is why it is an extremely bad idea to put RC filters directly on the input, in a dim-witted attempt to stop ultrasonic interference. The R is sometimes as high as 10 kΩ, which will play merry heck with the distortion and hum performance of most amplifiers.

Simple emitter degeneration with resistors provided all the linearity required at the power levels used here, but more complex and linear input stages are also examined in the chapter. CFP input stages are a definite possibility, but the cross-quad and cascomp configurations described [3] do not give an improvement in linearity consistent, with their complexity, and the cross-quad has also a nasty habit of latching-up solid unless handled very carefully. There are several other methods of improving linearity by adding two or more transistors to extend the region over which the input stage is linear, for example the multi-tanh approach. Some of these look quite promising, but I have yet to explore them in any detail.

Up to the time of writing, the simple degenerated input pair with current-mirror has still proved adequate for all requirements. If and when output stage distortion is reduced substantially, this may no longer be the case, and there is, as usual, no room for complacency.

References

1. Self, D., *Audio Power Amplifier Design*, 6th edition, pp. 140–142. Focal Press 2013. ISBN 978-0-240-52613-3.

2. Ibid., pp. 142–150.

3. Ibid., pp. 135–138.

DISTORTION IN POWER AMPLIFIERS

September 1993

The input stage of an amplifier performs the critical duty of subtracting the feedback signal from the input, to generate the error signal that drives the output. It is almost invariably a differential transconductance stage; a voltage-difference input results in a current output that is essentially insensitive to the voltage at the output port. Its design is also frequently neglected, as it is assumed that the signals involved must be small, and that its linearity can therefore be taken lightly compared with that of the voltage amplifier stage (VAS) or the output stage. This is quite wrong, for a misconceived or even mildly wayward input stage can easily dominate HF distortion performance.

The input transconductance is one of the two parameters setting HF open-loop (o/l) gain, and thus has a powerful influence on stability and transient behaviour as well as distortion. Ideally the designer should set out with some notion of how much o/l gain at 20 kHz will be safe when driving worst-case reactive loads—a precise measurement method of open-loop gain was outlined last month—and from this a suitable combination of input transconductance and dominant-pole Miller capacitance can be chosen.

Many of the performance graphs shown here are taken from a model (small-signal stages only) amplifier with a Class-A emitter-follower output, at +16 dBu on ±15 V rails. However, since the output from the input pair is in current form, the rail voltage in itself has no significant effect on the linearity of the input stage. It is the current swing at its output that is the crucial factor.

Vive la differential

The primary motivation for using a differential pair as the input stage of an amplifier is usually its low DC offset. Apart from its inherently lower offset due to the cancellation of the V_{be} voltages, it has the added advantage that its standing current does not have to flow through the feedback network. However a second powerful reason is that its linearity is far superior to single-transistor input stages. Figure 23.1 shows three versions, in increasing order of sophistication. The resistor-tail version in Figure 23.1(a) has poor CMRR and PSRR and is

generally a false economy; it will not be further considered. The mirrored version in Figure 23.1(c) has the best balance, as well as twice the transconductance of that in Figure 23.1(b).

Intuitively, the input stage should generate a minimal proportion of the overall distortion because the voltage signals it handles are very small, appearing as they do upstream of the VAS that provides almost all the voltage gain. However, above the first pole frequency P1, the current required to drive C_{dom} dominates the proceedings, and this remorselessly doubles with each octave, thus:

$$I_{pk} = 2\pi F \cdot C_{dom} \cdot V_{pk} \qquad (1)$$

For example the current required at 100 W, 8 Ω and 20 kHz, with a 100 pF C_{dom} is 0.5 mA peak, which may be a large proportion of the input standing current, and so the linearity of transconductance for large current excursions will be of the first importance if we want low distortion at high frequencies.

Figure 23.2, *curve A*, shows the distortion plot for a model amplifier (at +16 dBu output) designed so that the distortion from all other sources is negligible compared with that from the carefully balanced input stage. With a small-signal class A stage this essentially reduces to making sure that the VAS is properly linearised. Plots are shown for both 80 kHz and 500 kHz measurement bandwidths to show both HF behaviour and LF distortion. It demonstrates that the distortion is below the noise floor until 10 kHz, when it emerges and heaves upwards at a precipitous 18 dB/octave.

This rapid increase is due to the input stage signal current doubling with every octave to drive C_{dom}; this means that the associated third harmonic distortion will quadruple with every octave increase. Simultaneously

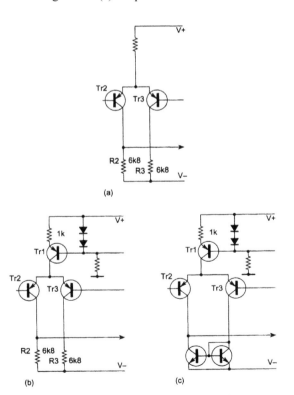

(a)

(b)   (c)

Figure 23.1 Three versions of an input pair: (a) Simple tail resistor; (b) Tail currentsource; (c) With collector current-mirror to give inherently good Ic balance.

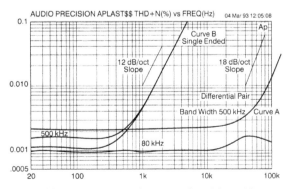

Figure 23.2 Distortion performance of model amplifier differential pair at A compared with singleton input at B. The singleton generates copious second-harmonic distortion.

the overall NFB available to linearise this distortion is falling at 6 dB/octave since we are almost certainly above the dominant pole frequency P1. The combined effect is an 18 dB/octave rise. If the VAS or the output stage were generating distortion, this would be rising at only 6 dB/octave and would look quite different on the plot.

This form of non-linearity, which depends on the rate-of-change of the output voltage, is the nearest thing to what we normally call TID, an acronym that now seems to be falling out of fashion. Slew-induced-distortion SID is a better description of the effect.

If the input pair is *not* accurately balanced, then the situation is more complex. Second as well as third harmonic distortion is now generated, and by the same reasoning this has a slope of closer to 12 dB/octave. This vital point requires examination.

Input stage in isolation

The use of a single input transistor (Figure 23.3(a)) sometimes seems attractive, where the amplifier is capacitor-coupled or has a separate DC servo; it at least promises strict economy. However, the snag is that this singleton configuration has no way to cancel the second-harmonics generated by its strongly-curved exponential V_{in}/I_{out} characteristic.[1] The result is shown in Figure 23.2 curve B, where the distortion is much higher, though rising at the slower rate of 12 dB/octave.

Although the slope of the distortion plot for the whole amplifier tells much, measurement of input-stage nonlinearity in isolation tells more. This may be done with the test circuit of Figure 23.4. The op-amp uses shunt feedback to generate an appropriate AC virtual earth at the input-pair output. Note that this current-to-voltage conversion op-amp requires a third −30 V rail to allow the i/p pair collectors to work at a realistic DC voltage—i.e. about one diode's-worth above the −15 V rail. R_f can be scaled to stop op-amp clipping without effect to the input stage. The DC balance of the pair may be manipulated by VR_1: it is instructive to see the THD residual diminish as balance is approached until, at its minimum amplitude, it is almost pure third harmonic.

The differential pair has the great advantage that its transfer characteristic is mathematically highly predictable.[2] The output current is related to the differential input voltage V_{in} by:

$$I_{out} = Ic \cdot \tanh(-V_{in}/2V_t) \tag{2}$$

where V_t is the usual 'thermal voltage' of about 26 mV at 25 °C and I_e the tail current.

This equation demonstrates that the transconductance, g_m, is highest at $V_{in} = 0$ when the two collector currents are equal, and that that the value of this maximum is proportional to the tail current, I_e. Note also that beta does not figure in the equation, and that the performance of the input pair is not significantly affected by transistor type.

Figure 23.5(a) shows the linearising effect of local feedback or degeneration on the voltage-in/current-out law. Figure 23.5(b) plots transconductance against input

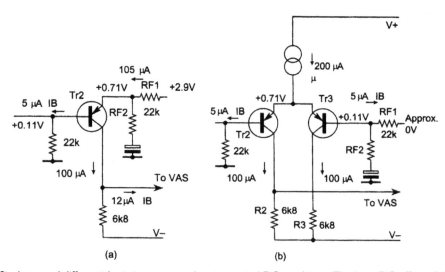

(a)   (b)

Figure 23.3 Singleton and differential pair input stages showing typical DC conditions. The large DC offset of the singleton (2.8 V) is largely due to all the stage current flowing through the feedback resistor RF1.

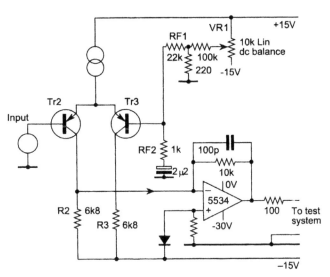

Figure 23.4 Test circuit for examining input stage distortion in isolation. The shunt-feedback opamp is biased to provide the right DC conditions for Tr2.

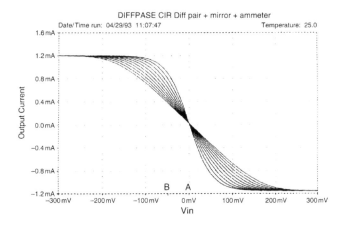

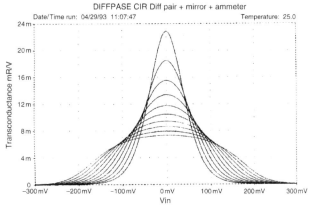

Figure 23.5 Effect of degeneration on input pair V/I law, showing how transconductance is sacrificed in favour of linearity (SPICE simulation).

voltage and demonstrates a reduced peak transconductance value but with the curve made flatter and more linear over a wider operating range. Adding emitter degeneration markedly improves input stage linearity at the expense of noise performance. Overall amplifier feedback factor is also reduced since the HF closed-loop gain is determined solely by the input transconductance and the value of the dominant-pole capacitor.

Input stage balance

One relatively unknown property of the differential pair in power amplifiers is its sensitivity to exact DC balance. Minor deviations from equality of I_c in the pair seriously upset the second-harmonic cancellation by moving the operating point from A in Figure 23.5(a) to B. Since the average slope of the characteristic is greatest at A, serious imbalance also reduces the open-loop gain. The effect of small amounts of imbalance is shown in Figure 23.6 and Table 23.1: for an input of −45 dBu a collector current imbalance of only 2% increases THD from 0.10% to 0.16%; for 10% imbalance this deteriorates to 0.55%. Unsurprisingly, imbalance in the other direction ($I_{c1} > I_{c2}$) gives similar results.

This gives insight [4] into the complex changes that accompany the simple changing the value of R_2. For example, we might design an input stage as per Figure 23.7(a),

where R_1 has been selected as 1 kΩ by uninspired guesswork and R_2 made highish at 10 kΩ in a plausible but misguided attempt to maximise o/l gain by minimising loading on Tr_1 collector. R_3 is also made 10 kΩ to give the stage a notional 'balance', though unhappily this is a visual rather than electrical balance. The asymmetry is shown in the resulting collector currents: this design will generate avoidable second harmonic distortion, displayed in the 10 kΩ curve of Figure 23.8.

However, recognising the importance of DC balancing, the circuit can be rethought as per Figure 23.7(b). If the collector currents are to be roughly balanced, then R_2 must be about $2 \times R_1$, as both have about 0.6 V across them. The effect of this change is shown in the 2.2 kΩ curve of Figure 23.8. The improvement is accentuated as the o/l gain has also increased by some 7 dB, though this has only a minor effect on the closed-loop linearity compared with the improved balance of the input pair. R_3 has been excised as it contributes little to stage balance.

The joy of current mirrors

While the input pair can be approximately balanced by the correct choice of R_1 and R_2, other circuit tolerances are significant and Figure 23.6 shows that balance is critical, needing to be accurate to at least 1% for optimal linearity. The standard current-mirror configuration

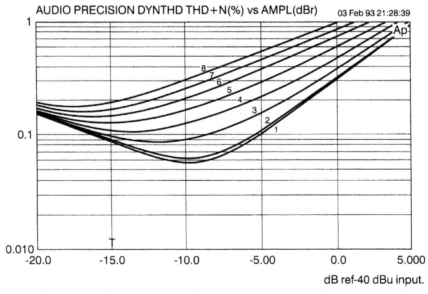

Figure 23.6 Effect of collector-current imbalance on an isolated input pair; the second harmonic rises well above the level of the third if the pair moves away from balance by as little as 2%.

Table 23.1 Key to Figure 23.6

Curve No.	I_c Imbalance (%)
1	0
2	0.5
3	2.2
4	3.6
5	5.4
6	6.9
7	8.5
8	10

Imbalance defined as deviation of I_c (per device) from that value which gives equal currents in the pair.

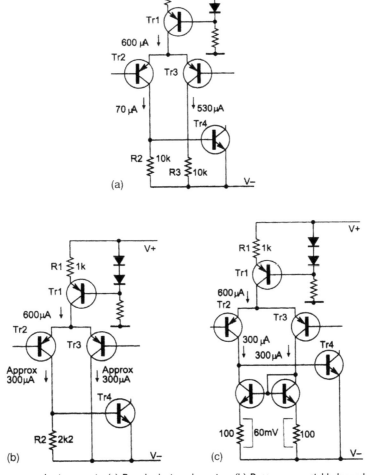

Figure 23.7 Improvements to the input pair: (a) Poorly designed version; (b) Better ... partial balance by correct choice of R2. (c) Best ... near-perfect Ic balance enforced by mirror.

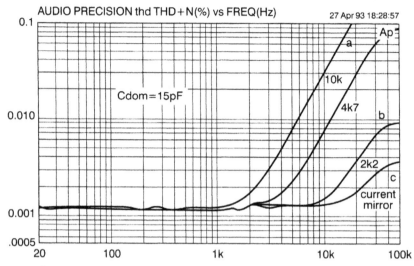

Figure 23.8 Distortion of model amplifier: (a) Unbalanced with R2 = 10 kΩ; (b) Partially balanced with R = 2.2 kΩ; (c) Accurately balanced by current-mirror.

shown in Figure 23.7(c) forces the two collector currents very close to equality, giving proper cancellation of second harmonic. The resulting improvement shows up in the current-mirror curve of Figure 23.8. There is also less DC offset due to unequal base currents flowing through input and feedback resistances; we often find that a power-amplifier improvement gives at least two separate benefits. This simple mirror has its own residual base current errors but they are not large enough to affect distortion.

The hyperbolic tangent law also holds for the mirrored pair, [3] though the output current swing is twice as great for the same input voltage as the resistor-loaded version. This doubled output occurs at the same distortion level as for the single-ended version, as linearity depends on the input voltage, which has not changed. Alternatively, to get the same output we can halve the input which, with a properly balanced pair generating only third harmonic, will produce just one-quarter the distortion, a pleasing result.

A low cost mirror made from discrete transistors forgoes the V_{be} matching available to IC designers, and so requires its own emitter degeneration for good current-matching. A voltage drop across the mirror emitter resistors in the range 30–60 mV will be enough to make the effect of V_{be} tolerances on distortion negligible If degeneration is omitted, there is significant variation in HF distortion performance with different specimens of the same transistor type. Adding a current mirror to

a reasonably well balanced input stage will increase the total o/l gain by at least 6 dB, and by up to 15 dB if the stage was previously poorly balanced. This needs to be taken into account in setting the compensation. Another happy consequence is that the slew-rate will be roughly doubled, as the input stage can now source and sink current into C_{dom} without wasting it in a collector load. If C_{dom} is 100 pF, the slewrate of Figure 23.7(b) is about 2.8 V/μs up and down, while Figure 23.7(c) gives 5.6 V/μs. The unbalanced pair in Figure 23.7(a) displays further vices by giving 0.7 V/μs positive-going and 5 V/μs negative-going.

Improving linearity

Now that the input pair has been fitted with a mirror, we may still feel that the HF distortion needs further reduction; after all, once it emerges from the noise floor it goes up eight times with each doubling of frequency, and so it is well worth pushing the turn point as far as possible up the frequency range. The input pair shown has a conventional value of tail-current. We have seen that the stage transconductance increases with I_c, and so it is possible to increase the g_m by increasing the tail-current, and then return it to its previous value (otherwise C_{dom} would have to be increased proportionately to maintain stability margins) by applying local NFB in the form of emitter-degeneration resistors. This ruse powerfully improves input linearity despite its rather unsettling flavour of something-for-nothing. The transistor nonlinearity can here be regarded as an internal nonlinear emitter resistance r_e, and what we have done is to reduce the value of this (by increasing I_c) and replace the missing part of it with a linear external resistor, R_e.

For a single device, the value of r_e can be approximated by:

$$r_e = 25/I_c \ \Omega \text{ (for } I_c \text{ in mA).} \tag{3}$$

Our original stage at Figure 23.9(a) has a perdevice I_c of 600 μA, giving a differential (i.e. mirrored) g_m of 23 mA/V and $r_e = 41.6 \ \Omega$. The improved version at Figure 23.9(b) has $I_c = 1.35$ mA and so $r_e = 18.6 \ \Omega$. Emitter degeneration resistors of 22 Ω are required to reduce the g_m back to its original value, as $18.6 + 22 = 41.6$. The distortion measured by the circuit of Figure 23.4 for a −40 dBu input voltage is reduced from 0.32% to 0.032%, which is an extremely valuable linearisation, and will translate into a distortion reduction at HF of about five times for a complete amplifier. For reasons that will emerge later the full advantage is rarely gained. The distortion remains a visually pure third harmonic so long as the input pair remains balanced. Clearly this sort of thing can only be pushed so far, as the reciprocal-law

reduction of r_e is limited by practical values of tail current. A name for this technique seems to be lacking; 'constant-g_m degeneration' is descriptive but rather a mouthful.

Since the standing current is roughly doubled so has the slew rate: 10 V/μs to 20 V/μs. Once again we gain two benefits for the price of one modification.

For still better linearity, various techniques exist. When circuit linearity needs a lift, it is often a good approach to increase the *local* feedback factor, because if this operates in a tight local NFB loop there is often little effect on the overall global-loop stability. A reliable method is to replace the input transistors with complementary-feedback (CFP or Sziklai) pairs, as shown in the stage of Figure 23.10(a). If an isolated input stage is measured using the test circuit of Figure 23.4, the constant g_m degenerated version shown in Figure 23.9(b) yields 0.35% third-harmonic distortion for a −30 dBu input voltage, while the CFP version gives 0.045%. Note that the input level here is 10 dB up on the previous example to get well clear of the noise floor. When this stage is put to work in a model amplifier, the third-harmonic distortion at a given frequency is roughly halved, assuming other distortion sources have been appropriately minimised. However, given the steep slope of input stage

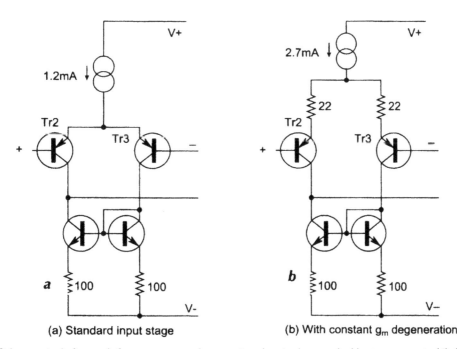

(a) Standard input stage   (b) With constant g_m degeneration

Figure 23.9 Input pairs before and after constant-gm degeneration showing how to double stage current while keeping transconductance constant: distortion is reduced by about ten times.

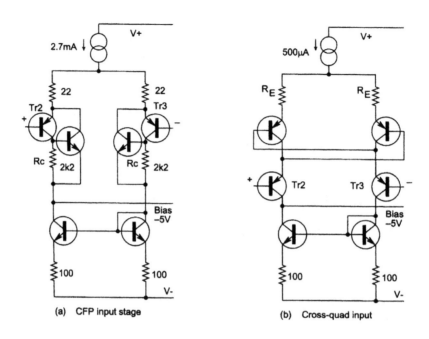

(a)   CFP input stage

(b)   Cross-quad input

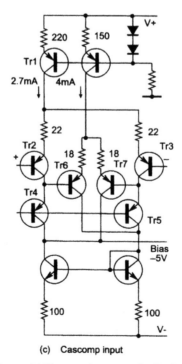

(c)   Cascomp input

Figure 23.10 Some enhanced differential pairs: (a) The complementary feedback pair; (b) The cross-quad; (c) The cascomp.

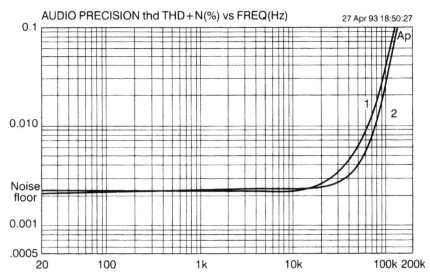

Figure 23.11 Whole-amplifier THD with normal and CFP input stages; input stage distortion only shows above noise floor at 20 kHz, so improvement occurs above this frequency. The noise floor appears high as the measurement bandwidth is 500 kHz.

distortion, this extends the low distortion regime up in frequency by less than an octave. See Figure 23.11.

The CFP circuit does require a compromise on the value of R_c, which sets the proportion of the standing current that goes through the NPN and PNP devices on each side of the stage. In general, a higher value of R_c gives better linearity, but more noise, due to the lower I_c in the NPN devices that are the inputs of the input stage, as it were, causing them to match less well the relatively low source resistances. 2.2 kΩ is a reasonable compromise.

Other elaborations of the basic input pair are possible. Power amp design can live with a restricted common-mode range in the input stage that would be unusable in an op-amp, and this gives the designer great scope. Complexity in itself is not a serious disadvantage as the small-signal stages of the typical amplifier are of almost negligible cost compared with mains transformers, heatsinks, etc.

Two established methods to produce a linear input transconductance stage (often referred to in opamp literature simply as a transconductor) are the cross-quad [5] and the cascomp [6] configurations. The cross-quad (Figure 23.10(b)) gives a useful reduction in input distortion when operated in isolation but is hard to incorporate in a practical amplifier because it relies on very low source resistance to tame the negative conductances inherent in its operation. The cross-quad works by imposing the

input voltage to each half across two base-emitter junctions in series, one in each arm of the circuit. In theory the errors due to non-linear r_e of the transistors is divided by beta, but in practice things seem less rosy.

The cascomp (Figure 23.10(c)) does not have this snag, though it is significantly more complex to design. Tr_2, Tr_3 are the main input pair as before, delivering current through cascode transistors Tr_4, Tr_5 (this does not in itself affect linearity) which, since they carry almost the same current as Tr_2, Tr_3 duplicate the input V_{bev} errors at their emitters. This is sensed by error diff-amp Tr_6, Tr_7 whose output currents are summed with the main output in the correct phase for error-correction. By careful optimisation of the (many) circuit variables, distortion at -30 dBu input can be reduced to about 0.016% with the circuit values shown. Sadly, this effort provides very little further improvement in whole-amplifier HF distortion over the simpler CFP input, as other distortion mechanisms are coming into play—for instance the finite ability of the VAS to source current into the other end of C_{dom}.

Power amplifiers with pretensions to sophistication sometimes add cascoding to the standard input differential amplifier. This does nothing to improve input stage linearity as there is no appreciable voltage swing on the input collectors; its main advantage is reduction of the high V_{ce} that the input devices work at. This allows cooler running, and therefore possibly improved thermal balance; a V_{ce} of 5 V usually works well. Isolating the input collector capacitance from the VAS input often allows C_{dom} to be somewhat reduced for the same stability margins, but it is doubtful if the advantages really outweigh the increased complexity.

Other considerations

As might be expected, the noise performance of a power amplifier is set by the input stage, and so it is briefly examined here. Power amp noise is not an irrelevance: a powerful amplifier is bound to have a reasonably high voltage gain and this can easily result in a faint but irritating hiss from efficient loudspeakers even when the volume control is fully retarded. In the design being evolved here the EIN has been measured at −120 dBu, which is only 7 or 8 dB inferior to a first-class microphone preamplifier. The inferiority is largely due to the source resistances seen by the input devices being higher than the usual 150 Ω microphone impedance. For example, halving the impedance of the feedback network shown in ptl (22 kΩ and 1 kΩ) reduces the EIN by approx 2 dB.

Slew rate is another parameter usually set by the input stage, and has a close association with HF distortion. The amplifier slew rate is proportional to the input's maximum-current capability, most circuit configurations being limited to switching the whole of the tail current to one side or the other. The usual differential pair can only manage half of this, as with the output slewing negatively half the tail-current is wasted in the input collector load R_2. The addition of an input current-mirror, as advocated, will double the slew rate in both directions. With a tail current of 1.2 mA, the slew rate is improved from about 5 V/μs to 10 V/μs. (for C_{dom} = 100 pF) The constant g_m degeneration method of linearity enhancement in Figure 23.9 further increases it to 20 V/μs. The mathematics of voltage-slewing is simple:

Slew rate = I/C_{dom} in V/μs for maximum I in μA, C_{dom} in pF.

The maximum output frequency for a given slew rate and voltage is:

$$F_{max} = S_r/2\pi V_{pk} = S_r/(2\pi \cdot \sqrt{2} \cdot V_{rms}) \qquad (4)$$

Likewise, a sinewave of given amplitude has a maximum slew-rate (at zero-crossing) of:

$$S_{r\,max} = dV/dt = \omega_{max} \cdot V_{pk} = 2\pi F V_{pk} \qquad (5)$$

So, for example, with a slew rate of 20 V/μs the maximum frequency at which 35 V r.m.s. can be sustained is 64 kHz, and if C_{dom} is 100 pF, then the input stage must be able to source and sink 2 mA peak.

A vital point is that the current flowing through C_{dom} must be sourced/sunk by the VAS as well as the input pair. Sinking is usually no problem, as the VAS common-emitter transistor can be turned on as hard as required. The current source or bootstrap at the VAS collector will however have a limited sourcing ability, and this can often turn out to be an unexpected limitation on the positive-going slew rate.

References

1. Gray and Meyer, *Analysis & Design of Analog Integrated Circuits*, Wiley 1984, p 172 (*exponential law of singleton*).

2. Gray and Meyer, *Analysis and Design of Analog Integrated Circuits*, p 194 (*tanh law of simple pair*).

3. Gray and Meyer, *Analysis and Design of Analog Integrated Circuits*, p 256 (*tanh law of current-mirror pair*).

4. Self, *Sound Mosfet Design Electronics & Wireless World*, September 1990.

5. Feucht, *Handbook of Analog Circuit Design*, Academic Press 1990, p 432.

6. Quinn, IEEE International Solid-State Circuits Conference, THPM 14.5, p 188 (Cascomp).

Chapter 24

Distortion in power amplifiers

Part III: the voltage-amplifier stage
(Electronics World, October 1993)

Unlike the input stage, the Voltage Amplifier Stage or VAS does have to handle big signals as its output swing is slightly larger than that of the output stage, and various distortion mechanisms that are of no importance in the input stage become very significant. The main reason why this stage works as well as it does is that it normally has a Miller dominant-pole capacitor connected between its input and output. As the frequency rises, this component smoothly transfers the global negative feedback right round the amplifier to a small stable loop around the VAS, providing it with a high level of local feedback and reducing its distortion.

Attempts at SPICE simulation of the VAS were not found to be very helpful, as it requires some sort of biasing system to make the collector sit at the right quiescent level, and DC simulations ignore important dynamic mechanisms that affect linearity.

The section on the various types of balanced VAS—which nowadays I prefer to call a push-pull VAS, as this makes it clearer what is its special property actually—was really little more than a placeholder, as space was limited and the topic is a big one. In the latest edition of *Audio Power Amplifier Design* [1], it has grown into a complete chapter of twenty-eight pages. Regrettably, the only conclusion that can be reached at the end of that chapter is that while a push-pull VAS as usually presented may *look* more sophisticated and apparently promise better linearity, the promise is worthless and the distortion is actually greater than for a single-ended VAS with a current-source load. There are all sorts of interesting extra problems; the accurate balance of the input stage collector currents, which is vital for optimal linearity is now not only not guaranteed but drifts disconcertingly as the transistors warm up after switch-on. In the current state of knowledge, the simplest approach is the best, which is not helpful to anyone hoping to show off some virtuoso circuit design skills.

The major omission from this chapter is the new information in the latest edition of *Audio Power Amplifier Design*. Experiments inspired and aided by Samuel Groner showed that in a simple VAS, the low-frequency distortion is due to Early effect [2], while the high-frequency distortion is due to local feedback through the transistor base-collector capacitance Cbc which varies with Vce and is therefore non-linear [3]. Adding an emitter–follower inside the VAS local loop disables both sources of distortion and allows excellent linearity. It does *not* do this primarily by enhancing the amount of local NFB around the VAS, despite the statement to that effect in this chapter, though its presence does increase low-frequency open-loop gain considerably, by about 30 dB. The emitter–follower disables the Cbc distortion because the distorted current required to drive the non-linear capacitance is now supplied by it, rather than the input of the VAS stage. Like the input stage current-mirror, the VAS emitter-follower is a hard-working circuit element that works in a quite subtle manner, costs very little and conveys more than one benefit.

References

1. Self, D., *Audio Power Amplifier Design*, 6th edition, pp. 201–229. Focal Press 2013. ISBN 978-0-240-52613-3.

2. Ibid., pp. 175–180.

3. Ibid., pp. 171–175.

DISTORTION IN POWER AMPLIFIERS

October 1993

The voltage-amplifier stage (or VAS) has often been regarded as the most critical part of a power-amplifier, since it not only provides all the voltage gain but also must deliver the full output voltage swing. This is in contrast to the input stage which may give substantial transconductance gain, but the output is in the form of a current. But as is common in audio design, all is not quite as it appears. A well-designed voltage amplifier stage will contribute relatively little to the overall distortion total of an amplifier, and if even the simplest steps are taken to linearise it further, its contribution sinks out of sight.

As a starting point, Figure 24.1 shows the distortion plot of a model amplifier with a Class-A output (±15 V rails, +16 dBu out). The model is as described in previous

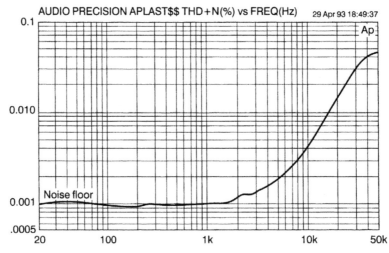

Figure 24.1 THD plot for model amp showing distortion below noise floor at low frequency, and increasing from 2 kHz to 20 kHz. The ultimate roll-off is due to the 80 kHz measurement bandwidth.

articles. No special precautions have been taken to linearise the input stage or the VAS and output stage distortion is negligible. It can be seen that the distortion is below the noise floor at low frequencies; the distortion slowly rising from about 1 kHz is coming from the voltage amplifier stage. At higher frequencies, where the VAS 6 dB/octave rise becomes combined with the 12 or 18 dB/octave rise of input stage distortion, we can see the accelerating distortion slope typical of many amplifier designs.

The main reason why the voltage amplifier stage generates relatively little distortion is because at LF, global feedback linearises the whole amplifier, while at HF the voltage amplifier stage is linearised by local negative feedback through C_{dom}.

Examining the mechanism

Isolating the voltage amplifier stage distortion for study requires the input pair to be specially linearised, or else its steeply rising distortion characteristic will swamp the VAS contribution. This is most easily done by degenerating the input stage which also reduces the open-loop gain. The reduced feedback factor mercilessly exposes voltage amplifier stage nonlinearity. This is shown in Figure 24.2, where the 6 dB/octave slope suggests origination in the VAS, and increases with frequency solely because the compensation is rolling-off the global feedback factor.

Confirming that this distortion is due solely to the voltage amplifier stage requires varying VAS linearity experimentally while leaving other circuit parameters unchanged. Figure 24.3 achieves this by varying the VAS negative rail voltage; this varies the proportion of its characteristic over which the voltage amplifier stage swings, and thus only alters the effective VAS linearity, as the important input stage conditions remain unchanged. The current-mirror must go up and down with the VAS emitter for correct operation, and so the V_{ce} of the input devices also varies, but this has no significant effect as can be proved by the unchanged behaviour on inserting cascode stages in the input transistor collectors.

The typical topology as shown in Figure 24.4(a) is a classical common emitter voltage amplifier stage with a current-drive input into the base. The small-signal characteristics, which set open-loop gain and so on, can be usefully simulated by the spice model shown in Figure 24.5, of a VAS reduced to its conceptual essentials. G is a current source whose value is controlled by the voltage-difference between R_{in} and R_{f2}, and represents the differential transconductance input stage. F represents the voltage amplifier stage transistor, and is a current source yielding a current of beta times that sensed flowing through ammeter V which, by spice convention, is a voltage source set to 0 V.

The value of beta, representing current-gain, models the relationship between VAS collector current and base current. R_c represents the total stage collector impedance, a typical real value being 22 kΩ. With suitable parameter values, this simple model provides a useful demonstration of relationships between gain, dominant-pole

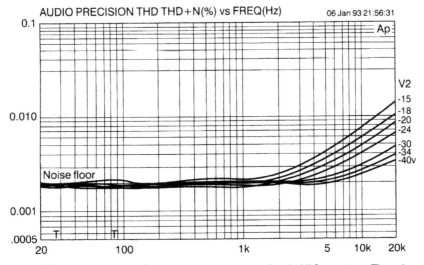

Figure 24.2 The change in HF distortion resulting from varying the negative rail in the VAS test circuit. The voltage amplifier stage distortion is only revealed by degenerating the input stage with 100 Ω resistors.

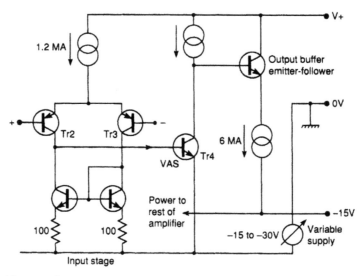

Figure 24.3 Voltage amplifier stage distortion test circuit. Although the input pair mirror moves up and down with the VAS emitter, the only significant parameter being varied is the available voltage swing at the collector.

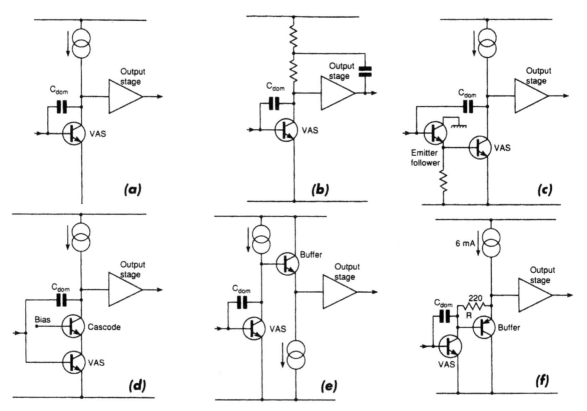

Figure 24.4 Six variations on a voltage amplifier stage: (a) conventional current source VAS, (b) conventional bootstrapped VAS, (c) increase in local NFB by adding emitter follower, (d) increase in local NFB by cascoding, (e) one method of buffering VAS collector from output stage, (f) alternative buffering arrangement uses bootstrapping resistor.

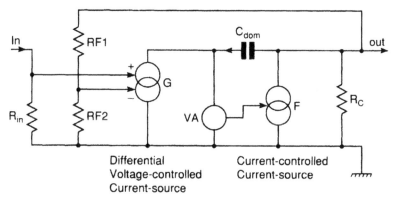

Figure 24.5 Conceptual spice model of differential input stage (G) and VAS (F). The current in F is beta times the current in G.

frequency, and input stage current outlined in the first article in this series. Injecting a small signal current into the output node from an extra current source also allows the fall of impedance with frequency to be examined.

The overall voltage gain clearly depends linearly on beta, which in real transistors may vary widely. Working on the trusty engineering principle that what cannot be controlled must be made irrelevant, local shunt NFB through C_{dom} sets the crucial HF gain that controls Nyquist stability. The LF gain below the dominant pole frequency P1 remains variable (and therefore so does P1) but is ultimately of little importance; if there is an adequate NFB factor for overall linearisation at HF then there are unlikely to be problems at LF where gain is highest. As for the input stage, the linearity of the voltage amplifier stage is not greatly affected by transistor type, given a reasonably high beta value.

Stage distortion

Voltage amplifier stage distortion arises from a curved transfer characteristic of the common-emitter amplifier, a small portion of an exponential.[1] This characteristic generates predominantly second-harmonic distortion, which, in a closed-loop amplifier, will increase at 6 dB/octave with frequency.

Distortion does not get worse for more powerful amplifiers as the stage traverses a constant proportion of its characteristic as the supply-rails are increased. This is not true of the input stage: increasing output swing increases the demands on the transconductance amp as the current to drive C_{dom} increases. The increased V_{ce} of the input devices does not measurably affect their linearity.

It seems ironic that VAS distortion only becomes clearly visible when the input pair is excessively degenerated—a pious intention to 'linearise before applying feedback' can make the closed loop distortion worse by reducing the open loop gain and hence the NFB factor available to linearise the VAS. In a real (non-model) amplifier with a distortive output stage, the deterioration will be worse.

The local open-loop gain of the VAS (that existing inside the local feedback loop closed by C_{dom}) should be high, so that the voltage amplifier stage can be linearised. This precludes a simple resistive load. Increasing the value of R_c will decrease the collector current of the transistor reducing its transconductance. This reduces voltage gain to the starting value.

One way to ensure sufficient gain is to use an active load. Either bootstrapping or a current source will do this effectively, though the current source is perhaps more dependable and is the usual choice for hi-fi or professional amplifiers.

The bootstrap promises more output swing as the collector of Tr_4 can soar above the positive rail. This suits applications such as automotive power amps that must make the best possible use of a restricted supply voltage.[2]

These two active-load techniques also ensure enough current to drive the upper half of the output stage in a positive direction right up to the supply rail. If the collector load were a simple resistor, this capability would certainly be lacking.

Checking the effectiveness of these measures is straightforward. The collector impedance may be determined by shunting the collector node to ground with decreasing resistance until the open loop gain falls by 6 dB indicating that the collector impedance is equal to the current value of the test resistor.

The popular current source version is shown in Figure 24.4(a). This works well, though the collector impedance is limited by the effective output resistance R_o of the voltage amplifier stage and the current source transistors [3] which is another way of saying that the improvement is limited by Early effect.

It is often stated that this topology provides current-drive to the output stage; this is only partly true. It is important to realise that once the local NFB loop has been closed by adding C_{dom} the impedance at the VAS output falls at 6 dB/octave for frequencies above P1. The impedance is only a few kΩ at 10 kHz, and this hardly qualifies as current-drive at all.

Bootstrapping (Figure 24.4(b)) works in most respects as well as a current source load, for all its old-fashioned flavour. The method has been criticised for prolonging recovery from clipping. I have no evidence to offer on this myself, but I can state that a subtle drawback definitely exists: LF open loop gain is dependent on amplifier output loading. The effectiveness of bootstrapping depends crucially on the output stage gain being unity or very close to it. However the presence of the output transistor emitter resistors means that there will be a load-dependant gain loss in the output stage significantly altering the amount by which the VAS collector impedance is increased. Hence the LF feedback factor is dynamically altered by the impedance characteristics of the loudspeaker load and the spectral distribution of the source material.

This has significance if the load is a quality speaker with impedance modulus down to 2 Ω, in which case the gain loss is serious. If anyone needs a new audio-impairment mechanism to fret about, then I offer this one in the confident belief that its effects, while measurable, are not of audible significance.

The standing d.c. current also varies with rail voltage. Since accurate setting and maintaining of quiescent current is difficult enough, an extra source of possible variation is decidedly unwelcome.

A less well known but more dependable form of bootstrapping is available if the amplifier incorporates a unity gain buffer between the VAS collector and the output stage as shown in Figure 24.4(f), where R_c is the collector load, defining the VAS collector current by establishing the V_{be} of the buffer transistor across itself. This is constant, and R_c is therefore bootstrapped and appears to the VAS collector as a constant current source.

In this sort of topology a voltage amplifier stage current of 3 mA is quite sufficient, compared with the 6 mA standing current in the buffer stage. The voltage amplifier stage would in fact work well with collector currents down to 1 mA, but this tends to compromise linearity at the high-frequency, high-voltage corner of the operating envelope, as the VAS collector current is the only source for driving current into C_{dom}.

Voltage stage enhancements

Figure 24.2, which shows only VAS distortion, clearly indicates the need for further improvement over that given inherently by the presence of C_{dom} if an amplifier is to avoid distortion. While the virtuous approach might be an attempt to straighten the curved voltage amplifier stage characteristic, in practice the simplest method is to increase the amount of local negative feedback through this capacitance. Equation 1 in the first article shows that the LF gain (i.e. the gain before C_{dom} is connected) is the product of input stage transconductance, Tr_4 beta and the collector impedance R_c. The last two factors represent the VAS gain and therefore the amount of local NFB can be augmented by increasing either. Note that so long as the value of C_{dom} remains the same, the global feedback factor at HF is unchanged and so stability is not affected.

The effective beta of the stage can be substantially increased by replacing the VAS transistor with a Darlington, Figure 24.4(c). Adding an extra stage to a feedback amplifier always requires thought because, if significant additional phase-shift is introduced, the global loop stability may suffer. In this case the new stage is inside the Miller loop and so there is little likelihood of trouble. The function of such an emitter follower is sometimes described as 'buffering the input stage from the VAS' but its true function is linearisation by enhancement of local NFB.

Alternatively the stage collector impedance may be increased for higher local gain. This is could be done with a cascode configuration (Figure 24.4(d)) but the technique is only useful when driving a linear impedance rather than a Class-B output stage with its non-linear input impedance.

Assuming for the moment that this problem is dealt with, either by use of a Class-A output or by VAS-buffering, the drop in distortion is dramatic as is the beta-enhancement method. The gain increase is ultimately limited by Early effect in the cascode and current source transistors, and more seriously by the loading effect of the next stage. But it is of the order of 10 times and gives a useful improvement.

This is shown by curves A, B in Figure 24.6 where the input stage of a model amplifier has been over-degenerated

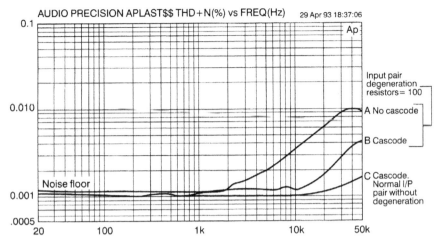

Figure 24.6 Showing the reduction of VAS distortion possible by cascoding. The results from adding an emitter follower to the voltage amplifier stage, as an alternative method of increasing local voltage amplifier stage feedback, are very similar.

with 100 Ω emitter resistors to bring out the voltage amplifier stage distortion more clearly.

Note that in both cases the slope of the distortion increase is 6 dB/octave. Curve C shows the result when a standard undegenerated input pair is combined with the cascoded VAS; the distortion is submerged in the noise floor for most of the audio band, being well below 0.001%.

This justifies my assertion that input stage and VAS distortion need not be a problem; we have all but eliminated distortions 1 and 2 from the list of seven given in the first article.

A cascode transistor also allows the use of a high-beta transistor for the voltage amplifier stage; these typically have a limited V_{ceo} that cannot withstand the high rail voltages of a high-power amplifier. There is a small loss of available voltage swing, but only about 300 mV, which is usually tolerable. Experiment shows that there is nothing to be gained by cascoding the current source collector load.

A cascode topology is often used to improve frequency response by isolating the upper collector from the C_{bc} of the lower transistor. In this case the frequency response is deliberately defined by a well defined passive component.

It is hard to say which technique is preferable; the emitter follower circuit is slightly simpler than the cascode version, which requires extra bias components, but the cost difference is minimal. When wrestling with these kind of financial decisions it as well to remember that the cost of a small-signal transistor

is often less than a fiftieth of that of an output device, and the entire small-signal section of an amplifier usually represents less than 1% of the total cost, when heavy metal such as the mains transformer and heatsinks are included.

Benefits of voltage drive

The fundamentals of linear voltage amplifier stage operation require that the collector impedance is high, and not subject to external perturbations. Thus a Class-B output stage, with large input impedance variations around the crossover point, is the worst possible load. The 'standard' amplifier configuration deserves tribute that it can handle this internal unpleasantness gracefully, 100 W/8 Ω distortion typically degrading only from 0.0008% to 0.0017% at 1 kHz assuming that the avoidable distortions have been eliminated. Note however that the effect becomes greater as the global feedback factor is reduced. There is little deterioration at HF, where other distortions dominate.[4]

The VAS buffer is most useful when LF distortion is already low, as it removes Distortion 4, which is—or should be—only visible when grosser non-linearities have been seen to. Two equally effective ways of buffering are shown in Figures 24.4(e) and 24.4(f).

There are other potential benefits to VAS buffering. The effect of beta mismatches in the output stage halves is minimised.[5] Voltage drive also promises

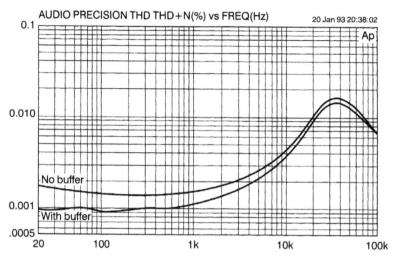

Figure 24.7 The beneficial effect of using a VAS buffer in a full scale Class B amplifier. Note that the distortion needs to be low already for the benefit to be significant.

the highest f_T from the output devices, and therefore potentially greater stability, though I have no data of my own to offer on this point. It is right and proper to feel trepidation about inserting another stage in an amplifier with global feedback, but since this is an emitter follower its phase shift is minimal and it works well in practice.

A VAS buffer put the right way up can implement a form of d.c. coupled bootstrapping that is electrically very similar to providing the voltage amplifier stage with a separate current-source.

The use of a buffer is essential if a VAS cascode is to do some good. Figure 24.7 shows before/after distortion for a full-scale power amplifier with cascode VAS driving 100 W into 8 Ω.

Balanced voltage amplifier stage

When linearising an amplifier before adding negative feedback one of the few specific recommendations made is usually the use of a balanced voltage amplifier stage—sometimes combined with a double input stage consisting of two differential amplifiers, one complementary to the other. The latter seems to have little to recommend it, as you cannot balance a stage that is already balanced, but a balanced (and, by implication, more linear) voltage amplifier stage has its attractions. However, as explained above, the distortion contribution from a properly designed VAS is negligible under

most circumstances, and therefore there seems to be little to be gained.

Two possible versions are shown in Figure 24.8; The first type gives approximately 10 dB more open loop gain than the standard, but this naturally requires an increase in C_{dom} if the same stability margins are to be maintained. In a model amplifier, any improvement in linearity can be wholly explained by this o/l gain increase, so this seems (not unexpectedly) an unpromising approach. Also, as John Linsley Hood has pointed out, [6] the standing current through the bias generator is ill-defined compared with the usual current source VAS. Similarly the balance of the input pair is likely to be poor compared with the current-mirror version. Two signal paths from the input stage to the VAS output must have the same bandwidth; if they do not then a pole-zero doublet is generated in the open-loop gain characteristic that will markedly increase settling-time after a transient. This seems likely to apply to all balanced voltage amplifier stage configurations. The second type is attributed by Borbely to Lender, [7] Figure 24.8 shows one version, with a quasi-balanced drive to the VAS transistor, via both base and emitter. This configuration does not give good balance of the input pair since it depends on the tolerances of R_2, R_3, the V_{be} of the voltage amplifier stage, and so on. Borbely has advocated using two complementary versions of this giving a third type, but this would not seem to overcome the objections and the increase in complexity is significant.

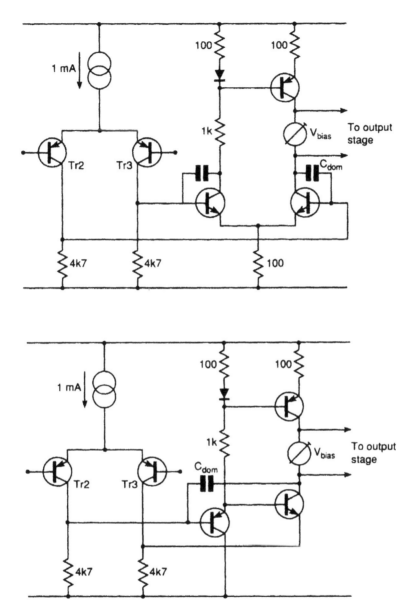

Figure 24.8 Two kinds of balanced voltage amplifier stage: Type I gives more open loop gain, but no better open loop linearity. Type 2 the circuit originated by Lender.

All balanced voltage amplifier stages seem to be open to the objection that the vital balance of the input pair is not guaranteed, and that the current through the bias generator is not well-defined. However one advantage would seem to be the potential for sourcing and sinking large currents into C_{dom}, which might improve the ultimate slew-rate and HF linearity of a very fast amplifier.

Open loop bandwidth

Acute marketing men will appreciate that reducing the LF open loop gain, leaving HF gain unchanged, must move the P1 frequency upwards, as shown in Figure 24.9. Open loop gain held constant up to 2 kHz appears so much better than open loop bandwidth restricted to 20 Hz. These two statements could describe near

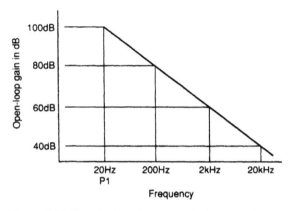

Figure 24.9 Showing how dominant pole frequency P1 can be altered by changing the LF open loop gain. The gain at HF, which determines Nyquist stability and HF distortion, is unaffected.

identical amplifiers, except that the first has plenty of open loop gain at LF while the second has even more. Both amplifiers have the same feedback factor at HF, where the amount available has a direct effect on distortion performance, and could easily have the same slew rate. Nonetheless the second amplifier somehow reads as sluggish and indolent, even when the truth of the matter is known.

Reducing low frequency open loop gain may be of interest to commercial practitioners but it also has its place in the dogma of the subjectivist. Consider it this way: firstly there is no engineering justification for it and, secondly, reducing the NFB factor will reveal more of the output stage distortion. NFB is the only weapon available to deal with this second item so blunting its edge seems ill-advised.

It is of course simple to reduce open loop gain by degenerating the input pair, but this diminishes it at HF as well as LF. To alter it at LF only requires engineering changes at the VAS. Figure 24.10 shows two ways. Figure 24.10(a) reduces gain by reducing the value of the collector impedance, having previously raised it with the use of a current source collector load. This is no way to treat a gain stage: loading resistors low enough to have a significant effect cause unwanted current variations in the VAS as well as shunting its high collector impedance, and serious LF distortion appears. While this sort of practice has been advocated in E&WW in the past, [8] it seems to have nothing to recommend it. Figure 24.10(b) also reduces overall open loop gain by adding a frequency insensitive component to the local shunt feedback around the voltage amplifier stage. The value of R_{NFB} is too high to load the collector significantly and therefore the full gain is available for local feedback at LF, even before C_{dom} comes into action. Figure 24.11 shows the effect on the open loop gain of a model amplifier for several values of R_{NFB}; this plot is in the format described in the first part of this series where error voltage is plotted rather than gain so the curve appears upside down compared with the usual presentation. Note that the dominant pole frequency is increased from 800 Hz to above 20 kHz by using a 220 kΩ value for R_{NFB}; however the gain at higher frequencies is unaffected and so is the stability. Although the amount of feedback available at 1 kHz has been decreased by nearly 20 dB, the distortion at +16 dBu output is only increased from below 0.001% to 0.0013%. Most of the reading is due to noise.

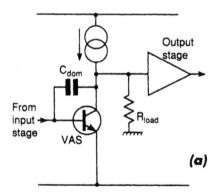

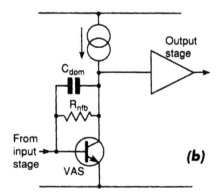

Figure 24.10 Two ways to reduce open loop gain: (a) simply loading down the collector. This is a cruel way to treat a VAS since current variations cause extra distortion. (b) local NFB with a resistor in parallel with Cdom. This looks crude, but actually works very well.

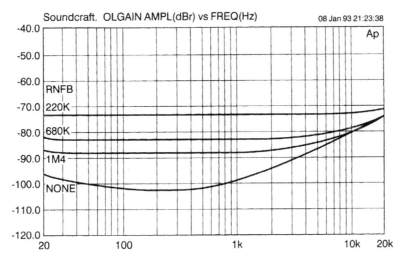

Figure 24.11 The result of voltage amplifier stage gain reduction by local feedback; the dominant pole frequency is increased from about 800 Hz to about 20 kHz, with high frequency gain hardly affected.

In contrast, reducing the open loop gain by just 10 dB through loading the VAS collector to ground requires a load of 4.7 kΩ which, under the same conditions, yields distortion of more than 0.01%.

It might seem that the stage which provides all the voltage gain and swing in an amplifier is a prime suspect for generating the major part of its non-linearity. In actual fact, this is unlikely to be true, particularly with a cascode VAS/current source collector load buffered from the output stage. Number 2 in the distortion list can usually be forgotten.

References

1. Gray & Meyer *Analysis & Design of Analog Integrated Circuits*, Wiley 1984, p 251 (VAS transfer characteristic).

2. Antognetti (Ed) *Power Integrated Circuits*, McGraw-Hill 1986, Section 9.3.2.

3. Gray & Meyer *Analysis & Design of Analog Integrated Circuits*, Wiley 1984, p 252 (Rco limit on VAS gain).

4. Self *Sound Mosfet Design*, *EW + WW*, September 1990, p 760.

5. Oliver *Distortion In Complementary-Pair Class-B Amplifiers, Hewlett-Packard Journal,* Feb 1971, p 11.

6. Linsley-Hood *Solid State Audio Power-3*, *EW + WW*, January 1990, p 16.

7. Borbely *A 60 W MOSFET Power Amplifier, The Audio Amateur,* Issue 2, 1982, p 9.

8. Hefley High Fidelity, Low Feedback, 200 W, *EW + WW*, June 92 p 45.

Chapter 25

Distortion in power amplifiers

Part IV: the power amplifier stages
(Electronics World, November 1993)

This article tackled the difficult subject of output stage design. It began, in Figure 25.1, with a demonstration of something that was dimly understood at the time but is now considered conventional wisdom—the fact that the distortion generated by Class-AB is relatively as bad in its harmonic structure as crossover distortion, and the best thing that can be said about it is that does not (unlike crossover distortion) occur at all output levels. The spectrum analysis was done with a huge Hewlett-Packard instrument on a trolley that was normally used for EMC RF measurements. My current digital scope can do all that for itself; of such is progress.

In Figure 25.6 and onwards, I published the first of the 'wingspread' incremental gain plots produced with SPICE. As I have explained elsewhere, I did not invent this extremely useful sort of visual aid, but I suspect I have made more use of them than anyone else.

In any article written years ago, there will be things that need correcting in the light of later knowledge. Without descending to nit-picking, here are a couple of them.

In the second section of the article I said '. . . there seems to be a clear case for avoiding Class-AB altogether', by which I meant avoiding it in amplifiers that are nominally Class-B. There, cranking up the quiescent current gives an area of Class-A operation at small signal levels, the downside being the rather large amount of distortion generated by the gain changes in the output stage when you leave the Class-A region. This distortion has a harmonic structure not very different from crossover distortion—see Figure 25.1c in the article—and is well worth avoiding. In contrast, the later article on Class-A amplifiers and the Trimodal design used Class-AB solely as a better expedient than hard clipping, if a low-impedance load means you run out of Class-A quiescent current.

At the end of the section of the article entitled 'The emitter-follower output', I said 'The (crossover) region extends over about ±5V, independent of load resistance'. This statement depends on how you define the crossover region, but is in general not true as crossover distortion is affected by the load on the amplifier output.

The time-stamp on Figure 25.8 in the article speaks of the long hours required of those who seek to understand power amplifiers. Six minutes to midnight . . .

SPICE simulation of the various output stages proved to be very useful. Each near-unity-gain stage can stand alone without any support circuitry, and the dynamic effects are relatively minor in practice (see the section on switch-off distortion). It is therefore very practicable to run a DC simulation and display the result as incremental gain versus input voltage, by a simple process of differentiation. It may be that not all simulators have this facility in their graphical post-processors, but certainly PSPICE and B2SPICE do. The only significant deviation from reality here is that the source impedance from which the output stages are driven is usually zero in the simulation, while in a complete amplifier it is driven from the VAS collector, which has distinctly non-zero impedance that falls as frequency rises, due to increasing negative feedback through the Miller compensation capacitor. I have to admit that I have spent very little time on this issue, but I can assure you that using a zero impedance for simulation does give very useful results. There is more on the impedance at the VAS collector in Chapter 27 of this book.

DISTORTION IN POWER AMPLIFIERS

November 1993

The almost universal choice in semiconductor power amplifiers is for a unity gain output stage, and specifically a voltage follower. Output stages with gain are not unknown, [1] but they are not common. Most designers feel that controlling distortion while handling large currents is hard enough without trying to generate gain at the same time.

The first three parts of this series have dealt with one kind of distortion at a time, due to the monotonic transfer characteristics of small signal stages, which usually, but not invariably, work in class A.[2] Economic and thermal realities mean that most output stages are class B, and so we must now consider crossover distortion, which remains the thorniest problem in power amplifier design, and HF switchoff effects.

It is now also necessary to consider what kind of active device is to be used; jfets offer few if any advantages in the small current stages, but power fets are a real possibility, providing that the extra cost brings with it tangible benefit.

The class war

The fundamental factor in determining output stage distortion is the class of operation. Apart from its inherent inefficiency, class A is the ideal operating mode, because there can be no crossover or switchoff distortion. However, of those designs which have been published or

reviewed, it is notable that the large signal distortion produced is still significant. This looks like an opportunity lost, as of the distortion mechanisms discussed in the first part of this series, we now only have to deal with Distortion 1 (input stage), Distortion 2 (VAS), and Distortion 3 (output stage large signal nonlinearity). Distortions 4, 5, 6 and 7, as mentioned earlier, are direct results of class B operation and therefore can be thankfully disregarded in a class A design. However, class B is overwhelmingly of the greater importance, and is therefore dealt with in detail.

Class B is subject to misunderstanding. The statement is often made that a pair of output transistors operated without any bias are working in 'class B', and therefore 'generate severe crossover distortion'. In fact, with no bias each output device is operating for slightly less than half the time, and the question arises as to whether it would not be more accurate to call this class C and reserve class B for that condition of quiescent current which eliminates, or rather minimises, the crossover artifacts.

A further complication exists; it is not generally appreciated that moving into what is usually called class AB, by increasing the quiescent current, does not make things better. In fact, the THD reading will increase as the bias control is advanced, with what is usually known as 'g_m doubling' (i.e. a voltage gain increase caused by both devices conducting simultaneously in the centre of the output voltage range) putting edges into the distortion residual that generate high order harmonics in much the same way that underbiasing does. This important fact seems almost unknown, presumably because the g_m doubling distortion is at a relatively low level and is completely obscured in most amplifiers by other distortion mechanisms.

The phenomenon is demonstrated in Figures 25.1(a), (b), (c) which shows spectrum analysis of the distortion residuals for under biasing, optimal, and over biasing of a 150 W/8 Ω amplifier at 1 kHz.

As before, all non-linearities except the unavoidable Distortion 3 (output stage) have been effectively eliminated. The over biased case had its quiescent current increased until the g_m doubling edges in the residual had an approximately 50:50 mark/space ratio, and so was in class A about half the time which represents a rather generous amount of quiescent for class AB. Nonetheless, the higher order odd harmonics in Figure 25.1(c) are at least 10 dB greater in amplitude than those for the optimal class B case, and the third harmonic is actually higher than for the under-biased case as well. However the under biased amplifier, generating the familiar sharp spikes on the residual, has a generally greater level of high-order odd harmonics above the 5th; about 8 dB higher than the AB case.

Bearing in mind that high order odd harmonics are generally considered to be the most unpleasant, there seems to be a clear case for avoiding Class AB altogether, as it will always be less efficient and generate more high order distortion than the equivalent class B circuit, class distinction therefore seems to resolve itself into a binary choice between A or B.

It must be emphasised that these effects can only be seen in an amplifier where the other forms of distortion have been properly minimised. The r.m.s. THD reading for case **1a** was 0.00151%, for case **1b** 0.00103%, and for case **1c** 0.00153%. The tests were repeated at the 40 W power level with very similar results. The spike just below 16 kHz is interference from the test gear VDU.

This may seem complicated enough, but there are other and deeper subtleties in class B.

Distortions of the output

I have designated the distortion produced directly by output stages as Distortion 3 (see Part 1); this subdivides into three categories. Mechanism 3a describes the large signal distortion produced by both class A and B, ultimately because of the large current swings in the active devices. In bipolars, but not fets, large collector currents reduce the beta leading to drooping gain at large output excursions. I shall use the term 'LSN' for large signal nonlinearity, as opposed to crossover and switchoff phenomena that cause trouble at all output levels.

The other two contributions to Distortion 3 are associated with class B only; Distortion 3b the classic

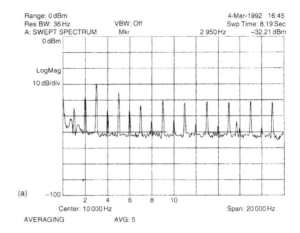

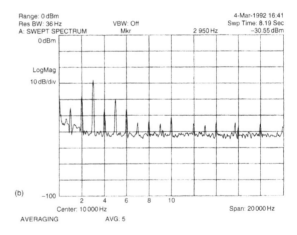

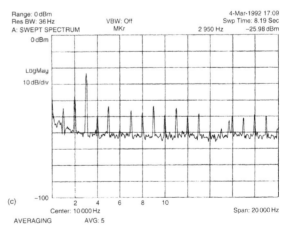

Figure 25.1 Spectrum analysis of class B &AB distortion residual. 1(a) Underbiased class B; 1(b) Optimal class B; 1(c) class AB.

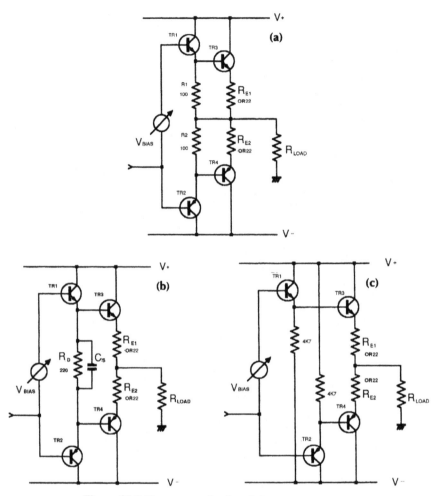

Figure 25.2 Three types of emitter follower output stages.

crossover distortion resulting from the non-conjugate nature of the output characteristics, and is essentially frequency independent.

In contrast, Distortion 3c is switchoff distortion generated by the output devices failing to turn off quickly and cleanly at high frequencies. This mechanism is strongly frequency dependent. It is sometimes called switching distortion, but this allows room for confusion, as some writers use 'switching distortion' to cover crossover distortion as well. I refer specifically to charge storage turn off troubles.

One of my aims for this series has been to show how to isolate individual distortion mechanisms. To examine output behaviour, it is perfectly practical to drive output stages open loop providing the driving source impedance

is properly specified; this is difficult with a conventional amplifier, as it means the output must be driven from a frequency dependant impedance simulating that at the VAS collector with some sort of feedback mechanism incorporated to keep the drive voltage constant.

However, if the VAS is buffered from the output stage by some form of emitter follower, as described in the last part, it makes things much simpler, a straightforward low impedance source (e.g. 50 Ω) providing a good approximation of a VAS-buffered closed loop amplifier. The VAS buffer makes the system more designable by eliminating two variables—the VAS collector impedance at LF, and the frequency at which it starts to decrease due to local feedback through C_{dom}. This markedly simplifies the study of output stage behaviour.

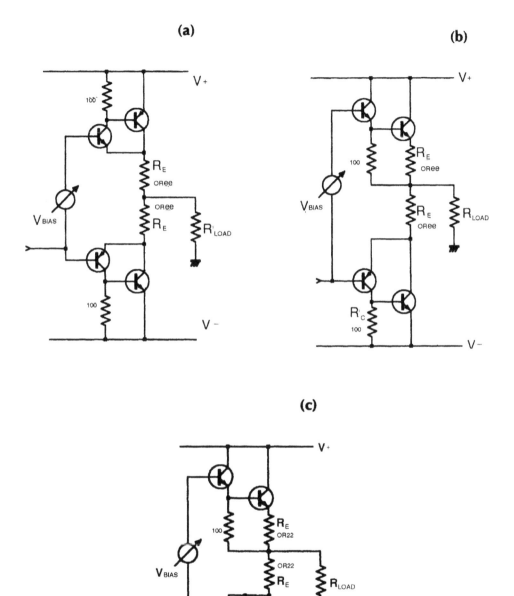

Figure 25.3 CFP circuit and quasi-complementary stages.

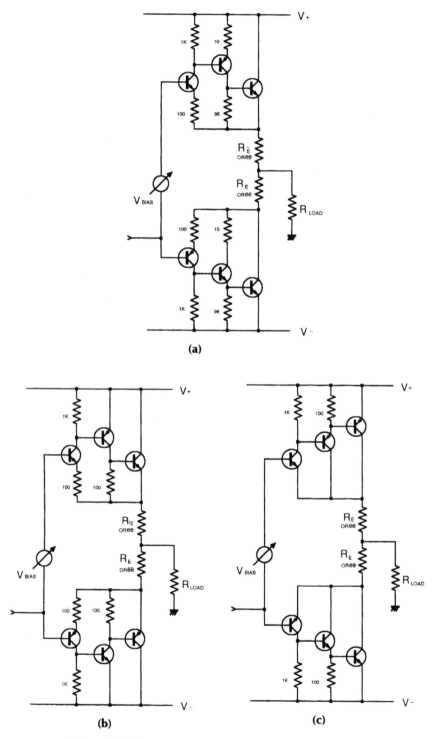

Figure 25.4 Three of the possible output triple configurations.

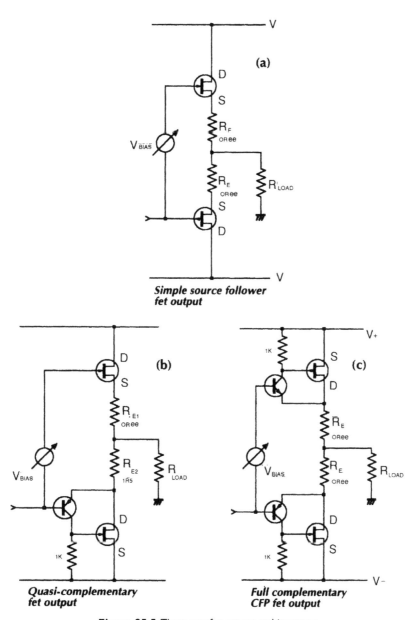

Figure 25.5 Three mosfet output architectures.

The large signal linearity of various kinds of open loop output stage with typical values are shown in Figures 25.6–25.15 These diagrams were all generated by spice simulation, and are plotted as incremental output gain against output voltage, with the load resistance stepped from 16 Ω to 2 Ω. The power devices are *MJ802* and *MJ4502*, which are more complementary than many transistor pairs, and minimise distracting large signal asymmetry. The quiescent current is in each case set to minimise the peak deviations of gain around the crossover point for 8 Ω loading; for the moment it is assumed that you can set this accurately and keep it where you want it. The difficulties in actually doing this will be examined later.

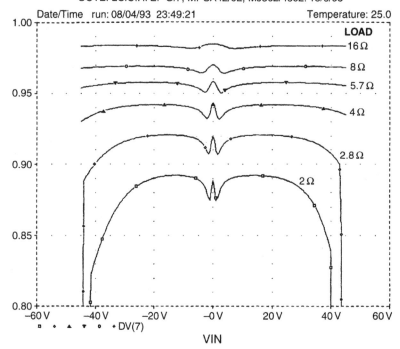

Figure 25.6 Emitter follower large signal gain vs output.

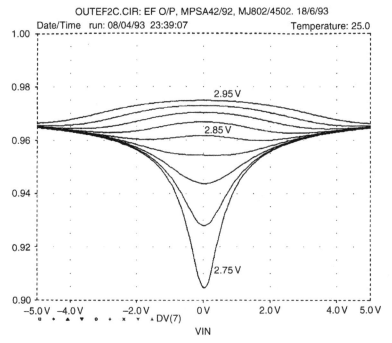

Figure 25.7 Emitter follower crossover region gain deviations, ±5 V range.

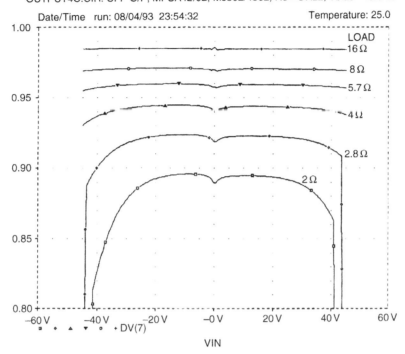

Figure 25.8 Complementary feedback pair gain versus output.

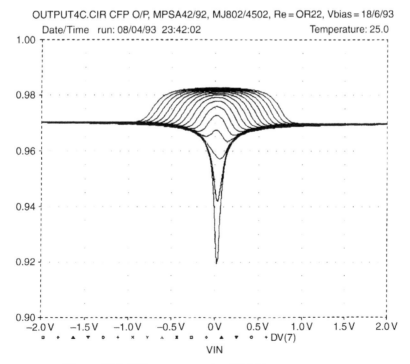

Figure 25.9 CFP crossover region ±2 V, Vbias as a parameter.

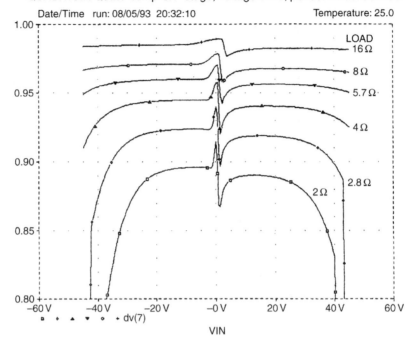

QUASI1.CIR Quasi–comp O/P stage, voltage drive; perfect Vbias. 30/4/93

Figure 25.10 Quasi complementary large signal gain vs output load resistance.

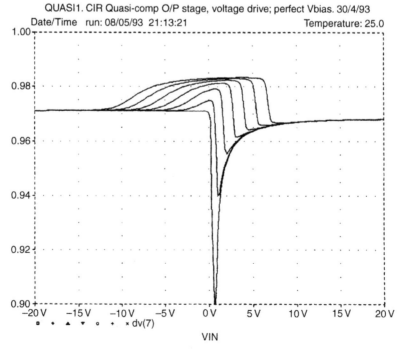

QUASI1. CIR Quasi-comp O/P stage, voltage drive; perfect Vbias. 30/4/93

Figure 25.11 Quasi crossover region ±20 V, Vbias as parameter.

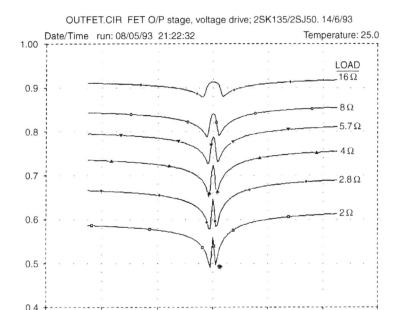

Figure 25.12 Source follower FET large signal gain vs output.

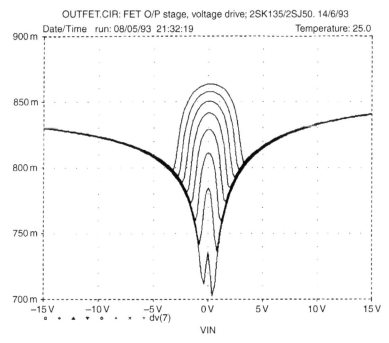

Figure 25.13 Source follower FET crossover region ±15 V range.

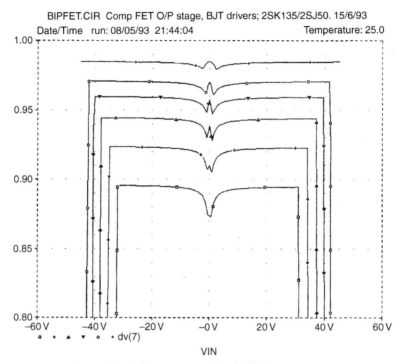

Figure 25.14 Complementary bipolar FET gain vs output.

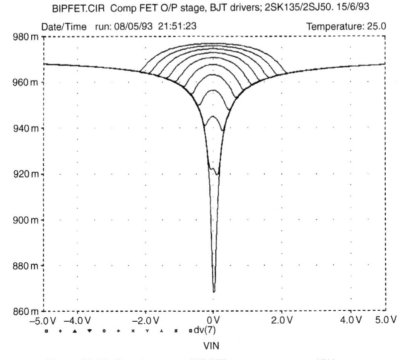

Figure 25.15 Complementary BJT FET crossover region ±15 V range.

There are at least 16 distinct configurations in straightforward output stages not including error correcting, [3] current dumping [4] or Blomley [5] types. These are as follows:

Emitter Follower	3 types	Figure 25.2
Complementary Feedback Pair	1 type	Figure 25.3
Quasicomplementary	2 types	Figure 25.3
Output Triples	At least 7 types	Figure 25.4
Power FET	3 types	Figure 25.5

The emitter follower output

Figure 25.2 shows three versions of the most common type of output stage; the double-emitter follower where the first follower acts as driver to the second (output) device. I have deliberately called this an emitter follower rather than a Darlington configuration, as this latter implies an integrated device with associated resistors. As for all the circuitry here, the component values are representative of real practice.

Two important attributes of this topology are:

- The input is transferred to the output via two base emitter junctions in series, with no local feedback around the stage (apart from the very local 100% voltage feedback that makes an emitter follower what it is);
- There are two dissimilar base emitter junctions between the bias voltage and the emitter resistor R_e, carrying different currents and at different temperatures. The bias generator must attempt to compensate for both at once, though it can only be thermally coupled to one. The output devices have substantial thermal inertia and thus thermal compensation represents a time average of the preceding conditions. Figure 25.2(a) shows the most prevalent version (type I) which has its driver emitter resistors connected to the output rail.

The type II configuration in Figure 25.2(b) is at first sight merely a pointless variation on type I, but in fact it has a valuable extra property. The shared driver emitter resistor R_d, with no output rail connection, allows the drivers to reverse bias the base emitter junction of the output device being turned off.

Assume that the output voltage is heading downwards through the crossover region; the current through R_{e1} has dropped to zero, but that through R_{e2} is increasing, giving a voltage drop across it, so Tr_4 base is caused to go more negative to get the output to the right voltage. This negative excursion is coupled to Tr_3 base through R_d, and with the values shown can reverse bias it by up to 0.5 V, increasing to 1.6 V with a 4 Ω load. The speed up capacitor C_s markedly improves this action, preventing the charge suckout rate being limited by the resistance of R_d. While the type I circuit has a similar voltage drop across R_{e2}, the connection of the mid point of R_1, R_2 to the output rail prevents this from reaching Tr_3 base; instead Tr_1 base is reverse biased as the output moves negative, and since charge storage in the drivers is usually not a problem, this does little good. In the type II circuit the drivers are never reverse biased, though they do turn off. The important issue of output turn off and switching distortion is further examined in the next part of this series.

The type III topology shown in Figure 25.2(c) maintains the drivers in class A by connecting the driver emitter resistors to the opposite supply rail rather than the output rail. It is a common misconception [6] that class A drivers somehow maintain better low frequency control over the output devices, but I have yet to substantiate any advantage myself. The driver dissipation is of course substantially increased, and nothing seems to be gained at LF as far as the output transistors are concerned, for in both type I and type II the drivers are still conducting at the moment the outputs turn off, and are back in conduction before the outputs turn on, which would seem to be all that matters.

Type III is equally good as type II in reverse biasing the output bases, and may give even cleaner HF turn off as the carriers are being swept from the bases by a higher resistance terminated in a higher voltage approximating constant current drive; I have yet to try this.

The large signal linearity of the three versions is virtually identical—all have the same feature of two base emitter junctions in series between input and load.

The gain/output voltage plot is shown at Figure 25.6; with BJTs the gain reduction with increasing loading is largely due to the emitter resistors. Note that the crossover region appears as a relatively smooth wobble rather than a jagged shape. Another major feature is the gain droop at high output voltages and low loads indicating that high collector currents are the fundamental cause of this.

A close up of the crossover region gain for 8 Ω loading only is shown in Figure 25.7; note that no V_{bias} setting can be found to give a constant or even monotonic gain; the double dip and central gain peak are characteristic of optimal adjustment. The region extends over about ±5 V, independent of load resistance.

Complementary feedback output

The other major type of bipolar output is the complementary feedback pair (CFP) sometimes called the Sziklai Pair, Figure 25.3(a). There seems to be only one popular configuration, though versions with gain are possible. The drivers are now placed so that they compare output voltage with that at the input. Wrapping the outputs in a local negative feedback loop promises better linearity than emitter follower versions with 100% feedback applied separately to driver and output transistors.

This topology also has better thermal stability, because the V_{be} of the output devices is inside the local feedback loop, and only the driver V_{be} affects the quiescent current. It is usually simple to keep drivers cool, and thermal feedback from them to the V_{bias} generator transistor can be much faster and mechanically simpler.

Like emitter follower outputs, the drivers are conducting whenever the outputs are, and so special arrangements to keep them in class A seem pointless. This stage, like emitter follower type I, can only reverse bias the driver bases rather than the outputs, unless extra voltage rails outside the main ones are provided.

The output gain plot is shown in Figure 25.8. Fourier analysis shows that the CFP generates less than half the large signal distortion of an emitter follower stage. (See Table 25.1) Given also the greater quiescent stability, it is hard to see why this topology is not more popular.

The crossover region is much narrower, at about ±0.3 V (Figure 25.9). When under biased, this shows up on the distortion residual as narrower spikes than an emitter follower output gives.

The bad effects of g_m doubling as V_{bias} increases above optimal (here 1.296 V) can be seen in the slopes moving outwards from the centre.

Quasicomplementary outputs

The original quasicomplementary configuration [7] was almost mandatory, as it was a long time before pnp silicon power transistors matched the performance of the npn versions. The standard version shown at Figure 25.3(b) is well known for poor symmetry around the crossover region, as shown at Figure 25.10. A close-up of the crossover region (Figure 25.11) reveals an unhappy hybrid of the emitter follower and CFP, as might be expected, and that no setting of bias voltage can remove the sharp edge in the gain plot.

A major improvement to symmetry may be made by using a Baxandall diode [8] as shown in Figure 25.3(c). This stratagem yields gain plots very similar to those for the true complementary emitter follower at Figures 25.6,

25.7, though in practice the crossover distortion seems rather higher. When a quasiBaxandall stage is used closed loop in an amplifier in which distortion mechanisms 1 and 2, and 4 to 7 have been properly eliminated, it is capable of better performance than is commonly believed. For example, 0.0015% (1 kHz) and 0.015% (10 kHz) at 100 W is straightforward to obtain from an amplifier with a negative feedback factor of about 34 dB at 20 kHz.

The best reason to use the quasiBaxandall approach today is to save money on output devices, as pnp power transistors remain somewhat pricier than npns. Given the tiny cost of a Baxandall diode, and the absolutely dependable improvement it gives, there seems no reason why anyone should ever use the standard quasi circuit. My experiments show that the value of R_1 in Figure 25.3(c) is not critical; making it about the same as R_c seems to work well.

Triples

With three rather than two bipolar transistors in each half of an output stage the number of circuit permutations possible leaps upwards. There are two main advantages if output triples are used correctly: better linearity at high output voltages and currents; and more stable quiescent setting as the predrivers can be arranged to handle very little power, and remain almost cold in use.

However, triples do not automatically reduce crossover distortion, and they are, as usually configured, incapable of reverse biasing the output bases to improve switch-off. Figure 25.4 shows three ways to make a triple output stage—all of those shown (with the possible exception of Figure 25.4(c), which I have just made up) have been used in commercial designs. The circuit of 4a is the Quad *303* quasicomplementary triple. The design of triples demands care, as the possibility of local HF instability in each output half is very real.

Power FET outputs

Power mosfets are often claimed to be a solution to all amplifier problems, but they have their drawbacks: poor linearity and a high on-resistance that makes output efficiency mediocre. The high frequency response is better, implying that the second pole P2 of the amplifier response will be higher, allowing the dominant pole P1 be raised with the same stability margin, and in turn allowing more overall feedback to reduce distortion. However, the extra feedback (if it proves available in practice) is needed to correct the higher open loop distortion.

To complicate matters, the compensation cannot necessarily be lighter because the higher output resistance makes the lowering of the output pole by capacitive loading more likely.

The extended frequency response creates its own problems; the HF capabilities mean that rigorous care must be taken to prevent parasitic oscillation, as this is often promptly followed by an explosion of disconcerting violence. Fets should at least give freedom from switchoff troubles as they do not suffer from charge storage effects.

Three types of FET output stage are shown in Figure 25.5. Figures 25.12 to 25.15 show spice gain plots, using *2SK135/2SJ50* devices.

Most FET amplifiers use the simple source follower configuration in Figure 25.5(a); the large signal gain plot at Figure 25.12 shows that the gain for a given load is lower, (0.83 rather than 0.97 for bipolar, at 8 Ω) because of low g_m. This, with the high on resistance, noticeably reduces output efficiency.

Open loop distortion is markedly higher; however large signal nonlinearity does not increase with heavier loading, there being no equivalent of 'bipolar gain droop'. The crossover region has sharper and larger gain deviations than a bipolar stage, and generally looks pretty nasty; Figure 25.13 shows the difficulty of finding a 'correct' V_{bias} setting.

Figure 25.5(b) shows a hybrid (i.e. bipolar/FET) quasi complementary output stage [9]. The stage is intended to maximise economy rather than performance, once the decision has been made (probably for marketing reasons) to use fets, by making both output devices cheap n-channel devices; complementary mosfet pairs remain relatively rare and expensive.

The basic configuration is badly asymmetrical, the hybrid lower half having a higher and more constant gain than the source follower lower upper half. Increasing the value of R_{e2} gives a reasonable match between the gains of the two halves, but leaves a daunting crossover discontinuity.

The hybrid full complementary stage in Figure 25.5(c) was conceived [10] to maximise performance by linearising the output devices with local feedback and reducing I_q variations due to the low power dissipation of the bipolar drivers. It is highly linear, showing no gain droop at heavier loadings (Figure 25.14) and promises freedom from switchoff distortion. But, as shown, it is rather inefficient in voltage swing. The crossover region (Figure 25.15) still has some dubious sharp corners, but the total crossover gain deviation (0.96–0.97 at 8 Ω is much smaller than for the quasi hybrid (0.78–0.90) and so less high order harmonic energy is generated.

References

1. Mann, R. 'The Texan 20 + 20 watt stereo amplifier'. *Practical Wireless*, May 1972, p 48.

2. Takahashi. 'Design & construction of high slew rate amplifiers', Preprint No. 1348 (A 4) for *60th AES Convention*, 1978.

3. Hawksford, M.O. 'Distortion correction in audio power amplifiers', *JAES* January/February 1981 p 27 (Error correction).

4. Walker, P. 'Current dumping audio amplifier', *Wireless World*, 1975 pp 60/62.

5. Blomley. 'New approach to class B'. *Wireless World*, February 1971, p 57 and March 1971, p 127/131.

6. Otala. 'An audio power amplifier for ultimate quality requirements', *IEEE Trans on Audio & Electroacoustics*, December 1973, p 548.

7. Lin, H. *Electronics*, September 1956 pp 173/175.

8. Baxandall, P. 'Symmetry in class B. Letters', *Wireless World*, September 1969 p 416.

9. Self, D. 'Sound mosfet design'. *Electronics & Wireless World*, September 1990, p 760.

10. Self, D. 'Mosfet audio output', letter, *Electronics & Wireless World*, May 1989, p 524 (see also Ref. 9).

Chapter 26

Distortion in power amplifiers

Part V: output stages
(Electronics World, December 1993)

This article carried on the investigation into amplifier output stages begun in the previous one.

In the first section of the article, there is the suggestion that the linearity of the VAS can be effectively eliminated by cascoding. With the wisdom of hindsight, 'eliminated' was probably a bit optimistic. Later experience has convinced me that adding an emitter-follower within the VAS Miller dominant-pole loop is a more effective and dependable way of implementing this linearisation; no components are required for biasing a cascode transistor, and less voltage is lost on negative swings. I have used the emitter-follower method in all my recent commercial designs.

I have to confess that I have done very little work on the issue of switching distortion since this article appeared. At the time I suggested it could be improved by using the version of the EF output circuit that has a single resistor between the driver emitters, because this slightly reverse-biases the output bases at turn-off, improving the speed at which the charge-carriers are swept away. This is still entirely sound, and I always use this version of the EF output stage—and it saves a resistor! But . . . my further addition of a large (1uF) speed-up capacitor across this resistor was perhaps a less happy notion. It works effectively to reduce distortion with steady sine-wave drive, but as Pete Baxandall pointed out [1], things might go wrong with large asymmetrical signals, interfering with the output bias conditions. This appears to be a rare case where sinewave testing alone will not show up a problem. I don't think I have ever tried the use of a speed-up capacitor since, and I have certainly never used one in any of my commercial designs. More well-funded research needed, clearly.

Peter went on to explore the business of turning-off output devices in great detail [2], including the use of dedicated pull-off transistors to speed things up. I have never had the opportunity to test these ideas, but they are most certainly worth reading.

The section on selecting an output stage reflects the available knowledge at the time, but it could do with considerable expansion. An important consideration is the number of parallel output devices you plan to use. With an EF output stage, the more output transistors you put in parallel the better, assuming each pair has its own emitter resistors. As the number of pairs increases, the gain gets closer to unity and the gain-wobble that causes crossover distortion becomes flatter and more spread out [3]. A CFP stage arranged in the usual way (with one pair of shared emitter resistors) acts completely differently; going from one to two pairs of output devices causes a significant increase in distortion, and it gets worse as more pairs are added [4]. When only one pair of output devices is used, the CFP is noticeably more linear than the EF, by about a factor of two. The break-even point is usually two parallel output devices.

References

1. Baxandall, P., and Self, D., Baxandall and Self on Audio Power. *Linear Audio* 2011, p. 96. Editor: Jan Didden. ISBN 978-94-90929-03-9.

2. Ibid., pp. 97–104.

3. Self, D., *Audio Power Amplifier Design*, 6th edition, p. 269. Focal Press 2013. ISBN 978-0-240-52613-3.

4. Ibid., pp. 240–241.

DISTORTION IN POWER AMPLIFIERS

December 1993

From earlier work in this series, distortion from the small-signal stages may be kept to levels that will prove negligible compared with distortion from a closed-loop output stage. Similarly, future work in this series will show that distortion mechanisms 4 to 7 from my original list (*EW + WW, July 93*) can be effectively eliminated by lesser-known but straightforward methods. This leaves the third mechanism in its three components as the only distortion that is in any sense unavoidable: Class-B stages free from crossover artifacts are not exactly commonplace.

This is a good place to introduce the concept of a blameless amplifier, one designed so that all the easily-defeated distortion mechanisms have been rendered negligible. The word *blameless* has been carefully chosen to not imply perfection.

The first distortion, non-linearity in the input stage, cannot be totally eradicated but its onset can be pushed well above 20 kHz. The second distortion, non-linearity in the voltage amplifier stage, can be effectively eliminated by cascoding. Distortion mechanisms 4 to 7, concerned with such things as earth return loops, power supply impedance and non-linear loading, can be made negligible by simple measures to be described later.

Large-signal distortion

The large-signal nonlinearity performance of all the bipolar junction transistor stages outlined in the previous part of this series have these features in common:

Large-signal nonlinearity increases as load impedance decreases. In a typical output stage loaded with 8 Ω, closed-loop LSN is usually negligible, the THD residual being dominated by high-order crossover artefacts that are reduced less by negative feedback. At lower impedances, such as 4 Ω, relatively pure third harmonic becomes obvious in the residual.

LSN worsens as the driver emitter or collector resistances are reduced, because the driver current swings are larger. On the other hand, this reduction improves output device turn-off, and will so decrease switchoff distortion; the usual compromise is around 47–100 Ω.

The BJT output gain plots in the previous article reveal that the LSN is compressive, the voltage gain falling off with higher output currents. It is roughly symmetrical, generating third-harmonic, and is much greater at the very lowest load impedances; this is more of an issue now that 2 Ω-capable (for a few minutes, anyway) amplifiers are considered macho, and some speaker designers are happy with 2 Ω impedance troughs.

I suggest that the fundamental reason for this gain droop is the fall in output transistor beta as collector current increases, due to the onset of high level injection effects [1]. In the emitter follower topology, this fall in beta draws more output transistor base current from the driver emitter, pulling its gain down further from unity; this is the change in gain that affects the overall transfer ratio.

The output device gain is not directly affected, as beta does not appear in the classical expression for emitter follower gain, providing the source impedance is negligibly low. This assertion has been verified by altering an output stage simulated in Spice such that the output bases are driven directly from zero impedance voltage sources rather than drivers; this abolishes the gain droop effect, so it must be in the drivers rather than the output transistors.

Further evidence for this view is that in Spice simulation, the output device Ebers-Moll model can be altered so that beta does not drop with I_c (simply increase the value of the parameter IKF) and once more the gain droop does not occur, even with drivers. Here is one of the best uses of circuit simulation tweaking the untweakable. Gain droop does not affect FET outputs, which have no equivalent beta loss mechanism.

It used to be commonplace for output transistors to be sold in pairs roughly matched for beta, allegedly to minimise distortion; this practice seems to have been abandoned. Simulation shows that beta mismatch produces an unbalanced gain droop that markedly increases low order harmonics without much effect on the higher ones. Modern amplifiers with adequate feedback factors will linearise this effectively. This appears to be why the practice has ceased.

Improving large signal linearity

It will be suggested that, in a closed loop blameless amplifier, the large signal nonlinearity contribution to total distortion (for 8 Ω loading) is actually very small compared with that from crossover and switchoff. This is no longer true at 4 Ω and still less so for lower load impedances. Thus ways of reducing this mechanism will still be useful.

The best precaution is to choose the most linear output topology; The previous article suggested that the open loop complementary feedback pair output is at least twice as linear as its nearest competitor, (the emitter follower output) and so the CFP is usually the best choice unless the design emphasis is on minimising switchoff distortion.

In the small signal stages, we could virtually eliminate distortion. If the linearity of the input or voltage amplifier stage was inadequate, it was possible to come up with several ways in which it could be dramatically improved. A Class B output stage is a tougher proposition. In particular we must avoid complications to the forward path that lower the second amplifier pole P2, as this would reduce the amount of feedback that can be safely applied.

Several authors [2,3] have tried to show that the output emitter resistors of bipolar outputs can be fine tuned in value to minimise large signal distortion, the rationale being that the current dependent internal r_e of the output transistors will tend to cause the gain to rise at high currents, and that this gain variation can be minimised by appropriate choice of the external R_e. This is not true in practical output stages whose gain behaviour tends to be dominated by beta loss and its effect on the drivers. In any case the resistor values suggested are such tiny fractions of an ohm that quiescent stability would be perilous.

In real life the R_e of a CFP output stage can be varied between 0.5 and 0.2 Ω without significantly affecting linearity; 0.22 Ω is a good compromise between efficiency and stability.

The gain droop at high I_cs can be partly cancelled by a simple but effective feedforward mechanism. The emitter resistors R_e are shunted with silicon power diodes, which with typical, circuit values will only conduct when 4 Ω loads (or less) are driven. This causes a slight gain increase that works against the beta loss droop. The modest but dependable improvement can be seen in Figure 26.1, measured with a 2.7 Ω load.

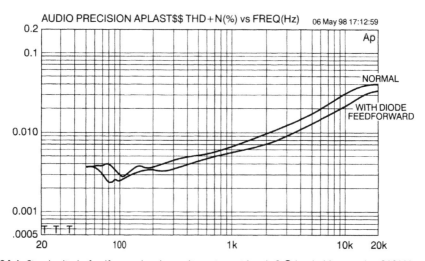

Figure 26.1 Simple diode feedforward reduces distortion with sub-8 Ω loads. Measured at 210 W into 2.7 Ω.

If a 100 W/8 Ω amplifier is required to drive 4 Ω loads then it will need paralleled output devices to cope with the power dissipation. Perhaps surprisingly, the paralleling of output BJTs (driven as usual from a single driver) has little effect on linearity, given elementary precautions to ensure current sharing. However, for the 2 Ω case there is a definite linearity improvement on resorting to tripled output devices; this is consistent with the theory that LSN results from beta loss at high collector currents.

Crossover distortion

The worst problem in Class B is the crossover region, where control of the output voltage must be transferred from one device to another. Crossover distortion generates unpleasant high order harmonics with the potential to increase in percentage as signal level falls. There is a consensus that crossover caused the transistor sound of the 1960's, though to the best of my knowledge this has never actually been confirmed by the double blind testing of vintage equipment.

The $V_{bc}-I_c$ characteristic of a bipolar transistor is initially exponential, blending into linear as the emitter resistance R_e comes to dominate the transconductance. The usual Class B stage puts two of these curves back to back, and Peter Blomley has shown that these curves are non-conjugate, [4] i.e. there is no way they can be rearranged to sum to a completely linear transfer characteristic, whatever the offset imposed by the bias voltage.

This can be demonstrated quickly and easily by Spice simulation; see Figure 26.2. There is at first sight not much you can do except maintain the bias voltage, and hence quiescent current, at some optimal level for minimum gain deviation at crossover; quiescent current control is a topic that could fill a book in itself, and cannot be considered properly here.

It should be said that the crossover distortion levels generated in a blameless amplifier can be low up to 1 kHz, being barely visible in residual noise and only measurable with a spectrum analyser. For example, if a blameless closed-loop Class B amplifier is driven through a *TL072* unity gain buffer the added noise from this op-amp will usually submerge the 1 kHz crossover artifacts into the noise floor. (It is most important to note that distortion

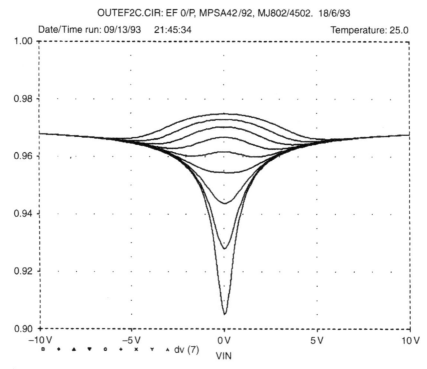

Figure 26.2 Gain/output-voltage Spice plot for an emitter follower output shows how non-conjugate transistor characteristics at the crossover region cannot be blended into a flat line at any bias voltage setting. Bias varies 2.75 to 2.95 V in 25 mV steps, from too little to too much quiescent.

mechanisms 4 to 7 create disturbances of the THD residual at the zero crossing point that can be easily mistaken for crossover distortion, but the actual mechanisms are quite different.) However, the crossover distortion becomes obvious as the frequency increases, and the high order harmonics benefit less from NFB. See text panel *Harmonic generation by crossover distortion* (p. 197).

It will be seen later that in a blameless amplifier the linearity is dominated by crossover distortion, even with a well designed and optimally biased output stage. There is an obvious incentive to minimise it, but there seems no obvious way to reduce crossover gain deviations by tinkering with any of the relatively conventional stages considered so far. Significant improvement is only likely through application of one of the following techniques:

- The use of Class AB stages where the handover from one output device to the other is genuinely gradual, and not subject to the g_m doubling effects that an over biased Class B stage shows. One possibility is the so called Harmonic AB mode [5].
- Non-switching output stages where the output devices are clamped to prevent turn off, and thus hopefully avoiding the worst part of the V_{bc}-I_c curve [6].
- Error correcting output stages implementing either error feedforward or error feedback. The latter is not the same thing as global NFB, being instead a form of cancellation [7].

Once more, these will have to be examined in the future.

Switching distortion

This depends on several variables, notably the speed characteristics of the output devices and the output topology. Leaving aside the semiconductor physics and concentrating on the topology, the critical factor is whether or not the output stage can reverse bias the output device base emitter junctions to maximise the speed at which carriers are sucked out, so the device is turned off quickly.

The only conventional configuration that can reverse bias the output base emitter junctions is the emitter follower *type II*, described in the previous article. A second influence is the value of the driver emitter or collector resistors; the lower they are the faster the stored charge can be removed.

Applying these criteria can reduce HF distortion markedly, but it is equally important that it minimises output conduction overlap at high frequencies. If unchecked, overlap results in an inefficient and potentially destructive increase in supply current [8]. Illustrating this, Figure 26.3 shows current consumption vs frequency for varying driver collector resistance, for a CFP type output.

Figure 26.4 shows how HF THD is reduced by adding a speed-up capacitor over the common driver resistor of a *type II* emitter follower. At LF the difference is small, but at 40 kHz THD is halved, indicating much cleaner switch-off. There is also a small benefit over the range 300 Hz to 8 kHz.

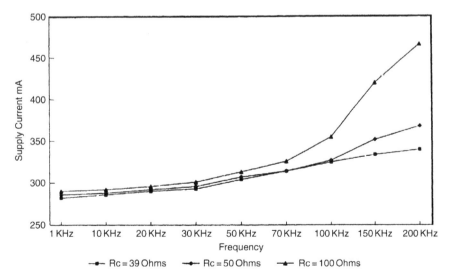

Figure 26.3 Power supply current versus frequency, for a CFP output with the driver collector resistors varied. There is little to be gained from reducing Rc below 50 Ω.

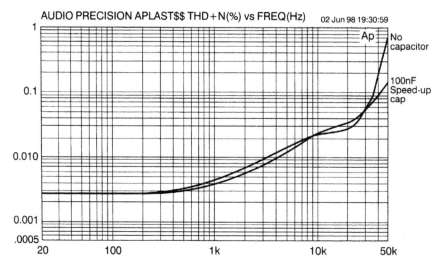

Figure 26.4 HF THD reduction by adding speed-up capacitance across the common driver resistance of a Type 11 emitter follower output stage. Taken at 30 W/8 Ω.

Selecting an output stage

Even if we stick to the most conventional of output stages, there are still an embarrassingly large number to choose from. The cost of a complementary pair of power fets is currently at least twice that of roughly equivalent BJTs, and taken with the poor linearity and low efficiency of these devices, the use of them may require a marketing rather than a technical motivation.

Turning to BJTs, and taking the material in this article with that in *Part 4*, I conclude that these are the following candidates for best output stage:

- The emitter follower *type II* output stage is the best at coping with switchoff distortion but the quiescent current stability is not of the best;
- The CFP topology has good quiescent stability and low LSN; its worst drawback is that reverse biasing the output bases for fast switchoff is impossible without additional HT rails;
- The quasi-complementary with Baxandall diode stage comes close to mimicking the emitter follower type stages in linearity, with a potential for cost saving on output devices. Quiescent stability is not as good as the CFP.

Closing the loop

In chapters 17 and 18, it was shown how relatively simple design rules could ensure that the THD of the small signal stages alone could be reduced to less than 0.001% across the audio band, in a repeatable fashion, and without using frightening amounts of negative feedback. Combining this subsystem with one of the more linear output stages such as the CFP version which gives 0.014% THD open loop, and having a feedback factor of at least 70 times across the band, it seems we have the ingredients for a virtually distortionless power amplifier, with THD below 0.001% from 10 Hz to 20 kHz. However, life is rarely so simple. . . .

The seven main sources of distortion

It is one of the central themes of this series that the primary sources of power amplifier distortion are seven-fold:

1. Nonlinearity in the input stage. For a well balanced differential pair distortion rises at 18 dB/octave, and is third harmonic. When unbalanced, HF distortion is higher and rises at 12 dB/octave, being mostly second harmonic.
2. Nonlinearity of the voltage amplifier stage (VAS), second harmonic, rising at 6 dB/octave.
3. Nonlinearity of the output stage. In Class B this may be a mix of large signal distortion and crossover effects, in general rising at 6 dB/octave as the amount of NFB decreases; worsens with heavier loads.
4. Nonlinear loading of the VAS by the nonlinear input impedance of the output stage. Magnitude is essentially constant with frequency.

5. Nonlinearity caused by large rail decoupling capacitors feeding the distorted supply rail signals into the signal ground.
6. Nonlinearity caused by induction of Class B supply currents into the output, ground, or negative feedback lines.
7. Nonlinearity resulting from taking the NFB feed incorrectly.

Table 26.1 Summary of closed loop amp THD performance

	1 kHz (%)	10 kHz (%)	
Emitter follower	0.0019	0.013	Figure 26.5
CFP	0.0008	0.005	Figure 26.6
Quasi Bax	0.0015	0.015	Figure 26.7

Figure 26.5 shows the distortion performance of such a closed loop amplifier with an emitter follower output stage, Figure 26.6 showing the same with a CFP output stage. Figure 26.7 shows the THD of a quasi-complementary stage with Baxandall diode [9]. In each case distortion mechanisms 1, 2 and 4–7 have been eliminated by methods described in past and future chapters, to make the amplifier blameless.

It will be seen at once that these amplifiers are definitely not distortionless, though the performance is markedly superior to the usual run of hardware. THD in the LF region is very low, well below a noise floor of 0.0007%, and the usual rise below 100 Hz is very small indeed. However, above 2 kHz, THD rises with frequency at between 6 to 12 dB/octave, and the residual in this region is clearly time aligned with the crossover region, and consists of high order harmonics rather than second or third.

It is intriguing to note that the quasi-Bax output gives about the same HF THD as the emitter follower topology, confirming the statement that the addition of a Baxandall diode turns a conventional quasi-complementary stage with serious crossover asymmetry into a reasonable emulation of a complementary emitter follower stage.

There is significantly less HF THD with a CFP output; this cannot be due to large signal nonlinearity as this is negligible with an 8 Ω load for all three stages, and must result from lower levels of high order crossover products.

Despite the promising ingredients, a distortionless amplifier has failed to materialise, so we had better find out why?

When an amplifier with a frequency dependent NFB factor produces distortion, the reduction is not due to the NFB factor at the fundamental frequency, but the amount available at the frequency of the harmonic in question.

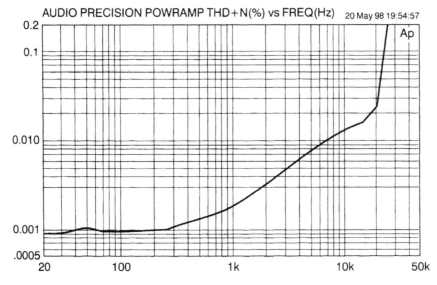

Figure 26.5 Closed-loop amplifier performance with emitter follower output stage. 100 W into 8 Ω.

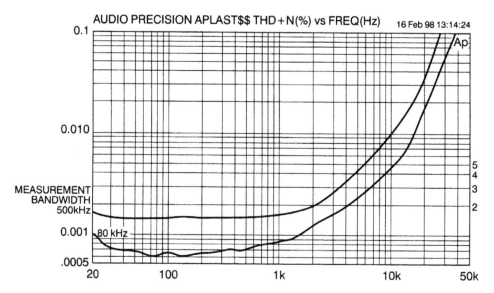

Figure 26.6 Closed-loop amplifier performance with CFP output. 100 W into 8 Ω.

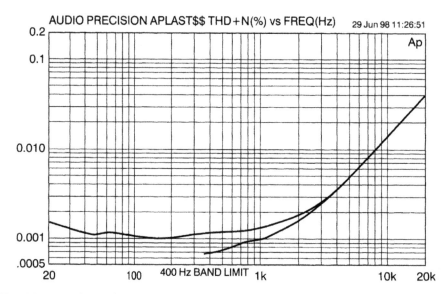

Figure 26.7 Closed-loop amplifier performance; quasi-complementary output stage with Baxandall diode. 100 W into 8 Ω. AP plots in Figures 26.5 to 26.7 were taken at 100 Wrms/8 Ω, from an amplifier with an input error of −70 dB at 10 kHz and c/1 gain of 27 dB, giving a feedback factor of 43 dB at this frequency. This is well above the dominant pole frequency, so the NFB factor is dropping at 6 dB/octave and will be down to 37 dB (or 70x) at 20 kHz. My experience suggests that this is about as much NFB as is safe for general use, assuming an output inductor to improve stability with capacitive loads. Sadly, published data on this touchy topic seems non-existent.

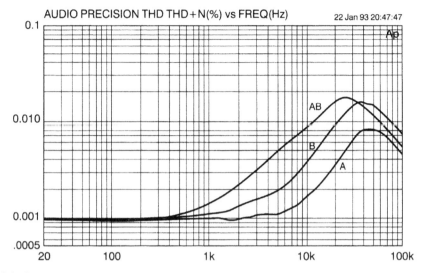

Figure 26.8 Closed-loop CFP amp. Setting quiescent for Class AB gives more HF THD than either Class A or B.

Harmonic generation by crossover distortion

The usual nonlinear distortions generate most of their unwanted energy in low order harmonics that NFB can deal with effectively. However, crossover and switching distortions that warp only a small part of the output swing tend to push energy into high order harmonics, and this important process is demonstrated here, by Fourier analysis of a Spice waveform.

Take a sinewave fundamental, and treat the distortion as an added error signal E, letting the ratio WR describe the proportion of the cycle where E > 0. If this error is a triangle wave extending over the whole cycle (WR = 1) this would represent large signal nonlinearity, and Figure 26.9 shows that most of the harmonic energy goes into the third and fifth harmonics; the even harmonics are all zero due to the symmetry of the waveform.

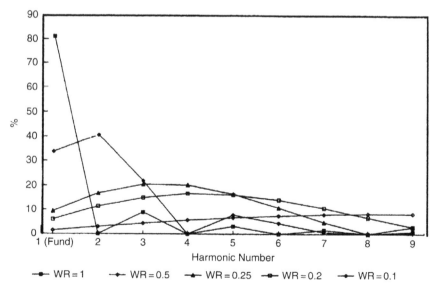

Figure 26.9 The amplitude of each harmonic changes with WR; as the error waveform gets narrower, energy is transferred to the higher harmonics.

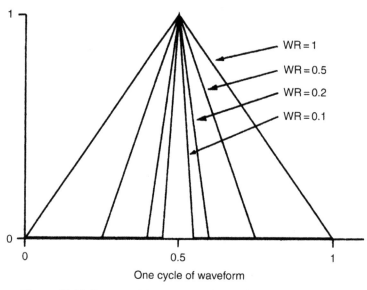

Figure 26.10 Diagram of the error waveform E for some values of WR.

Figure 26.10 shows how the situation is made more like crossover or switching distortion by squeezing the triangular error into the centre of the cycle so that its value is zero elsewhere; now $E > 0$ for only half the cycle (denoted by $WR = 0.5$) and Figure 26.9 shows that the even harmonics are no longer absent. As WR is further decreased, the energy is pushed into higher order harmonics, the amplitude of the lower harmonics falling.

These high harmonics have roughly equal amplitude, spectrum analysis confirming that even in a blameless amplifier driven at 1 kHz, harmonics are freely generated from the 7th to the 19th at a level within a dB or so. The 19th harmonic is only 10 dB below the third.

Thus, in an amplifier with crossover distortion, the order of the harmonics will decrease as signal amplitude reduces, and WR increases; their lower frequencies allow them to be better corrected by the frequency dependant negative feedback. This effect seems to work against the commonly assumed rise of percentage crossover distortion as level is reduced.

A typical amplifier with open loop gain rolling off at 6 dB/octave will be half as effective at reducing fourth-harmonic distortion as it is at reducing the second harmonic. LSN is largely third (and possibly second) harmonic, and so NFB will deal with this effectively. However, both crossover and switchoff distortions generate high-order harmonics significant up to at least the 19th and these receive much less linearisation. As the fundamental moves up in frequency the harmonics do too, and get even less feedback. This is the reason for the differentiated look to many distortion residuals; higher harmonics are emphasised at the rate of 6 dB/octave.

Here is a real example of the inability of NFB to cure all amplifier ills. To reduce this HF distortion we must reduce the crossover gain deviations of the output stage before closing the loop. There seems no obvious way to do this by minor modifications to any of the conventional output stages; we can only optimise the quiescent current.

Increasing the quiescent current will do no good for, as outlined in the previous chapter, Class AB is generally Not A Good Thing, producing more distortion than Class B, not less. Figure 26.8 makes this painfully clear for the closed-loop case; Class AB clearly gives the worst performance. (As before, the AB quiescent was set for 50:50 m/s ratio of the g_m doubling artifacts on the residual).

In this case the closed loop distortion is much greater than that from the small signal stages alone; however this is not automatic, and if the input pair is badly designed its HF distortion can easily exceed that caused by the output stage.

The distortion figures given in this article are rather better than the usual run. I must emphasise that these are not freakish or unrepeatable figures. They are simply the result of attending to all seven of the major sources of distortion rather than just one or two. I have so far built 12 CFP amplifiers, and performance shows little variation.

Conclusions

Taking this and the previous chapter together, we can summarise. Class AB is best avoided. Use pure Class A or B, as AB will always have more distortion than either. Fet outputs offer freedom from some BJT problems, but in general have poorer linearity, lower efficiency, and cost more.

Distortion generated by a blameless amplifier driving an 8 Ω load is almost entirely due to crossover effects and switching distortion. This does not hold for 4 Ω or lower loads where third harmonic on the residual shows the presence of large signal nonlinearity caused by beta loss at high output currents.

References

1. Gray and Meyer *Analysis & Design of Analog Integrated* Circuits, Wiley 1984, p 172.

2. Crecraft *et al. Electronics,* Chapman & Hall 1993, p 538.

3. Oliver 'Distortion in complementary pair class B amps', *Hewlett Packard Journal,* February 1971, p 11.

4. Blomley, P. 'New approach to class B', *Wireless World,* February 1971, p 57.

5. Thus, F.A. 'Compact bipolar class AB output stage using 1 V power supply', *IEEE Journal of Solid State Circuits,* December 1992, p 1718 (Harmonic AB).

6. Tanaka, S.A. 'New biasing circuit for class B operation', *JAES,* January/February 1981, p 27.

7. Hawksford, M. 'Distortion correction in audio power amplifiers', *JAES,* March 1981, p 148.

8. Alves, J. 'Power bandwidth limitations in audio amplifiers', *IEEE Trans on Broadcast & TV,* March 1973, p 79.

9. Baxandall, P. 'Symmetry in class B', *Letters, Wireless World,* September 1969, p 416 (Baxandall diode).

Chapter 27

Distortion in power amplifiers

Part VI: the remaining distortions
(Electronics World, January 1994)

The first part of the article dealt with the vital matter of control of the quiescent conditions in a Class-B output stage. This is a big subject and only the briefest of summaries was possible; it was much expanded into an entire chapter called 'Thermal Dynamics' in *Audio Power Amplifier Design*. The two most important issues described were firstly, the almost constant dissipation and hence temperature of the drivers in the EF output stage, making its quiescent conditions more stable than would otherwise be the case. Secondly, the most useful and perhaps non-intuitive finding that the top of an output transistor package gets hotter than anywhere else (much hotter than an adjacent bit of heatsink) and provides the nearest thing possible to sensing the device junction temperature of a conventional transistor. As far as I am aware, I was the first person to observe this; it certainly seems to have caught on as a popular piece of amplifier technology.

Distortion 4 (the nonlinear loading of the VAS by the output stage) is not a subject I have put much time into over the years since this article was written. Figure 27.4 shows considerable reduction in LF distortion from the use of a buffering emitter-follower between the VAS and the output stage, and looks rather like the benefit gained by adding an emitter-follower inside the Miller compensation loop; however, in the latter case there is also a dramatic decrease in HF distortion. This leads me to think that the amplifier tested in Figure 27.4 had only a simple VAS. Adding a VAS buffer to an EF-VAS may well not show much improvement. Digging out my notes for this test, I am unable to confirm if a simple VAS was used, but I can fill in some more details. The test power was 100W/8Ω, the input stage was not degenerated, had no mirror, and the Cdom capacitor was only 47pF. The next test with mirror added and Cdom = 100pF gave much lower HF distortion with the VAS buffer fitted.

This article was not as useful as it might have been on the subject of how to avoid induction distortion from the supply cables to the signal circuitry (Distortion 6). The advice to minimise the loop areas at both the transmitting and receiving ends, and to keep those areas as far apart as possible is of course sound, but it is rather non-specific. The trouble is that each amplifier has its own physical layout, so coming up with more detailed instructions is difficult because the results depend so much on the details of supply-rail tracking around the output devices. Since that can only be changed by PCB revisions, experimentation is going to be slow and expensive.

From my subsequent experience, there are some more firm statements that can be made. The risk of trouble is minimised if the positive and negative supply rails to the output devices do *not* trail their way around the PCB, but dive in from overhead on separate cables. They connect only to the DC fuseholders, decouplers, and output devices by PCB tracks, and these can be kept very short and loop-free. This introduces the third dimension, and makes it much easier to keep things apart. If inductive distortion is present, with its characteristic saw-tooth shape on the THD residual, then moving the supply cables about will alter its magnitude and give a quick diagnosis.

Another consideration is the placement of the output inductor. This may have up to twenty turns and makes an efficient way of coupling inductive distortion into the output path. With modern amplifiers and testgear this can easily place a limit on the distortion performance. The apparently logical place for this inductor is close to the output stage, but this is of course the worst place possible for inductive pickup. Some improvement can be got by moving the inductor away from the output stage and into the middle of the small-signal stages, which may sound iffy but in my experience works fine. The low closed-loop gain means that there is no possibility of instability due to coupling between output and input. My current policy is that the output inductor should not be on the power amplifier PCB at all, but next to the output terminals, on a small sub-PCB if need be. This should put enough distance between the coil and the Class-B currents.

Note that this article was the first to introduce the concept of the Blameless amplifier, in the panel towards the beginning.

DISTORTION IN POWER AMPLIFIERS

January 1994

Distortion 3: quiescent current control

An optimised amplifier requires minimisation of output stage gain irregularities around the crossover point by holding the quiescent current I_q at its optimal value. Increasing I_q to move into Class-AB makes the distortion worse, not better, as g_m-doubling artifacts are generated.

The initial setting of quiescent current is simple, given a distortion analyser to get a good view of the residual; keeping that setting under varying operating conditions is a much greater problem because I_q depends on small voltages established across low value resistors by power devices with thermally dependant V_{be} drops.

How accurately does quiescent current need to be maintained? I wish I could be more specific on this. Some informal experiments with Blameless CFP type outputs at 1 kHz indicate that crossover artifacts on THD residual seem to stay at roughly the same level, partly submerged in the noise, over an I_q range of about 2:1, the centre of this region being around 20 mA. Results may well be different for emitter follower type outputs.

Blameless amplifiers

I have adopted the term *blameless* to describe a Class-B amplifier designed in accordance with the philosophy of this series, with the use of simple circuit enhancements to minimise distortions 1,2 and 4, and correct layout to prevent distortions 5,6 and 7. Such a device will still suffer from output stage distortion 3, and so exhibit measurable distortion at high frequencies due to the difficulty that NFB has in dealing with the high order crossover distortion products generated by a conventional (but well designed) output stage. Distortion will usually be greater when driving loads below 8 Ω.

The word is specifically chosen to imply the avoidance of error but not perfection.

This may seem a wide enough target, but given that junction temperature of power devices may vary over a 100°C range, this is not so. Some kinds of amplifier (e.g. current dumping types) manage to evade the problem altogether, but in general the solution is thermal compensation: the output stage bias voltage is set by a temperature sensor (usually a V_{be} multiplier transistor) coupled as closely as possible to the power devices.

There are inherent inaccuracies and thermal lags in this sort of arrangement leading to programme dependency of I_q. A sudden period of high power dissipation will begin with the bias current increasing above optimum, as the junctions will heat up very quickly. Eventually the thermal mass of the heatsink will respond, and the bias voltage will be reduced. When the power dissipation falls again, the bias voltage will now be too low to match the cooling junctions and the amplifier will be under biased, producing crossover spikes that may persist for some minutes. This is well illustrated in an important paper by Sato [1].

Emitter follower outputs

The major drawback of emitter follower output stages is thermal stabilisation. This can cause production problems in initial setting up since any drift of quiescent current will be very slow as a lot of metal must warm up.

For EF outputs, the bias generator must attempt to establish an output bias voltage that is a summation of four driver and output V_{be}'s. These do not vary in the same way. It seems at first a bit of a mystery how the EF stage, which still seems to be the most popular output topology, works as well as it does. The probable answer is Figure 27.1, which shows how driver dissipation (averaged over a complete cycle) varies with peak output level for the three kinds of EF output, and for the CFP configuration. The Spice simulations used to generate this graph used a triangle waveform to give a slightly closer approximation to the peak-average ratio of real waveforms. The rails were ±50 V, and the load 8 Ω.

It is clear that the driver dissipation for the EF types is relatively constant with power output, while the CFP driver dissipation, although generally lower, varies strongly. This is a consequence of the different operation of these two kinds of output. In general, the drivers of an EF output remain conducting to some degree for most or all of a cycle, although the output devices are certainly off half the time.

In the CFP, however, the drivers turn off almost in synchrony with the outputs, dissipating an amount of power that varies much more with output. This implies that EF drivers will work at roughly the same temperature, and can be neglected in arranging thermal compensation; the temperature dependent element is

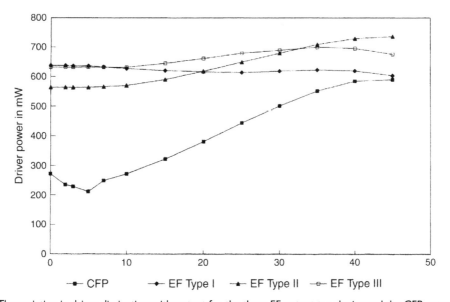

Figure 27.1 The variation in driver dissipation with output for the three EF output topologies and the CFP output. All three EF types keep driver power fairly constant, simplifying the thermal compensation problem.

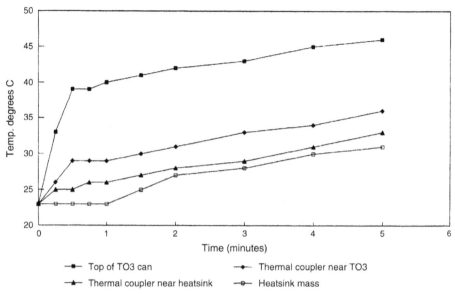

Figure 27.2 Thermal response of a TO3 coupled to a large heatsink when power is abruptly applied. The top of the TO3 can responds most rapidly.

usually attached to the heatsink to compensate for the junction temperature of the output devices alone. The Type I EF output keeps its drivers at the most constant temperature.

The above does not apply to integrated Darlington outputs, with drivers and assorted emitter resistors combined in one ill-conceived package where the driver sections are directly heated by the output junctions. This works directly against quiescent stability.

The drawback with most thermal compensation schemes is the slow response of the heatsink mass to thermal transients. The obvious solution is to find some way of getting the sensor closer to one of the output junctions. If TO3 devices are used, then the flange on which the actual transistor is mounted is as close as one can get without a hacksaw. This is however clamped to the heatsink, and almost inaccessible, though it might be possible to hold a sensor under one of the mounting bolts. A simpler solution is to mount the sensor on the top of the TO3 can. This is probably not as accurate an estimate of junction temperature as the flange would give, but measurement shows the top gets much hotter much faster than the heatsink mass, so while it may appear unconventional, it is probably the best sensor position for an EF output stage.

Figure 27.2 shows the results of an experiment designed to test this. A TO3 device was mounted on a thick aluminium L-section thermal coupler in turn clamped to a heatsink; this construction represents many typical designs. Dissipation equivalent to 100 W/8 Ω was suddenly initiated, and the temperature of the various parts monitored with thermocouples. The graph clearly shows that the top of the TO3 responds much faster, and with a larger temperature change, though after the first two minutes the temperatures are all increasing at the same rate. The whole assembly took more than an hour to asymptote to thermal equilibrium.

The CFP output

In the CFP configuration, the output devices are inside a local feedback loop, and play no significant part in setting I_q, which is affected only by thermal changes in the drivers' V_{be}. Such stages are virtually immune to thermal runaway; I have found that assaulting the output devices with a powerful heat gun induces only insignificant I_q changes. Thermal compensation is mechanically simpler as the V_{be} multiplier transistor is simply mounted on one of the driver heatsinks, where it aspires to mimic the driver junction temperature. It is now practical to make the bias transistor of the same type as the drivers, which should give the best matching of V_{be}, though how important this is in practice I wouldn't like to say [2].

Because driver heatsinks are much smaller than the main heatsink, the thermal compensation time constant is now measured in tens of seconds rather than tens of minutes, and should give much shorter periods of non optimal quiescent current than the EF output topology.

Distortion 4: nonlinear loading of the voltage amplifier stage by the nonlinear impedance of the output stage

This distortion mechanism was examined in Chapter 18 from the point of view of the voltage amplifier stage (VAS). Essentially, since the VAS provides all the voltage gain, its collector impedance tends to be made high. This renders it vulnerable to nonlinear loading unless it is buffered.

Making a linear VAS is most easily done by applying a healthy amount of local negative feedback via the dominant pole Miller capacitor, and if VAS distortion needs further reduction, then the open loop gain of the VAS stage must be raised to increase this local feedback. The direct connection of a Class-B output can make this difficult for, if the gain increase is attempted

by cascoding with intent to raise the impedance at the VAS collector, the output stage loading will render this almost completely ineffective. The use of a VAS buffer eliminates this effect.

As explained previously, the collector impedance, while high at LF compared with other circuit nodes, falls with frequency as soon as C_{dom} starts to take effect, and so the fourth distortion mechanism is usually only visible at LF. It is also masked by the increase in output stage distortion above dominant pole frequency P1 as the amount of global NFB reduces.

The fall in VAS impedance with frequency is demonstrated in Figure 27.3, obtained from the Spice conceptual model outlined previously, with real life values. The LF impedance is basically that of the VAS collector resistance, but halves with each octave once P1 is reached. By 3 kHz it is down to 1 kΩ and still falling. Nevertheless, it can remain high enough for the input impedance of a Class-B output stage to significantly degrade linearity, the actual effect being shown in Figure 27.4.

An alternative to cascoding for VAS linearisation is to add an emitter follower within the VAS local feedback loop, increasing the local NFB factor by raising effective

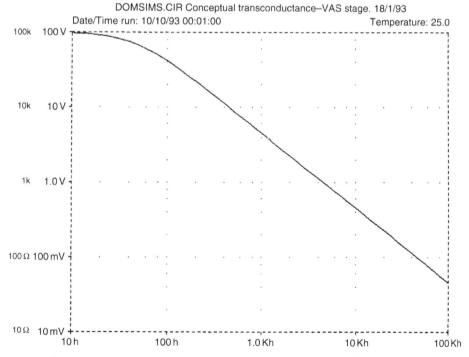

Figure 27.3 Distortion 4. The impedance at the VAS collector falls at 6 dB/octave with frequency.

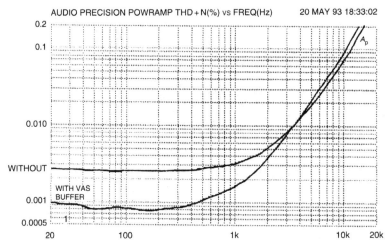

Figure 27.4 Distortion 4 in action. The lower trace shows the result of its elimination by the use of a VAS buffer.

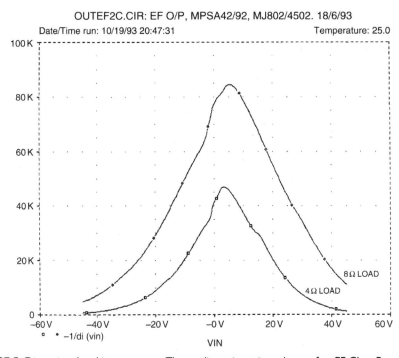

Figure 27.5 Distortion 4 and its root cause. The nonlinear input impedance of an EF Class B output stage.

beta rather than the collector impedance. Preliminary tests show that as well as providing good VAS linearity, it establishes a lower VAS collector impedance across the audio band. It should be more resistant to this type of distortion than the cascode version.

Figure 27.5 confirms that the input impedance of a conventional EF Type I output stage is anything but linear; the data is derived from a Spice output stage simulation with optimal I_q. Even with an undemanding 8 Ω load, the impedance varies by 10:1 over the output voltage swing. Interestingly, the Type II EF output (using a shared drive emitter resistance) has a 50% higher impedance around crossover, but the variation ratio is rather greater. CFP output stages have a more complex variation that includes a precipitous drop to less than 20 kΩ around the crossover point. With all types under biasing produces additional sharp impedance changes at crossover.

Distortion 5: supply ground loops

Virtually all amplifiers include some form of rail decoupling apart from the main reservoir capacitors; this is usually required to improve HF stability. The standard decoupling arrangements include small to medium sized electrolytics (say 10–1000 μF) connected between each rail and ground, and an inevitable consequence is that voltage variations on the rails cause current to flow into the ground connection chosen. This is just one mechanism that defines the power supply rejection ratio (PSRR) of an amplifier, but it is one that can do serious damage to linearity. If we assume a simple unregulated power supply, (and there are excellent reasons for using such a supply[3]) then these rails have a significant A.C. impedance and superimposed voltage will be due to amplifier load currents as well as 100 Hz ripple. In Class-B, these supply rail currents are halfwave rectified sine pulses with strong harmonic content, and if they contaminate the signal, then distortion will degrade badly. A common route for interaction is via decoupling

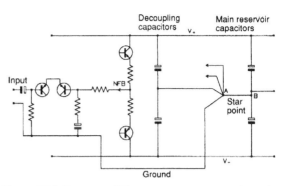

Figure 27.6 Distortion 5. The correct way to route decouple grounding to the star point.

grounds shared with input or feedback networks, and a completely separate decoupler ground usually effects a total cure. This point is easy to overlook, and attempts to improve amplifier linearity by labouring on the input pair, VAS, etc., are doomed to failure unless this distortion mechanism is eliminated first.

As a rule it is simply necessary to take the decoupling ground separately back to the ground star point, as shown in Figure 27.6. Note that the star point A is defined on a short spur from the heavy connection joining the reservoirs; trying to use B as the star point will introduce ripple due to the large reservoir charging current pulses passing through it.

Figure 27.7 shows the effect on an otherwise optimised amplifier delivering 60 W/8 Ω with 220 μF

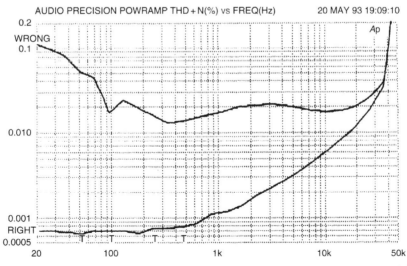

Figure 27.7 Distortion 5 in action. The upper trace was produced simply by taking the decoupler ground from the star point and connecting it via the input ground line instead.

rail decoupling capacitors. At 1 kHz distortion has increased by more than ten times, which is quite bad enough. However, at 20 Hz the THD has increased at least 100 fold, turning a very good amplifier into a profoundly mediocre one with a single misconceived connection.

If the residual on the supply rails is examined, the ripple amplitude will usually be found to exceed the pulses due to Class-B signal current, and so some of the 'distortion' on the upper curve of the plot is actually due to ripple injection. This is hinted at by the phase crevasse at 100 Hz, where ripple partly cancelled the signal at the instant of measurement. Below 100 Hz the—curve rises as greater demands are made on the reservoirs, the signal voltage on the rails increases, and so more distorted current is forced into the ground system.

Generally, if an amplifier is made free from ripple injection under drive conditions, shown by a THD residual without ripple components, there will be no distortion from the supply rails and the complications and inefficiency of high current rail regulators are unnecessary.

There has been much discussion of PSRR induced distortion in *EW + WW* recently, led by Ben Duncan [4] and Greg Ball [5]. I part company with Ben Duncan on this issue where he assumes that a power amplifier is likely to have 25 dB PSRR, making expensive high power DC regulators the only answer. He agrees that this sort of PSRR is highly unlikely with the relatively conventional amplifier topologies I have been considering [6].

Greg Ball also initially assumes that a power amp has the same PSRR characteristics as an op-amp, i.e. falling steadily at 6 dB/octave. There is absolutely no need for this to be so, given a little RC decoupling, and Ball states at the end of his article that 'a more elegant solution . . . is to depend on a high PSRR in the amplifier proper.'

Power supply rejection

For low noise and distortion, all the obvious methods of rail injection must be attended to as a matter of routine. I therefore give here some guidelines that I have found effective with unregulated supplies:

- The input pair must have a tail current source. A tail made of two resistors decoupled mid way is simply not adequate.
- This tail source will probably be biased by a pair of diodes or a led fed from a resistor to ground. This resistor should be split and the midpoint decoupled with an electrolytic of about 10 µF to the appropriate rail.

- If a cascode transistor is used in the VAS, then its base will need to be biased about 1.2 V above whichever rail the VAS emitter sits on; if this is implemented with a pair of diodes then further decoupling seems unnecessary.
- Having taken care of the above, the PSRR will now be limited by injection from the negative rail by a mechanism that is not yet fully clear. RC decoupling can however reduce this to negligible levels.

This is not the whole story on power rail rejection, but it does provide a starting point.

Distortion 6: induced output current coupling

This distortion mechanism, like the previous case, stems directly from the Class-B nature of the output stage. Assuming a sine input, the output hopefully carries a good sinewave, but the supply rail currents are halfwave rectified sine pulses, which are quite capable of inductive crosstalk into sensitive parts of the circuit. This can be very damaging to the distortion performance, as Figure 27.8 shows.

The distortion signal may intrude into the input circuitry, the feedback path, or even the cables to the output terminals. The result is a kind of sawtooth on the distortion residual that is very distinctive, an extra distortion component which rises at 6 dB/octave with frequency.

This effect appears to have been first publicised by Cherry, [7] in a paper that deserves much more attention than it appears to have got. Having examined many power amplifiers, I feel that this effect is probably the most widespread cause of unnecessary distortion.

Effects of this distortion mechanism can be reduced below the measurement threshold by taking care over supply rail cabling layout relative to signal leads, and avoiding loops that will induce or pick up magnetic fields. There are no precise rules for layout that would guarantee freedom from rail induction since each amplifier has its own physical layout and the cabling topology needs to take this into account. All I can do is give guidelines:

- Firstly, implement rigorous minimisation of loop area in the input and feedback circuitry; keep each signal line as close to its ground return as possible.
- Secondly, minimise the ability of the supply wiring to create magnetic fields.
- Thirdly, put as much distance between these two areas as you can. Fresh air beats shielding on price every time. Figure 27.9(a) shows one straightforward approach to tackling the problem; the supply and

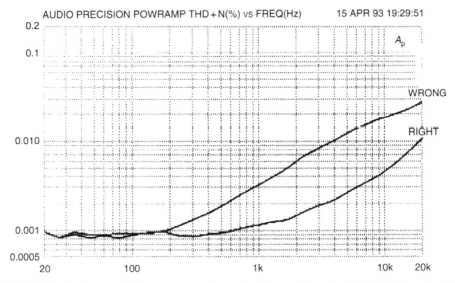

AUDIO PRECISION POWRAMP THD + N(%) vs FREQ(Hz)   15 APR 93 19:29:51

Figure 27.8 Distortion 6 exposed. The upper trace shows the effects of Class B rail induction into signal circuitry.

ground wires are tightly twisted together to reduce radiation. In practice this doesn't seem too effective for reasons that are not wholly clear, but appear to involve the difficulty of ensuring exactly equal coupling between three twisted conductors.

In Figure 27.9(b), the supply rails are twisted together but kept well away from the ground return. This allows field generation, but if currents in the two rails butt together to make a sinewave at the output, they should do the same when the magnetic fields from each rail sum. There is an obvious risk of interchannel crosstalk with this approach in a stereo amplifier, but it does seem to deal most effectively with the induced distortion problem.

Distortion 7: nonlinearity from incorrect NFB connection point

Negative feedback is a powerful technique and must be used with care. Designers are repeatedly told that too much feedback can affect slew rate. Possibly true, though the greater danger is that an excess amplifier may produce tweeter-frying HF instability.

However, there is another and more subtle danger. Class-B output stages are a hotbed of high amplitude halfwave rectified currents, and if the feedback takeoff

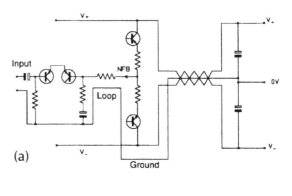

(a)

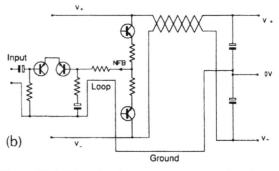

(b)

Figure 27.9 Distortion 6. Countermeasures against the induction of distortion from the supply rails. 27.9(b) is usually more effective.

point is even slightly asymmetric, these will contaminate the feedback signal making it an inaccurate representation of the output voltage. This will manifest itself as distortion, Figure 27.10.

At the current levels in question, all wires and PCB tracks must be treated as resistances, and it follows that point C is not at the same potential as point D whenever TR_1 conducts. If feedback is taken from D, then a clean signal will be established here, but the signal at output point C will have a half wave rectified sinewave added to it, due to the resistance $C–D$. The output will be distorted but the feedback loop will do nothing about it as it does not know about the error.

Figure 27.11 shows the practical result for an amplifier driving 100 W into 8 Ω, with the extra distortion shadowing the original curve as it rises with frequency. Resistive path $C–D$ that did the damage was a mere 6 mm length of heavy gauge wirewound resistor lead.

Elimination of this distortion is easy, once you know the danger. Connecting the feedback arm to D is not advisable as it will not be a mathematical point, but will have a physical extent inside which the current

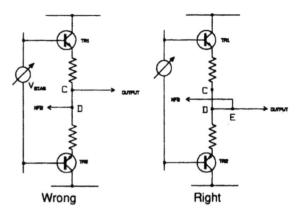

Figure 27.10 Distortion 7. Wrong and right ways of arranging the critical negative feedback takeoff point.

distribution is unknown. Point E on the output line is much better, as the half wave currents do not flow through this arm of the circuit.

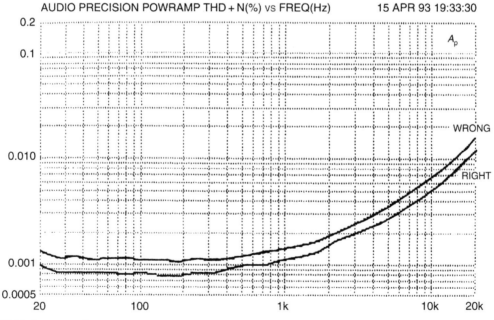

Figure 27.11 Distortion 7 at work. The upper trace shows the result of a mere 6 mm of heavy gauge wire between the output and the feedback point.

References

1. Sato 'Amplifier transiant grossover distortion'. AES Preprint for 72nd Convention, October 1982.

2. Evans, J 'Audio amplifier bias current', Letters, *Electronics + Wireless World,* January 1991, p 53.

3. Self, D 'Sound mosfet design', *Electronics + Wireless World,* September 1990, p 760.

4. Duncan, B 'PSU regulation boosts audio performance', *Electronics + Wireless World*, October 92, p 818.

5. Ball, G 'Distorting power supplies', *Electronics + Wireless World,* December 1990, p 1084.

6. Duncan, B, Private communication, October 93.

7. Cherry 'A New distortion mechanism in class-B amplifiers', *JAES* May 1981, p 327.

Chapter 28

Distortion in power amplifiers

Part VII: frequency compensation and real designs
(Electronics World, February 1994)

This crowded article attempted to deal with the extensive subject of compensation for stability, to showcase a complete 50W/8Ω power amplifier, and demonstrate the distortion performance of that amplifier with various compensation schemes. It was difficult to pack everything in, as the eighth and final article of the series was reserved for Class-A, and perhaps inevitably, there were a few rough edges.

The first section on dominant pole compensation was of course written before the role of the non-linear Cbc had been properly explored. (See Chapter 24.) I say in the article that the added Miller capacitor swamps the effect of the non-linear Cbc, implying it is rendered harmless. I confess got this plausible but wholly wrong notion from other writers on the subject, and it demonstrates that you should never take anything in the amplifier business as true until you have tested it yourself. In fact, Cbc can cause a lot of distortion in a simple VAS, even if it has a Miller capacitor of more than ten times its value connected across it [1]. Adding an emitter follower to turn the simple VAS into an EF-VAS completely suppresses the effect of Cbc.

The article does not look into two-pole compensation in any depth, as space was in short supply. At one point I say: 'The open-loop gain peak at 8kHz looks extremely dubious, but I have so far failed to detect any resulting ill-effects . . .' as indeed I hadn't. My confidence was bolstered by Peter Baxandall's theoretical demonstration that the closed-loop response was flat, and his comment that 'any notion of a high-Q resonance disappears' [2]. However . . . simulation of a complete amplifier shows something different, with the closed-loop response peaking by up to 0.7 dB around 100 kHz, as described in [3]. This is well outside the audio band, but it puts a visible overshoot on fast square-waves. Methods for suppressing this peak are also described in [3].

There is a definite infelicity in Figure 27.1d. Peter Baxandall pointed out to me [4] that the capacitor values are not optimal, as the larger capacitor Cp2 is loading the VAS collector via Rp; increasing Cp1 to 1nF and reducing Cp2 to 100pF instead gives exactly the same gain response but much reduces the loading. I do express my concern in the text about the loading effects of Rp, but regrettably, I did not investigate further at the time. Peter was of course quite right, and I corrected the capacitor values in [5] and [3]. I reported that in tests on an experimental amplifier based on the Load Invariant design at 25W/8Ω, the original values gave THD at 10 kHz of 0.0043%, but swapping them dropped it to 0.00317%. It is difficult to see that this useful improvement could be due to anything except reduced VAS loading.

The section on quiescent current stability is OK as far as it goes in that all the information presented is correct. The method shown will much improve the stability of the voltage set up by the bias generator. However, a constant voltage bias is not actually what we want, and this is a good example of the need to look at a whole system rather than just one part. If an amplifier with a truly constant-voltage bias generator has its supply voltage altered by a variable-voltage transformer, the quiescent voltages and currents increase with the supply voltage, due to Early effect in the output devices. What we really need is a bias generator that cancels out this effect. Increasing R14 considerably (typically to 100Ω) can do this but the substantial voltage drop across it may impair the positive voltage swing of the amplifier. To make this work really consistently it might be necessary to design a special VAS current-source that had an accurate and dependable relationship with the supply voltage, and did not depend on uncertainties in beta, Early effect, and so on. Even this might not be enough; I have no data on how the Early effect in output devices is likely to vary, and it might be necessary to provide a preset adjustment to allow for this factor. This would be set-and-forget (unless you had to change the output devices) and well worthwhile if it gave a useful increase in the stability of the quiescent voltages and currents in the output stage.

The problem naturally goes away if regulated DC supply rails are used, but so many new and intractable problems appear at the same time that this is really not a promising route.

References

1. Self, D., *Audio Power Amplifier Design*, 6th edition, pp. 169–171. Focal Press 2013. ISBN 978-0-240-52613-3.

2. Baxandall, P., and Self, D., Baxandall and Self on Audio Power, *Linear Audio* 2011, p. 114. Editor: Jan Didden. ISBN 978-94-90929-03-9.

3. Self, pp. 345.

4. Baxandall and Self, 2011, p. 112.

5. Self, D., *Audio Power Amplifier Design*, 5th edition, p. 221. Focal Press 2009. ISBN 978-0-240-52162-6.

DISTORTION IN POWER AMPLIFIERS

February 1994

Making a pole dominant

Dominant pole compensation is the simplest kind, though its implementation involves subtlety. Simply take the lowest pole to hand (PI), and make it dominant, i.e. so much lower in frequency than the next pole P2 that the total loop gain (the open loop gain as reduced by the attenuation in the feedback network) falls below unity before enough phase shift accumulates to cause HF oscillation. With a single pole, the gain must fall at 6 dB/octave, corresponding to a constant 90° phase shift. Thus the phase margin will be 90° giving good stability.

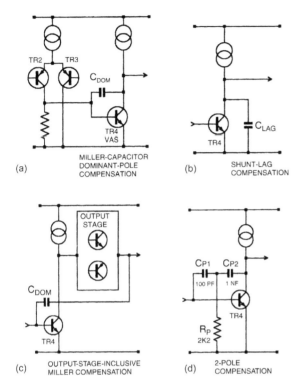

(a) MILLER-CAPACITOR DOMINANT-POLE COMPENSATION

(b) SHUNT-LAG COMPENSATION

(c) OUTPUT-STAGE-INCLUSIVE MILLER COMPENSATION

(d) 2-POLE COMPENSATION

Figure 28.1 Implementing dominant-pole compensation. (a) Miller capacitor, (b) Shunt-lag circuit (c) Output-stage Inclusive Miller compensation. (d) How to implement 2-pole compensation.

Figure 28.1(a) shows the traditional Miller method of making a dominant pole. The collector pole of Tr_4 is lowered by adding the Miller capacitance C_{dom} to that which unavoidably exists as the C_{bc} of the VAS transistor. However there are other beneficial effects; C_{dom} causes 'pole splitting', in which the pole at Tr_2 collector is pushed up in frequency as P1 moves down—most desirable for stability. Simultaneously the local NFB through C_{dom} linearises the VAS.

Assuming that input stage transconductance is set to a plausible 5 mA/V, and stability considerations set the maximal 20 kHz open loop gain to 50 dB, then from the equations in Part 1, C_{dom} must be 125 pF. This is more than enough to swamp the internal capacitances of the VAS transistor, and is a realistic value.

The peak current that flows in and out of this capacitor for an output of 20 V r.m.s., 20 kHz, is 447 μA. Recalling that the input stage must sink C_{dom} current while the VAS collector load sources it, and likewise the input stage must source it while the VAS sinks it, there are four possible places in which slew rate might

be limited by inadequate current capacity. If the input stage is properly designed then the usual limiting factor is VAS current sourcing. In this example a peak current of less than 0.5 mA should be easy to deal with, and the maximum frequency for unslewed output will be comfortably above 20 kHz.

Figure 28.1(b) shows a much less satisfactory method—the addition of capacitance to ground from the VAS collector. This is usually called shunt lag compensation, and as Peter Baxandall aptly put it, 'The technique is in all respects sub-optimal'.[2]

We have already seen in Part 3 that loading the VAS collector resistively to ground is a very poor option for reducing LF open loop gain, and a similar argument shows that capacitive loading to ground for compensation purposes is an even worse idea. To reduce open loop gain at 20 kHz to 50 dB as before, the shunt capacitor C_{lag} must be 43.6 nF, which is a whole different order of things from 125 pF. The current flowing in C_{lag} at 20 V r.m.s., 20 kHz, is 155 mA peak, which is going to require some serious electronics to provide it. This important result can be derived by simple calculation, and I have confirmed it with Spice simulation. The input stage no longer constrains the slew rate limits, which now depend entirely on the VAS.

A VAS working under these conditions is almost certain to have poor linearity. The current variations in the stage, caused by the extra loading, produces more distortion and there is now no local NFB through a Miller capacitor to correct it. To make matters worse, the dominant pole P1 will probably need to be set to a lower frequency than for the Miller case, to maintain the same stability margins, as there is now no pole splitting to raise the pole at the input stage collector. Hence C_{lag} may have to be even larger, and require even higher peak currents. Takahashi has produced a fascinating paper on this approach, [3] showing one way of heaving about the enormous compensation currents required for good slew rates. The only thing missing is an explanation of why shunt compensation was chosen in the first place.

Including the output stage

Miller capacitor compensation elegantly solves several problems at once, and the decision to use it is not hard. However the question of whether to include the output stage in the Miller feedback loop is less easy. Such inclusion (see Figure 28.1(c)) presents the desirable possibility that local feedback could linearise both the VAS and the output stage, with just the input stage left out in the cold as frequency rises and global NFB falls. This idea is most attractive as it would greatly increase the feedback available to linearise a Class B output stage.

There is certainly *some* truth in this where applying C_{dom} around the output as well as the V_{as} reduced the peak 1 kHz THD from 0.05% to 0.02%.[4] However, it should be pointed out that the output stage was deliberately under biased to induce crossover spikes because, with optimal bias, the improvement was too small to be either convincing or worthwhile. Also, this demonstration used a model amplifier with TO-92 'output' transistors. In my experience this technique just does not work with real power bipolars because it induces intractable HF oscillation.

The use of local NFB to linearise the VAS demands a tight loop with minimal extra phase shift beyond that inherent in the C_{dom} dominant pole. It is permissible to insert a cascode or a small signal emitter follower into this local loop, but a sluggish output stage seems to be pushing the phase margin too far; the output stage poles are now included in the loop, which loses its dependable HF stability. Bob Widlar has stated that output stage

behaviour must be well controlled up to 100 MHz for the technique to be reliable.[5] This would appear to be virtually impossible for discrete power stages with varying loads.

While I have so far not found 'Inclusive Miller compensation' to be workable myself, others may know different. If anyone can shed further light I would be most interested.

Nested feedback loops

Nested feedback is a way to apply more NFB around the output stage without increasing the global feedback factor. The output has an extra voltage gain stage bolted on, and a local feedback loop is closed around these two stages. This NFB around the composite block reduces output stage distortion and increases frequency response, to make it safe to include in the global NFB loop.

Suppose that block A_1 (Figure 28.2(a)) is a distortionless small signal amplifier providing all the open loop gain and so including the dominant pole.

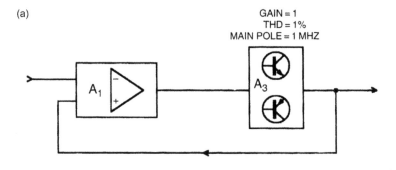

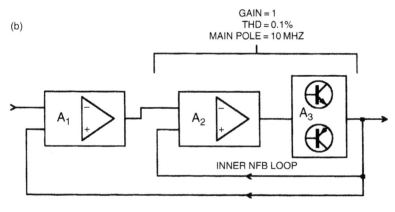

Figure 28.2 The principle of nested feedback loops.

A_3 is a unity gain output stage with its own main pole at 1 MHz and distortion of 1% under given conditions: this 1 MHz pole puts a firm limit on the amount of global NFB that can be safely applied.

Figure 28.2(b) shows a nested feedback version; an extra gain block A_2 has been added, with local feedback around the output stage. A_3 has the modest gain of 20 dB so there is a good chance of stability when this loop is closed to bring the gain of $A_3 + A_2$ back to unity. A_2 now experiences 20 dB of NFB, bringing the distortion down to 0.1%, and raising the main pole to 10 MHz, which should allow the application of 20 dB more global NFB around the overall loop that includes A_1. We have thus decreased the distortion that exists before global NFB is applied, and simultaneously increased the amount of NFB that can be safely used, promising that the final linearity could be very good indeed. For another theoretical example see Ref. 6.

Real life examples of this technique in power amps are not easy to find, but a variation is widely used in op-amps. Many of us were long puzzled by the way that the much loved *5534* maintained such low THD up to high frequencies. Contemplation of its entrails appears to reveal a three-gain stage design with an inner Miller loop around the third stage, and an outer Miller loop around the second and third stages; global NFB is then applied externally around the whole lot. Nested Miller compensation has reached its apotheosis in cmos op-amps—the present record appears to be three nested Miller loops plus the global NFB.[7] Don't try this one at home.

Two pole compensation

Two pole compensation is a mildly obscure technique for squeezing the best performance from an op-amp, [8,9] but it has rarely been applied to power amplifiers. I know of only one example.[5] An extra HF time constant is inserted in the C_{dom} path, giving an open loop gain curve that initially falls at almost 12 dB/octave, but which gradually reverts to 6 dB/octave as frequency continues to increase. This reversion is arranged to happen well before the unity loop gain line is reached, and so stability should be the same as for the conventional dominant pole scheme, but with increased negative feedback over part of the operational frequency range. The faster gain roll off means that the maximum amount of feedback can be maintained up to a higher frequency. There is no measurable mid band peak in the closed loop response.

One should be cautious about any circuit arrangement which increases the NFB factor. Power amplifiers face loads that vary widely: it is difficult to be sure that a design will always be stable under all circumstances. This makes designers rather conservative about compensation, and I approached this technique with some trepidation. However, results were excellent with no obvious reduction in stability. Figure 28.7 shows the result of applying this technique to the Class B amplifier described below.

The simplest way to implement two pole compensation is shown in Figure 28.1(d), with typical values. C_{p1} should have the same value as it would for stable single pole compensation, and C_{P2} should be at least twice as big; R_p is usually in the region 1 k–10 k. At intermediate frequencies C_{P2} has an impedance comparable with R_p, and the resulting extra time constant causes the local feedback around the VAS to increase more rapidly with frequency, reducing the open loop gain at almost 12 dB/octave.

At HF the impedance of R_p is high compared with C_{P2}, the gain slope asymptotes back to 6 dB/octave, and then operation is the same as conventional dominant pole, with C_{dom} equal to the series capacitance combination. So long as the slope returns to 6 dB/octave before the unity loop gain crossing occurs, there seems no obvious reason why the Nyquist stability should be impaired.

Figure 28.3 shows a simulated open loop gain plot for realistic component values; C_{P2} should be at least twice C_{P1} so the gain falls back to the 6 dB/octave line before the unity loop gain line is crossed. The potential feedback factor has been increased by more than 20 dB from 3 kHz to 30 kHz, a region where THD tends to increase due to falling NFB. The open loop gain peak at 8 kHz looks extremely dubious, but I have so far failed to detect any resulting ill effects in the closed loop behaviour.

There is however a snag to the simple approach shown here, which reduces the linearity improvement. Two-pole compensation may decrease open loop linearity at the same time as it raises the feedback factor that strives to correct it. At HF, C_{P2} has low impedance and allows R_p to directly load the VAS collector to ground. This worsens VAS linearity as we have seen. However, if C_{P2} and R_p are correctly proportioned the overall reduction in distortion is dramatic and extremely valuable. When two pole compensation was added to Figure 28.4, the crossover glitches on the THD residual almost disappeared, being partially replaced by low level 2nd harmonic which almost certainly results from VAS loading. The positive slew rate will also be slightly reduced.

This looks like an attractive technique, as it can be simply applied to an existing design by adding two

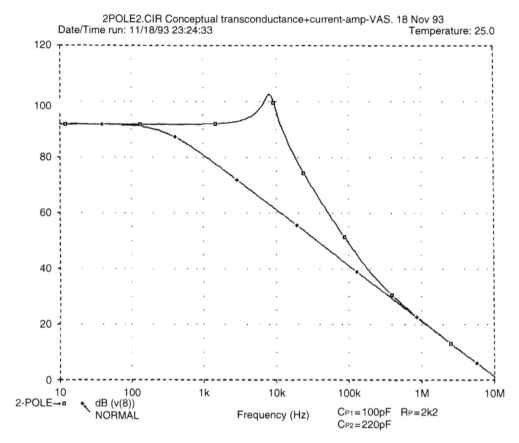

Figure 28.3 J3 Spice plot of the open-loop gain of a 2-pole compensated amplifier. The difference between the two plots shows the amount of extra NFB possible.

inexpensive components. If C_{p2} is much larger than C_{P1}, then adding/removing R_p allows instant comparison between the two kinds of compensation. Be warned that if an amplifier is prone to HF parasitics then this kind of compensation may exacerbate them.

Design example: a 50 W Class B amplifier

Figure 28.4 shows a design example of a Class B amplifier intended for domestic audio. Despite its conventional appearance, the circuit delivers a far better distortion performance than that normally associated with the arrangement.

With the supply voltages and values shown it delivers 50 W/8 Ω from IV r.m.s. input. In previous articles I have used the word *blameless* to describe amplifiers in which all distortion mechanisms, except the apparently

unavoidable ones due to Class B, have been rendered negligible. This circuit has the potential to be *blameless*, but achieving this depends on care in cabling and layout. It does not aim to be a cookbook project; for example, overcurrent and DC offset protection are omitted.

The investigation presented in chapters 19 and 20 concluded that power fets were expensive, inefficient and non linear. Bipolars make good output devices. The best BJT configurations were the emitter follower type II, with least output switch-off distortion, and the complementary feedback pair (CFP) giving best basic linearity and quiescent stability.

I assume that domestic ambient temperatures will be benign, and the duty moderate, so that adequate quiescent stability can be attained by suitable heatsinking and thermal compensation. The configuration chosen is therefore emitter follower type II, which has the

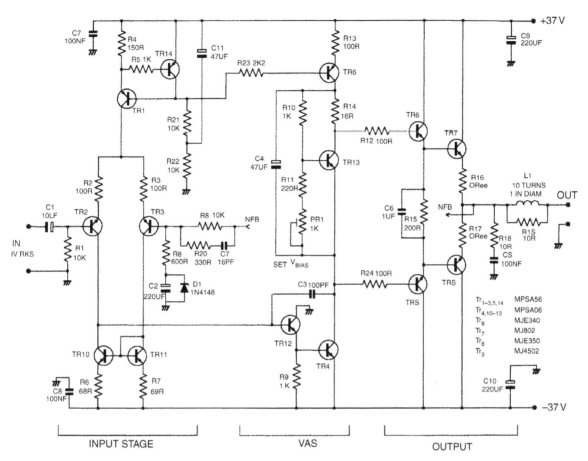

Figure 28.4 50 W Class B amplifier circuit diagram. Transistor numbers correspond with the generic amplifier in chapter 16.

advantage of reducing switch-off nonlinearities (Distortion 3(c)) due to the action of R_{15} in reverse biasing the output base emitter junctions as they turn off. The disadvantage is that quiescent stability is worse than for the CFP output topology, as there is no local feedback loop to servo out V_{be} variations in the hot output devices.

The NFB factor was chosen as 30 dB at 20 kHz, which should give generous HF stability margins. The input stage (current source Tr_1, Tr_{14} and differential pair $Tr_{2,3}$) is heavily degenerated by R_2R_3 to delay the onset of third harmonic Distortion 1. To assist this the contribution of transistor internal r_e variation is minimised by using the unusually high tail current of 4 mA. $Tr_{10,11}$ form a degenerated current mirror that enforces accurate balance of the $Tr_{2,3}$ collector currents, preventing second harmonic distortion. Tail source $Tr_{1,14}$ has a basic PSRR 10 dB better than the usual two diode version, though this is academic when C_{11} is fitted.

Input resistor R_1 and feedback arm R_8 are made equal and kept as low as possible consistent with a reasonably high input impedance, so that base current mismatch caused by beta variations will give a minimal DC offset. This does not affect Tr_2-Tr_3V_{be} mismatches, which appear directly at the output, but these are much smaller than the effects of I_b. Even if $Tr_{2,3}$ are high voltage types with low beta, the output offset should be within ±50 mV, which should be quite adequate, and eliminates balance presets and DC servos. A low value for R_8 also gives a low value for R_9, which improves the noise performance.

The value of C_2 shown (220 µF) gives an LF roll off with R_9 that is −3 dB at 1.4 Hz. The aim is not an unreasonably extended sub-bass response, but to prevent an LF rise in distortion due to capacitor non linearity.

For example, 100 µF degraded the THD at 10 Hz from less than 0.0006% to 0.0011%. Band limiting should be done earlier, with non electrolytic capacitors. Protection diode D_1 prevents damage to C_2 if the amplifier suffers a fault that makes it saturate negatively; it looks unlikely but causes no measurable distortion.[10] C_7 provides some stabilising phase advance and limits the closed loop bandwidth; R_{20} prevents it upsetting Tr_3.

The VAS stage is enhanced by an emitter follower inside the Miller compensation loop, so that the local NFB which linearises the VAS is increased by augmenting total VAS beta, rather than by increasing the collector impedance by cascoding. The extra local NFB effectively eliminates VAS nonlinearity (Distortion 2).

Increasing VAS beta like this presents a much lower collector impedance than a cascode stage due to the greater local feedback. The improvement appears to make a VAS buffer to eliminate Distortion 4 (loading of VAS collector by the nonlinear input impedance of the output stage) unnecessary. C_{dom} is relatively high at 100 pF, to swamp transistor internal capacitances and circuit strays, and make the design predictable. The slew rate calculates as 40 V/µsec. The VAS collector load is a standard current source, to avoid the uncertainties of bootstrapping.

Quiescent current stability

Since almost all the THD from a *blameless* amplifier is crossover, keeping the quiescent optimal is essential. Quiescent stability requires the bias generator to cancel out the V_{be} variations of four junctions in series; those of two drivers and two output devices. Bias generator Tr_{13} is the standard V_{be} multiplier, modified to make its voltage more stable against variations in the current through it. These occur because the biasing of Tr_5 does not completely reject rail variations: its output current drifts initially due to heating thus changing its V_{be}. Keeping a Class B quiescent stable is hard enough at the best of times, and so it makes sense to keep these extra factors out of the equation.

The basic V_{be} multiplier has an incremental resistance of about 20 Ω; in other words its voltage changes by 1 mV for a 50 µA drift in standing current. Adding R_{14} converts this to a gently peaking characteristic that can be made perfectly flat at one chosen current; see Figure 28.5. Setting R_{14} to 22 Ω makes the voltage peak at 6 mA, and standing current now must deviate from this value by more than 500 µA for a 1 mV bias change. The R_{14}

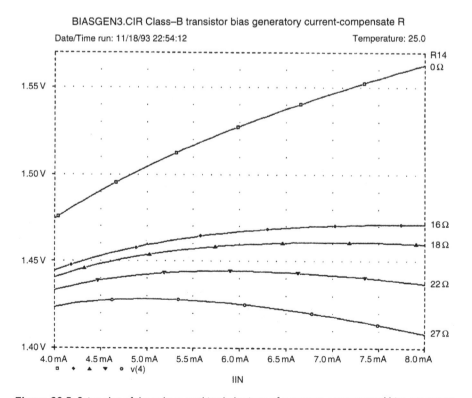

Figure 28.5 Spice plot of the voltage-peaking behaviour of a current-compensated bias generator.

value needs to be altered if Tr_5 is run at a different current. For example, 16 Ω makes the voltage peak at 8 mA instead. If TO3 outputs are used, the bias generator should be in contact with the top or can of one of the output devices, rather than the heatsink, as this is the fastest and least attenuated source for thermal feedback.

Output stage

The output stage is a standard double emitter follower apart from the connection of R_{15} between the driver emitters without connection to the output rail. This gives quicker and cleaner switch-off of the outputs at high frequencies; this may be significant from 10 kHz upwards dependent on transistor type. Speed up capacitor C_5 improves the switch-off action. C_6, R_{18} form the Zobel network while L_1, damped by R_{19}, isolates the amplifier from load capacitance.

Figure 28.6 shows the 50 W/8 Ω distortion performance, about 0.001% at 1 kHz, and 0.006% at 10 kHz. The measurement bandwidth makes a big difference to the appearance, because what little distortion is present is crossover derived, and so high order. It rises at 6 dB/octave, the rate at which feedback factor falls. The crossover glitches emerge from the noise, like Grendel from the marsh, as the test frequency increases above 1 kHz. There is no precipitous THD rise in the ultrasonic region, and so I suppose this might be called a high speed amplifier.

Note that the zigzags on the LF end of the plot are measurement artifacts, apparently caused by the Audio Precision system trying to winkle distortion from visually pure white noise. Below 700 Hz the residual was pure noise with a level equivalent to approx 0.0006% at 30 kHz bandwidth. The actual THD here must be microscopic.

This performance can only be obtained if all seven of the distortion mechanisms are properly addressed; Distortions 1–4 are determined by the circuit design, but the remaining three depend critically on physical layout and grounding topology.

Figure 28.7 shows the startling results of applying two pole compensation to the amplifier. C_3 remains 100 pF, while C_{p2} was 220 pF and R_p1 kΩ. The extra NFB does its work extremely well, the 10 kHz THD dropping to 0.0015%, while the 1 kHz figure can only be guessed at. There were no unusual signs of instability on the bench, but I have not tried a wide range of loads.

This experimental amplifier was rebuilt with three alternative output stages: the simple quasi-complementary, the quasi-Baxandall and the CFP. The results for both single and two pole compensation are shown in Figures 28.8, 28.9, and 28.10. The simple quasi-complementary generates more crossover distortion, as expected, and the quasi-Baxandall version is not a lot better, due to remaining asymmetry around the crossover region. The CFP gives even lower distortion than the original EF-II output. Figure 28.10 shows only the result for single pole

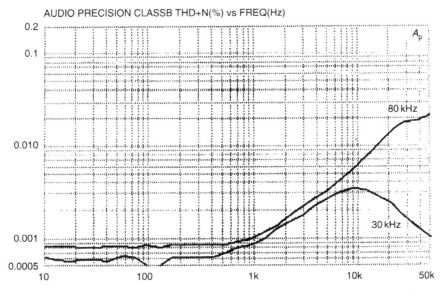

Figure 28.6 Class B amplifier: THD performance at 50 W/8-ohm; measurement bandwidths 30 kHz and 80 kHz.

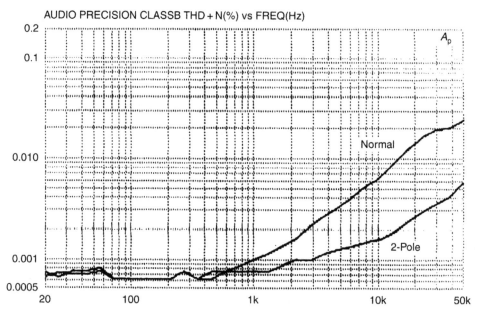

Figure 28.7 The dramatic THD improvement obtained by converting the Class B amplifier to two pole.

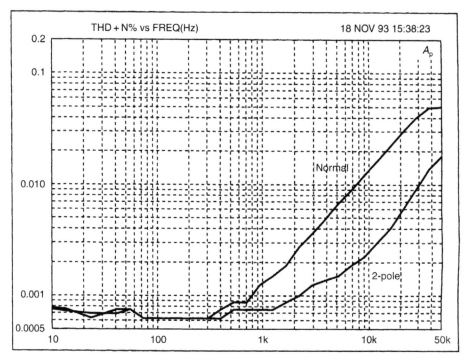

Figure 28.8 Class B amplifier with simple quasi-complementary output. Lower trace is for two pole compensation (80 kHz bandwidth).

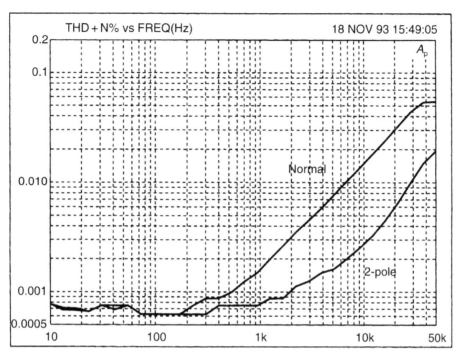

Figure 28.9 Class B amplifier with quasi-complementary plus Baxandall diode output. Lower trace is the two pole case (80 kHz bandwidth).

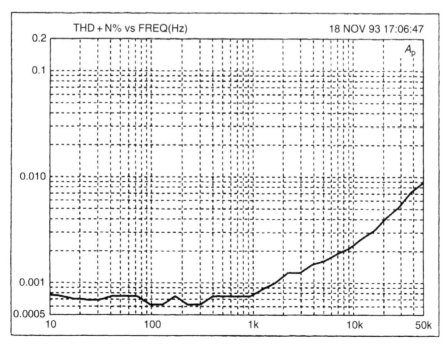

Figure 28.10 Class B amplifier with complementary feedback pair (CFP) output stage.

compensation; in this case the improvement with two pole was marginal and the trace is omitted for clarity.

The AP plots in earlier parts of this series were mostly done with an amplifier similar to Figure 28.4, though of higher power. Main differences were the use of a cascode-VAS with a buffer, and a CFP output to minimise distracting quiescent variations. Measurements at powers above 100 W/8 Ω used a version with two paralleled output devices.

References

1. Otala, 'An audio power amplifier for ultimate quality requirements', *IEEE Trans on Audio & Electroacoustics,* Vol AU-21, No. 6, December 1973.

2. Baxandall, 'Audio power amplifier design. Part 4', *Wireless World,* July 1978, p 76.

3. Takahashi *et al.* 'Design & construction of high slew-rate amplifiers', *AES* 60th Convention, Preprint No. 1348 (A-4) 1978.

4. Self, 'Crossover Distortion & Compensation', Letters, *Electronics & Wireless World,* August 1992, p 657.

5. Widlar, 'A monolithic power op-amp', *IEEE Journal of Solid-State Circuits,* Vol 23, No 2, April 1988.

6. Bonello, 'Advanced negative feedback design for high performance amplifiers', *AES* 67th Convention, Preprint No. 1706 (D-5) 1980.

7. Pernici, 'A CMOS low-distortion amplifier with double-nested et al miller compensation', *IEEE Journal of Solid-State Circuits,* July 1993, p 758.

8. 'Fast compensation extends power bandwidth', Linear Brief 4, National Semiconductor Linear Applications Handbook 1991.

9. Feucht, *Handbook of Analog Circuit Design,* Academic Press 1990, p 264.

10. Self, 'An Advanced Preamplifier', Wireless World, November 76, p 43.

Chapter 29

Distortion in power amplifiers

Part VIII: class A amplifiers
(Electronics World, March 1994)

The class A amplifier described in this chapter has now been in production by The Signal Transfer Company [1] for twenty years without so much as a component value change, which I think indicates it is pretty sound. I have to admit that I was initially nervous about having the voltage reference and quiescent-control subtractor sailing up and down *en bloc* with the amplifier output, but there was never a hint of a problem. This is a lesson to anyone nervously contemplating a complicated output stage, but one thing you must never omit is practical measurements on a real physical amplifier.

I have never been tempted to upgrade its power capability, because as I show in Chapters 40 and 41 of this book, the power efficiency of a push-pull Class-A amplifier reproducing real music as opposed to test sine waves is not 50%, but something more like 1% at maximum level, defined as clipping occasionally.

The canonical series of Class-A power amplifiers traditionally runs from the resistor loaded version with 12.5% efficiency, through the constant-current-source loaded type with 25% efficiency, to the full push-pull version with 50%. These well-known percentages must be treated with caution, for they apply only to sinewaves driven into a resistive load at the maximum amplitude permitted by the rail voltages. Lower amplitudes and reactive loads give even lower efficiencies. Therefore waveforms like music, with a high peak/average ratio, spend almost all their time in the low-efficiency area, and hence the 1%.

The canonical series in the article confined itself to resistors and semiconductors as amplifier elements, and so omitted at least one type of Class-A amplifier. An inductor (what used to be called a choke) can be used as an emitter or collector load in the same way as a resistor, with the added advantage that since current cannot rapidly change in the inductor, all the signal current must pass into the external load, and the maximum efficiency is once more increased to 50%. Also, it can give a greater output voltage swing than the supply voltage. A complication is that the DC resistance of the inductor is low, and if it is used to set the quiescent current simply by putting half the supply

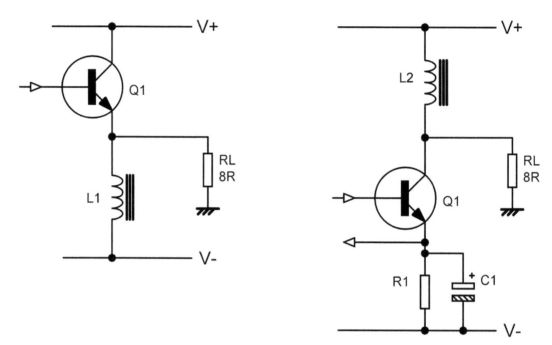

Preface Figure 29.1 Two ways of making a Class-A amplifier with inductive loading. In (b) the quiescent current can be set to any value as it is not affected by the DC resistance of the inductor.

voltage across it, as in Preface Figure 29.1a, the current is very likely to be excessive. A common solution to this problem is shown in Preface Figure 29.1b, where the average value of the voltage drop across low-value R1 is sensed and used to set the quiescent current and hold it constant by negative feedback.

Inductively loaded amplifiers like this had a niche application in car radios during the 1960s because of their voltage efficiency, and a classic version of it by Mullard is described in [1]. This design has an inductor load that is centre-tapped to drive a 3Ω loudspeaker, i.e., it is working as an auto-transformer; the output device therefore saw an apparent load of 12Ω. There is a separate secondary winding that is used solely for negative feedback, which may have been a clever way to extend the LF response, or reduce the effect of inductor non-linearity. In operation the output node voltage can swing either side of the top rail by 9 Volts, giving a an output swing of 18 Vpk-pk from a 14V supply rail, and this over-unity voltage efficiency shows why inductive loading was once popular for car-battery operation. In the Mullard design, R1 was not decoupled (probably because a large and expensive capacitor would have been required), and the voltage drop across it limits the negative output swing.

The Mullard design specifies a 30 mH choke, which must be able to pass the 900 mA quiescent current without saturation or showing lesser signs of non-linearity. Forgetting the 3Ω centre-tap, if the amplifier was driving 8Ω directly, the LF rolloff is −3 dB at 40 Hz by simulation. This agrees with the turnover frequency of 42 Hz, calculated from the simple 30 mH and 8Ω combination in the usual way.

A design using a mains transformer as an inductor was published in *Electronics World* in 1999 [2], specified to give 32W/8Ω from a single 12V supply rail. Since this requires a peak-to peak output of 23V, there is clearly something interesting going on with the voltage efficiency. In this case, the output inductor was centre-tapped and the outside connections driven in push-pull. The tapped inductor was the primary of a mains transformer, as these are much easier to obtain than hefty chokes. Chokes can of course be custom-made, but at a price. Mains transformers appear to generally work well, but you must be wary of dangerous voltages if there is an unconnected secondary winding. The design was inspired by an earlier germanium version of the Mullard 5W amplifier.

If over-unity voltage efficiency from a low supply rail is not an issue, the use of inductive loading has quite serious drawbacks. A suitably sized inductor for a good bass response is going to be a large, heavy, and expensive component, even if you can get it off the shelf in the guise of a mains transformer. The linearity will be poor at low frequencies, and some power will inevitably be lost in the DC resistance of the windings. In a stereo amplifier, you will have to be aware of the possibility of magnetic crosstalk between the two inductors.

References

1. Mullard, *Transistor Audio and Radio Circuits*, Mullard Ltd 1972, Chapter 6, pp. 112–114.

2. Burfoot, R., 32W Class A from a 12V rail, *Electronics World*, Nov. 1999, pp. 934–936.

DISTORTION IN POWER AMPLIFIERS

March 1994

The art of compromise

The only real disadvantage of class A is inefficiency, so inevitably efforts have been made to compromise between A and B. As compromises go, traditional class AB is not a happy one because, when the AB region is entered, the step change in gain generates significantly greater high order distortion than that from optimally biased class B. However, a well-designed AB amplifier will give pure class A performance below the AB threshold, something a class B amp cannot do.

Another compromise is the so-called non-switching amplifier, with its output devices clamped to pass a minimum current. However, it is not immediately obvious that a sudden halt in current-change as opposed to complete turn-off makes for a better crossover region. Those residual oscillograms that have been published seem to show that some kind of discontinuity still exists at crossover.[2]

A potential problem is the presence of maximum ripple on the supply rails at zero signal output; the PSRR must be taken seriously if good noise and ripple figures are to be obtained. This problem can be simply solved by the measures proposed for class B designs.

There is a kind of canonical sequence of efficiency improvement in class A amplifiers. The simplest kind is single-ended and resistively loaded, as at Figure 29.1(a).

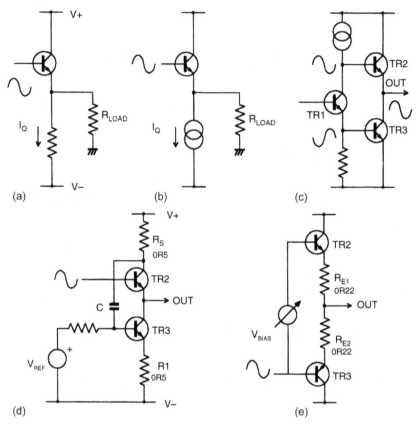

Figure 29.1 The major class A configurations. Ic, Id and Ie are push-pull variants, Ie being simply a class B stage with higher V_{bias}.

When it sinks output current, there is an inevitable voltage drop across the emitter resistance, limiting the negative output capability, and resulting in an efficiency of 12.5% (erroneously quoted in at least one textbook as 25%, apparently on the grounds that power not dissipated in silicon doesn't count) This would be of purely theoretical interest—and not much of that—except that a single ended design has recently appeared. This reportedly produces a 10 W output for a dissipation of 120 W, with output swing predictably curtailed in one direction.[3]

A better method—constant current class A—is shown in Figure 29.1(b). The current sunk by the lower constant current source is no longer related to the voltage across it, and so the output voltage can approach the negative rail with a practicable quiescent current. (Hereafter shortened to 'I_q') Maximum efficiency is doubled to 25% at maximum output; for an example with 20 W output (and a big fan) see Ref.

4. Some versions (Krell) make the current source value switchable, controlling it with a kind of noise gate.

Push-pull operation once more doubles full-power efficiency, producing a more practical 50%; most commercial class A amplifiers have been of this type. Both output halves now swing from zero to twice the I_q, and least voltage corresponds with maximum current, reducing dissipation. There is also the intriguing prospect of cancelling the even-order harmonics generated by the output devices.

There are several ways to induce push-pull action. Figures 29.1(c), (d) show the lower constant current source replaced by a voltage controlled current source. This can be driven directly by the amplifier forward path, as in Figure 29.1(c), [5] or by a current control negative feedback loop, as at Figure 29.1(d).[6] The first of these methods has the drawback that the stage generates gain, phase splitter

Tr_1 doubling as the VAS; hence there is no circuit node that can be treated as the input to a unity gain output stage, making the circuit hard to analyse, as VAS distortion cannot be separated from output stage non-linearity. There is also no guarantee that upper and lower output devices will be driven appropriately for class A if the effective quiescent varies by more than 10% over the cycle [5]

The second push-pull method in 29.1(d) is more dependable, and I can vouch that it works well. The disadvantage with the simple form shown is that a regulated supply is required to prevent rail ripple from disrupting the current loop control. Designs of this type have a limited current control range. In Figure 29.1(d), Tr_3 cannot be turned on further once the upper device is fully off—so the voltage controlled current source will not be able to respond to an unforeseen increase in the output loading. If this happens there is no way of resorting to class AB to keep the show going and the amplifier will show some form of asymmetrical hard clipping.

The best push-pull stage seems to be that in Figure 29.1(e), which probably looks rather familiar. Like all the conventional class B stages examined in Chapter 19, this one will operate effectively in push-pull class A if the bias voltage is sufficiently increased; the increase over class B is typically 700 mV, dependant on the value of the emitter resistors. For an example of high biased class B see Ref. 7. This topology has the great advantage that, when confronted with an unexpectedly low load impedance, it will operate in class AB. The distortion performance will be inferior not only to class A but also to optimally biased class B, once above the AB transition level, but can still be made very low by proper design.

Although the push-pull concept has a maximum efficiency of 50%, this is only true at maximum sine-wave output. Due to the high peak/average ratio of music, the true average efficiency probably does not exceed 10%, even at maximum volume before obvious clipping.

Other possibilities are signal controlled variation of the class A amplifier rail voltages, either by a separate class B amplifier, or a modulated switch mode supply. Both approaches are capable of high power output, but involve extensive extra circuitry, and present sent some daunting design problems.

A class B amplifier has a limited voltage output capability, but can be flexible about load impedances, as more current will be simply turned on when required. However, class A has also a current limitation, after which it enters class AB, and so loses

its *raison d'etre*. The choice of quiescent value has a major effect on thermal design and parts cost so a clear idea of load impedance is important. The calculations to determine the required I_q are straightforward, though lengthy if supply ripple, $V_{ce(sat)}$, and R_e losses, etc., are all considered, so I just give the results here. An unregulated supply with 10,000 μF reservoirs is assumed.

A 20 W/8 Ω amplifier will require rails of approx ±24 V and a quiescent of 1.15A. If this is extended to give roughly the same voltage swing into 4 Ω, then the output power becomes 37 W, and to deliver this in class A the quiescent must increase to 2.16A, almost doubling dissipation. If however full voltage swing into 6 Ω will do, (which it will for many reputable speakers) then the quiescent only needs to increase to 1.5A; from here on I assume a quiescent of 1.6A to give a margin of safety.

The class A output stage

I consider here only the high biased class B topology, because it is probably the most popular approach, effectively solving the problems presented by the others. Figure 29.2 shows a Spice simulation of the collector currents in the output devices versus output voltage for the emitter follower configuration, and also the sum of these currents. This sum of device currents is, in principle, constant, but need not be so for low THD. The output signal is the difference of device currents and is not inherently related to the sum. However, a large deviation from this constant sum condition means inefficiency, as the stage is conducting more quiescent than it needs to for some part of the cycle. The constancy of this sum is important because it shows that the voltage measured across R_{e1} and R_{e2} together is also effectively constant so long as the amplifier stays in class A. This in turn means that I_q can be simply set with a constant voltage bias generator, in very much the same way as class B.

Figures 29.3, 29.4, 29.5 show Spice gain plots for open loop output stages, with 8 Ω loading and 1.6 A quiescent; the circuitry is exactly as for class B in Part 4. The upper traces show class A gain, and the lower traces gain under optimal class B bias for comparison. Figure 29.3 shows an emitter follower output, Figure 29.4(a) simple quasi complementary stage, and Figure 29.5(a) CFP output.

We would expect class A stages to be more linear than B, and they are Harmonic and THD figures for the three configurations, at 20 V peak, are shown in

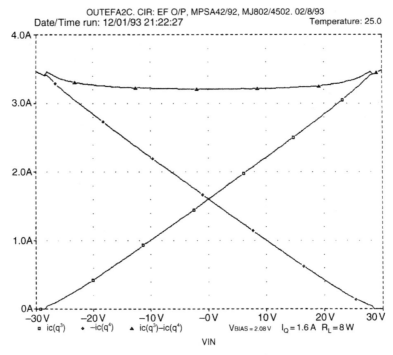

Figure 29.2 How output device current varies in push-pull class A. The sum of the currents is near-constant, simplifying biasing.

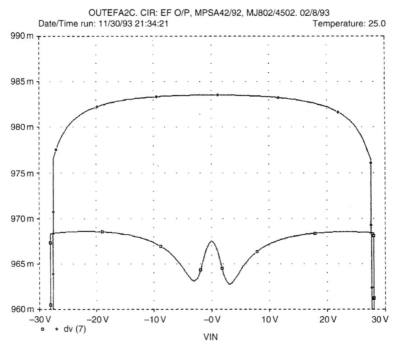

Figure 29.3 Gain linearity of the class A emitter-follower output stage. Load is 8 Ω, and quiescent current (Iq) is 1.6A. Upper trace class A, lower trace optimal class B.

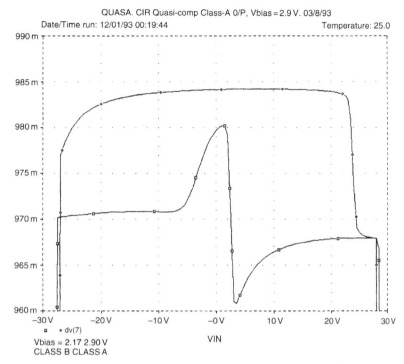

Figure 29.4 Gain linearity of the class A quasi-complementary output stage. Conditions as in Figure 29.3. Upper trace class A, lower class B.

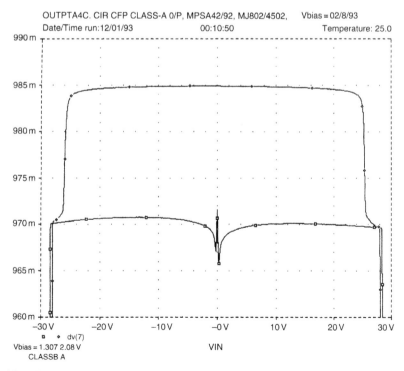

Figure 29.5 Gain linearity of the class A CFP output stage. Upper trace class A, lower trace class B.

Table 29.1

Harmonic	Emitter Follower (%)	Quasi-Comp (%)	CFP Output (%)
Second	0.00012	0.0118	0.00095
Third	0.0095	0.0064	0.0025
Fourth	0.00006	0.0011	0.00012
Fifth	0.00080	0.00058	0.00029
THD	0.0095	0.0135	0.0027

THD is calculated from the first nine harmonics, though levels above the fifth are very small

Table 29.1. There is absolutely no gain wobble around OV, and push-pull class A genuinely does cancel even order distortion. Class B only does this in the crossover region, in a partial and unsatisfactory way.

It is immediately clear that the emitter follower has more gain variation, and therefore worse linearity, than the CFP, while the quasi complementary circuit shows an interesting mix of the two. The more curved side of the quasi gain plot is on the negative side, where the CFP half of the quasi circuit is passing most of the current. However we know by comparing Figure 29.3 and Figure 29.5 that the CFP is the more linear structure. Therefore it appears that the shape of the gain curve is determined by the output half that is turning off, presumably because this shows the biggest g_m changes. The CFP structure maintains g_m better as current decreases, and so gives a flatter gain curve with less rounding of the extremes.

The gain behaviour of these stages is reflected in their harmonic generation; Table 29.1 reveals that the two symmetrical topologies give mostly odd order harmonics as expected. The asymmetry of the quasi comp version causes a large increase in even order harmonics, and this is reflected in the higher THD figure. Nonetheless the THD figures are still two to three times lower than for their class B equivalents.

If this factor of improvement seems a poor return for the extra dissipation of class A, this is not so. The crucial point about the distortion from a class A output stage is not just that is low, but that it is low order, and so benefits much more from a typical NFB factor that falls with frequency than does high order crossover distortion.

The choice of class A output topology is now simple. For best performance, use the CFP. Apart from greater basic linearity, the effects of output device temperature on I_q are servoed out by local feedback, as in class B. For utmost economy, use the quasi complementary with two NPN devices: these need only a low $V_{ce(max)}$ for a typical

class A amp, so here is an opportunity to recoup some of the money spent on heatsinking.

The rules are different from class B; the simple quasi configuration will give first class results with moderate NFB, and adding a Baxandall diode to simulate a complementary emitter follower stage makes little difference to linearity.[7]

It is sometimes assumed that the different mode of operation of class A makes it inherently short circuit proof. This may be true with some configurations, but the high biased type shown here will continue delivering current until it bursts. Overload protection is no less necessary.

Quiescent control systems

Unlike class B, precise control of quiescent current is not required to optimise distortion. For good linearity there just has to be enough of it. However, I_q must be under some control to prevent thermal runaway, particularly if the emitter follower output is used, and an ill conceived controller can ruin the THD. There is also the point that a precisely held standing current is considered the mark of a well bred class A amplifier; a quiescent that lurches around like a drunken sailor does not inspire confidence.

Thermal feedback from the output stage to a standard V_{be} multiplier bias generator will work,[8] and should be sufficient to prevent run-away. However, unlike class B, class A gives the opportunity of tightly controlling I_q by negative feedback. This is profoundly ironic because now that we can precisely control I_q, it is no longer critical. Nevertheless it seems churlish to ignore the opportunity.

There are two basic methods of feedback current control. In the first, the current in one output device is monitored, either by measuring the voltage across *one* emitter resistor, (R_s in Figure 29.6(a)), or by a collector sensing resistor. The second method monitors the sum of the device currents, which as described above, is constant in class A.

The first method as implemented in Figure 29.6(a) [7] compares the V_{be} of Tr_4 with the voltage across R_s, with filtering by R_F, C_F. If quiescent is excessive, then Tr_4 conducts more, turning on Tr_5 and reducing the bias voltage between points A and B.

In Figure 29.6(b), which uses the voltage controlled current source approach, the voltage across collector sensing resistor R_s is compared with V_{ref} by Tr_4, the value of V_{ref} being chosen to allow for Tr_4 V_{be}.[9] Filtering is once more by R_F, C_F.

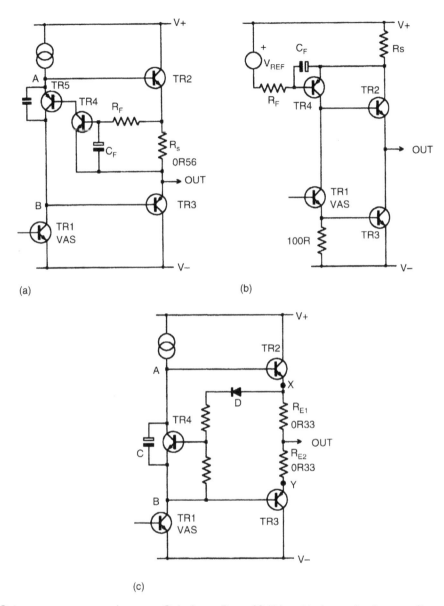

(a)

(b)

(c)

Figure 29.6 Quiescent current-control systems. Only that at Figure 29.6(c) avoids the need to low pass filter the control signal; C simply provides feed forward to speed up signal transfer to Tr2.

For either Figure 29.6(a) or 29.6(b), the current being monitored contains large amounts of signal, and must be low pass filtered before being used for control purposes. This is awkward as it adds one more time constant to worry about if the amplifier is driven into asymmetrical clipping, and implies the desirability of large electrolytic capacitors to minimise the a.c. voltage drop across the sense resistors. In the case of collector sensing there are unavoidable losses in the extra sense resistor. It is my experience that imperfect filtering can produce a serious rise in distortion.

The better way is to monitor current in *both* emitter resistors. As explained above, the voltage across both is very nearly constant, and in practice filtering is unnecessary. An example of this approach is shown in Figure 29.6(c), based on a concept originated by

Nelson Pass.[10] Here Tr_4 compares its own V_{be} with the voltage between X and B; excessive quiescent turns on Tr_4 and reduces the bias directly. Diode D is not essential to the concept, but usefully increases the current feedback loop gain; omitting it more than doubles I_q variation with Tr_7 temperature in the Pass circuit.

The trouble with this method is that Tr_3V_{be} directly affects the bias setting, but is outside the current control loop. A multiple of V_{be} is established between X and B, when what we really want to control is the voltage between X and Y. The temperature variations of Tr_4 and Tr_3V_{be} partly cancel, but only partly. This method is best used with a CFP or quasi output so that the difference between Y and B depends only on the driver temperature, which can be kept low. The 'reference' is $Tr_4 \, V_{be}$, which is itself temperature dependent. Even if it is kept away from the hot bits it will react to ambient temperature changes, and this explains the poor performance of the Pass method for global temperature changes (Table 29.2).

To solve this problem, I would like to introduce the novel control method in Figure 29.7. We need to compare the floating voltage between X and Y with a fixed reference, which sounds like a requirement for two differential amplifiers. This can be reduced to one by sitting the reference

Table 29.2 Iq change per °C change in temperature

	Changing Tr₇ temperature only	Changing Global temperature (%)
Quasi + V_{be} mult	+0.112%	−0.43
Pass: as Flg. 6c	+0.0257	−14.1
Pass: no dlode D	+0.0675	−10.7
New system:	+0.006%	−0.038

(assuming 0.22 Ω emitter resistors and 1.6A I_q.)

V_{ref} on point Y. This is a very low impedance point and can easily swallow a reference current of 1 mA or so. A simple differential pair $T_{r15,16}$ then compares the reference voltage with that at point X: excess quiescent turns on Tr_{16}, causing Tr_{13} to conduct more and reducing the bias voltage.

The circuitry looks enigmatic because of the high impedance of Tr_{13} collector would seem to prevent signal from reaching the upper half of the output stage; this is in essence true, but the vital point is that Tr_{13} is part of a NFB loop that establishes a voltage at A that will keep the bias voltage between A and B constant. This comes to the same thing as maintaining a constant V_{bias} across Tr_{13}. As might

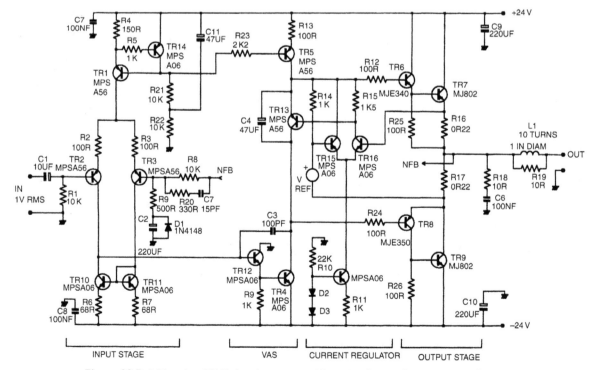

Figure 29.7 A Blameless 20 W class A power amplifier, using the novel current-control system.

be imagined, this loop does not shine at transferring signals quickly, and this duty is done by feedforward capacitor C_4.

Without it, the loop (rather surprisingly) works correctly, but HF oscillation at some part of the cycle is almost certain. With C_4 in place the current loop does not need to move quickly, since it is not required to transfer signal but rather to maintain a DC level.

The experimental study of I_q stability is not easy because of the inaccessibility of junction temperatures. Professional Spice implementations like *PSpice* allow both the global circuit temperature and the temperature of individual devices to be manipulated; this is another aspect where simulators shine. The exact relationships of component temperatures in an amplifier is hard to predict: I show here just the results of changing the global temperature of all devices, and changing the junction temp of Tr_7 alone (Figure 29.7) with different current controllers. Tr_7 will be one of the hottest transistors and unlike Tr_9 it is not in a local NFB loop, which would greatly reduce its thermal effects.

A new class A design

The full circuit diagram shows a 'blameless' 20 W/8 Ω class A power amplifier. This is as close as possible in operating parameters to the previous class B design to aid comparison. In particular the NFB factor remains 30 dB at 20 kHz. The front end is as for the class B version, which should not be surprising as it does exactly same job, input Distortion 1 being unaffected by output topology.

As before the input pair uses a high tail current, so that $R_{2,3}$ can be introduced to linearise the transfer characteristic and set the transconductance. Distortion 2 (VAS) is dealt with as before, the beta enhancer Tr_{12} increasing the local feedback through C_{dom}. There is no need to worry about Distortion 4 (non-linear loading by output stage) as the input impedance of a class A output, while not constant, does not have the sharp variations shown by class B.

The circuit uses a standard quasi output. This may be replaced by a CFP stage without problems. In both cases the distortion is extremely low but, gratifyingly, the CFP proves even better than the quasi, confirming the simulation results for output stages in isolation.

The operation of the current regulator $Tr_{13,15,16}$ has already-been described. Using a band gap reference, it holds a 1.6 A I_q to with in ±2 mA from a second or two after switch on. Looking at Table 29.2, there seems no doubt that the new system is effective.

As before an unregulated power supply with 10,000 µF reservoirs was used, and despite the higher prevailing ripple, no PSRR difficulties were encountered once the usual decoupling precautions were taken.

Performance

The closed loop distortion performance (with conventional compensation) is shown in Figure 29.8 for the quasi comp output stage, and in Figure 29.9 for a CFP

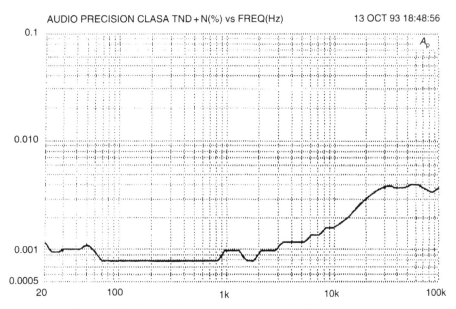

Figure 29.8 Class A distortion performance with CFP output stage.

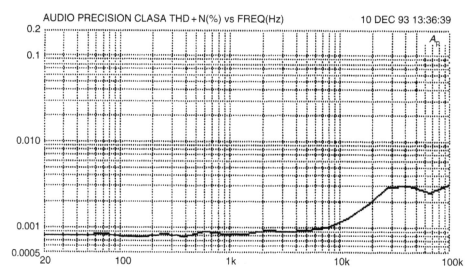

Figure 29.9 Class A amplifier THD performance with quasi-comp output stage. The steps in the LF portion of the trace are measurement artifacts.

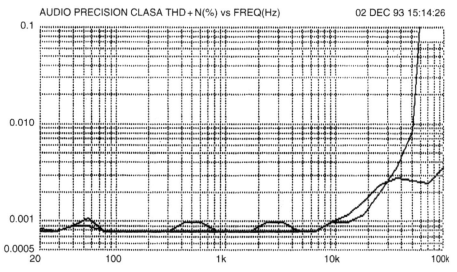

Figure 29.10 Distortion performance for CFP output stage with 2-pole compensation. The THD drops to 0.0012% at 20 kHz, but the extra VAS loading has compromised the positive-going slew capability, The 2-pole trace is shown moving off the graph at 50 kHz.

output version. The THD residual is pure noise for almost all of the audio spectrum, and only above 10 kHz do small amounts of third harmonic appear. The suspected source is the input pair, but this so far remains unconfirmed.

The distortion generated by the class B and A design examples is summarised in Table 29.3, which shows a pleasing reduction as various measures are taken to deal with it. As a final tweak, two pole compensation was applied to the most linear (CFP) of the class A versions, reducing distortion to 0.0012% at 20 kHz, at some cost in slew rate (Figure 29.10). While this may not be the fabled straight wire with gain, it must be a near relation . . .

Table 29.3

	I kHz	I0 kHz	20 kHz	Power
class B EF	<.0006%	.0060%	.012%	50W
class B CFP	<.0006%	.0022%	.0040%	50 W
class B EF 2-pole	<.0006%	.0015%	.0026%	50 W
class A quasl	<.0006%	.0017%	.0030%	50 W
class A CFP	<.0006%	.0010%	.0018%	20 W
class A CFP 2-pole	<.0006%	.0010%	.0012%	20 W

(All for 8 Ω loads and 80 kHz bandwidth. Single pole compensation unless otherwise stated.)

And finally

The techniques in this series have a relevance beyond power amplifiers. Applications obviously include discrete op-amp based preamplifiers[11] and extend to any amplifier offering static or dynamic precision. My philosophy is that all distortion is bad, and high order distortion is worse . . . $n^2/4$ worse, according to many authorities.[12] Digital audio routinely delivers the signal with less than 0.002% THD, and I can earnestly vouch for the fact that analogue console designers work hard to keep the distortion in long and complex signal paths down to similar levels. I think it an insult to allow the very last piece of electronics in the chain to make nonsense of these efforts.

I do not believe that an amplifier yielding 0.001% THD is going to sound much better than another generating 0.002%. However, if there is ever a doubt as to what level of distortion is perceptible, then using the techniques I have presented, it should be possible to reduce the THD below the level at which there can be any rational argument.

I am painfully aware of the school of thought that regards low distortion as inherently immoral, but this is to confuse electronics with religion. The implication is that very low THD can only be obtained by huge global NFB factors which in turn require heavy dominant pole compensation that severely degrades slew rate. The obvious flaw in this argument is that, once the compensation is applied, the amplifier no longer has a large global NFB factor: Its distortion performance presumably reverts to mediocrity, further burdened with a slew rate of 4 V per fortnight.

To me low distortion has its own aesthetic appeal. All of the linearity enhancing strategies examined in this series are of minimal incremental cost to existing designs with the possible exception of using class A. There seems to be no reason to not use them.

References

1. Moore, B.J. *An Introduction To The Psychology of Hearing,* Academic Press, 1982, pp 48–50.

2. Tanaka, S. 'A new biasing circuit for class B operation', *JAES* January/February 1981, p 27.

3. Fuller, S. Private communication.

4. Nelson Pass 'Build a class A amplifier', *Audio,* February 1977, p 28 (constant current).

5. Linsley Hood, J. 'Simple class A Amplifier', *Wireless World,* April 1969, p 148.

6. Self, D. 'High-performance preamplifier', *Wireless World,* February 1979, p 41.

7. Nelson-Jones, L. 'Ultra-low distortion class A amplifier', *Wireless World,* March 1970, p 98.

8. Giffard, T. 'Class A power amplifier', *Elektor,* November 1991, p 37.

9. Linsley-Hood, J. 'High-quality headphone amp', *HiFi News & RR,* January 1979, p 81.

10. Nelson Pass. 'The pass/A40 power amplifier', *The Audio Amateur,* #4, 1978, p 4 (Push-pull).

11. Self, D. 'Advanced Preamplifier Design', *Wireless World,* November 1976, p 41.

12. Moir, J. 'Just detectable distortion levels', *Wireless World,* February 1981, p 34.

Chapter 30

Power amplifier input currents and their troubles

(Electronics World, May 2003)

This article grew out of some work that I was doing for a well-known amplifier company. I had produced a nice low-distortion design, which was to a great extent a straightforward application of the Blameless amplifier design methodology described in the Distortion In Power Amplifier series. However, a late change to the specification of the product—a thing not wholly unknown in the world of audio engineering—meant that a resistive network had to be added immediately before the power amplifier stage. The effective source resistance of this network was, if memory serves, 2k2, and when you have perused the following article, you will understand that the bad effects on both hum and distortion performance were most unwelcome.

Adding a 5532 buffer stage between resistive network and amplifier would have been a quick fix, but running another opamp from the very limited amount of ±15 V power available was going to be awkward, and PCB area in the right place was also a very scarce resource. The 5532 is a low-noise opamp, but it is not as quiet as the pair of discrete transistors in the power amplifier input section, and the overall noise performance would definitely have suffered.

It was therefore time to look a little more closely at the exact mechanisms by which the source resistance was causing trouble, in the hope that more elegant ways of retrieving the original performance could be found. They were, and this chapter tells the story of how those mechanisms were uncovered, and rendered less troublesome. The hum issue was completely eliminated by cascoding the input pair tail source to prevent Early effect, but the extra distortion generated was a tougher nut to crack, and was only be somewhat reduced by using high-beta input devices.

Very little work has been done on this topic since this article was published, because the driving need is not there. In most cases, there is no difficulty in arranging that an opamp output drives the amplifier input directly, and this seems preferable to adding complications to the amplifier input stage that may have bad effects on other performance parameters. The only time this is difficult is when for reasons of quality or parsimony, the number of opamps in the signal is being strictly minimised. The most promising way to reduce input currents would seem to be to convert the single transistors in the input stage into CFP pairs. This does not, as far as I know at present, compromise any amplifier parameters; if it does, I suspect it would be in the area of noise. The CFP structure in fact improves the linearity of the input stage transconductance [1]. It is, however a relatively untried approach so far as I am concerned, and so some careful testing would be required before I felt confident enough to put it into production.

There are many reasons for making a circuit as simple as possible, and one of the less obvious ones is that the more complex it is, the greater the likelihood that it might have hidden in its intricacies some misbehaviour that is only triggered by a certain set of specific circumstances, and which might be overlooked in testing.

Reference

1. Self, D., *Audio Power Amplifier Design*, 6th edition, pp. 135–137. Focal Press 2013.

POWER AMPLIFIER INPUT CURRENTS AND THEIR TROUBLES

May 2003

When power amplifiers are measured, the input is normally driven from a low impedance signal generator. Some testgear, such as the much-loved Audio Precision System-1, has selectable output impedance options of 50, 150, and 600 Ω. The lowest value available is almost invariably used because (1) it minimises the Johnson noise from the source resistance; (2) it minimises level changes due to loading by the amplifier input impedance.

This is all very sensible, and exactly the way I do it myself—99% of the time. There is however two subtle effects that can be missed if the amplifier is always tested this way. These are: distortion caused by the non-linear input currents drawn by the typical power amplifier, and hum caused by ripple modulation of the same input currents.

Note that this is not the same effect as the excess distortion produced by FET-input opamps when driven from significant source impedances; this is due to their non-linear input capacitances to the IC substrate, and has no equivalent in power amplifiers made of discrete transistors.

Figure 30.1 shows both the effects. The amplifier under test was a conventional Blameless design with an EF output stage comprising a single pair of sustained-beta bipolar power transistors; see Figure 30.2 for the basic circuit. Output power was 50 W into 8 Ω. The bottom trace is the distortion + noise with the usual source impedance of 50 Ω, and the top one shows how much worse the THD is with a source impedance of 3.9 K. Intermediate traces are for 2.2 K and 1.1 K sources. The THD residual shows both second harmonic distortion and 100 Hz ripple components, the latter dominating at low frequencies, while at higher ones the reverse is true. The presence of ripple is signalled by the dip in the top trace at 100 Hz, where distortion products and ripple have partially cancelled. The amount of degradation is proportional to the source impedance.

This is not a problem in most cases, where the preamplifier is driven by an active preamplifier, or by a buffer internal to the power amplifier. Competent preamplifiers have a low output impedance, often around 50–100 Ω, to minimise high-frequency losses in cable capacitance. (I have just been hearing of a system with 10m of cable between preamp and power amp.)

However, there are two scenarios where the input source resistance is higher than this. If a so-called 'passive preamp' is used then the output impedance is both higher and volume-setting dependent. A 10 K volume potentiometer has a maximum output impedance of one-quarter the track resistance, i.e. 2.5 K, at its midpoint setting. It is also possible for significant source resistance to exist inside the power amplifier for example, there might be an balanced input amplifier, which

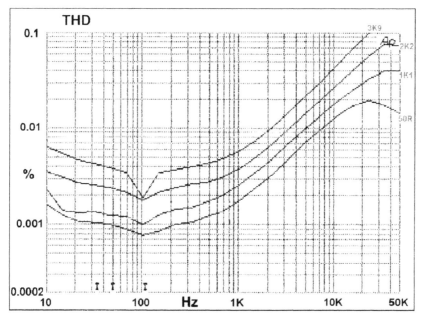

Figure 30.1 Second-harmonic distortion and 100 Hz ripple get worse as the source impedance rises from 50 Ω to 3.9 K. 50 W into 8 Ω.

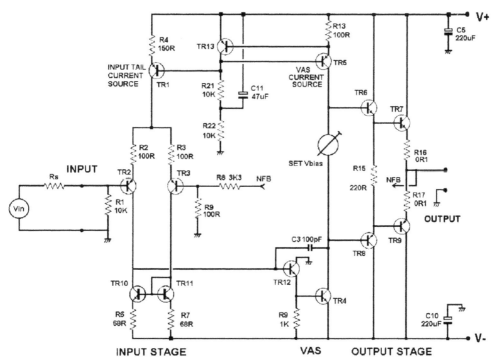

Figure 30.2 Simplified circuit of a typical Blameless power amplifier, with negative-feedback control of VAS current source TR5 by TR13. The bias voltage generated is also used by the input tail source TR1.

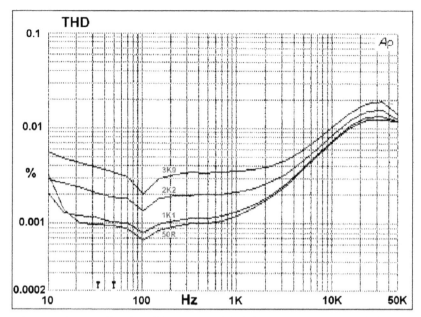

Figure 30.3 There is less introduction of ripple and distortion with high-beta input transistors and the same set of source resistances as Figure 30.1.

while it has a very low output impedance itself, may have a resistive gain control network between it and the power amp.

So—we have a problem, or rather two of them. It seems very likely that the input transistor base currents are to blame for both, hence an obvious option is to minimise these currents by using transistors with the highest available beta in the input pair. In this amplifier the input pair were originally ZTX753, with a beta range of 70–200. Replacing these with BC556B input devices (beta range 180–460) gives Figure 30.3 which shows a useful improvement in THD above 1 kHz; distortion at 10 kHz drops from 0.04% to 0.01%. Our theory that the base currents are to blame is clearly correct. The bottom trace is the reference 50 Ω source plot with the original ZTX753s, and this demonstrates that the problem has been reduced but certainly not eliminated.

The amplifier here is very linear with a low source impedance, and it might well be questioned as to why the input currents drawn are distorted if the output is beautifully distortion-free. The reason is of course that global negative feedback constrains the output to be linear because this is where the NFB is taken from

but the internal signals of the amplifier are whatever is required to keep the output linear. The VAS is known to be non-linear, so if the output is sinusoidal the collector currents of the input pair clearly are not. Even if they were, the beta of the input transistors is not constant so the base currents drawn by them would still be non-linear.

It is also possible to get a reduction in hum and distortion by reducing the input pair tail current, but this very important parameter also affects input stage linearity and the slew-rate of the whole amplifier. Figure 30.4 shows the result. The problem is reduced—though far from eliminated—but the high-frequency THD has actually got worse because of poorer linearity in the input stage. This is not a promising route to follow.

Both ripple and THD effects consequent on the base currents drawn could be eliminated by using FETs instead of bipolars in the input stage. The drawbacks are:

1. Poor Vgs matching, which means that a d.c. servo becomes essential to control the amplifier output d.c. offset. Dual FETs do exist but they are discouragingly expensive.

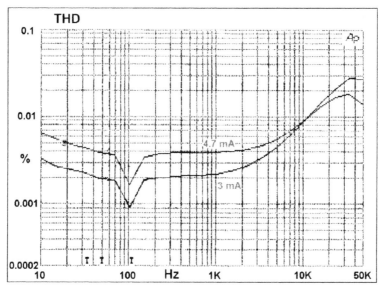

Figure 30.4 Reducing the tail current improves things at low frequencies but increases HF distortion above 10 kHz. The notches at 100 Hz indicate that the ripple content is still substantial.

2. Low transconductance, which means the stage cannot be linearised by local feedback as the raw gain is just not available.
3. Although there is no d.c. gate current, there might well be problems with non-linear input capacitance, as there are with FET-input op-amps.

Once again, not a promising route.

The distortion problem looks rather intractable; one possible total cure is to put a unity-gain buffer between input and amplifier. The snag (for those seeking the highest possible performance) is that any opamp will compromise the noise and distortion of a Blameless amplifier. It is quite correct to argue that this doesn't matter, as any preamp hooked up to the power amp will have opamps in it anyway, but the preamp is a different box, a different project, and possibly has a different designer, so philosophically this does not appeal to everyone. If a balanced input is required then an opamp stage is mandatory. (unless you prefer transformers, which of course have their own problems.)

The best choice for the opamp is either the commonplace but extremely capable 5532 (which is pretty much distortion-free, but not alas noise-free, though it is very quiet) or the rather expensive but very quiet AD797.

The ripple problem, however, has a more elegant solution. If there is ripple in the input base current, then clearly there is some ripple in the tail current. This is

not normally detectable because the balanced nature of the input stage cancels it out. A significant input source impedance upsets this balance, and the ripple appears.

The tail is fed from constant-current source TR1, and this is clearly not a mathematically perfect circuit element. Investigation showed that the cause of the tail-current ripple contamination is Early effect in this transistor, which is effectively fed with a constant bias voltage A tapped off from the VAS negative-feedback current source. (Early effect is the modulation of transistor collector current caused by changing the Vce; as a relatively minor aspect of bipolar transistor behaviour it is modelled by SPICE simulators in a rather simplistic way.) Note that this kind of negative-feedback current-source could control the tail current instead of the VAS current, which might well reduce the ripple problem, but is arranged this way as it gives better positive slewing. Another option is two separate negative-feedback current-sources.

The root cause of our hum problem is therefore the modulation of the Vce of TR1 by ripple on the positive rail, and this variation is easily eliminated by cascoding, as shown in Figure 30.5. This forces TR1 emitter and collector to move up and down together, preventing Vce variations. It completely eradicates the ripple components, but leaves the inputcurrent distortion unaltered, giving the results in Figure 30.6, where the upper trace is degraded only by the extra distortion introduced by a

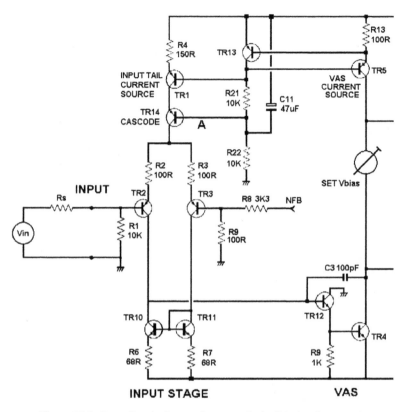

Figure 30.5 Cascoding the input tail; one method of biasing the cascode.

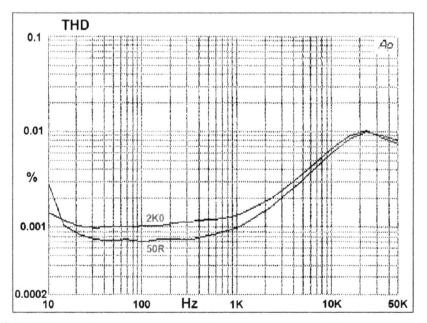

Figure 30.6 Cascoding the input tail removes the ripple problem, but not the extra distortion.

2 K source impedance; the 100 Hz cancellation notch has also disappeared. The reference 50 Ω source plot is below it.

The voltage at A that determines the Vce of TR1 is not critical. It must be sufficiently below the positive supply rail for TR1 to have enough Vce to conduct properly, and it must be sufficiently above ground to give the input pair enough common-mode range. I usually split the biasing chain R21, R22 in half, as shown, so C11 can be used to filter out rail noise and ripple, and biasing the cascode transistor from the mid-point works very well.

It may have occurred to the reader that simply balancing the impedances seen by the two inputs will cancel out the unwanted noise and distortion. This is not very practical as with discrete transistors there is no guarantee that the two input devices will have the same beta.

(I know there are such things as dual bipolars, but once more the cost is depressing.) This also implies that the feedback network will have to have its impedance raised to equal that at the input, which would give unnecessarily high levels of Johnson noise.

Conclusions

If the system design requires an opamp at the input, then both hum and distortion problems are removed with no further effort. If not, perhaps because the amplifier must be as quiet as possible, then cascoding the input pair tail cures the ripple problem but not the distortion. Using high-beta input transistors reduces both problems but does not eliminate them.

Chapter 31

Diagnosing distortions

(Electronics World, January 1998)

Intended title: *THD Testing & Amplifier Distortion Residuals*

This article focused on the visual appearance of the distortion residuals produced by THD analysis. The distortion products may be low-order, in which case, if they appear alone they are simply second or third harmonic, looking basically sinusoidal. If both second and third appear together, which is very common, it is hard to visually disentangle the two because of unknown phase shifts between them; spectrum analysis is useful here. Higher-order harmonics are signalled by angular features or glitches in the THD residual.

Crossover distortion, the greatest disturbance to the amplifier designer's peace of mind, is instantly recognizable not only by its shape but by its timing, coinciding as it does with the zero-crossings of the output waveform. The other distortions are somewhat harder to distinguish, as most have their origin in contamination of the signal with half-wave rectified sinewaves, and so their THD residuals look similar. Fortunately, it usually takes only one or two simple experiments to determine the root cause of the trouble. On several occasions I have watched people trying to impose their will on a recalcitrant design without looking at the distortion residual, an approach which leaves me shaking my head in bafflement. To me the THD analyser is as much a part of the amplifier designer's armoury as the doctor's stethoscope, the pathologist's microscope or even the geologist's hammer. Sometimes, even in electronics, a big hammer is what you need.

The obscure reference to the HP546100B digital oscilloscope is simply explained. It was the prize I won in the Electronics World/Hewlett-Packard Writers competition in June 1994. I still use it all the time, because it has a really lovely user interface with real physical knobbage. There is a downside in that graphics can only be exported via the Centronics connector.

I should have recorded the type of output stage used in the tests that generated the residual images. As might be deduced from the narrow crossover spikes in Figure 31.4, it was a CFP output stage rather than an EF type. Only normal Miller dominant-pole compensation was used.

The list of distortion mechanisms given in the article is now recognised to be incomplete. In the original amplifier articles in 1993, I listed an Unmagnificent Seven mechanisms, and by the time this article was written in 1998, I had added to them Distortion 8, caused by electrolytic capacitors and most likely to occur in the feedback network. When the fourth edition of *Audio Power Amplifier Design* came out in 2006, there were still only 8 distortion mechanisms given; by 2009 when the fifth edition appeared, there were 11 distortions on the list. The sixth and current edition (2013) stuck with 11 distortions, but I would now add Distortion 12 to the list. The later members are listed below. I have added the procedures for confirming that you really do suffer from the Distortion in question, and not something else; I should have emphasised that more in the original article.

Distortion 9. Magnetic distortion occurs when a loudspeaker-level signal goes through a piece of ferrous metal, such as a bolt or the iron frame of a relay [1]. It occurs to me that many TO-3 installations must have had the collector current passing through ferrous mounting bolts, and though the distance is short, and the current (presumably) split through both bolts, I suspect that measurable distortion may have occurred in some cases. This is now pretty much an historical issue, as TO-3 packages are obsolete, but it would be an interesting project to see if the effect was detectable with modern testgear.

Confirmation: Measure THD at both the output connector and at the actual output of the amplifier, i.e. the negative feedback node. If you get extra third harmonic in the first case, there is probably a ferrous bolt or relay frame in the path. You have Distortion 9.

The offending part can usually be detected with a strong permanent magnet.

Distortion 10. Input current distortion. The base of a power amplifier input transistor draws a non-linear current from the source impedance, even if the amplifier output is free of distortion, and if this impedance is significant (say more than 1 kΩ) then distortion occurs because of the voltage drop across that impedance [2]. Many amplifier designs have so-called TID filters, which are just an RC lowpass at the input. The R part will cause distortion and can cause hum.

Confirmation: There has to be significant impedance in series with the input in some form for this to happen. Find it and short it out. If the THD drops, you have Distortion 10.

Distortion 11. Premature overload protection distortion [3]. Commonly, SOAR protection circuitry begins to operate and degrade the amplifier linearity long before visually obvious limiting of the output amplitude occurs. One cure for this is latching snap-action protection circuitry that is either hard off or hard on, but this means hitting a reset switch for minor overloads that in non-latching circuitry would probably pass unnoticed. There should be no change in distortion at the lowest load impedance contemplated, with the protection circuitry either disabled or enabled. This can only be achieved by building adequate safety margins into the output stage. Simple SOAR protection circuitry is sensitive to internal temperature changes in the equipment, the protection threshold falling as temperature rises, and this must be allowed for. I once worked with someone who failed to check this, and 500 amplifiers had to be taken out of their packing, disassembled, and modified to prevent premature protection. Not sure what he's doing nowadays.

Confirmation: Usually the protection transistors have series diodes in their collectors to make sure they do not turn on when reverse-biased; disconnecting these disables the protection. With a double-sided PTH PCB (which most of them are these days) then it should be possible to lift one leg of the diode from the top side without having to disassemble anything. If the diode leg has been clinched over underneath the PCB during component auto-insertion, you will have to cut the diode leg as near the pad as possible, so it can be re-soldered easily. Remember to disable both positive and negative halves of the protection system, remember to be careful with an amplifier that has no short-circuit protection, and remember to re-enable it afterwards.

Distortion 12. Thermal distortion in the upper feedback resistor. This sounds highly specific, and it is, because this is the only place in which thermal distortion has definitely been found so far. This is the latest addition to the list, because its incidence has only been detected by the use of new and superior testgear, and high-power amplifiers having the right circuit conditions. There is more on this new development below.

Confirmation: At this point only one component is believed to show thermal distortion, the upper resistor in the negative feedback loop. If you replace it with a series-parallel combination of four of the same resistor type, which has the same total resistance, and only the LF distortion drops, you have thermal distortion. If distortion drops at all frequencies, voltage-coefficient distortion may be the problem (see below).

The procedures for confirming the presence of Distortions 1 to 8 are mostly less straightforward. Here are the simple ones:

Distortion 3a. Crossover distortion in output stage.

Confirmation: Very obvious changes in the distortion residual when quiescent conditions are altered using the bias adjustment. In a Class-B amplifier you don't really need to confirm it because you know it will be there to some extent.

Distortion 6. Induction from Class-B supply rails.

Confirmation: If possible move the power-supply cabling and see if the sawtooth on the residual varies. If the supply path is fixed as PCB tracks, this can't be done. One expedient is to cut a supply track, and bridge the gap with a piece of wire that can be moved about near the amplifier circuitry. If cancellation of the sawtooth is possible, you have identified Distortion 6. Good luck with sorting out the PCB.

Distortion 8. Non-linearity of feedback network capacitor (at LF only).

Confirmation: Put another capacitor of the same value across the feedback capacitor; if the LF distortion drops significantly, you have Distortion 8. Increase the capacitor size in the design.

For the remaining distortions, confirmation can only be obtained by effectively altering the amplifier design.

At the end of the list in the article, I put 'Non-existent or negligible distortions' in which I specifically mentioned common mode distortion in the input stage, and thermal distortion just about anywhere. This needs a bit of amendment now.

My casual dismissal of common mode distortion in the input stage was based on the fact that I was sure I had never in my amplifier investigations seen any distortion that looked like it. In this I was quite right. Later I investigated the common-mode issue properly, and demonstrated that it really was of unmeasurable and therefore negligible magnitude in a BJT input stage unless the amplifier was deliberately run at a very low closed-loop gains of two or less, to put a large common-mode voltage on the input stage [4]. This has nothing to with normal power amplifiers operating at a gain of twenty times or so. To date, I still consider common mode distortion to be something that can usually be ignored.

You often read about 'thermal distortion' in hi-fi magazines. This is supposed, insofar as any hypothesis is advanced at all, to be due to changes in device parameters over a cycle due to changes in power dissipation and heating over that cycle. I was unimpressed by the possibility of thermal distortion anywhere in the amplifiers I measured, reasoning that it would inevitably cause rising distortion with falling frequency, and this was not observable at the 0.001% level in any of my measurements. Changes in distortion over time due to long-term thermal changes in the output stage—measured in minutes rather than milliseconds—are another matter altogether, and clearly do exist [5].

Through the natural effluxion of time, and consequent improvements in testgear, the use of an Audio Precision SYS-2702 rather than a System-1 has significantly lowered the noise floor of THD measurements. When this is combined with a relatively high output power, of around 100W/8Ω, then it is possible to see very small amounts of distortion that increase as frequency falls. Preface Figure 31.1 demonstrates this; distortion is barely measurable at <0.0004% around the 300 Hz–1 kHz region. Above this band, the usual crossover distortion artefacts appear, while below it, some LF distortion appears as fairly pure third harmonic, increasing in level as frequency falls. No trace of this exists on the testgear output; see the flat genmon line at the bottom of the plot.

Preface Figure 31.1 shows that even radical changes in output loading have no effect on the LF distortion. Preface Figure 45.2 later in this book extends this to 100W/8Ω. A puzzling feature of this distortion is how slowly it increases as frequency falls. Most distortion mechanisms give more THD at HF, increasing at 6, 12, or 18 dB/octave with frequency; this one only doubles over the three octaves from 100 to 10 Hz. The only place I have seen such a rise before is when examining old carbon-film resistors for thermal distortion [6].

After a few more experiments, I could say that the LF distortion:

1. Rises very slowly as frequency falls, taking three octaves to double in amplitude

2. Increases with output voltage, increasing more steeply as output voltage increases

3. Is not a function of output current

4. Is not a function of supply voltage

5. The feedback capacitor is not involved. Doubling its size made no difference.

6. Is not affected by the compensation scheme used

Digesting those points, the source of the LF distortion must therefore be either the input circuitry, which is outside the negative feedback loop, or the negative feedback network. Signal levels in the input

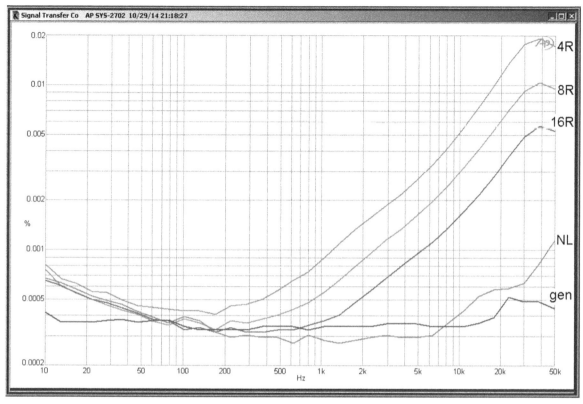

Preface Figure 31.1 THD of Compact Class-B CFP power amp: no-load, 20W/16Ω, 40W/8Ω and 80W/4Ω plus genmon (marked gen). Measurement bandwidth 10–80kHz.

circuitry, are low, maximally about 1.2 Vrms, and a few simple experiments disposed of this as a possibility.

This leaves the negative feedback network, and if your mind is open to the possibility of non-ideal resistors, there is a prime suspect in the shape of the upper feedback resistor. This component has almost the whole output voltage swing across it; 22/23 of it in my amplifiers, so there is about 27 Vrms across it at 100W/8Ω, and for my usual 2k2 upper resistor, the power dissipated is 330 mW, a significant amount. Even good-quality metal-film resistors change their value slightly when their temperature or the voltage across them changes, and this is reflected in their temperature coefficient and voltage coefficient. Regrettably the temperature coefficient is rarely quoted in any detail, and the voltage coefficient is rarely quoted at all.

In the amplifiers I tested, this the upper feedback resistor was a high-quality metal-film part of 750 mW rating, and so well within its limits. Its temperature coefficient was merely specified as ±100 ppm/°C; no voltage coefficient was given. Replacing the resistor with a series-parallel combination of four of them, to give the same total value, reduced the power dissipation in each resistor to a quarter, and the voltage across each one by half; this somewhat reduced the LF distortion, but definitely did not eliminate it. Since this test alters both power dissipation and voltage, it does not give a very clear idea as to which is causing the trouble.

An alternative experiment is to use two resistors the same type and of twice the value in parallel. This halves the power dissipation, and hence reduces temperature changes, but leaves the voltage across each resistor the same. Hence any improvement must be due to the reduction of thermal distortion. This was explored further by using four 9k1 resistors in parallel, and then eight 18 kΩ resistors in parallel; the LF distortion slowly decreased as the number of resistors (250 mW rated) increased, but it was not eliminated. Eight resistors would be quite practical on a PCB if a hairpin format was used. There is a non-obvious snag here in that there is no guarantee that a two resistors of the same type, but differing values will have identical coefficients.

In the next experiments, a single resistor of another type, though still metal-film and from the same respected manufacturer, was tried as it had an unusually low temperature coefficient specified as ±50 ppm/°C. I hoped this might tackle thermal distortion more directly. Preface Figure 31.2 shows that it did, up to a point, giving much better results than the series-parallel combination of four ±100 ppm/°C 750 mW resistors. The LF distortion is now reduced by a useful amount, despite the fact that the ±50 ppm/°C part was rated at only 500 mW rather than 750 mW. This appears to me to

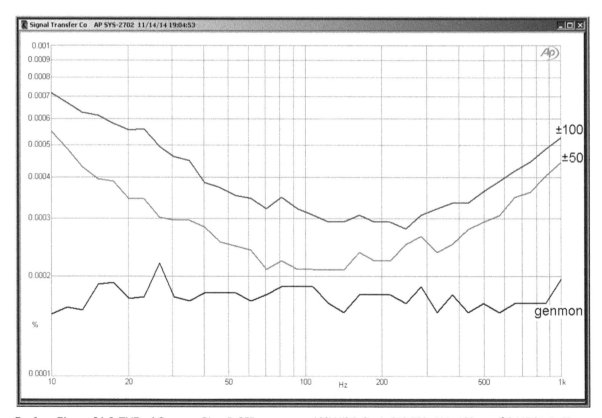

Preface Figure 31.2 THD of Compact Class-B CFP power amp 100W/8Ω. Single 2k2 750mW ±100 ppm/°C MF, Single PR 2k2 500mW ±50 ppm/°C MF resistors. Test gear output labelled 'genmon'. Measurement bandwidth 10–22kHz.

be a good example of thermal distortion found in the wild, and partly at least cured just by changing resistor type.

In the current state of knowledge, it could be argued that voltage-coefficient distortion is also probably involved, but this is a hard point to settle without any knowledge of the voltage coefficients for the two types of resistor. There is no reason why thermal and voltage-coefficient distortion cannot co-exist in the same resistor, which raises the question as to if they should be given different Distortion numbers; this may be resolved when further research is done.

In understanding distortion, it is extremely helpful to have a mathematical model of its mechanism, particularly where it is not possible to measure it directly. A good number of these are given in [7]. Voltage-coefficient distortion in metal-film resistors is too low for direct THD methods (some useful techniques are described in [8] and [9]), but a first-order mathematical model tells us that the level of THD is proportional to the signal level and that all odd harmonics are produced but only the third is of significant amplitude. A similar model for thermal distortion would be extremely useful for comparison, but is unfortunately not a simple thing to build. It is the temperature of the metal film itself that matters, but it is intimately attached to a the body of the resistor which much affects the thermal inertia. The thermal variations inside it will depend on the distance from the surface, and modelling the effects of this is probably going to require some sort of network of RC time-constants. I'm working on it.

In Preface Figure 31.2 there is still LF distortion visible as third harmonic below 100 Hz. The new resistor was tried in the series-parallel combination of four, but there was no further improvement, which seems to prove that upper-resistor-based non-linearity has been made negligible, but another source of it remains somewhere else in the amplifier, possibly in the forward path. Given that radical changes of loading, which would affect both the VAS and the output stage, have no effect, it is possible it is in the input stage. It certainly looks as though there is a Distortion 13 to be elucidated, and clearly more well-funded research is needed.

References

1. Self, D., *Audio Power Amplifier Design*, 6th edition, p. 119. Focal Press 2013. ISBN 978-0-240-52613-3.

2. Ibid., p. 142.

3. Ibid., p. 119.

4. Ibid., pp. 140–143.

5. Ibid., Chapter 22.

6. Ibid., pp. 45–48.

7. Ibid., pp. 25–44.

8. Simon, E., Resistor Non-linearity, *Linear Audio,* Volume 1, April 2011, p. 138.

9. Groner, S, Comments on Ed Simon article http://linearaudionet.solide-ict.nl/sites/linearaudio.net/files/volume1 ltees.pdf (accessed Feb. 2015)

DIAGNOSING DISTORTIONS

January 1998

In recent years, some audio commentators have been rudely dismissive of the simplest and most basic kind of distortion measurement—the totalharmonic distortion, or THD, test.

Because THD measurement has a long history, it is easy to imply that it is outdated and used only by the clueless. This is not so. Many other distortion tests exist, but none of them allows instant diagnosis of audio problems with one glance at an oscilloscope.

The test requires an oscillator with negligible distortion, feeding the unit under test. A notch-filter then completely suppresses the fundamental, to reveal the distortion products that have been generated. What remains after the fundamental is removed is not unnaturally called the THD residual.

The blameless amplifier concept

A Blameless amplifier results when the known distortions in the panel on distortion mechanisms have been either minimised or reduced to below visibility on the THD residual. It is so-called because it achieves its superb linearity not by startling innovation but simply by avoiding a series of possible errors. Avoiding them is straightforward once they are identified.

The concept of a Blameless amplifier has proved extremely useful. Such an amplifier has surprisingly low THD, despite its conventional-looking circuitry, but its greatest advantage is its defined performance, only weakly dependent on component characteristics.

If an amplifier does not perform to Blameless standards of linearity, then there is something fairly simple wrong with it, and to attempt to improve it by adding extra circuitry or turning up the bias into Class AB misses the point totally.

In several previous articles I have described the various distortions that afflict audio power amplifiers. In the generic circuit, these are relatively few in number as is evident from the panel entitled 'Distortion mechanisms'. Here I will show what some of these distortion residuals actually look like. Distortions 1, 2, 4 and 8 are not very informative visually, being essentially

second or third harmonic, so I have omitted them to make room for the more complex waveforms specific to Class B.

Making distortion measurements

Total harmonic distortion is the r.m.s. sum of all the distortion components generated by the path under test. It is usually quoted as a percentage of the total signal level.

The r.m.s. calculation—taking the square-root of the sum of the squares of the harmonics—emphasises spiky distortions, but whether this helps to mimic human perception of distortion is unclear. The peak capability of true-r.m.s. circuitry is limited, and this may well lead to under-reading of crossover spikes and such.

I hold that the best method is to observe the residual simultaneously, and time-aligned, with the output sinewave as in Figure 31.1. If you are testing similar pieces of equipment, then the gain of the oscilloscope's second channel for the residual can be kept at the same setting. This allows linearity to be assessed at a glance.

In contrast, it is wiser to connect the actual output to channel 1, rather than an auto-scaled version from the analyser, as this prevents parasitics, etc., from being filtered out by the analyser input circuitry.

The beauty of THD testing is that the error is isolated; in essence, the residual is the difference between perfection and reality. When viewed time-aligned with the output sinewave, crossover distortion can be diagnosed immediately as it occurs at the zero-crossings. On the other hand, non-linearity confined to one peak is probably due to something running out of voltage swing or current capability.

Two technical challenges

Figure 31.1 shows the basic THD measuring system. There are two major technical challenges to be overcome. The signal source must be extremely pure, as any oscillator distortion puts an immediate limit on the measurement floor; it must maintain superb performance at least over the range 10 Hz to 20 kHz. A balanced output is highly desirable.

In the analyser section a balanced input is essential. Very great attenuation of the fundamental is required—about 120 dB if you are going to measure down to 0.0005%, making notch tuning is extremely

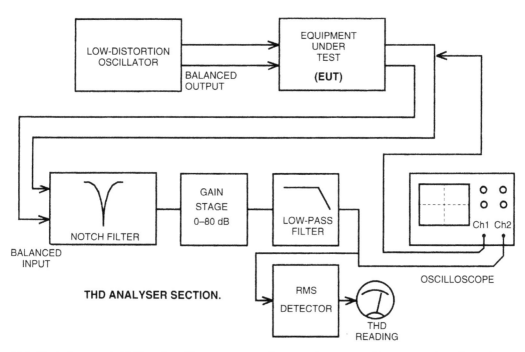

Figure 31.1 Block diagram of a THD analyser. The minimum reading is set by input amplifier noise and oscillator distortion rather than the filter auto-tuning.

critical. This cannot be attained by fixed-tuned filters, and manual tuning, requiring at least six controls, is about as much fun as picking oakum. In modern THD equipment both frequency and phase are continuously adjusted by a twin servo-loop that optimises the cancellation.

An additional low-pass filter defines the measurement bandwidth. Usually, 80 kHz is a good compromise, retaining most of the important harmonics while reducing noise. A switchable 400 Hz high-pass filter is often fitted, allowing measurements at 1 kHz and up, in the presence of hum. Such a filter should be used only in exceptional cases, for THD often rises sharply at low frequencies, and this would be missed.

While frequently advocated as a more searching examination of an audio path, twin-tone intermodulation tests are almost useless for circuit investigation. They give very little information about the source of the nonlinearity as the phase relationship between the test signal and the result is lost. It is often claimed they give a better measure of audible degradation in real use, but a test using two or three tones is still a long way away from music that has tens or hundreds of simultaneous

frequencies. Intermodulation tests can often dispense with very-low-THD oscillators, but this in itself is not much of a recommendation.

If real subjective degradation is the issue, a test signal much closer to reality is required. This can be either pseudorandom noise as in the Belcher test, [1] or real music, as in the Baxandall [2] and Hafler [3] cancellation tests.

Returning to harmonic distortion, much better correlation between THD measurements and subjective impairment is possible if the harmonics are weighted so that the higher order components are emphasised.

Weighting by $n^2/4$, so that the second harmonic is unchanged, the third increased by 9/4, and so on, is generally accepted to be roughly correct.[4,5] I was surprised to find that this approach goes back to 1937 and before.[6] I doubt however whether this can be applied to crossover distortion.

When the THD residual is displayed on an analogue oscilloscope, artifacts in the noise are easily detectable by the averaging processes of our vision, but they remain unavailable to conventional measurement. A digital scope can perform even more effective averaging by computation, making submerged distortion artifacts both visually clearer and readily measurable, though an r.m.s. mode may not be available.

If a noisy signal is averaged two times, by combining two sweeps, the coherent signal stays at the same level, while the uncorrelated noise decreases by 3 dB. Averaging 64 times performs this process six-fold, so noise is then reduced by 18 dB. The oscilloscope used here was a digital *HP54600B* 100 MHz digital storage; an excellent instrument. This choice will not come as a surprise to alert readers.

Distortion mechanisms

My original series on amplifiers[7] listed seven independent distortions inherent to the generic/ Lin Class-B amplifier, and whose existence is not dependant on circuit details. I have now increased this to eight.

Distortion one

Input-stage distortion. Non-linearity in the input stage. If this is a carefully-balanced differential pair then the distortion is typically only measurable at high frequencies, rises at 18 dB/octave, and is almost pure third harmonic.

If the input pair is unbalanced—which from published circuitry it usually is—then enough second harmonic is produced to swamp the third. Hence the h.f. distortion emerges from the noise at a lower frequency, rising at 12 dB/octave.

Distortion two

Voltage amplifier stage distortion. Surprisingly, non-linearity in the voltage-amplifier stage does not always contribute significantly in the total distortion. If it does, it remains constant until the dominant-pole frequency P1 is reached, and then rises at 6 dB/octave. In the generic configuration discussed here it is always second harmonic.

Distortion three

Output-stage distortion. Non-linearity in the output stage—the most obvious source. This has three components: crossover distortion (3a) usually dominates for Class-B into 8 Ω, generating high-order harmonics rising at 6 dB/octave as global negative feedback decreases. Low-order large-signal nonlinearity (3b)

---

appears with 4 Ω loads and worsens at 2 Ω. Distortion 3c stems from overlap of output device conduction and only appears at high frequencies.

Distortion four

Voltage-amplifier loading. Loading of the voltage-amplifier stage by the non-linear input impedance of the output stage.

Distortion five

Rail decoupling distortion. Non-linearity caused by large rail-decoupling capacitors feeding the distorted signals on the supply lines into the signal ground. This seems to be the reason that many amplifiers have rising THD at low frequencies.

Distortion six

Induction distortion. Induction of Class-B supply currents into the output, ground, or negative-feedback lines. Almost certainly the least understood and so must common distortion afflicting commercial amplifiers.

Distortion seven

Negative-feedback take-off distortion. Non-linearity resulting from taking the negative feedback feed from slightly the wrong place near the point where Class-B currents sum to form the output.

Distortion eight

Capacitor distortion. Rising as frequency falls, capacitor distortion is caused by non-linearity in the input d.c.-blocking capacitor or the feedback network capacitor. The latter is more likely.

Distortions X

Non existent or negligible distortions. Common-mode distortion in the input stage and thermal distortion in the output stage—or anywhere else.

Although sometimes invaluable, digital oscilloscopes are often not the best choice for audio THD testing and general amplifier work; in particular the problems of aliasing make the detection and cure of h.f. oscillations very difficult.

To create the residuals shown here, a Blameless amplifier was used essentially identical to that published in Ref. 7. Output was 25 W into 8 Ω, or 50 W into 4 Ω. The Blameless amplifier concept is outlined in a separate panel.

Crossover distortion

Crossover distortion is only one of the three components that make up Distortion 3 but is often the dominant one. Blameless amplifiers show only crossover distortion when driving 8 Ω or more, and at low and medium frequencies it should be below the noise. This remains true even if the amplifier noise is within a few decibels of the theoretical minimum from a 50 Ω source resistance.

Figure 31.2 shows the THD residual from such a Blameless power amplifier, with optimally biased in Class-B. Since this is a record of a single sweep, the residual appears to be almost wholly noise. The visual averaging process is absent and so the crossover artifacts are actually less visible than on an analogue scope in real time.

In Figure 31.3, 64 times digital averaging is applied, which makes the disturbances around crossover very

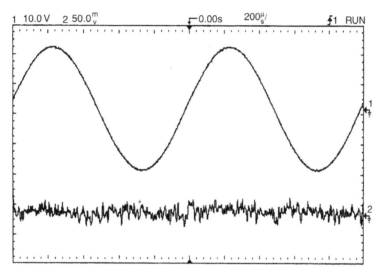

Figure 31.2 The THD residual from an optimally-biased Blameless power amplifier at 1 kHz, 25W/8 Ω is essentially white noise. There is some evidence of artifacts at the crossover point, but they are not measurable. THD 0.00097%, 80 kHz bandwidth.

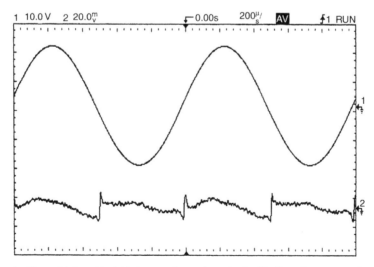

Figure 31.3 Averaging the Figure 31.2 residual 64 times reduces the noise by 18 dB, and crossover discontinuities are now obvious. The residual has been scaled up by 2.5 times from Figure 31.2 for greater clarity.

clear. A low-order component at roughly 0.0003% is also revealed, which is probably due to very small amounts of Distortion 6 that were not visible when the amplifier layout was optimised.

Figure 31.4 shows Class B mildly underbiased to generate crossover distortion. The crossover spikes are very sharp, and their height in the residual depends critically on measurement bandwidth. Their presence warns immediately of underbiasing and avoidable crossover distortion.

In Figure 31.5 an optimally-biased amplifier is tested at 10 kHz. The THD has increased to approx 0.004%,

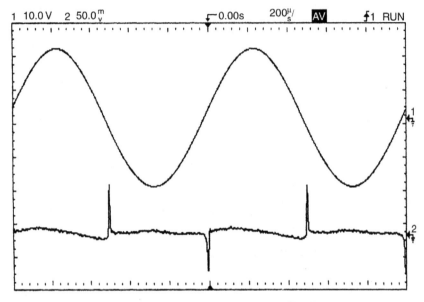

Figure 31.4 Results of mild underbias in Class B.

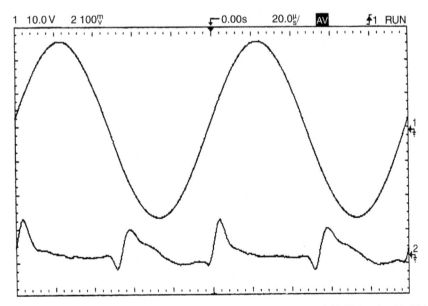

Figure 31.5 An optimally-biased Blameless power amplifier at 10 kHz. THD is around 0.004%, bandwidth 80 kHz. Averaged eight times.

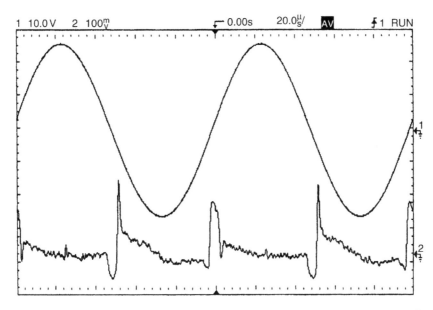

Figure 31.6 As Figure 31.5, but in 500 kHz bandwidth. The distortion products look quite different.

as the amount of global negative-feedback is 20 dB less than at 1 kHz. The crossover events appear wider than in Figure 31.3. The higher THD level is above the noise so the residual is averaged eight times only.

The measurement bandwidth is still 80 kHz, so harmonics above the eighth are lost. This is illustrated in Figure 31.6, which is Figure 31.5 rerun with a 500 kHz bandwidth. The distortion products look very different.

The 80 kHz cutoff point is something of *de facto* standard, which is reasonable as it seems highly unlikely that ultrasonic harmonics can detract from one's listening pleasure. This does not mean THD testing can stop at 10 kHz, as there might be an area of bad intermodulation in the top octave.

My practice is to test up to 50 kHz, to check that nothing awful is lurking just outside the audio band; this is safe for moderate powers, and short durations.

Classes B and AB

I showed in my series on power amplifier distortion[7] that Class AB is not a true compromise between Class A and Class B operation. If AB is used to trade off efficiency and linearity, its linearity is superior to B since below the AB transition level, it is pure Class A.

The Class-A region can—and should—have very low THD indeed, below 0.0006% up to 10 kHz, as demonstrated in Ref. 8. However, above the AB transition level THD abruptly worsens. This is due to what has been called 'g_m-doubling', but is better regarded as a step in the gain/output-voltage relationship. Linearity is then inferior not only to Class-A but also to optimal-bias Class-B.

It is possible to make Class AB distortion very low by proper design. Basically, this means using the lowest possible emitter resistors to reduce the size of the gain step.[9] Even so, THD remains at least twice as high as Class-B.

Tweaking up the bias of a Class-B amplifier most certainly does *not* offer a simple trade-off between power dissipation and overall linearity, despite the constant repetition this notion receives in some parts of the audio press. The real choice is: very low THD at low power and high THD at high power, or medium THD at all powers. The electricity bill is another issue.

Figure 31.7 shows the gain-step distortion introduced by Class AB. The undesirable edges are caused by gain changes that are no longer partially cancelled at the crossover; they are now displaced to either side of the zero-crossing. No averaging is used here as the THD is higher and well above the noise.

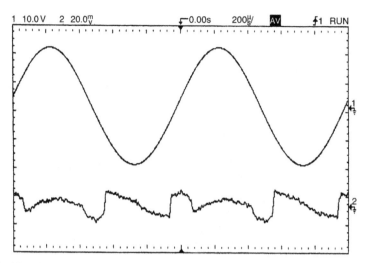

Figure 31.7 The gm-doubling distortion introduced by Class AB. The edges in the residual are larger and no longer at the zero-crossing, but displaced either side of it.

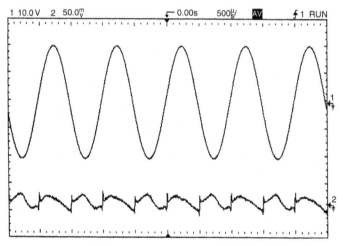

Figure 31.8 Large-signal nonlinearity, driving 50 W into 4 Ω, and averaged 64 times. The extra distortion appears to be a mixture of third harmonic—occurring as a consequence of the compressive nature of beta-loss—and second harmonic arising because the beta-loss is not perfectly symmetrical in the two halves of the output stage.

Large-signal non-linearity

When the load resistance falls below 8 Ω, extra low-order distortion components appear. This is true for most or all modern power bipolar junction transistors, but with old devices like 2N3055, some large-signal nonlinearity may appear at 8 Ω. This is a compressive non-linearity, i.e. gain falls as level increases, 'squashing' the signal, and is due to fall-off of transistor beta at high collector currents.

Figure 31.8 shows the typical appearance of large-signal non-linearity, driving 50 W into 4 Ω, and averaged 64 times. The extra distortion appears to be a mixture of

third harmonic, due to the basic symmetry of the output stage, with some second harmonic, because the beta-loss is component-dependant and not perfectly symmetrical in the two halves of the output.

Other distortions

Of the distortions that afflict generic Class-B power amplifiers, 5, 6 and 7 all look rather similar in the THD residual. This is perhaps not surprising since all result from adding half-wave disturbances to the signal.

Distortion 5 is usually easy to identify as it is accompanied by 100 Hz power-supply ripple; 6 and 7 introduce no ripple. Distortion 6 is easily identified if the d.c. power cables are movable, for altering their run will strongly affect the quantity generated.

Figure 31.9 shows Distortion 5, provoked by connecting the negative supply rail decoupling capacitor to the input ground instead of giving it its own return to the far side of the star point. Doing this increases THD from 0.00097% to 0.008%, mostly as second harmonic. Ripple contamination is significant and contributes to the THD figure. It could be easily filtered out to make the measurement, but this is just brushing the problem under the carpet.

Distortion 6 is displayed in Figure 31.10. The negative supply rail was run parallel to the negative-feedback

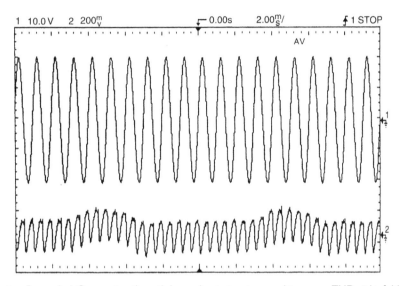

Figure 31.9 Distortion 5 revealed. Connecting the rail decoupler to input ground increases THD eight-fold from 0.00097% to 0.008%, mostly as second harmonic. 100 Hz ripple is also visible. No averaging.

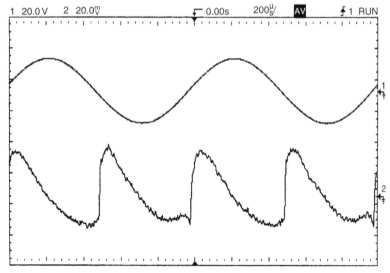

Figure 31.10 Distortion 6. Induction of half-wave signal from the negative supply rail into the negative feedback line increases THD to 0.0021% Averaged 64 times.

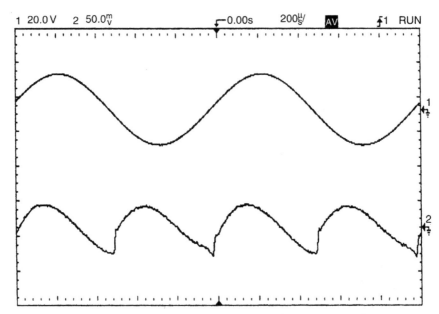

Figure 31.11 Distortion 7, caused by choosing an negative feedback take-off point inside the Class-B output stage rather than on the output line itself. THD increases from 0.00097% to 0.0027%, by taking the negative feedback from the wrong end of 10 mm of very thick resistor leg. Averaged 64 times.

line to produce this diagram. Although more than doubled, THD is still relatively low at 0.0021%, so 64-times averaging is used.

Figure 31.11 shows a case of Distortion 7, introduced by deliberately making a minor error in the negative feedback take-off point.

If it is attached to a part of the Class-B output stage so that half-wave currents flow through it, rather than being on the output line itself, THD is increased. Here it rose from 0.00097% to 0.0027%, caused by taking the negative feedback from the wrong end of the leg of one of the output emitter resistors, R_e.

Note this was at the right end of the resistor, otherwise THD would have been gross, but 10 mm along a very thick resistor leg from the output line junction. Truly, God is in the details.

Diagnosis

The rogue's gallery of real-life THD residuals portrayed here will hopefully help with the problem of identifying the distortion mechanism in a misbehaving amplifier. There is no reason why the generic/Lin configuration should give measurable THD at 1 kHz, or more than say, 0.004% at 10 kHz when driving 8 Ω.

It is important to be sure that you are measuring a real distortion mechanism, and not the results of parasitic oscillation upsetting circuit conditions; the oscillation itself may be outside the scope bandwidth. Parasitics usually vary greatly when a cautious finger is applied to the relevant section of the circuitry. Real distortion changes little, though the THD reading will probably be increased by the introduction of hum.

I hope I have shown that THD testing gives an immediate view into circuit operation that other methods do not, however useful they may be in other applications.

It cannot be stated too strongly that to attempt amplifier design and diagnosis without continuous visual observation of the THD residual is to work blind. You will proverbially fall into the ditch.

References

1. Belcher, R. 'A new distortion measurement', *Wireless World,* May 1978, pp 36–41.

2. Baxandall, P. 'Audible amplifier distortion is not a mystery', *Wireless World,* November 1977, pp 63–66.

3. Hafler, D. 'A listening test for amplifier distortion', *HiFi News & RR,* November 1986, pp 25–29.

4. Moir, J. 'Just detectable distortion levels', *Wireless World,* February 1981, p 34.

5. Shorter, D. 'Influence of high-order products in non-linear, distortion', *Electronic Engineering,* April 1950, p 152.

6. Callendar, M, Letter, *Electronic Engineering,* October 1950, p 443.

7. Self, D. 'Distortion in power amplifiers', *Electronics World,* August 1993 to March 1994.

8. Self, D. 'Distortion in power amplifiers, part 8', *Electronics & Wireless World,* March 1994, p 225 (Class A amp).

9. Self, D. 'Trimodal Audio Power', *Electronics World,* June 1995, p 462. (Effect of R_e in Class AB).

Chapter 32

Trimodal audio power, Part I

(Electronics World, June 1995)

Intended title: *A Trimodal Power Amplifier: Part 1*

This project had its roots in a large number of requests to *Electronics World* for a PCB for the Class-A amplifier described in the last part of the Distortion In Power Amplifiers series; see Chapter 29. Rather than just using the original circuit, I took the chance to look closely at output stage voltage efficiency, and also to re-examine the input stage balance and noise performance. I added a safety circuit to prevent catastrophe if the quiescent-current control arrangements went haywire. A little thought showed that this could very conveniently double as a bias voltage generator that would allow the amplifier to be switched into Class-B mode. Some very simple switching gave two kinds of amplifier for the price of one.

Like its progenitor, the original Class-A amplifier described in Chapter 29, the Trimodal had its quiescent current set at 1.5 Amps. In a push-pull design the peak current is twice the quiescent, so that allows the maximum rail-limited voltage swing to be driven into 6Ω. The voltage swing into 4Ω is limited to ±12V before the amplifier goes into Class-AB, so as it stands, the Trimodal is not well suited to nominal 4Ω loads, especially since there will almost certainly be impedance dips below that figure. Increasing the quiescent current to 2.25 Amps will give the full Class-A output swing into 4Ω, but no lesser impedance. This modification should be straightforward, requiring only an increase in the reference voltage of the current-control circuitry. Given the very low distortion this design shows when it moves into Class AB on encountering an unexpectedly low impedance in use, no one is ever going to know it happened unless they have some pretty sophisticated monitoring apparatus.

Likewise, it should be relatively simple to increase the power output by increasing the supply rails; this naturally requires increasing the quiescent current as well, to maintain the voltage-swing, so there is a double increase in heat dissipation. As I have noted elsewhere, I have little enthusiasm for Class-A power amplifiers of more than a very modest size, because of their absolutely deplorable efficiency with music signals. The overall efficiency is something like 1% if the output level is set to maximum, defining that as occasional clipping.

This proved a very popular design, and I had a lot of correspondence about it. People are still enthusiastically building them. Bare PCBs, kits, and complete tested versions of an updated Trimodal amplifier are available from The Signal Transfer Company [1].

A stereo pair of Signal Transfer Trimodal PCBs were put in a box, together with power supply, protection system, and ultra-low noise balanced input card, all by Signal Transfer. This amplifier was reviewed by TNT Audio [2] in 2010, and they liked it. It is worth noting that the power supply reservoir capacity was only 20,000 uF for each rail. Power-supply rejection is a big issue in Class-A amplifiers because there is maximum ripple on the supply rails when there is no signal to mask any hum that gets through, and this means that many designs have enormous amounts of reservoir capacity; the 20,000 uF used here is very modest, as power-supply rejection is obtained by more intelligent means. The V-rail filtering of the Trimodal amplifiers was slightly upgraded, and this was all that was required to sink the measured hum levels below the noise floor.

References

1. http://www.signaltransfer.freeuk.com/ (accessed Feb. 2015)

2. http://www.tnt-audio.com/ampli/trimodal_amp_e.html (accessed Dec. 2014)

TRIMODAL AUDIO POWER, PART I

June 1995

I present here my own contribution to global warming in the form of an improved Class-A amplifier that I believe is unique. It not only copes with load impedance dips by means of an unusually linear form of Class-AB, but will also operate as a 'blameless' Class-B engine. The power output in pure Class-A is 20 to 30 W into 8 Ω, depending on the exact supply rails chosen.

Initially, I simply intended to provide an updated version of the Class-A circuit published in reference 1, in response to requests for a PCB for the Class-A amplifier designed with my methodology. I decided to use a complementary-feedback-pair (CFP), output stage for best possible linearity, and some incremental improvements have been made to noise, slew rate and maximum d.c. offset.

Naturally, the Class-A circuit bears a very close resemblance to a 'blameless' Class-B amplifier. As a result, I decided to retain the Class-B V_{be} multiplier, and use it as a safety-circuit to prevent catastrophe if the relatively complex Class-A current-regulator failed. From this the idea arose of making the amplifier instantly switchable between Class-A/AB and Class-B modes. This gives two kinds of amplifier for the price of one, and permits of some interesting listening tests. Now you really can do an A/B comparison.

In the Class-B mode the amplifier has the usual negligible quiescent dissipation, but in Class-A the thermal efflux is naturally considerable. This is because true Class-A operation is extended down to 6 Ω resistive loads for the full output voltage swing, by suitable choice of the quiescent current.

With heavier loading the amplifier gracefully enters Class-AB, in which it will give full output down to 3 Ω before the safe-operating-area (SOAR), limiting begins to act. Output into 2 Ω is severely curtailed, as it must be with only one output pair, and this kind of load is not advisable.

In short, the amplifier allows a choice between being firstly very linear all the time—blameless Class-B—and secondly ultra-linear most of the time—Class-A—with occasional excursions into Class-AB.

The amplifier's AB mode is still extremely linear by current standards, though inherently it can never be as good as properly-handled Class-B, and nothing like as good as A. Since there are three possible classes of operation I have decided to call the design a Trimodal power amplifier. It is impossible to be sure that you have read all the literature on an area of technology; however, to the best of my knowledge this is the first ever Trimodal amplifier.

As I said earlier, designing a low-distortion Class-A amplifier is in general a good deal simpler than the same exercise for Class-B. All the difficulties of arranging the best possible crossover between the output devices disappear. Because of this it is hard to define exactly what 'blameless' means for a Class-A amplifier.

In Class-B the situation is quite different, and 'blameless' has a very specific meaning; when each of the eight or more distortion mechanisms has been minimised in effect, there always remains the crossover distortion inherent in Class-B. There appears to be no way to reduce it without departing radically from that might be called the generic Lin amplifier configuration. Therefore the 'blameless' state appears to represent some sort of theoretical limit for Class-B, but not for Class-A.

However, Class-B considerations cannot be ignored, even in a design intended to be Class-A only, because if the amplifier does find itself driving a lower load impedance than expected, it will move into Class-AB. In this case, all the additional Class-B requirements are just as significant as for a Class-B design proper. Class-AB can never give distortion as low as optimally-biased Class-B, but it can be made comparable if the extra distortion mechanisms are correctly handled.

My correspondence has made it abundantly clear that *EW* readers are not going to be satisfied with anything less than state-of-the-art linearity, and so the amplifier described here uses the CFP type of output stage, which has the lowest distortion due to the local feedback loops enclosing the output devices. It also has the advantage of better output efficiency than the emitter-follower version, and inherently superior quiescent current stability. It will shortly be seen that these are both important for this design.

Half-serious thought was given to labelling the Class-A mode 'distortionless' as the THD is completely unmeasurable across most of the audio band. However, detectable distortion products do exist above 10 kHz, so sadly, I abandoned this provocative idea.

Before putting cursor to CAD, it seemed appropriate to take another look at the Class-A design, to see if it could be inched a few steps nearer perfection. The result is a slight improvement in efficiency, and a 2 dB improvement in noise performance. In addition the expected range of output d.c. offset has been reduced from ±50 mV to ±15 mV, still without any adjustment.

The power and the glory

The amplifier is 4 Ω capable in both A/AB and B operating modes, though it is the nature of things that the distortion performance is not quite so good. All solid-state amplifiers—without qualification, as far as I am aware—are much happier with an 8 Ω load, both in terms of linearity and efficiency; loudspeaker designers please note.

With a 4 Ω load, Class-B operation gives better THD than Class-A/AB, because the latter will always be in AB mode, and therefore generating extra output stage distortion through g_m-doubling. This should really be called gain-deficit-halving, but somehow I don't see this term catching on. These not entirely obvious relationships are summarised on the right.

Figure 32.1 attempts to show diagrammatically just how power, load resistance, and operating mode are related. The rails have been set to ±20 V, which just allows 20 W into 8 Ω in Class-A. The curves are lines of constant power, i.e. $V \times I$ in the load, the upper horizontal line represents maximum voltage output, allowing for $V_{ce(sat)}$s, and the sloping line on the right is the SOAR protection locus; the output can never move outside this area in either mode. The intersection between the load resistance lines sloping up from the origin and the ultimate limits of voltage-clip and SOAR protection define which of the curved constant-power lines is reached.

In A/AB mode, the operating point must be left of the vertical push-pull current-limit line (at 3A, i.e. twice the quiescent current) for Class-A. If we move along one of the impedance lines, when we pass to the right of the push-pull limit the output devices will begin turning off for part of the cycle; this is the AB operation zone. In Class-B mode, the 3A line has no significance and the amplifier remains in optimal Class-B until clipping or SOAR limiting occurs. Note that the diagram axes represent instantaneous power in the load, but the curves show sine-wave r.m.s. power, and that is the reason for the apparent factor-of-two discrepancy between them.

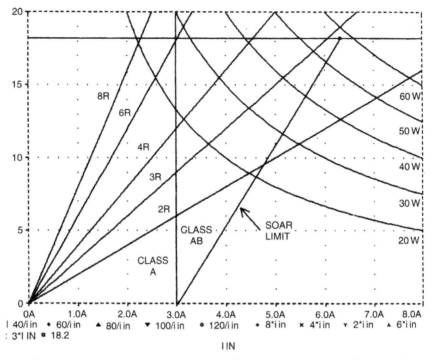

Figure 32.1 Relationships between load, mode, and power output. The intersection between the sloping load resistance lines and the ultimate limits of voltage-clipping and SOAR protection define which of the curved constant-power lines is reached. In A/AB mode, the operating point must be to the left of the vertical push-pull current-limit line for true Class-A.

Table 32.1

Load (Ω)	Mode	Distortion	Dissipation
8	A/AB	very low	high
4	A/AB	high	high
8	B	low	low
4	B	medium	medium

Note that in the context of this sort of amplifier, 'high' means about 0.002% THD at 1 kHz and 0.01% at 10 kHz.

Health and efficiency

Concern for efficiency in Class-A may seem paradoxical, but one way of looking at it is that Class-A watts are precious things, wrought in great heat and dissipation, and so for a given quiescent power it makes sense to ensure that the amplifier approaches its limited theoretical efficiency as closely as possible. I was confirmed in this course by reading of another recent design [2] which seems to throw efficiency to the winds by using a hybrid bjt/FET cascode output stage. The voltage losses inherent in this arrangement demand ±50 V rails and sixfold output devices for a 100 W Class-A capability; such rail voltages would give 156 W from a 100% efficient amplifier.

Voltage efficiency of a power amplifier is the fraction of the supply-rail voltage which can actually be delivered as peak-to-peak voltage swing into a specified load; efficiency is invariably less into 4 Ω due to the greater resistive voltage drops with increased current.

The Class-B amplifier I described in Ref. 3 has a voltage efficiency of 91.7% for positive swings, and 92.5% for negative, into 8 Ω. Amplifiers are not in general completely symmetrical, and so two figures need to be quoted; alternatively the lower of the two can be given as this defines the maximum undistorted sine-wave. These figures above are for an emitter-follower output stage, and a complementary-feedback pair output does better, the positive and negative efficiencies being 94.0% and 94.7% respectively.

The emitter follower version gives a lower output swing because it has two more V_{be} drops in series to be accommodated between the supply rails; the CFP is always more voltage-efficient, and so selecting it over the emitter follower for the current Class-A design is the first step in maximising efficiency.

Figure 32.2 shows the basic complementary-feedback pair output stage, together with its two biasing elements. In Class-A the quiescent current is rigidly controlled by negative-feedback; this is possible because in Class-A the total voltage across both emitter resistors R_e is constant throughout the cycle. In Class-B this is not the case, and we must rely on 'thermal feedback' from the output stage, though to be strictly accurate this is not 'feedback' at all, but a kind of feed-forward.

It is a big advantage of the CFP configuration that quiescent current, I_q depends only on driver temperature, and this is important in the Class-B mode, where true feedback control of quiescent current is not possible.

This has special force if low-value emitter resistors such as 0.1 Ω, are chosen, rather than the more usual 0.22 Ω; the motivation for doing this will soon become clear.

Voltage efficiency for the quasi-complementary Class-A circuit of Ref. 1 into 8 Ω is 89.8% positive and 92.2% negative. Converting this to the CFP output stage increases this to 92.9% positive and 93.6% negative. Note that a Class-A I_q of 1.5 A is assumed throughout; this allows 31 W into 8 Ω in push-pull, if the supply rails are adequately high. However the assumption that loudspeaker impedance never drops below 8 Ω is distinctly doubtful, to put it mildly, and so as before this design allows for full Class-A output voltage swing into loads down to 6 Ω.

So how else can we improve efficiency? The addition of extra and higher supply rails for the small-signal section of the amplifier surprisingly does not give a

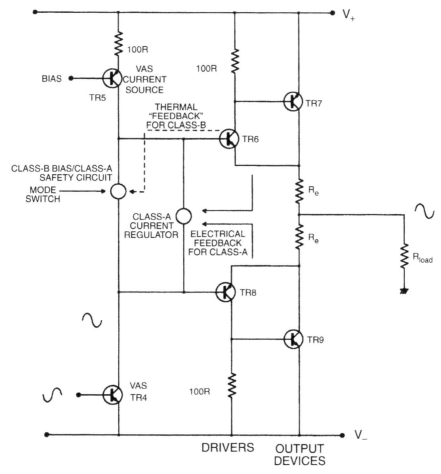

Figure 32.2 Basic current feedback output stage, equally suited to operating Class B, AB and A, depending the magnitude of Vbias. The resistors Re may be from 0.1 to 0.47 Ω.

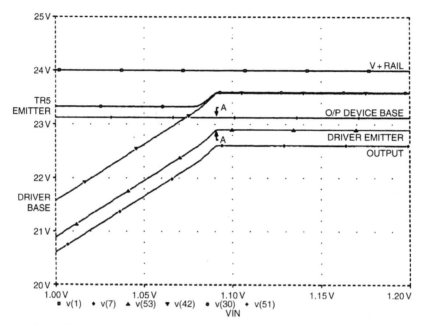

Figure 32.3 PSpice simulation showing how positive clipping occurs in the current feedback output. A higher sub-rail for the voltage amplifier cannot increase the output swing, as the limit is set by the minimum driver Vce and not the voltage amplifier output swing.

significant increase in output; examination of Figure 32.3 shows why. In this region of operation, the output device Tr_7 base is at a virtually constant 880 mV below the positive rail, and as Tr_6 driver base rises it passes this level, and keeps going up; clipping has not yet occurred.

The driver emitter follows the driver base up, until the voltage difference between this emitter and the output base, i.e. the driver V_{ce}, becomes too small to allow further conduction; this choke point is indicated by the arrows A-A. At this point the driver base is forced to level off, although it is still about 500 mV below the level of the positive rail. Note also how the voltage between the positive rail and Tr_5 emitter collapses. Thus a higher rail will give no extra voltage swing, which I must admit came as something of a surprise. Higher sub-rails for small-signal sections only come into their own in FET amplifiers, where the high V_{gs} for FET conduction (5 V or more) makes their use almost mandatory.

Efficiency figures given so far are all greater for negative rather than positive voltage swings. The approach to the rail for negative clipping is slightly closer because there is no equivalent to the 0.6 V bias established across R_{13}; however this advantage is absorbed by the need to lose a little voltage in the RC filtering of the negative supply to the current-mirror and voltage amplifier stage. This filtering is essential if really good ripple/hum performance is to be obtained[3].

In the quest for efficiency, an obvious variable is the value of the output emitter resistors R_e. The performance of the current-regulator described, especially when combined with a CFP output stage, is more than good enough to allow these resistors to be reduced while retaining first-class I_q stability. I took 0.1 Ω as the lowest practicable value, and even this is comparable with PCB track resistance, so some care in the exact details of physical layout is essential; in particular the emitter resistors must be treated as four-terminal components to exclude unwanted voltage drops in the tracks leading to the resistor pads.

If R_e is reduced from 0.22 Ω to 0.1 Ω then voltage efficiency improves from 92.9%/93.6%, to 94.2%/95.0%. Is this improvement worth having? Well, the voltage-limited power output into 8 Ω is increased from 31.2 to 32.2 W with ±24 V rails, at absolutely zero cost, but it would be idle to pretend that the resulting increase in sound-pressure level is highly significant. It does however provide the philosophical satisfaction faction that as much Class-A power as possible is being produced for a given dissipation; a delicate pleasure.

The linearity of the CFP output stage in Class-A is very slightly worse with 0.1 Ω emitter resistors, though the difference is small and only detectable open-loop; the simulated THD of an output stage alone (for 20 V pk-pk in 8 Ω) is only increased from 0.0027% to 0.0029%. This is probably due to simply to the slightly lower total resistance seen by the output stage.

However, at the same time reducing the emitter resistors to 0.1 Ω provides much lower distortion when the amplifier runs out of Class-A; it halves the size of the

step gain changes inherent in Class-AB, and so effectively reduces distortion into 4 Ω loads.

Figures 32.4 & 32.5 are output linearity simulations; the measured results from a real and 'blameless' Trimodal amplifier are shown in Figure 32.6, where it can be clearly seen that THD has been halved by this simple

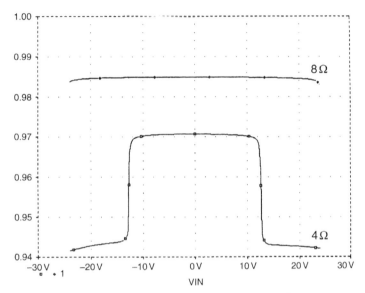

Figure 32.4 Complementary feedback pair output stage linearity with Re set at 0.22 Ω. Upper trace is Class-A into 8 Ω, lower is Class-AB operation into 4 Ω, showing step changes in gain of 0.024 units.

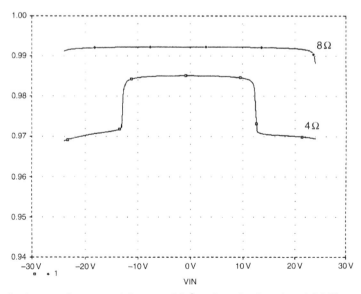

Figure 32.5 Current feedback output linearity with Re set at 0.1 Ω, re-biased to keep Iq at 1.5 A. There is slightly poorer linearity in the flat-topped Class-A region than for an Re of 0.22 Ω, but the 4 Ω AB steps are halved in size at .012 units. Note that both gains are now closer to unity; same scale as Figure 32.4.

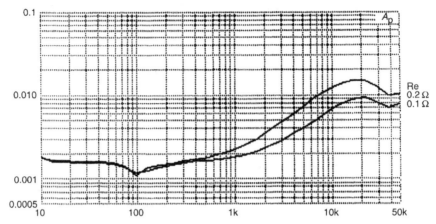

Figure 32.6 Proving that emitter resistor value really matters in Class-AB, Output was 20 W in 4 Ω, so amplifier was leaving Class-A for about 50% of the time. Changing emitter resistors from 0.2 to 0.1 Ω halves the distortion. Current Iq is 1.5 A for both cases.

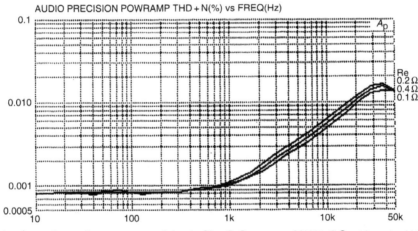

Figure 32.7 Proving that emitter resistors matter much less in Class-B. Output was 20 W in 8 Ω, with optimal bias. Interestingly, the bias does NOT need adjusting as the value of Re changes. Bandwidth 80 kHz.

change. To the best of my knowledge this is a new result; my conclusion is that if you must work in Class-AB, keep the emitter resistors as low as possible, to minimise the gain changes.

Having considered the linearity of Class-A and AB, we must not neglect what effect this radical Re change has on Class-B linearity. The answer is, not very much, but there is a slightly reduction in THD, Figure 32.7, where crossover distortion seems to be slightly higher with R_e at 0.2 Ω than for either 0.1 or 0.4 Ω. Whether this is a consistent effect—for complementary-feedback pair stages anyway—remains to be seen.

The detailed mechanisms of bias control and mode-switching are described in the second part of this article.

Improving noise performance

In a power amplifier, noise performance is not an irrelevance.[4] It is well worth examining just how good it can be. As in most amplifiers, noise is set here by a combination of the active devices at the input and the surrounding resistances.

Operating conditions of the input transistors themselves are set by the demands of linearity and slew-rate, and there is little freedom of design here; however the collector currents are already high enough to give near-optimal noise figures with the low source impedances—a few hundred ohms—that we have here, so this is not too great a problem. Also remember that noise figure is a weak function of I_c, so minor tweaking makes no detectable difference. We certainly have the choice of input device type; there are many more possibilities now that we have relatively low rail voltages. Noise performance is, however, closely bound up with source impedance, and we need to define this before device selection.

Looking therefore to the passives, there are several resistances generating Johnson noise in the input, and the only way to reduce this noise is to reduce them in value. The obvious candidates are input stage degeneration resistors $R_{2,3}$ and R_9, which determines the output impedance of the negative-feedback network. There is also another unseen component; the source resistance of the preamplifier or whatever upstream.

Even if this equipment were miraculously noise-free, its output resistance would still generate Johnson noise. If the preamplifier had, say, a 20 kΩ volume pot at its output—not a good idea, as this gives a poor gain structure and cable dependent h.f. losses, but that is another story[5] then the source resistance could be a maximum of 5 kΩ, which would almost certainly generate enough Johnson Noise to dominate the power-amplifier's noise behaviour. However, there is nothing that power-amp designers can do about this, so we must content ourselves with minimising the noise-generating resistances we do have control over.

The presence of input degeneration resistors $R_{2,3}$ is the price we pay for linearising the input stage by running it at a high current, and then bringing its transconductance down to a useable value by adding linearising local negative feedback. These resistors cannot be reduced, for if the h.f. negative-feed-back factor is then to remain constant, C_{dom} would have to be proportionally increaseed, with a consequent reduction in slew rate. Used with the original negative feedback network, these resistors degrade the noise performance by 1.7 dB. Like all the other noise measurements given here, this figure assumes a 50 Ω external source resistance.

If we cannot alter the input degeneration resistors, then the only course left is the reduction of the feedback network impedance, and this sets off a whole train of consequences. If R_8 is reduced to 2.2 kΩ, then R_9 becomes 110 Ω, and this reduces noise output from −93.5 dBu to −95.4 dBu. Note that if $R_{2,3}$ were not present, the respective figures would be −95.2 and −98.2 dBu. However, R_1

must also be reduced to 2.2 kΩ to maintain d.c. balance, and this is too low an input impedance for direct connection to the outside world.

If we accept that the basic amplifier will have a low input impedance, there are two ways to deal with it. The simplest is to decide that a balanced line input is essential; this puts an opamp stage before the amplifier proper, buffers the low input impedance, and can provide a fixed source impedance to allow the high and low-frequency bandwidths to be properly defined by an RC network using non-electrolytic capacitors. The common practice of slapping an RC network on an unbuffered amplifier input must be roundly condemned as the source impedance is unknown, and so therefore is the roll-off point. A major stumbling block for subjectivist reviewing, one would have thought.

The other approach is to have a low resistance d.c. path at the input but maintain a high a.c. impedance; in other words to use the fine old practice of input bootstrapping. Now this requires a low-impedance unity-gain-with-respect-to-input point to drive the bootstrap capacitor, and the only one available is at the amplifier inverting input, i.e. the base of Tr_3. While this node has historically been used for the purpose of input bootstrapping[6] it has only been done with simple circuitry employing very low feedback factors.

There is good reason to fear that any monkey business with the feedback point, at Tr_3's base, will add shunt capacitance, creating a feedback pole that will degrade h.f. stability. There is also the awkward question of what will happen if the input is left open-circuit.

Figure 32.8 shows how the input can be safely bootstrapped.

The total d.c. resistance of R_1 and R_{boot} equals R_8, and their centre point is driven by C_{boot}. Connecting C_{boot} directly to the feedback point did not produce gross instability, but it did seem to increase susceptibility to sporadic parasitic oscillation. Resistor R_{iso} was added to isolate the feedback point from stray capacitance: this seemed to effect a complete cure.

The input could be left open-circuit without any apparent ill-effects, though this is not exactly good practice if loud-speakers are connected. A value for R_{iso} of 220 Ω increases the input impedance to 7.5 kΩ, and 100 Ω raises it to 13.3 kΩ, safely above the 10 kΩ standard value for a bridging impedance. Despite successful tests, I must admit to a few lingering doubts about the high-frequency stability of this approach, and it might be as well to consider it as experimental until more experience is gained.

Another consequence of a low-impedance negative feedback network is the need for feedback capacitor C_2

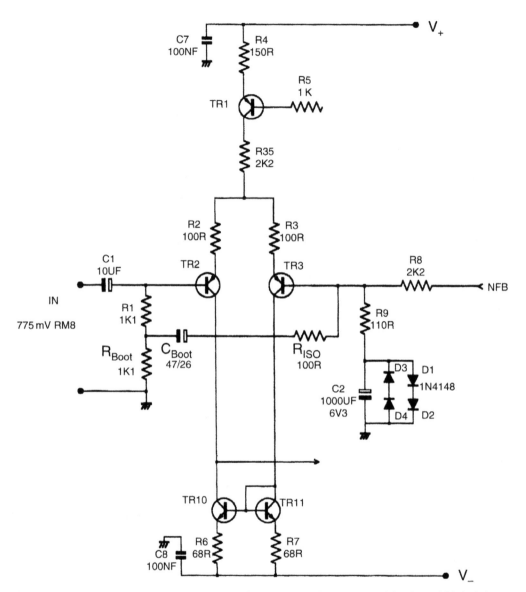

Figure 32.8 Method used for input bootstrapping from the feedback point. Riso is essential for dependable high-frequency stability; with it set to 100 Ω, input impedance is 13 kΩ.

to be proportionally increased to maintain the low-frequency response, and prevent capacitor distortion from causing a rise in THD at low frequencies; it is the latter constraint that determines the value. This is a separate distortion mechanism from the seven previously considered, and I think deserves the title Distortion 8. This criterion gives a value of 1000 μF, which necessitates a low rated voltage such as 6.3 V if the component is to be

of reasonable size. As a result, C_2 now needs protective shunt diodes in both directions, because if the amplifier fails it may saturate in either direction.

Close examination of the distortion residual shows that the onset of conduction of back-to-back diodes will cause a minor increase in THD at 10 Hz, from less than 0.001% to 0.002%, even at the low power of 20 W/8 Ω. It is not my practice to tolerate such gross non-linearity, and therefore four diodes are used in the final circuit, and this eliminates the distortion effect, Figure 32.8. It could be argued that a possible reverse-bias of 1.2 V does not protect C_2 very well, but at least there will be no explosion.

We can now consider alternative input devices to the *MPSA56,* which was never intended as a low-noise device. Several high-beta low-noise types such as *2SA970* give an improvement of about 1.8 dB with the low-impedance negative feedback network. Specialised low-R_b devices like *2SB737* give little further advantage—possibly 0.1 dB—and it is probably better to go for one of the high-beta types; the reason why will soon emerge.

It could be argued that the complications of a low-impedance negative feedback network are a high price to pay for a noise reduction of some 2 dB; however, there is a countervailing advantage, for the above negative feedback network modification significantly improves the output d.c. offset performance. The second and final part of this article shows how, and also gives full details of the mode-switching and bias control systems, and the performance of the complete amplifier.

References

1. Self, D. 'Distortion in power amplifiers', Part 8, *Electronics World & Wireless World,* March 1994, p 225.

2. Thagard, N. 'Build a 100 W Class-A Mono Amp', *Audio,* January 1995, p 43.

3. Self, D. 'Distortion in power amplifiers', Part 7, *Electronics World & Wireless World,* February 1994, p 137.

4. Self, D. 'Distortion in power amplifiers', Part 2, *Electronics World & Wireless World,* September 1993, p 736.

5. Self, D. 'A Precision Preamplifier', *Wireless World,* October 1983, p 31.

6. Mullard Ltd, *Transistor Audio & Radio Circuits,* Mullard Ltd. 1972, second edn., p 122.

Chapter 33

Trimodal audio power, Part II

(Electronics World, July 1995)

Intended title: *A Trimodal Power Amplifier: Part 2*

The Trimodal project created a lot of new information and required some relatively subtle concepts to be explained, so two quite lengthy articles in *Electronics World* were needed.

I must confess I have not yet got around to implementing the idea of the Adaptive Trimodal Amplifier, which would switch from Class-A to Class-B if the heatsink got too hot. It should be a straightforward bit of electronics using a temperature sensor, very possibly the LM35, and a latching comparator. A different approach would be to control the mode on the basis of the ambient temperature, so Class-A would supplement the central heating in winter, while Class-B would provide negligible dissipation in summer.

There were a few rough edges on the original article that I will now sand down:

In the section 'Class A/AB mode' I mention 0.22Ω emitter resistors. This is a mistake; the emitter resistors in the Trimodal are 0.1Ω, and this is important because it is the reason that the Class-AB mode is unusually linear. In fact, 0.6V, as referred to in the text, is no longer an appropriate voltage drop across two 0.1Ω emitter resistors, as that would set up a quiescent current of 3 Amps, which is twice what is required.

In the section 'The complete circuit', there is a reference to the generic/Lin amplifier configuration. Since the Lin design is frequently referred to in the literature, but its circuitry apparently never looked at, it is worth pointing out that what I call a "generic" amplifier is a three-stage configuration with the same general arrangements as a Blameless amplifier. In contrast the Lin is a two-stage amplifier, and its internal workings are quite different. I was clumsily attempting to refer to the basic structure of the output stages of the two amplifier types, which is basically the same. Wherever the Lin amplifier is referred to in these articles, please remember that it is a two-stage amplifier and not a three-stage amplifier like the Blameless configuration.

Also in this section, I say that output emitter resistors of only 0.1Ω are likely to have inadequate stability of quiescent current when used with an EF output stage. Experience has shown that this is absolutely not the case; I have found them perfectly safe with any reasonable size of heatsink.

I have however yet to make any serious trials of emitter resistors any lower in value than 0.1Ω. Such resistors promise greater efficiency and lower distortion, but quiescent stability will have to be assessed very carefully.

In July 1995, *Electronics World* were advertising a stereo pair of Trimodal amplifier PCBs for £49.48.

The Trimodal design is still very much in production at The Signal Transfer Company, with very few changes. This is pleasing as I was initially rather nervous about having quite complicated negative-feedback quiescent-control circuitry moving up and down with the output, at up to 50 kHz under test conditions. However, this approach proved totally docile and has never given a moment's anxiety. With suitably careful design, more complex circuitry than that could be made to operate in the same way.

PCBs, kits, or complete tested versions of the Trimodal amplifier are available from the Signal Transfer Company [1].

Reference

1. http://www.signaltransfer.freeuk.com/ (accessed Feb. 2015)

TRIMODAL AUDIO POWER, PART II

July 1995

The same components that dominate amplifier noise performance also determine the output d.c. offset; if R_9 is reduced to minimise the source resistance seen by Tr_3, then the value of R_8 is scaled to preserve the same closed-loop gain, and this reduces the voltage drops caused by input transistor base currents.

My previous amplifier designs assumed that a ± 50 mV output d.c. offset is acceptable. This allowed d.c. trimming, offset servos, etc. to be gratefully dispensed with. However, it is not in my nature to leave well enough alone, and it could be argued that ± 50 mV is on the high side for a top-flight amplifier. For this reason, I have reduced this range as much as possible without resorting to a servo; the required changes were already made when impedance of the feedback network was reduced to minimise Johnson noise. There were details on this in Part I of this article.

With the usual range of component values, the d.c. offset is determined not so much by input transistor V_{be} mismatch, which tends to be only 5 mV or so, but more by a second mechanism—imbalance in beta. This causes imbalance of base currents, I_b, drawn thorough input bias resistor R_1 and feedback resistor R_8. Cancellation of the voltage-drops across these components is therefore compromised.

A third source of d.c. offset is non-ideal matching of input degeneration resistors $R_{2,3}$. Here they are 100 Ω, with 300 mV dropped across each, so two 1% components at opposite ends of their tolerance bands could give a maximum offset of 6 mV. In practice, it is unlikely that the error from this source will exceed 2 mV.

There are several ways to reduce d.c. offset. Firstly, a class-a amplifier with a single output pair must be run from modest ht rails, so the requirement for high-V_{ce} input transistors is relaxed. This allows higher beta devices to be used, directly reducing I_b. The *2SA970* devices used in this design have a beta range of 350 to 700, compared with 100 or less for *MPSA06/56*. Note the pinout is *not* the same.

In Chapter 33, we reduced the impedance of the feedback network by a factor of 4.5, and the offset component due to I_b imbalance is reduced by the same ratio. We might therefore hope to keep the d.c. output offset for the improved amplifier to within ± 15 mV without trimming or servos. Using high-beta input devices, the I_b errors did not exceed ± 15 mV for ten sample pairs—*not* all from the same batch—and only three pairs exceeded ± 10 mV. Errors in I_b are now reduced to the same order of magnitude as V_{be} mismatches, and so no great improvement

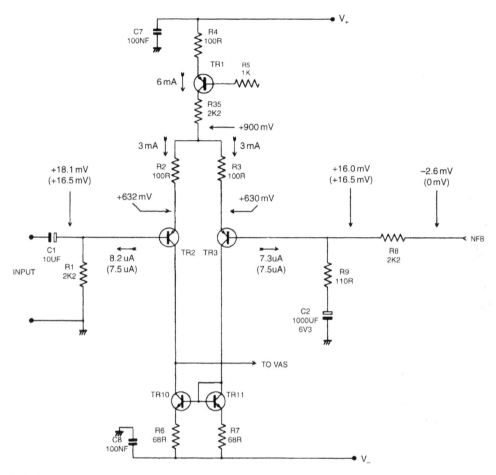

Figure 33.1 A close look at input stage balance. Circuit conditions shown here are a real example. Ideal conditions for β = 400 are shown in brackets. All voltages measured to ground.

can be expected from further reduction of circuit resistances. Drift over time was measured at less than 1 mV, and this seems to be entirely a function of temperature equality in the input pair.

Figure 33.1 shows the ideal d.c. conditions in a perfectly-balanced input stage, assuming a β of 400, compared with a set of real voltages and currents from the prototype amplifier. In the latter case, there is a typical partial cancellation of offsets from the three different mechanisms, resulting in a creditable output offset of −2.6 mV.

Biasing for three modes

Figure 33.2 shows a simplified rendering of the Trimodal biasing system; the full version appears in Figure 33.3. The voltage between points A and B is determined by

one of which can be in command at a time. Since both are basically shunt voltage regulators sitting between A and B, the result is that the lowest voltage wins. The novel Class-A current-controller introduced in the original article [1] is used here adapted for 0.1 Ω emitter resistors, mainly by reducing the reference voltage to 300 mV, which gives a quiescent current (I_q) of 1.5 A when established across the total emitter resistance of 0.2 Ω.

In parallel with the current-controller is the V_{be} multiplier Tr_{13}. In Class-B mode, the current-controller is disabled, and critical biasing for minimal crossover distortion is provided in the usual way by adjusting preset Pr_1 to set the voltage across Tr_{13}. In Class-A/AB mode, the voltage Tr_{13} attempts to establish is increased (by shorting out Pr_1) to a value greater than that required for Class-A. The current-controller therefore takes charge

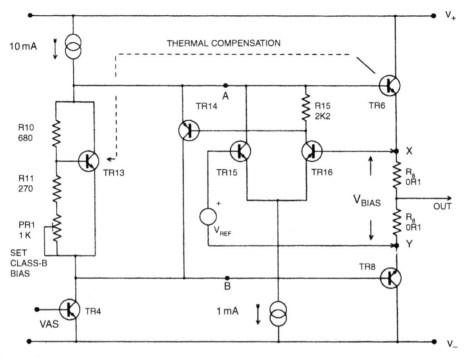

Figure 33.2 Simplified current-controller in action, showing typical d.c. voltages in class-A. Points A, B, X and Y are the same as in the original class-A article. The grey panel on the left is the Vbe multiplier, Class-B biasing and Class-A safety circuit. Panel in the middle is the Class-A current regulator. Voltage over points A, B is 1.5 V while over X, Y, i.e. Vbias, there is 300 mV.

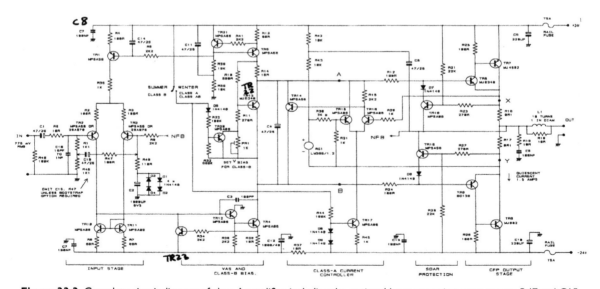

Figure 33.3 Complete circuit diagram of class-A amplifier, including the optional boot-strapping components, R47 and C15.

of the voltage between X and Y, and unless it fails Tr_{13} does not conduct. Points A B X Y are the same circuit nodes as in reference 1.

Class A/AB mode

In Class-A/AB mode, the current-controller, comprising $Tr_{14,15,16}$ in Figure 33.2, is active and Tr_{13} is off, as Tr_{20} has shorted out Pr_1. Transistors $Tr_{15,16}$ form a simple differential amplifier that compares the reference voltage across R_{31} with the V_{bias} voltage across output emitter resistors R_{16} and R_{17}; as explained in Ref. 1, for Class-A this voltage remains constant despite delivery of current into the load. If the voltage across $R_{16,17}$ tends to rise, then Tr_{16} conducts more, turning Tr_{14} more on and reducing the voltage between A and B. $Tr_{14,15,16}$ all move up and down with the amplifier output, and so a tail current-source Tr_{17} is used.

I am aware that the current-controller is more complex than the simple V_{be} multiplier used in most Class-B designs. There is an obvious risk that an assembly error could cause a massive current that would prompt the output devices to lay down their lives to save the rail fuses. The tail-source Tr_{17} is particularly vulnerable because any fault that extinguishes the tail current removes the drive to Tr_{14}, the controller is disabled, and the current in the output stage will be very large. In Figure 33.2 the V_{be}-multiplier Tr_{13} acts as a safety-circuit which limits V_{bias} to about 600 mV rather than the normal 300 mV, even if the current-controller is completely non-functional and Tr_{14} fully off. This gives a 'quiescent' of 3A, and I can testify this is a survivable experience for the output devices in the short-term; however they may eventually fail from overheating if the condition is allowed to persist.

There are important points about the current-controller. The entire tail-current for the error-amplifier, determined by Tr_{17}, is syphoned off from the voltage amplifier stage current source Tr_5. This must be taken into account when ensuring that the upper output half gets enough drive current.

There must be enough tail current available to turn on Tr_{14}, remembering that most of Tr_{16} collector-current flows through R_{15}, to keep the pair roughly balanced. If you feel moved to alter the voltage-amplifier stage current, remember also that the base current for driver Tr_6 is higher in Class-A than Class-B, so the positive slew-rate is slightly reduced in going from Class-B to A.

I must admit that the details of the voltage reference were rather glossed over in Ref. 1, because space was running out fast. The original amplifier shown last month used a National *LM385/1.2*, its output voltage fixed at 1.223 V nominal; this was reduced to approx 0.6 V by a 1 kΩ/1 kΩ divider.

The circuit also worked well with V_{ref} provided by a silicon diode, 0.6 V being an appropriate bias voltage drop across two 0.22 Ω output emitter resistors. This is simple, and retains the immunity of I_q to heatsink and output device temperatures, but it does sacrifice the total immunity to ambient temperature that a band-gap reference gives.

The *LM385/1.2* is the lowest voltage band-gap reference commonly available; however, the voltages shown in Figure 33.2 reveal a difficulty with the new lower V_{bias} value and the complementary feedback pair stage; points A and Y are now only 960 mV apart, which does not give the reference room to work in if powered from node A, as in the original circuit.

The solution is to power the reference from the positive rail, via $R_{42,43}$. The midpoint of these two resistors is boot-strapped from the amplifier output rail by C_5, keeping the voltage across R_{43} effectively constant. Alternatively, a current-source could be used, but this might reduce positive headroom. Since there is no longer a strict upper limit on the reference voltage, a more easily obtainable 2.56 V device could be used providing R_{30} is suitably increased to 7 kΩ to maintain V_{ref} at 300 mV across R_{31}.

In practice, stability of I_q is very good, staying within 1% for long periods. The most obvious limitation on stability is differential heating of $Tr_{15,16}$ due to the main heatsink. Transistor Tr_{14} should also be sited with this in mind, as heating it will increase its beta and slightly imbalance $Tr_{15,16}$.

Class-B mode

In Class-B mode, the current-controller is disabled, by turning off tail-source Tr_{17} so Tr_{14} is firmly off, and critical biasing for minimal crossover distortion is provided as usual by V_{be}-multiplier Tr_{13}. With 0.1 Ω emitter resistors V_{bias} (between X and Y) is approx 10 mV. I would emphasise that in Class-B this design, if constructed correctly, will be as 'blameless' as a purpose-built Class-B amplifier. No compromises have been made in adding the mode-switching.

As in the previous Class-B design, the addition of R_{14} to the V_{be}-multiplier compensates against drift of the voltage amplifier stage current-source Tr_5. To make an old but much-neglected point, the preset potentiometer should always be in the bottom arm of the V_{be} divider $R_{10,11}$ because when presets fail it is usually by the wiper going open; in the bottom arm this gives minimum bias voltage, but in the upper arm it would give maximum.

In Class-B, temperature compensation for changes in driver dissipation remains vital. Thermal runaway with the complementary feedback pair is most unlikely, but accurate quiescent setting is the only away to minimise cross-over distortion. Tr_{13} is therefore mounted on the same small heatsink as driver Tr_6. This is often called thermal feedback, but it is no such thing as Tr_{13} in no way controls the temperature of Tr_6; 'thermal feedforward' would be a more accurate term.

Switching modes

The dual nature of the biasing system means Class-A/Class-B switching is easily implemented, as in Figure 33.3. A Class-A amplifier is an uneasy companion in hot weather, and so I was unable to resist the temptation to sub-title the mode switch 'Summer/Winter', by analogy with a car air intake.

Switchover is d.c.-controlled, as it is not desirable to have more signal than necessary running around inside the box, possibly compromising inter-channel crosstalk. In Class-A/AB mode, S_1 is closed, so Tr_{17} is biased normally by $D_{5,6}$, and Tr_{20} is held on via R_{33}, shorting out present Pr_1 and setting Tr_{13} to safety mode, maintaining a maximum V_{bias} limit of 600 mV. For Class-B, S_1 is opened, turning off Tr_{17} and therefore $Tr_{15,16}$ and Tr_{14}. Transistor Tr_{20} also ceases to conduct, protected against reverse-bias by D_9, and reduces the voltage set by Tr_{13} to a suitable level for Class-B. The two control pins of a stereo amplifier can be connected together, and the switching performed with a single-pole switch, without interaction or increased crosstalk.

Mode-switching affects the current flowing in the output devices, but the output voltage is controlled by the global feedback loop, and switching is completely silent in operation. The mode is switchable while the amplifier is handling audio, allowing some interesting 'A/B' listening tests.

It may be questioned why it is necessary to explicitly disable the current-controller in Class-B; Tr_{13} is establishing a lower voltage than the current-controller which latter subsystem will therefore turn Tr_{14} off as it strives futilely to increase V_{bias}. This is true for 8 Ω loads, but 4 Ω impedances increase the currents flowing in $R_{16,17}$ so they are transiently greater than the Class-A I_q, and the controller will therefore intermittently take control in an attempt to reduce the average current to 1.5A. Disabling the controller by turning off Tr_{17} via R_{44} prevents this.

No warm up

Audio magazines often state that semiconductor amplifiers sound better after hours of warm-up. If this is true—in most cased it almost certainly isn't—the admission represents truly spectacular design incompetence. Accusations of this type are applied with particular venom to class-A designs, because it is obvious that the large heat sinks required take time to reach final temperature. So it is important to record that in class-A operation this design stabilises its electrical operating conditions in less than a second, giving the full intended performance.

No 'warm-up time' beyond this is required.

Obviously the heat sinks take time to reach thermal equilibrium. But as already described, measures have been taken to ensure that component temperature has no significant effect on operating conditions or performance.

Supplying power

Regulated supplies are quite unnecessary, and are virtually certain to do more harm than a good unregulated power supply (Figure 33.4).

The supply must be designed for continuous operation at maximum current, so the bridge rectifier should be properly heat-sunk, and careful consideration given to the ripplecurrent ratings of the reservoirs. This is one reason why reservoir capacitance has been doubled to 20,000 µF per rail: the ripple voltage is halved, improving voltage efficiency as it is the ripple troughs that determine clipping onset. But the ripple current, although unchanged in total value, is now split between two components. (The capacitance was not increased to reduce ripple injection. This is dealt with far more efficiently and economically by making amplifier psrr high.[3])

Do not omit the secondary fuses. Even in these modern times rectifiers do fail, and transformers are horribly expensive.

Test mode

If the Class-A controller is enabled, but preset Pr_1 is left in circuit, (e.g. by shorting Tr_{20} base-emitter) we have a test mode which allows suitably cautious testing; current I_q is zero with the preset fully down, as Tr_{13} overrides the current-controller, but increases steadily as Pr_1 is advanced, until it suddenly locks at the desired quiescent current. If the current-controller is faulty then I_q continues to increase to the defined maximum of 3A.

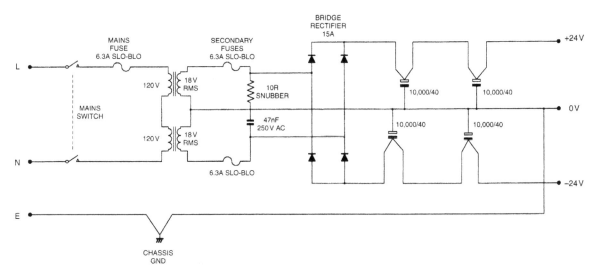

Figure 33.4 Power supply circuit diagram.

Thermal design

Class-A amplifiers are hot almost by definition, and careful thermal design is needed if they are to be reliable, and not take the varnish off the Sheraton. Since the internal dissipation of the amplifier is maximal with no signal, simply turning on the prototype and leaving it to idle for several hours will give an excellent idea of worst-case component temperatures. In Class-B the power dissipation is very programme-dependant, and estimates of actual device temperatures in realistic use are notoriously difficult.

Table 33.1 shows the output power available in the various modes, with typical transformer regulation, etc; the output mode diagram in Part 1, Figure 33.1, showed exactly how the amplifier changes mode from A to AB with decreasing load resistance. Remember that in this context 'high distortion' means 0.002% at 1 kHz. This diagram was produced in the analysis section of *PSpice*

Table 33.1 Power capability of the Trimodal power amplifier

	W	W	W	Distortion
Load resistance (Ω)	8	6	4	
Class A	20	27	15	low
Class AB	n/a	n/a	39	high
Class B	21	28	39	medium

simply by typing in equations, and without actually simulating anything at all.

The most important thermal decision is the size of the heatsink; it is going to be expensive, so there is a powerful incentive to make it no bigger than necessary. I have ruled out fan cooling as it tends to make concern for ultra-low electrical noise look rather foolish; let us rather spend the cost of the fan on extra cooling fins and convect in ghostly silence. The exact thermal design calculations are simple but tedious, with many parameters to enter; the perfect job for a spreadsheet. The final answer is the margin between the predicted junction temperatures and the rated maximum.

Once power output and impedance range is decided, the heatsink thermal resistance to ambient is the main variable to manipulate; and this is a compromise between coolness and cost, for high junction temperatures always reduce semi-conductor reliability, Table 33.2.

This shows that the transistor junctions will be 80 °C above ambient, i.e. at around 100 °C; the rated junction maximum is 200 °C, but it isn't wise to get anywhere close to this very real limit. Note the *Case-Sink* thermal washers are made from high-efficiency material. Standard versions have a slightly higher thermal resistance.

The heatsinks used in the prototype had a thermal resistance of 0.65 °C/W per channel. This is a substantial piece of metal, and is expensive.

Table 33.2 Temperature considerations

	Thermal resist °C/W	heat flow W	temp rise °C	temp °C
Juncn to to 3 case	0.7	36	25	100 junction
Case to sink	0.23	36	8	75 TO3 case
Sink to air	0.65	72	47	67 heatsink
Total			80	20 ambient

The complete circuit

The complete Class-A amplifier is shown in Figure 33.3, complete with optional input bootstrapping but omitting any balanced-line input amplifier or gain control. The circuitry may look a little complex at first, but we have only added four low-cost transistors to realise a high-accuracy Class-A quiescent controller, and one more to implement mode-switching. Since the biasing system has been described above, only the remaining amplifier subsystems are dealt with here.

The input stage follows my design methodology in running at a high tail current to maximise transconductance, and then linearizing it by adding input degeneration resistors $R_{2,3}$. These reduce the final transconductance to a suitable level. Current-mirror $Tr_{10,11}$ forces the collector currents of the two input devices $Tr_{2,3}$ to be equal, balancing the input stage to prevent the generation of second-harmonic distortion. The mirror is degenerated by $R_{6,7}$ to eliminate the effects of V_{be} mismatches in $Tr_{10,11}$.

With some misgivings I added the input network R_9, C_{15}, which is definitely *not* intended to define the system bandwidth, unless fed from a buffer stage; with practical values the h.f. roll off could vary widely with the source impedance driving the amplifier. It is intended rather to give the possibility of dealing with rf interference without having to cut tracks. Resistor R_9 could be increased for bandwidth definition if the source impedance is known, fixed, and taken into account when choosing R_9; bear in mind that any value over 47 Ω will measurably degrade the noise performance. The values given roll off above 150 MHz to keep out uhf.

As a result of insights gained while studying the slewing behaviour of the generic/Lin configuration, I have increased the input-stage tail current from 4 to 6 mA, and increased the voltage amplifier stage standing current from 6 to 10 mA over the original circuit. This increases the maximum positive and negative slew rates from the basic +21, –48 V/μs of reference 4 to +37, –52 V/μs; as described elsewhere [2] this amplifier architecture is always assymetrical in slew rate. One

reason is feedthrough in the voltage amplifier current source; in the original circuit an unexpected slew rate limit was set by fast edges coupling through the current source c-b capacitance to reduce the bias voltage during positive slewing. This effect is minimised here by using the negative-feedback type of current source bias generator, with voltage amplifier collector current chosen as the controlled variable.

Transistor Tr_{21} senses the voltage across R_{13}, and if it attempts to exceed V_{be}, turns on further to pull up the bases of Tr_1 and Tr_5. Capacitor C_{11} filters the d.c. supply to this circuit and prevents ripple injection from the positive rail. Capacitor C_{14}, with R_5, provides decoupling. Increasing input tail-current also mildly improves input-stage linearity, as it raises the basic transistor g_m and allows $R_{2,3}$ to apply more local feedback.

The voltage amplifier stage is linearised by beta-enhancing stage Tr_{12}, which increases the amount of local feedback through Miller dominant-pole capacitor C_3, often referred to as C_{dom}. Resistor R_{36} has been increased to 2.2 kΩ to minimise power dissipation, as there seems to be no significant effect on linearity or slewing. Do not, however, attempt to omit it altogether, or linearity *will* be affected and slewing much compromised.

As described in Ref. 3, the simplest way to prevent ripple from entering the voltage amplifier via the negative rail is old-fashioned RC decoupling, with a small R and a big C. We have some 200 mV in hand (Chapter 26) in the negative direction, compared with the positive, and expending this as the voltage-drop through the RC decoupling will give symmetrical clipping. R_{37} and C_{12} perform this function; the low rail voltages in this design allow the 1000 μF capacitor C_{12} to be a fairly compact component.

The output stage is of the complementary feedback pair (CFP) type. As described in Chapter 26, this gives the best linearity and quiescent stability, due to the two local negative feedback loops around driver and output device. Quiescent stability is particularly important with $R_{16,17}$ as low as 0.1 Ω, and this low value would probably be rather dicey in a double emitter-follower output stage.

Voltage efficiency of the complementary feedback pair is also higher than the emitter follower version. Resistor $R_{25,26}$ define a suitable quiescent collector current for the drivers $Tr_{6,8}$, and pull charge carriers from the output device bases when they are turning off. The lower driver is now a *BD136;* this has a higher f_T than the *MJE350,* and seems to be more immune to odd parasitics at negative clipping.

The new lower values for the output emitter resistors $R_{16,17}$ halve the distortion in Class-AB. This is equally effective when in Class-A with too low a load impedance, or in Class-B but with I_q maladjusted too high. It is now true in the latter case that too much I_q really is better than too little—but not much better, and AB still comes a poor third in linearity to Classes A and B.

An adaptive Trimodal design?

One interesting extension of the ideas presented here is the adaptive Trimodal amplifier. This would switch into class-B on detecting device or heat-sink over-temperature, and would be a unique example of an amplifier that changed mode to suit the operating conditions.

Thermal protection would need to be latching as flipping from class-A to class-B every few minutes would subject the output devices to unnecessary thermal cycling.

Safe operating area protection is given by the networks around $Tr_{18,19}$. This is a single-slope safe operating area system that is simpler than two-slope safe area, and therefore somewhat less efficient in terms of snuggling the limiting characteristic up to the true safe operating area of the output transistor. However, in this application, with low rail voltages, maximum utilisation of the transistor safe area is not really an issue; the important thing is to observe maximum junction temperatures in the A/AB mode.

The global negative-feedback factor is 32 dB at 20 kHz, and this should give a good margin of safety against Nyquist-type oscillation. Global negative feedback increases at 6 dB/octave with decreasing frequency to a plateau of around 64 dB, the corner being at a rather ill-defined 300 Hz; this is then maintained down to 10 Hz. It is fortunate that magnitude and frequency here are non-critical, as they depend on transistor beta and other doubtful parameters.

Performance

The performance of a properly-designed Class-A amplifier challenges the ability of even the Audio Precision measurement system. To give some perspective on this, Figure 33.5 shows the distortion of the AP oscillator driving the analyser section directly for various bandwidths. There appear to be internal mode changes at 2 kHz and 20 kHz, causing step increases in oscillator distortion content; these are just visible in the THD plots for Class-A mode.

Figure 33.6 shows Class-B distortion for 20 W into 8 and 4 Ω, while Figure 33.7 shows the same in Class-A/AB.

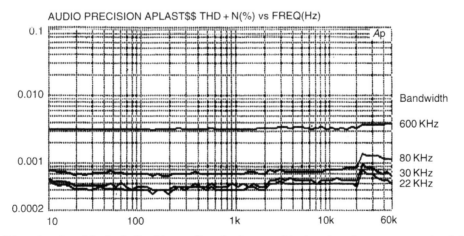

AUDIO PRECISION APLAST$$ THD + N(%) vs FREQ(Hz)

Figure 33.5 Distortion plot of the Audio Precision oscillator/analyser combination alone, for measurement bandwidths of 500, 80, 30 and 22 kHz. The saw-teeth below 1 kHz are artifacts. The residual appears to be pure noise.

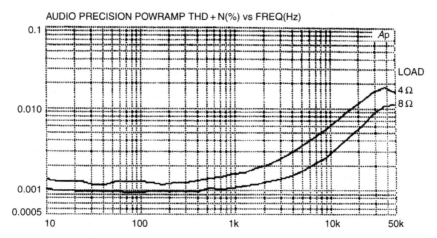

Figure 33.6 Distortion in class-B (summer) mode. Distortion into 4 Ω is always worse. Power was 20 W in 8 Ω and 40 W in 4 Ω, bandwidth 80 kHz.

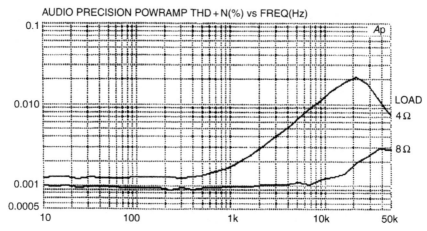

Figure 33.7 Distortion in class-A/AB (winter) mode, same power and bandwidth. The amplifier is in AB mode for the 4 Ω case, and so distortion is higher than for class-BΩ. At 80 kHz bandwidth, the class-A plot below 10 kHz merely shows the noise floor.

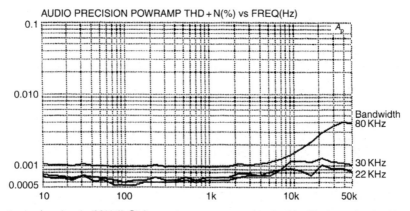

Figure 33.8 Distortion in class-A only (20 W/8 Ω) for varying measurement bandwidths. The lower bandwidths ignore h.f. distortion, but give a much clearer view of the excellent linearity below 10 kHz.

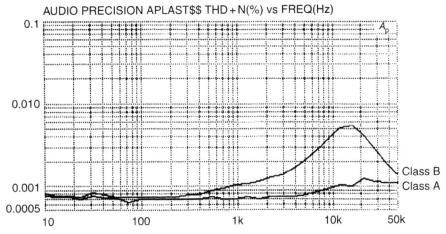

Figure 33.9 Direct comparison of classes A and B (20 W/8 Ω) at 30 kHz bandwidth. The h.f. rise for B is due to the inability of negative feedback that falls with frequency to linearise the high-order crossover distortion in the output stage.

I would like to acknowledge the invaluable help and encouragement of Gareth Connor. Credit goes to him for the tricky task of PCB layout—and not me, as previous adverts have implied.

References

1. Self, D. 'Distortion in power amplifiers; part 8', *Electronics World & Wireless World*, March 94, p 225.

2. Self, D. 'High speed audio power', *Electronics World & Wireless World*, September 1994, p 760.

3. Self, D. 'Off the rails', *Electronics World & Wireless World*, March 1995, p 201.

4. Self, D. 'Distortion in power amplifiers; part 7', *Electronics World & Wireless World*, February 1994, p 139.

Erratum

Regrettably, a couple of errors crept into the original article on Class-A.[1] On page 229, second column: '$Tr_{15,16}$ then compares the reference voltage with that at point Y' should read 'at point X'. On page 229, third column: 'This comes to the same thing as maintaining a constant V_{bias} across Tr_5' should read 'across Tr_{13}'. This is nobody's fault but mine, and I humbly apologise as it cannot have made understanding the current-controller action any easier. (Author's note: these corrections have been made in this edition.)

Chapter 34

Load-Invariant audio power

(Electronics World, January 1997)

Intended title: *A Load-Invariant Power Amplifier*

It has been my experience that power amplifiers always, without exception, give more distortion with heavier loading, i.e., with a lower load impedance. This article looked into why that happens and tried to fix it.

To see if this extra distortion could be eliminated, or at any rate reduced with respect to its 'normal' levels, I conducted a small research program, and this chapter was the result. It was not possible to make the amplifier totally Load-Invariant, i.e., with the same THD at 4Ω as 8Ω, but I think I got pretty close.

In Table 34.1, I tried to categorise the named distortion mechanisms, which at this point in time numbered eight, into those inherent in the design—in the sense that every input stage will have *some* non-linearity—and those which could be utterly eliminated by avoiding errors. Avoiding Distortions 5 and 7 is very simple when you know what they are; it's simply a matter of making connections to the right point. In contrast, Distortion 6 (induction from supply rail currents) is more of a challenge. It would be nice to give simple and foolproof design rules for avoiding it, but given the variations in the mechanical topology of power amplifiers, this is hard to do.

As explained in much more detail in the Preface to Chapter 31, the list of eight distortions in Table 34.1 has grown to twelve in the intervening years, by the addition of:

Distortion 9 Magnetic distortion. It's hard to call this either inherent or topological—it's certainly not inherent because ferrous metal connections are not inherent in amplifiers. Nor is it topological, as it's not a matter of the routing of connections; it's a matter of what the connections are made of.

Distortion 10 Input current distortion. This one is inherent insofar as input currents will always be drawn, but there is nothing inherent about the significant source impedance that must also be present for this distortion to occur.

Distortion 11 Premature overload protection distortion. Likewise, it's hard to call this either inherent or topological. I am concluding that trying to make this a binary distinction is not very useful.

Distortion 12 Thermal distortion in upper feedback resistor. This could be called inherent, in
   that presumably no resistor in existence has a temperature coefficient which is absolutely and
   exactly zero. It is certainly not a topological distortion.

In the caption to Figure 34.1 of the article, I should have specified that a CFP output stage was used
in the test amplifier. This could be deduced from the 8Ω distortion at 1 kHz, which is about 0.003%.
With an EF stage, the figure would be more like 0.006%. The CFP stage is more linear than the EF
for a single pair of output devices; with more output pairs the situation reverses.

The section 'Feeding forward' refers to putting power diodes across the output emitter resistors to
counteract loss of output stage gain at currents, and so reduce Large Signal Nonlinearity (LSN),
which becomes significant with sub-8Ω loads. The technique was initially described in Part 5 of the
original 1993 articles—see Chapter 26 in this book. Recently more research on this has been done,
and a typical result is shown in Preface Figure 34.1. The upper trace is an unmodified EF output
stage driving 8Ω. The lower traces are an identical EF stage driving 4Ω, with and without 20 Amp
Schottky diodes across the 0.1Ω output emitter resistors. Things have gone quite well on the
negative side, the flat gain range being extended by 10 Volts. Things are not so good on the positive
side, where the gain at high output voltages is greater than it is at the central crossover point, and
there is a slope between +25 and +35Volts that will surely generate extra distortion.

The only difference between the negative and positive regions of operation are the transistors, and
these are nominally complementary pairs: MJE340/350 drivers and MJ15022/23 output devices.
This indicates that the differing gain results seen are due to interactions between the Schottky diodes
and quite small differences in the transistor model characteristics. Investigations are continuing into
the details of the models, but if precise matching between the diodes and the transistors is required
for the plan to work, this is not looking good for quantity production.

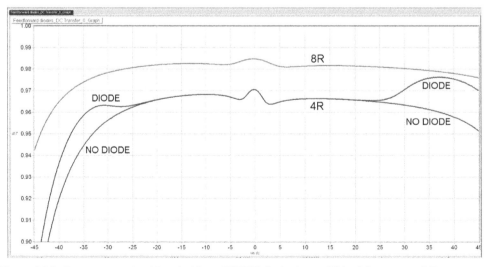

Preface Figure 34.1 Output stage gain with and without 20ETS12 diodes into a 4Ω load (lower traces). Top trace with 8Ω load
and no diodes for reference. Gain scale 0.90–1. ±50V rails.

LOAD-INVARIANT AUDIO POWER

January 1997

My investigations into power amplifiers have so far largely concentrated on 8 Ω resistive loading. This is open to criticism, as loudspeaker impedance dips to 4 Ω or less are not uncommon. Solid-state amplifiers always give more distortion with heavier loading, without exception so far as I am aware.

While it would be highly desirable from the amplifier designer's point of view for the loudspeaker designer to strive for a reasonably flat impedance, it has to be accepted that electronic problems are much easier to solve than electromechanical ones. It follows that it is reasonable for amplifiers to accommodate themselves to loudspeakers rather than the other way around. Thus an amplifier must be able to cope gracefully with impedance dips to 4 Ω or lower.

Such dips tend to be localised in frequency, so music does not often dwell in them. An amplifier should be capable of driving half the nominal load impedance at almost the full voltage swing, though not necessarily for more than a minute or so.

Contemporary power amplifier ratings tend to be presented in the format '*X* watts into 8 Ω, *Y* watts into 4 Ω' from which we presumably may deduce:

- The amplifier will deliver sustained power into 4 Ω.
- Since 2 Ω loads are not explicitly mentioned, they cannot be driven in a sustained fashion.

It may also be assumed, but with much less certainty, that,

- The amplifier will cope with short-term 2 Ω impedance dips; i.e. half the lowest nominal load quoted.
- The overload protection—if it exists at all—activates below 2 Ω. Note that no minimum load impedance is specified.

Output loading and distortion

A 'Blameless' Class-B power amplifier is one wherein all the distortion mechanisms shown in Table 34.1 have been eliminated or reduced to below the noise floor, except for the intractable Distortion 3 in its three subcategories. I have produced a slim monograph which describes the philosophy and practicalities of this in greater detail than *EW* articles permit.[1]

A Blameless design gives a distortion performance into 8 Ω that depends very little on variable transistor characteristics such as beta. This is because at this load impedance the output stage nonlinearity is almost all crossover distortion, which is primarily a voltage-domain effect.

Note that for optimal crossover behaviour the quantity to be set is V_q, the voltage across the two output emitter resistors R_e, and the actual value of the resulting I_q is incidental.[2] Mercifully, in Class-B the same V_q remains optimal whatever the load impedance; if it did not the extra complications would be serious.

Table 34.1 Characteristics of distortion mechanisms.

No.	Mechanism	Category	Component sensitive?
1	Input V_{in}/I_{out} nonlinearity	Inherent	No
2	VAS I_{in}/V_{out} nonlinearity	Inherent	Yes?
3	Output stage distortions:		
	a) Large-signal nonlinearity	Inherent	Yes
	b) Crossover distortion	Inherent	No?
	c) Switch-off distortion	Inherent	Yes
4	Non-linear voltage-amplifier stage loading	Inherent	Yes
5	Rail decouple grounding	Topological	No
6	Rail current induction	Topological	No
7	Error in negative-feedback take-off-point	Topological	No
8	Feedback cap distortion	Inherent	Yes

As the load impedance of a Blameless Class-B amplifier is decreased from infinite to 4 Ω, distortion increases in an intriguing manner. Unloaded, the THD is not much greater than that from the Audio Precision test oscillator, but with loading crossover distortion increases steadily, Figure 34.1.

When the load impedance falls below about 8 Ω, a new distortion begins to appear, overlaid on the existing crossover nonlinearities. It is low-order, and essentially third-harmonic. In Figure 34.2 the upper 4 Ω THD trace is consistently twice that for 8 Ω, once it clears the noise floor.

In Chapter 20, I labelled this as Distortion 3a, or large signal nonlinearity. The word 'large' refers to currents rather than voltages. Unlike crossover distortion 3b, the amount of LSN produced is significantly dependant on device characteristics.[3] The distortion residual is essentially third-order due to the symmetric and compressive nature of the output stage gain characteristic, but its appearance on a scope can be complicated by

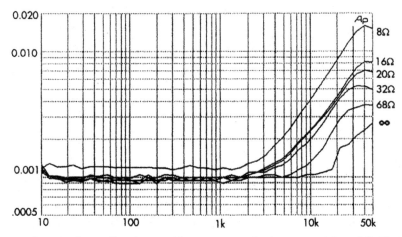

Figure 34.1 Crossover distortion from a Blameless amplifier increases as load resistance falls to 8 Ω. All plots at 80 kHz bandwidth.

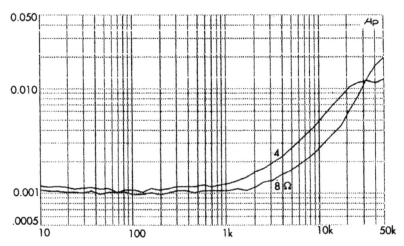

Figure 34.2 Upper trace shows distortion increase due to large-signal nonlinearity as load goes from 8 to 4 Ω. Blameless amplifier at 25 W/8 Ω.

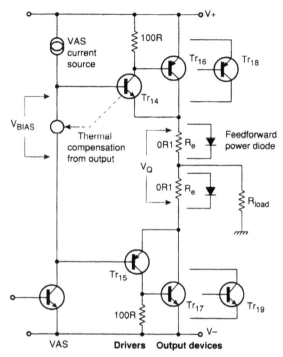

Figure 34.3 Complementary feedback pair output stage, showing how extra devices are added in parallel, and where feedforward diodes would be fitted.

different amounts of nonlinearity in the upper and lower output stage halves.

Large signal nonlinearity occurs in both emitter-follower and complementary feedback pair output configurations; this chapter concentrates on the complementary feedback pair, as in Figure 34.3. Incremental gain of a simulated complementary feedback pair output stage for 8 and 4 Ω is shown in Figure 34.4; the lower 4 Ω trace has greater downward curvature, i.e. a greater fall off of gain with increasing current. Simulated emitter follower behaviour is similar.

As it happens, an 8 Ω nominal impedance is a pretty good match for standard power bipolar junction transistors, though 16 Ω might be better for minimising large-signal nonlinearity—loudspeaker technology permitting. It is presumably coincidental that the 8 Ω nominal impedance corresponds approximately with the heaviest load that can be driven without large signal nonlinearity appearing.

Since large signal nonlinearity is an extra distortion component laid on top of others, and usually dominating them in amplitude, it is obviously simplest to minimise the 8 Ω distortion first, so that 4 Ω effects can be seen more or less in isolation when they appear.

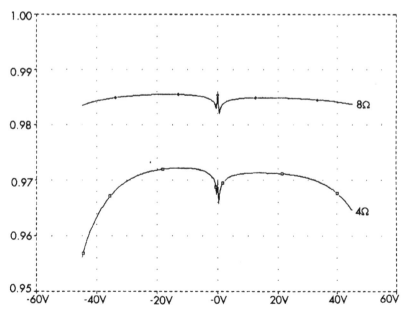

Figure 34.4 Incremental gain of a standard complementary feedback pair output stage. The 4 Ω trace droops much more as the gain falls off at higher currents. (PSpice)

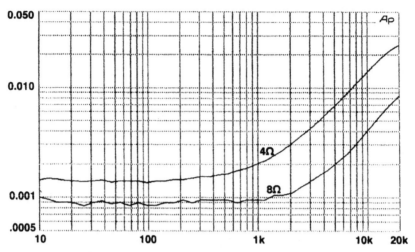

Figure 34.5 Distortion with 4 Ω load is 3 × greater than 8 Ω for 2N3055/2955 output devices. Compare Figure 34.2.

The typical result of 4 Ω amplifier loading was shown in Figure 34.2, for the relatively modern *MJ15024/25* complementary pair from Motorola. Figure 34.5 shows the same for one of the oldest silicon complementary pairs, the *2N3055/2955*, unfortunately on a slightly different frequency scale. The 8 Ω distortion is similar for the different devices, but the 4 Ω THD is 3.0 times worse for the venerable *2N3055/2955.* Such is progress.

Such experiments with different output devices throw useful light on the Blameless concept. From various types tried so far it can be said that Blameless performance, independent of output device type, should not exceed 0.001% at 1 kHz and 0.006% at 10 kHz, into 8 Ω. All the components existed to build sub-0.001% THD amplifiers in mid-1969—if only we had known how to do it.

Low-impedance loads have other implications beyond worsening the THD. The requirements for long-term 4 Ω operation are severe, demanding significantly more heatsinking and power supply capacity if reliability is to be maintained.

For economic reasons the peak-average ratio of music is usually fully exploited, though this can cause real problems on extended tests, such as the FTC 40%-power-for-an-hour preconditioning procedure.

The main subject of this article is the extra distortion generated in the output stage itself by increased loading, but there are other ways in which the total amplifier distortion may be degraded by the increased currents flowing.

Table 34.1 shows the main distortion mechanisms in a power amplifier; Distortions 1, 2, and 8 are unaffected by output stage conditions. Distortion 4 might be expected to increase, as the increased loading on the output stage is reflected in increased voltage amplifier stage loading.[4] However, both the beta-enhanced emitter-follower and buffered-cascode methods seem to cope effectively with sub–8 Ω loads.

The greater supply currents drawn could increase the rail ripple, which will worsen Distortion 5 if it exists. But since the supply reservoir capacitance must also be increased to permit greater power delivery, ripple will be reduced again and this tends to cancel out. If the rail ripple does increase, the usual *RC* filtering of bias supplies [5] deals with it effectively, preventing it getting in via the input pair tail, etc.

Distortion 6 may be more difficult to eliminate as the halfwave currents flowing in the output circuitry are twice as large, with no counteracting mechanism. Distortion 7, if present, will be worsened due the increased load currents flowing in the output stage wiring resistances.

Of those mechanisms above, Distortion 4 is inherent in the circuit configuration—though not a problem in practice—while 5, 6, and 7 are topological, in that they depend on the spatial and geometrical arrangements of components and wiring. The latter three can therefore be completely eliminated in both theory and practice. This leaves us with only the large signal nonlinearity component of Distortion 3 to grapple with.

The Load-Invariant concept

Ideally, the extra distortion component large signal non-linearity would not exist. Such an amplifier would give no more distortion into 4 Ω than 8, and I call it 'Load-Invariant to 4 Ω'. The loading qualification is required because, as you will see, the lower the impedance, the greater the difficulties in aspiring to load-invariance.

I am assuming that we start out with an amplifier that is Blameless at 8 Ω; it would be logical but pointless to apply the term 'Load-Invariant' to an ill-conceived amplifier delivering 1% THD into both 8 and 4 Ω.

Large signal nonlinearity

Large signal nonlinearity is clearly a current-domain effect, dependent on the magnitude of the signal currents flowing in drivers and output devices, as the voltage conditions are unchanged.

A 4 Ω load doubles the output device currents, but this does not in itself generate significant extra distortion. The crucial factor appears to be that the current drawn from the drivers by the output device bases *more* than doubles, due to beta fall-off in the output devices with increasing collector current. It is this *extra* increase of current due to beta-droop that causes almost all the additional distortion.

The exact details of how this works are not completely clear, but seems to be because the 'extra current' due to beta fall-off varies very non linearly with output voltage. It appears that the non linear extra current combines with driver nonlinearity in a particularly pernicious way. Beta-droop is ultimately due to what are called high-level injection effects. These vary with device type, so device characteristics now matter.

As I stated in my original power-amplifier series, [6] there is good simulator evidence that large signal non-linearity is entirely due to the beta-droop causing extra current to be drawn from the drivers. To recapitulate:

- Simulated output stages built from output devices modified to have no beta-droop (by increasing Spice model parameter IKF) have no large signal nonlinearity. It seems to be specifically the extra current taken due to beta-droop that causes the trouble.
- Simulated output devices driven with zero-impedance voltage sources instead of transistor drivers show no large signal nonlinearity. This shows that such non-linearity does not occur in the outputs themselves, but in the driver transistors.
- Output stage distortion can be regarded as an error voltage between input and output. The double

emitter-follower emitter-follower stage error is driver V_{be} + output V_{be} + R_e drop. A simulated emitter-follower output stage with the usual drivers shows that it is primarily nonlinearity in the driver V_{be} that increases as the load resistance reduces, rather than in the output V_{be}. The drop across R_e is essentially linear.

These three results have naturally been rechecked for this article.

Knowing that beta-droop caused by increased output device I_c is at the root of the problem leads to some solutions. Firstly, the per-device I_c can be reduced by using parallel output devices. Alternatively I_c can be left unchanged and output device types selected for the least beta droop.

Feedforward diodes across the emitter resistors sometimes help, but they treat the symptoms—by attempting distortion cancellation—rather the root cause, so it is not surprising this method is much less effective.

Doubled output devices

The basic philosophy here, indicated above, is that the output devices are doubled even though this is quite unnecessary for handling the power output required.

The fall-off of beta depends on collector current. If two output devices are connected in parallel, the collector current divides in two between them, and beta-droop is much reduced. From the above evidence, I predicted that this ought to reduce large-signal nonlinearity and when measured, indeed it does.

This sort of reality-check must never be neglected when you are using simulations. Figure 34.6 compares 4 Ω THD at 60 W for single and doubled output devices, showing that doubling reduces distortion by about 1.9 times; well worthwhile. The output transistors were standard power devices, in this case Motorola *MJ15024/15025*.

The *2N3055/2955* complementary pair give a similar halving of largesignal nonlinearity on being doubled, though the initial distortion is three times higher into 4 Ω. Those *2N3055s* with an H suffix are markedly worse than those without.

No current-sharing precautions were taken when doubling the devices, and this lack seemed to have no effect on large-signal nonlinearity reduction. There was no evidence of current-hogging.

Doubling the power devices naturally increases the power output capability, though if this is fully exploited large-signal nonlinearity will tend to rise again, and you are back where you started. It will also be necessary to

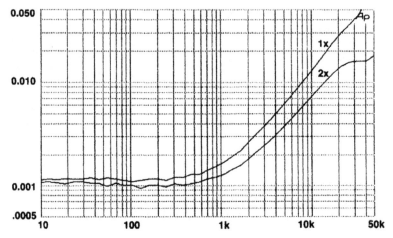

Figure 34.6 Distortion with 4 Ω load is reduced by 1.9 × upon doubling standard MJ15024/15025 output transistors 30 W/8 Ω.

uprate the power supply and so on. The essence of this technique is to use parallel devices to reduce distortion long before power handling alone compels you to do so.

Better output devices

The TO3P-packaged *2SC3281* and *2SA1302* complementary pair has a reputation in the hi-fi world for being 'more linear' than the run of transistors. This is the sort of vague claim that arouses the deepest of suspicions, and is comparable with the many assertions of superior linearity in power fets, which is the exact opposite of reality.[7]

In this case however, the kernel of truth is that the *2SC3281* and *2SA1302* show much less beta-droop than average power transistors. These devices were introduced by Toshiba; Motorola versions are *MJL3281A* and *MJL1302A*, also in TO3P. Figure 34.7 shows beta-droop, for the various devices discussed here, and it is clear that more droop means more large-signal nonlinearity.

The *3281/1302* pair is clearly in a different class from more conventional transistors as regards maintenance of beta with increasing collector current. There seems to be no special name for this class of bipolar junction transistors, so I have called them 'sustained-beta' devices here.

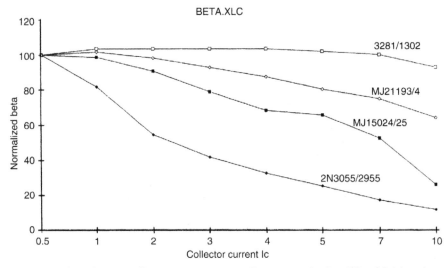

Figure 34.7 Power transistor beta-droop as collector current increases. Beta is normalised to 100 at 0.5 A based on manufacturers' data sheets.

Into 4 and 8 Ω, the THD for single *3281/1302* devices is shown in Figure 34.8. Distortion is reduced by about 1.4 times compared with the standard devices of Figure 34.2, over 2–8 kHz. Several pairs of *3281/1302* have been tested and the 4 Ω improvement is consistent and repeatable.

The obvious next step is to combine the two techniques by using double sustained-beta devices. Doubled device results are shown in Figure 34.9 where the distortion at 80 W/4 Ω (15 kHz) is reduced from 0.009% in Figure 34.8 to 0.0045%—in other words halved. The 8 and 4 Ω traces are now very close, the 4 Ω THD being only 1.2 times higher than the 8 Ω case.

Some similar devices exist. Other devices showing less beta-droop than standard are *MJ21193, MJ21194*, in TO3 packaging, and *MJL21193, MJL21194* in TO3P, also from Motorola. These devices show beta-maintenance intermediate between the 'super' *3281/1302* and 'ordinary' *MJ15024/25*, so it seemed likely that they would give less large-signal nonlinearity than ordinary power devices, but more than the *3281/1302*. This prediction was happily fulfilled.

It could be argued that multiplying output transistors is an expensive way to solve a problem. To give this perspective, in a typical stereo power amplifier, including heatsink, metal work and mains transformer, doubling

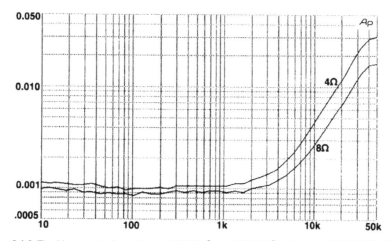

Figure 34.8 Total harmonic distortion at 40 W/8 Ω and 80W/4 Ω with single 3281/1302 devices.

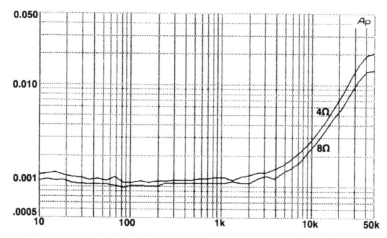

Figure 34.9 At 40 W/8 Ω and 80 W/4 Ω with doubled 3281/1302 output transistors, the total harmonic distortion looks like this. 4 Ω THD has been halved compared with Figure 34.8.

the output devices will only increase the total cost by about 5%.

Feeding forward

In the Distortion in Power Amplifiers series, the only technique I could offer for improving large-signal non-linearity was the use of power diodes across 0.22 Ω output emitter resistors.[8] The improvement was only significant for high power into less-than 3 Ω loading, and was of doubtful utility for hifi.

It is now my practice to make output emitter resistors R_e.0.1 Ω, rather than the more usual 0.22 Ω. This both improves voltage-swing efficiency and reduces the extra distortion generated if the amplifier is erroneously biased into Class AB.[8] Thus even with low-impedance loads the R_e voltage drop is very small, and insufficient to turn on a silicon power diode at realistic output levels.

Schottky diodes have a much lower forward voltage drop and might be useful here. Tests with 50 A diodes have been made but have so far not been encouraging in the distortion reduction achieved. A suitable Schottky diode costs at least as much as an output transistor, and two will be needed.

The trouble with triples

In electronics, there is often a choice between applying brawn—in this case using multiple power devices—or brains to solve a given problem. The 'brains' option would be represented by a clever circuit configuration that gave the same results without replication of expensive power silicon.

The obvious place to start looking is the various output-triple topologies that have occasionally been used. Note that 'output-triples' here refers to pre-driver, driver, and output device all in a local negative-feedback loop, rather than three identical output devices in parallel, which I would call 'tripled outputs'. Nomenclature is a problem.

In simulation, output-triple configurations do indeed reduce the gain-droop that causes large-signal nonlinearity. There are many different ways to configure output-triples. They vary in their general linearity and effectiveness at minimising large-signal nonlinearity.

The real difficulty with this approach is that three transistors in a local loop are very prone to parasitic and local oscillations. This is exacerbated by reducing the load impedances, presumably because the higher collector currents lead to increased device transconductance. This sort of problem can be very hard to deal with, and in some configurations appears almost insoluble. I have not studied this approach further.

Loads below 4 Ω

So far I have concentrated on 4 Ω loads; loudspeaker impedances can sink lower than this, so I pursued the matter down to 3 Ω. One pair of *3281/1302* devices will deliver 50 W into 3 Ω for THD of 0.006% at 10 kHz, Figure 34.10. Two pairs of *3281/1302*s reduce this to 0.003% at 10 kHz, Figure 34.11. This is a very good

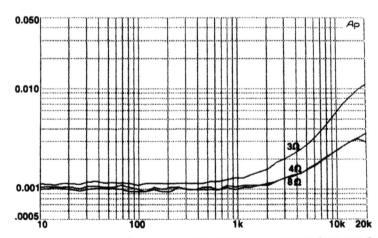

Figure 34.10 Distortion for 3, 4 and 8 Ω loads, single 3281/1302 devices, 20 W/8 Ω, 40 W/4 Ω and 60 W/3 Ω.

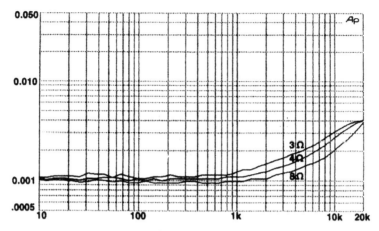

Figure 34.11 Distortion for 3, 4 and 8 Ω load, double 3281/1302 devices. Power as Figure 34.10.

result for such simple circuitry, and may be something of a record for 3 Ω linearity.

At this point it seems that whatever the device type, doubling the outputs halves the THD percentage for 4 Ω loading. The principle can be extended down to 2 Ω operation, but tripled devices are required for sustained operation at significant powers. Resistive losses are serious, so 2 Ω power output may be little greater than that into 4 Ω.

Improved 8 Ω performance

It was wholly unexpected that the sustained-beta devices would also show lower crossover distortion at 8 Ω—but they do. What is more, the effect is again repeatable.

Possibly, whatever improves the beta characteristics has also somewhat altered the turn-on law so that crossover distortion is reduced; alternatively traces of large-signal nonlinearity, not visible in the THD residual, may have been eliminated.

Plot Figure 34.11 shows the improvement over the *MJ15024/25* pair; the 8 Ω THD at 10 kHz is reduced from 0.003% to 0.002%, and with correct bias adjustment, crossover artifacts are simply not visible on the 1 kHz THD residual.

The artifacts are only just visible in the 4 Ω case. To get a feel for the distortion being produced, and to set the bias optimally, it is necessary to test at 5 kHz into 4 Ω.

Implementing the Load-Invariant concept

Figure 34.12 shows the circuit of a practical Load-Invariant amplifier intended for 8 Ω nominal loads

with 4 Ω impedance dips. Its distortion performance is shown in Figures 34.6–11, depending on the output devices fitted.

Apart from load-invariance, this design also incorporates two new techniques from the thermal dynamics series.

The first technique greatly reduces time-lag in the thermal compensation. With a complementary-feedback pair output stage, the bias generator aims to shadow driver junction temperature rather than the outputs. A much faster response to power dissipation changes is obtained by mounting the bias generator transistor Tr_8 on top of driver Tr_{14}, rather than on the other side of the heat-sink. Driver heat-sink mass is thus largely decoupled from the thermal compensation system, speeding up the response by at least two orders of magnitude.[9]

The second new technique is the use of a bias generator with an increased temperature coefficient, to reduce the static errors introduced by thermal losses between the driver and the sensor. Temperature coefficient is increased to −4.0 mV/°C.[10] Diode D_5 also compensates for the effect of ambient temperature changes.

The design is not described in detail because much of it closely follows the Blameless Class-B amplifier described Refs 1 and 11. Some features are derived from the Trimodal amplifier.[8] Most notable of these is the low-noise feedback network, with its requirement for input boot-strapping if a 10 kΩ input impedance is required. Single-slope *VI* limiting is incorporated for overload protection; see $Tr_{12,13}$.

As usual the global negative feedback factor is a modest 30 dB at 20 kHz.

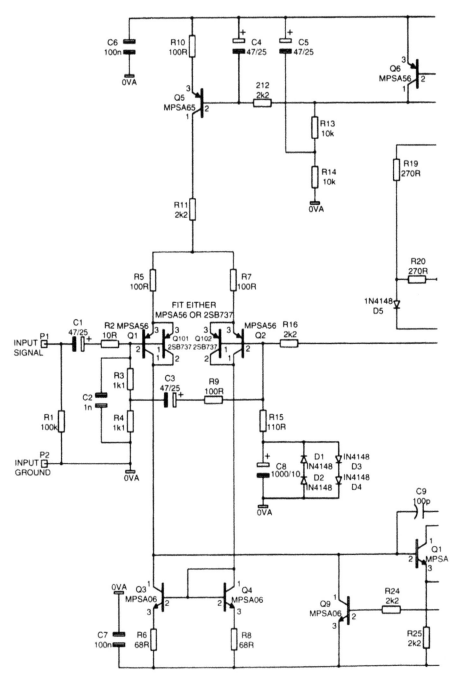

Figure 34.12 This Load-Invariant power amplifier is designed to keep performance constantly high as the loudspeaker impedance rises and falls with frequency. It is intended for 8 Ω nominal loads with 4 Ω impedance dips. Distortion, Figures 34.6–34.11, depends on output devices fitted.

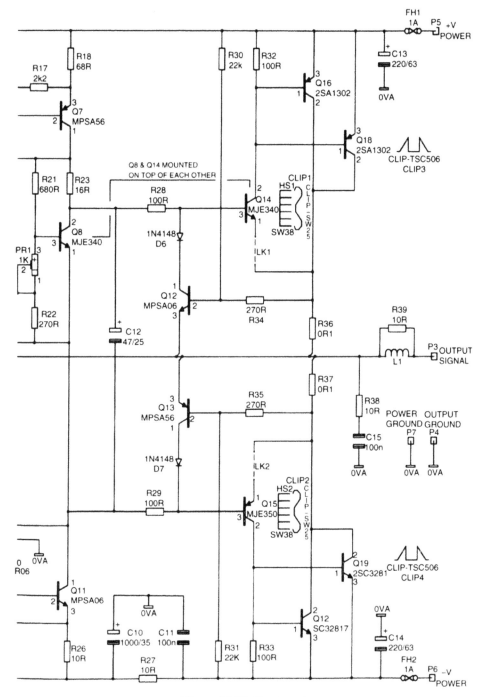

Figure 34.12 (Continued)

A point of departure

The improvements described here fit neatly into the philosophy of Blameless power amplifiers. The fundamental principle of the Blameless concept is that Distortion 3 should be the only significant distortion remaining. Distortions 1, 2 and 4–8 can all be reduced to negligible levels in straightforward ways.

For 8 Ω operation, the main nonlinearity left is crossover distortion, which seems to vary only very slightly with output transistor type.

As I hoped, the concept of a Blameless power amplifier is proving extremely useful as a defined point of departure for new amplifier techniques. Starting from the standard Blameless Class-B amplifier, I have derived:

- The pure Class-A power amplifier [12]
- The Trimodal A/AB/B amplifier [8]
- The Load-Invariant amplifier described here
- A further new design to be announced.

Note that Trimodal and new Load-Invariant amplifier are simple add-ons to the basic Blameless Class-B configuration. The Trimodal design adds a Class-A biasing subsystem, and the new amplifier grafts on extra—or improved—output devices.

In summary

This study is incomplete in that the details of the large-signal nonlinearity mechanism remain incompletely understood, even though several practical methods for reducing it now exist. A detailed mathematical analysis would probably get to the bottom of it, but a foot-long equation usually gives little physical insight.

My initial thoughts were that an amplifier could be considered as Load-Invariant if the rise in THD from 8 Ω to 4 Ω was less than some given ratio. For normal amplifiers the THD increase factor is from two to three times. The actual figure attained by the amplifier presented here is 1.2 times. I, for one, am prepared to classify this as 'Load-Invariant'. The ratio could probably be made even closer to unity by tripling the outputs.

Remember that this amplifier is designed for 8 Ω nominal loads, and their accompanying impedance dips; it is not intended for speakers that start out at 4 Ω nominal and plummet from there. Nonetheless, I hope it is some progress towards load-invariance, and that power amplifier design might have taken another small step forward.

References

1. Self, D. '*Audio Power Amplifier Design Handbook*', Newnes, ISBN 0–7506–2788–3 (The Blameless amplifier concept, and more).

2. Self, D. 'Night thoughts on crossover distortion', *Electronics World*, November 1996, p 858.

3. Self, D. 'Distortion in power amplifiers: part 4', *Electronics World*, November 1993, p 929 (Distortion Number 3a: large-signal nonlinearity).

4. Self, D. 'Distortion in power amplifiers: part 6', *Electronics World*, January 1994, p 43, 44 (Distortion Number 4: VAS loading).

5. Self, D. 'Distortion off the rails', *Electronics World*, March 1995, p 201 (Supply rejection in power amplifiers).

6. Self, D. 'Distortion in power amplifiers: part 5', *Electronics & Wireless World*, December 1993, p 1010 (Large-signal nonlinearity and the Beta-droop effect. Feedforward diodes).

7. Self, D. 'FETs vs BJTs—the linearity competition', *Electronics & Wireless World*, May 1995, p 387 (The poor linearity of power FETs).

8. Self, D. 'Trimodal audio power: part I', *Electronics World*, June 1995, p 465 (Reduction of THD in Class AB by low R_e's).

9. Self, D. 'Thermal dynamics in audio power: part II', *Electronics World*, June 1996, p 484 (Reduction of sensor delay by mounting on driver).

10. Self, D. 'Thermal dynamics in audio power: part III', *Electronics World*, October 1996, p 755 (Bias generator with increased temperature coefficient and ambient compensation).

11. Self, D. 'Distortion in power amplifiers: part 6', *Electronics & Wireless World*, January 1994, p 41 (The Blameless amplifier concept).

12. Self, D. 'Distortion in power amplifiers: part 8', *Electronics & Wireless World*, March 1994, p 229 (A Class-A power amplifier).

Chapter 35

Common-emitter power amplifiers

A different perception?
(Electronics World, July 1994)

Intended title: *Common-Emitter Power Amplifiers*

In those days, articles often appeared in *Electronics World* that provoked serious thought. The sum-of-squares principle of Michael Williams, published in 1994 [1], sounded most intriguing, so I ran a few simulations to see if the linearity it could provide was an improvement over conventionality.

This proved not to be the case, but it did lead to an interesting intellectual journey. As usually happens when I am evaluating an idea, enough work had been done to write an article on the topic, and here it is.

Looking back, there isn't really a lot more I have to say about this article in itself. The notion of synthesising a beautifully linear transfer characteristic by a combination of square laws founders immediately on the fact that there are no accurate square-law devices. FETs are often sloppily described as square-law devices, but their operation does not give a good square-law and can only give a rough square law over a limited range of drain currents. If you did have a magic three-terminal device where current was strictly proportional to the square of the input voltage, when you configured it as some sort of emitter-follower, as is usually done in power amplifier output stages, the 100% voltage feedback would make its operation approximate to linear.

The idea of combining square laws to get a linear result surfaced again in Ian Hegglun's design for *Linear Audio* [2]. The article referenced Bengt Olsson's article in *Electronics World* in 1995, (see Chapter 36) but rather surprisingly, not that of Michael Williams. It is accepted that the power FETs used are not square-law, so it is claimed a square law is approximated by having two pairs of output devices with different bias voltages. The value of the idea is hard to determine because the implementation is just an output stage without negative feedback wrapped round it, and the adoption of a floating-amplifier topology does not make things any clearer. The best distortion result quoted is 0.1% at 10W/8Ω at an unspecified frequency, which it is difficult to get excited about; it is not stated if that figure comes from simulation or measurement.

A later design by Ian Hegglun, which appeared in *Linear Audio* in 2014, uses a cubic law and claims to give Class-A performance at one-quarter of the usual quiescent dissipation [3]. THD figures mentioned include 0.04% at 30W/8Ω, which isn't my idea of what you should get from a Class-A amplifier. This is stated to be 'mostly second-harmonic from the signal generator', which does not help in the evaluation of the idea.

Another approach to curvilinear operation that is worth consulting was published by Wim de Jager and Ed van Tuyl in *Electronics World* in 1999 [4]. There the THD data quoted is an order of magnitude higher than for a Blameless amplifier with standard compensation.

It seems to me that this progression from a square-law to a cube law, if it is susceptible to being made to work linearly, could be extended to quartic (fourth-power) or quintic (fifth-power) laws, still with both output devices conducting all the time. The more steeply rising the curve, the greater reduction in quiescent current there should be. This process can be considered to reach its limit in the optimally biased Class-B amplifier, where the exponential curves around the crossover point are very steep. In this case of course, both output devices are not conducting all the time. See the Preface in Chapter 41 for more on this.

References

1. Williams, M., Square-Law amplifiers, *Electronics World + Wireless World,* Jan. 1994, p. 82.

2. Hegglun, I., Square-law Class-A: a family of efficient Class-A amplifiers, *Linear Audio,* Volume 1, April 2011, p. 6.

3. Hegglun, I., A Cube-law Audio Power Amplifier, *Linear Audio,* Volume 8, Sept. 2014, p. 149.

4. de Jager, W., and van Tuyl, E., A New Class-AB Design, *Electronics World + Wireless World,* Dec. 1994, pp. 988–992.

COMMON-EMITTER POWER AMPLIFIERS

July 1994

When I read Michael William's intriguing article *Making a Linear Difference to Square-Law fets*, [1] I was attracted by the prospect of applying it to an audio power output stage. I found the phrase 'curvilinear class A' particularly appealing.

The basic concept of the difference-of-squares is not new, as several correspondents to *EW + WW* have pointed out.[2,3] Another early reference (1949) to the quarter-squares principle can be found in the monumental MIT Radiation Lab series on radar techniques.

Mr William's basic circuit is shown in Figure 35.1, and the first problem to overcome in applying it for audio power is that the wanted output is the difference of two currents whereas hard-bitten amplifier designers are more used to a low impedance voltage output. Note that with the usual enhancement-mode power fets, if V_1, V_2 are a.c. sources only, and carry no d.c. bias, then V_b will have to establish point M some volts below ground. No doubt something could be done with industrial-sized current-mirrors, but it struck me that the circuit could be rearranged as Figure 35.2, by making use of complementary devices. We now need two bias voltages V_{b1}, V_{b2}, and the positioning of the two signal sources V_1, V_2 on opposite rails looks a little awkward,

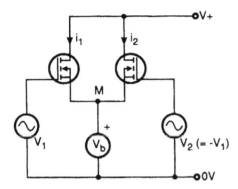

Figure 35.1 The original Williams circuit; the output required is the difference between i_1 and i_2.

but at least the current-difference will be mathematically perfect, if Kirchhoff has anything to say on the matter.

So far so good. We now have a single current output i_{out}. But is this any use for driving loudspeakers? I am assuming that current-drive of speakers is not the final goal; I appreciate that this can be made to work, and promises some tempting advantages in terms of reducing bass-unit distortion.[4] My immediate reaction to Figure 35.2 was no, it can't work, because with a high impedance output, the output stage gain will vary wildly with load impedance making the amount of NFB applied a highly variable quantity. It would also appear that any capacitive loading of this high-impedance node would generate an immediate output pole that would make stable compensation a waking nightmare.

However, just as I was discarding the notion, it occurred to me that the structure in Figure 35.2 looks very much like the bipolar common emitter (CE) stage in Figure 35.3. This is widely used in low voltage opamps because the low saturation voltage allows a close approach to the rails.[5] The more usual emitter follower type of opamp output is usually called a CC or common-collector stage. It is highly probable that the widest application of these voltage-efficient CE configurations is in the headphone amplifiers of personal stereos.

At about the same time I encountered a paper by Cherry [6] which pointed out that, so long as NFB is applied, the output impedance of such a stage can be as low as for the usual voltage follower type output. Cherry's paper is dauntingly mathematical, so I will summarise it thus. The vital point about using NFB to reduce the output impedance of an amplifier is that the amount of NFB applied must be calculated *assuming that the*

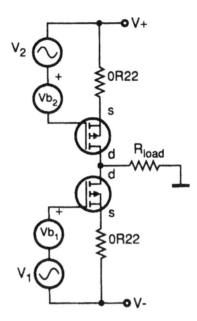

Figure 35.2 The i_1, i_2 subtraction carried out by inverting the polarity of one of the FETs. Two bias voltage generators are now needed.

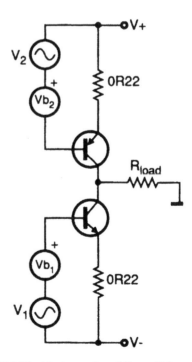

Figure 35.3 The bipolar version of Figure 35.2, as used in many low-voltage opamps and Walkman output amplifiers.

open-loop case is unloaded. This condition looks unfamiliar, because the average amplifier usually has a fairly low output resistance even when open-loop, due to its output follower configuration, and so the loaded/unloaded distinction makes only a negligible difference when calculating the reduction of output resistance by NFB.

Using this condition, Cherry shows that output impedance of a CE stage should be exactly equivalent to the usual CC stage, when the global NFB is applied. I appreciate that this result is counter-intuitive; it looks as though the current output version must have a higher output impedance, even with NFB, but it appears not to be so. Doubters who are unafraid of matrix algebra should consult Cherry's paper.

Topology to the test

Nonetheless, before reaching for the power fets, I felt the need for further reassurance that a CE output stage was workable. There are several low voltage opamps that use the CE output topology, so it seemed instructive to provoke one of these with some output capacitance and see what happens. A suitable candidate is the Analog Devices *AD820*, which has a BJT output stage looking like Figure 35.3 and provides all you need for CE experimentation in one 8-pin package.[7]

My practical findings were that the opamp works well, and while THD may not be up to the very best standards, it was happy with varying load resistances, proved stable with capacitors hung directly on the output, and was relaxed about rail decoupling. Once again, so far, so good.

By this stage, the quarter-squares principle was slipping somewhat into the background. My attention was focusing on the possibilities of a BJT power output stage something like Figure 35.4, which shows the addition of drivers and emitter resistors to make the circuit more practical. A good output swing is facilitated by the inward-facing driver arrangement. In a conventional emitter follower output the need to leave the drivers room to work in further reduces output swing.

Figure 35.4 could be configured into something like a normal Class-B amp, except that the novel use of a CE output stage would allow greater efficiency than usual because there would be the low $V_{ce(sat)}$ drops mentioned above. Also the crossover behaviour would presumably be different from a normal CC output, and quite possibly better, or at least more easily manipulated.

In a previous article [8] I tried to demonstrate that for an amplifier in which all the easily manipulated distortion mechanisms had been suitably dealt with, the low frequency THD was below the noise when driving an 8 Ω load. This without large global feedback factors: 30 dB at 20 kHz is quite adequate.

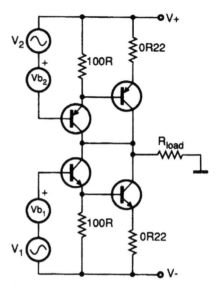

Figure 35.4 A practical circuit based on Figure 35.3. Drivers and emitter-resistors have been added.

At high frequencies (say above 2 kHz) the distortion is easily measurable, and almost all of its results from crossover effects in the output stage. Since NFB typically falls with frequency, these high-order harmonics receive much less linearisation. This is why any technique that promises a reduction in basic crossover nonlinearity is of immediate interest to those concerned with power amplifier design.

I began to think that Mr Williams had opened up a whole new field of audio amplification; each conventional CC output stage would have its dual in CE topology, perhaps with new and exciting characteristics.

The next stage of the investigation was more sobering. There was a familiarity about CE output stages. Readers old enough to recall paying 30 shillings for their first OC72 will recognise Figure 35.5 as the configuration used almost universally for low power audio output for many years when there was no such thing as a complementary device. Transformers provide one way to make a push-pull output. At first sight bias voltage V_b looks as if it will be far too low but bear in mind these are germanium transistors. Note the upside-down format of the circuit which is typical of the period. The circuit values are appropriate for an output of about 500 mW.

While it is perhaps not obvious, this is the equivalent of Figure 35.3. The need for an npn is avoided by using phase inversions in the transformers. So clearly CE output stages were not as rare and specialised as I thought; however they might still have handy distortion

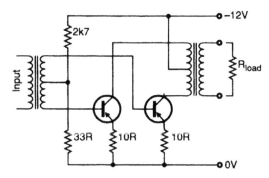

Figure 35.5 A rather old-fashioned CE amplifier: the transformers are expensive but avoid the need for complementary devices.

properties that were not obvious in the long-gone days of transformer coupling.

Adding spice to the investigation

The next step was Spice simulation of the practical BJT output circuit in Figure 35.4: Figure 35.6 shows how the device currents vary in a relationship that looks

ominously like classic Class-B. Somehow I was expecting more overlap of conduction. The linearity results are presented in Figure 35.7 as a plot of incremental gain versus output voltage for varying loads, as in the *Distortion In Power Amplifiers* series.[8]

The first obvious difference is that stage gain, instead of staying close to unity, varies hugely with load impedance—pretty much what we expect from a CE stage operating open-loop. Note that the X-axis is V_1 ($V_2 = -V_1$ to induce push-pull operation) and so represents the input voltage only rather than both input and output as before. Multiplying this input voltage by the gain taken from the Y-axis gives the peak output voltage swing. The vertical gain drop-offs that indicate clipping move inwards with higher load impedances because of the greater output gain rather than through any hidden limitation on output swing.

Figure 35.8 shows the effect of varying the bias, and hence quiescent current, for an 8-Ω load.

This circuit certainly works, but somehow the linearity results seem depressingly familiar. There is the same gain-wobble at crossover we have seen *ad nauseam* with CC output stages, and once again there is no bias setting that removes or significantly smooths it out. As before, the usual falling-with-frequency NFB will not deal with

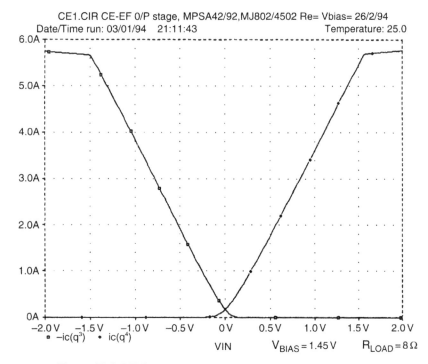

Figure 35.6 BJT Collector currents in Figure 35.4 driving an 8 Ω load.

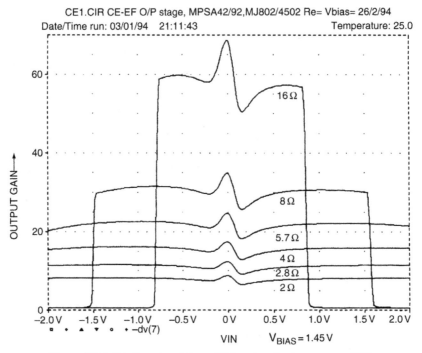

Figure 35.7 Gain linearity of Figure 35.4, various load resistances (BJT).

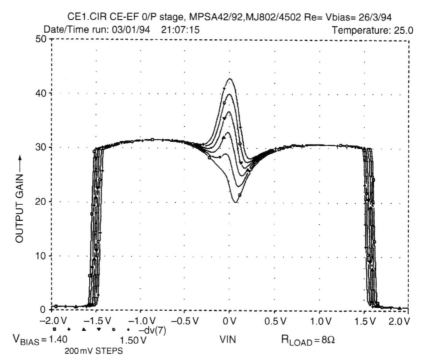

Figure 35.8 Gain linearity of Figure 35.4 for various bias voltages, load is 8 Ω (BJT).

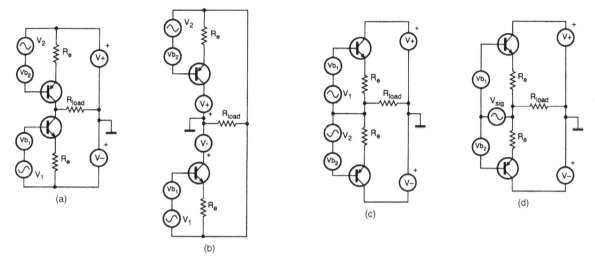

Figure 35.9 Showing how our experimental CE amplifier turns out to be a more or less conventional CC amp when turned inside out.

this sort of high-order distortion very effectively, leading to a rise in THD above the noise in the upper audio band.

In fact, the characteristics look so suspiciously similar to the standard emitter-follower CC stage, that it began to belatedly dawn on me they might actually be the same thing . . .

Figure 35.9 shows the final stages of this conceptual hejira. Figure 35.9(a) shows the simplified circuit of Figure 35.3 with the power supplies V+, Vincluded; they no doubt come from a mains transformer so we can float them at will, and it seems quite in order to pluck them from their present position and put them in the collectors of the output devices instead. All the other supplies shown are equally without ties forming an independent unit with the associated transistor and emitter resistor R_e. Thus they cannot affect device currents. Since there is only one ground reference in the circuit, it is also a legitimate gambit to put it wherever we like, which in this case is now the opposite end of the load R_1. (See Ref. 9 for another example of this manoeuvre). This gives us the unlikely looking but functionally equivalent circuit in Figure 35.9(b).

A purely cosmetic rearrangement of Figure 35.9(b) produces Figure 35.9(c), which is topologically identical, and reveals that the new output stage is . . . a CC stage after all. Figure 35.9(d) shows the standard output.

The only true difference between the 'CE' stage and the traditional CC stage is the arrangement of the two bias voltages V_{b1}, V_{b2}. In a conventional CC stage, the output bases or gates are held apart by a single fixed voltage,

shown here as V_{b1} and V_{b2} connected together. This rigid 'unit' can be regarded as driven with respect to the output rail by the signal source V_{sig}, representing the difference between input and output of the stage. Normally, of course, it is more useful to regard the earlier circuitry as generating a signal voltage with respect to ground.

In contrast to Figure 35.9(d), Figure 35.9(c) has two bias voltage generators, and the consequence of this is that voltage drops in the emitter resistors R_e are not coupled across to the opposite device by the bias voltage. This does not seem to offer immediately any magical stratagems for reducing the gain deviation around crossover, and creates the need for two drive voltages referenced about the output rail. This should be fairly easy to contrive, but is bound to be more complex than the traditional method.

Squaring the circle

Having gone through these manipulations, it is time to reconsider fets and the quarter-squares approach, knowing now that we are dealing with something very close to a standard power-amp configuration. To underline the point. Figure 35.10 shows the gain characteristics for the circuit of Figure 35.2, using *2SK135/2SJ50* power fets. Note the very close resemblance to a conventional source follower.[8]

As Mr Williams points out, the V_{gs}/I_d characteristic curve for power fets may follow a square law at low currents, but it is more or less linear at high ones, and

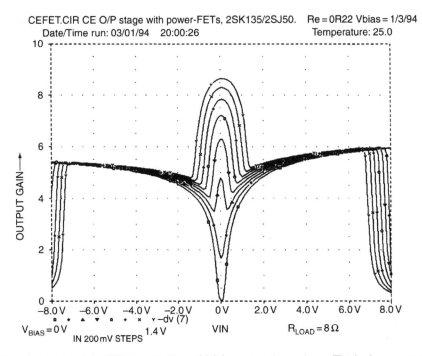

Figure 35.10 The gain linearity of the FET circuit in Figure 35.2 for various bias voltages. This looks very similar to a conventional source-follower output stage.

this appears to rule out any simple approach to 'curvi-linear class A'. For the fets I used, the 'square lawish' region is actually tiny, being roughly between 0 to 80 mA which is of limited use for a power stage. In so far as second-harmonic cancellation occurs at all, it is in the crossover region where, without this effect, the central gain deviations would probably be greater than they are.

As I can see, the quarter-squares concept is already in use in most FET power amplifiers in heavy disguise but only operational in the crossover region. If this idea is to be pursued further, we need a true square-law output device. Since there is no such thing, it would need to be realised by some kind of law-synthesis circuitry. If amplifier distortion needs reducing below the tiny levels possible with relatively conventional techniques, there are probably better avenues to explore.

References

1. Williams, M. 'Making a linear difference to square-law fets', *EW + WW*, January 94, p 82.

2. Brown, S F. Letters, *EW + WW*, March 94, p 247.

3. Owen, D, Letters, *EW + WW*, March 94.

4. Mills and Hawksford, 'Transconductance power amplifier systems, current-driven loudspeakers', *JAES* Vol 37 (10), October 1989.

5. Fonderie and Huijsing, *Design of Low-Voltage Bipolar Operational Amplifiers*, Kluwer Academic Publishers, 1993, Chapter 3.

6. Cherry and Cambrell, '*Output Resistance & Intermodulation Distortion of Feedback Amplifiers*', *JAES*, Vol 30 (4), April 82.

7. Jung and Wong, 'High-performance ICs in single-supply analog circuits', *Analog Dialogue* #27–2 (1993) p 16.

8. Self, D. 'Distortion in power amplifiers: Part 4'. *EW + WW*, November 93, p 932.

9. Baxandall, P. 'Symmetry in Class B', Letter, *Wireless World*, September 1969, p 416.

Chapter **36**

Few compliments for non-complements

(Electronics World, September 1995)

Intended title: *Two-Stage Amplifiers and the Olsson Output*

Another wince-inducing title from the editor of *Electronics World.*

A further article in December 1994 by another author, Mr Bengt Olsson [1], also caused me to begin my own investigations, not least because few performance details were given in the original. The concept might have worked, but it did not, so all I could really do was say so, adding some more positive material on two-stage power amplifiers in general. They turned out to be rather awkward things.

I want to stress that the last thing intended in these 'reactive' articles, like this one and the previous piece on common-emitter power amplifiers, is discourtesy to the original authors, though I doubt if they were very enthusiastic about my investigative activities. Presented with an ingenious idea, anyone with a spark of curiosity would want to find the truth about. Does it really work better than conventional methods? Does it work at all? The Truth has proverbially no respect for persons, and I also understand from Jack Nicholson that some people can't handle it.

It is a sad fact that when people are showcasing what they consider a new and useful idea in power amplifier design, they often cannot resist the temptation to put in more than one innovation at a time, which makes it hard to discern the exact derivation of any improvements that may be seen. I was inclined to think that Mr Olsson was guilty of this in using a two-stage configuration, but as the Olsson output stage inherently includes a device performing a VAS-like role, and so a two-stage amplifier, composed of differential input stage and output stage, results almost automatically. The famous Lin power amplifier [2] was a two-stage design, though with what is now considered a conventional quasi-complementary BJT output stage. The Lin design has very little in common with the Olsson circuit.

It may not be intuitively obvious, but three-stage amplifiers are much more tractable and easier to handle than two-stage designs. The root of the problem is that in a two-stage design, the signal from the input stage to the output stage has to be passed as a voltage, rather than as a current as it is in Blameless three-stage designs. This means that pole-splitting does not occur to improve the input stage frequency response, and it is harder to apply Miller compensation to reduce the output impedance of the 'VAS', not least because it now has two outputs rather than one. This is all fully explained in the article.

It sounds as though three-stage amplifiers should have inevitably more parts than two-stage ones, but this is not necessarily the case, and even if it is so the savings will be in the small-signal circuitry, which makes up only a small percentage of the cost of a power amplifier. The expensive parts are the output devices and the heatsink. If the cost of the power supply, with its transformer, rectifier and reservoir capacitors is included in the reckoning then saving a couple of small-signal transistors is going to have very little effect on the result.

I have no plans at present to invest any more time researching the troubles of two-stage power amplifiers.

Bengt Olsson's original article in *Electronics World* was referenced by Ian Hegglun in his article on square-law Class-A amplification in *Linear Audio* [3].

References

1. Olsson, B., Better audio from non-complements? *Electronics World + Wireless World,* Dec. 1994, pp. 988–992.

2. Lin, H.C., Quasi-Complementary Transistor Amplifier, *Electronics,* Sept. 1956, p. 173ff.

3. Hegglun, I., Square-law Class-A: a family of efficient Class-A amplifiers, *Linear Audio,* Volume 1, April 2011, p. 6.

FEW COMPLIMENTS FOR NON-COMPLEMENTS

September 1995

Bengt Olsson's most interesting article on quasi-complementary FET output stages [1] prompted me to examine how his proposed configuration works. Investigations showed that his scheme changes not just the output stage but the entire structure of the amplifier, and it presents some intriguing new design problems.

An alternative architecture

Nearly all audio amplifiers use the conventional architecture I have analysed previously.[2] There are three stages, the first being a transconductance stage, i.e. differential voltage in/current out, the second is a transimpedance stage i.e. current in/voltage out and lastly a unity-gain output stage, Figure 36.1(a).

Clearly, the second stage has to provide all the voltage gain and is therefore formally named the voltage amplifier stage (VAS). This architecture has several advantages. A main benefit is that it is straightforward to arrange things so that the interaction between stages is negligible. For example, there is very little signal voltage at the input to the second stage, due to its current-input nature. This results in very little voltage on the first stage output, which in turn minimises phase-shift and possible Early effect.

In contrast, the architecture presented by Olsson is a two stage amplifier, Figure 36.1(b). The first stage is once more a transconductance stage, though now without a guaranteed low impedance to accept its output current. The second combines voltage amplifier stage and output stage in one block. It is inherent in this scheme

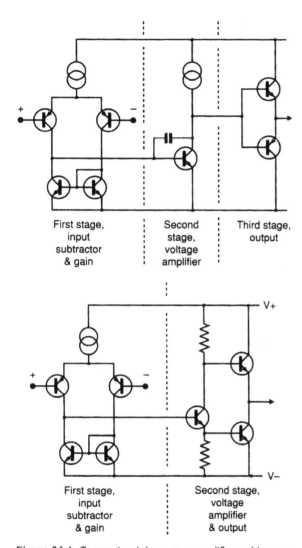

First stage,
input
subtractor
& gain

Second
stage,
voltage
amplifier

Third stage,
output

First stage,
input
subtractor
& gain

Second stage,
voltage
amplifier
& output

Figure 36.1 Conventional three-stage amplifier architecture, and two-stage architecture advocated by Olsson.

has a unity-gain output stage—unless you really want to make life difficult for yourself. As a result the total forward gain is simply the product of the transconductance of the input stage and the transimpedance of the voltage amplifier stage. Transimpedance is determined solely by the Miller capacitor C_{dom}, except at very low frequencies.[3]

Typically, the feedback factor at 20 kHz will be 25–40 dB. It will increase at 6 dB per octave with falling frequency until it reaches the dominant pole frequency P_1, when it flattens out. What matters for the control of distortion is the amount of negative feedback, NFB, available, rather than the open-loop bandwidth, to which it has no direct relationship.

In my *EW + WW* Class-B design, input stage g_m is about 9 mA/V, and C_{dom} is 100 pF, giving a feedback factor of 31 dB at 20 kHz. In other designs I have used as little as 26 dB at the same frequency with good results.

Arranging the compensation of a three-stage amplifier can be relatively simple. Since the pole at the voltage-amplifier stage is already dominant, it can be easily increased to lower the h.f. negative-feedback factor to whatever level is considered safe. The local negative feedback working on the voltage amplifier has an invaluable linearising effect. I am aware that some consider there are better ways to perform this sort of compensation, but the Miller approach is so far the most stable method in my experience.

Fewer stages, more complexity?

Paradoxically, a two-stage amplifier may be more complex in its gain structure than a three-stage. Forward gain depends on the input-stage g_m, the input-stage collector load, and the gain of the output stage, which will be seen to vary in a most unsettling manner with bias and loading. Input-stage collector loading plays a part here since the input stage cannot be assumed to be feeding a virtual earth.

Choosing the compensation is also more complex for a two-stage amplifier. The voltage-amplifier/phase-splitter has a significant signal voltage on its input. Usually, the pole-splitting mechanism enhances Nyquist stability by increasing the pole frequency associated with the input stage collector. But because of the relatively high voltage on the voltage-amplifier/phase-splitter, the pole-splitting mechanism is no longer effective.

This may be why Olsson's circuit uses a cascoded input stage comprising $Tr_{6,7}$ in his original circuit.

that the voltage amplifier must double as a phase-splitter. This results in two dissimilar signal paths to the output, and it is not at all clear that trying to break this block down further will assist a linearity analysis. The use of a phase-splitting stage harks back to valve amplifier days, when it was essential due to the lack of complementary valve technology.

Since the amount of linearising global feedback available depends upon amplifier open-loop gain, the way in which the stages contribute to this is are of great interest. The normal three-stage architecture always

This presents the input device collectors with a low impedance and prevents a significant collector pole. Another valid reason is that it also allows the use of high-beta low-V_{ce} input transistors, which minimise output d.c. offset due to base current mismatch. This is usually much larger than the d.c. offset due to V_{be} mismatch.

Such an input cascode can also improve power-supply rejection as it prevents Early effect from modulating the subtractive action of the input pair.

Simple calculation gives the g_m of Olsson's amplifier as 16 mA/V, but the effective gain of the next stage seems much more difficult to equate. A full *PSpice* simulation of the complete amplifier with an 8 Ω load shows that the feedback factor is 36 dB up to 300 Hz. It then rolls off at the usual 6 dB/octave, until it passes through 0dB at about 20 kHz. This 36 dB represents much less feedback than the three-stage version. It indicates that C_{dom}, notionally connected between drain and gate of M_3, must be comparatively very large at approximately 3 nF.

Specified internal capacitances of M_3 are certainly orders of magnitude larger than those of an equivalent bipolar device—they vary from sample to sample and also with operating conditions such as V_{ds}. These unwanted variations would appear to make stable and reliable compensation a difficult business.

The low-frequency feedback factor is about 6 dB less with a 4 Ω load, due to lower gain in the output stage. However, this variation is much reduced above the dominant pole frequency, as there is then increasing local negative feedback acting in the output stage.

Devices and desires

In his opening paragraph, Olsson says that the symmetry of complementary transistor output stages is theoretical rather than practical. Presumably he is referring only to power-fets, as suitable pairs of bipolar devices, such as Motorola *MJ802/MJ4502*—old favourites of mine—exhibit excellent symmetry.[4] Admittedly, the two devices are not exact mirror-images, but the asymmetries are small enough for even-order harmonic generation in the output stage to be negligible. This is surely what counts.

This symmetry does not hold for power-fets however, and so it may be that some of Olsson's concern with symmetry flows from an initial decision to use fets. I find it difficult to understand why power fets in particular suffer from so many mis-statements. It is still confidently held that fets are more linear than bipolars, although the opposite is certainly the case when the two types of device are used in normal Class-B output stages.

Similarly, FET robustness is often exaggerated, the devices being prone to summary explosion under serious parasitic oscillation. Mercifully bipolars are not. In particular, I find it hard to understand Olsson's contention that FET parameters are predictable—they are notorious for being anything but.

From one manufacturer's data, namely Harris, the V_{gs} for the *IRF240* varies between 2 and 4 V for an I_d of 250 µA—a range of two to one. In contrast the V_{be}/I_c relation in bipolars is fixed by a mathematical equation for a given transistor type. The exponential relationship may be regarded as more non-linear than the partially-square-law FET V_{gs}/I_d relationship but it is dependable, and gives a much higher transconductance. This can always be traded for linearity by introducing local negative feedback.

Output considerations

Figure 36.2 shows the basic output configuration I have investigated—I have not examined the 'anti-saturation' schemes intended to provide the output fets with extra gate drive.

My first discovery is that the voltage-amplifier stage/phase-splitter does not have to be a FET. Replacing it with a bipolar junction transistor, for example *MPSA92*, gives almost identical results. I have used $M_{1,2}$ etc. rather than $Tr_{1,2}$ for FET designations as this preserves consistency with the *PSpice* output.

This output stage configuration is totally different in operation from the conventional Class-B stages discussed in Ref. 1. It is a hybrid common-drain/common-source configuration, or, in bipolar speak, a common-collector/common-emitter (cc–ce) stage. In this sort of output, the upper emitter-follower has a common-emitter active-load. This load may or may not deliver an appropriate current into the node it shares with the upper device.

The input-voltage/output-current relationship for the upper and lower devices will be different, as a result of the two dissimilar paths to the output. This means that while such a stage can always be biased into Class-A by increasing the quiescent enough[5] there is every likelihood that it will be an inefficient kind of Class-A. It will deviate seriously from the constant-sum-of-currents condition that distinguishes classical Class-A.[6]

Lower quiescents tend to give a depressingly non-linear and asymmetrical Class-AB. In general there is no equivalent at all to standard Class-B, where the symmetry of the configuration—rather than anything else—allows both output devices to be biased to the edge of conduction simultaneously.

In general, cc–ce stages generate large amounts of even-order distortion, due to their inherent asymmetry. I appreciate however that avoiding this is one of the prime purposes of Olsson's circuit.

In Figure 36.2, the position of V_{bias} causes a form of bootstrapping, and I can confirm that driving it from the output is essential to make the scheme work.

Figure 36.3 shows drain current in each output FET when driving an 8 Ω load, with V_{bias} stepped, as simulated by *PSpice.* Compared with conventional Class-B,

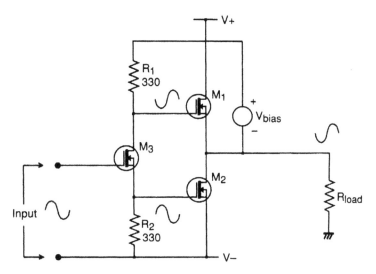

Figure 36.2 Basic FET output stage used for simulations of Figures 36.3, 36.4 and 36.5. Transistors M1,2 are IRF240 and M3 is IRF710.

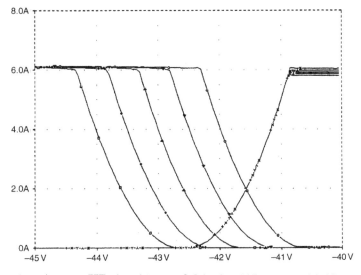

Figure 36.3 Drain current in each output FET when driving an 8 Ω load, with Vbias stepped. As Vbias increases the Id(M1) line moves to the right and overlaps more with the Id(M2) line. In each case symmetry exists about the intersection of the two Id lines. Values for Vbias are 7.5, 8, 8.5, 9 and 9.5 V.

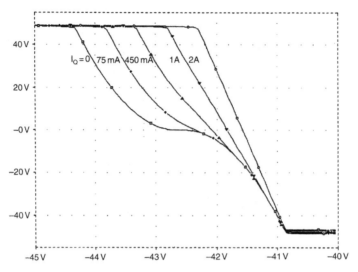

Figure 36.4 Output transfer functions for stepped bias voltage. For lower bias, the characteristic is sigmoidal—S-shaped, unlike a conventional Class-B stage. Values for Vbias are 7.5, 8, 8.5, 9 and 9.5 V.

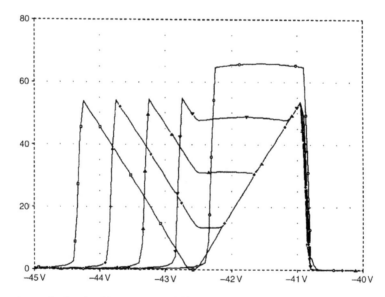

Figure 36.5 Incremental gain plot for the Olsson stage, with Vbias stepped. The—AB mode (curve B) shows serious gain variations and therefore poor linearity.

the drain currents cross over smoothly. This seems intuitively a good idea, and has been recommended by several writers, but in fact a smooth-looking current crossover does not guarantee a linear composite gain characteristic.

Figure 36.4 shows the input/output characteristic for the stage. You can see that the lower V_{bias} levels produce a sigmoidal transfer function, with gain falling off in the crossover region. This gross output distortion

is much greater than that given by a normal Class-AB stage. It suggests that it is only practical to run the stage in full Class-A, i.e., the straight line at the right of the plot. I do not recall a mention of this point in the original article.

Quiescent current needed to achieve this is about 2 A. The desirability of Class-A operation is reinforced by the incremental gain plot in Figure 36.5. It is clear that the gain variations are serious for lower V_{bias}, and do not augur well for the closed-loop distortion performance. Only the rightmost Class-A gain characteristic has a clear flat portion over its operational range.

It is true that the drain currents in this stage are symmetrical—but the quiescent required to remove the sharp-eared 'Batman' effect in Figure 36.5 is so high that the amplifier is working entirely in Class-A. The symmetry of the circuit means only that when distortion is produced, it will be predominantly odd-order, which is not normally considered a good thing from the subjective point of view. To keep the stage linear into 4 Ω loads would demand a quiescent of 2.9A. It is interesting that this is not twice the current for the 8 Ω case.

There are other significant differences from the usual voltage-follower configuration, which if nothing else has a stage gain reliably close to unity. Olsson's output stage gain varies with V_{bias} adjustment—even when in Class-A—and also varies strongly with load impedance. This would seem to make reliable compensation a difficult business, but in the complete amplifier this variation is probably only significant below the dominant-pole frequency P_1.

Figure 36.6 is a comparison between the Olsson configuration and three conventional stages, all biased to drive an 8 Ω load in Class-A. Traces 1 and 2, at the top, are bipolar-emitter follower and complementary-feedback-pair stages. These produce the usual linearity and close approach to unity gain.

The curved lower trace, 4, is a conventional complementary FET output. Trace 3 is the Olsson stage, with its gain of about 65 times normalised to fit in between the other curves. It shows stronger curvature than the bipolar stages, and despite everything, is actually less symmetrical than the usual output stages. I think this is inherent in the circuit's lack of symmetry about the output line.

I hope this article is a fair analysis of the proposed configuration, and that I have not made any serious misinterpretations of Mr Olsson's intentions. I also trust that it will not be taken as purely destructive criticism, for that is not my intention. I have to conclude that the configuration appears to require a very high quiescent current for linear operation, and has only a limited amount of negative feedback available to correct output stage distortions. Any deeper investigation would need to be encouraged by some promise that the Olsson configuration can deliver substantial benefits, and as far as my analysis goes, this does not seem likely.

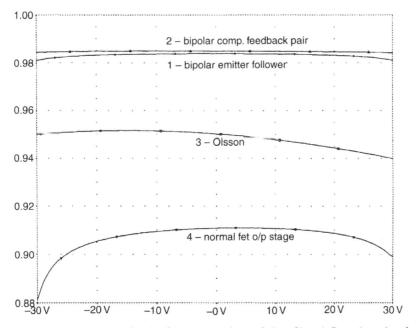

Figure 36.6 Curve 3 shows incremental gain for the Olsson stage, driving 8 Ω in Class-A. Equivalent plots for conventional Class-A bipolar emitter follower (curve 1), bipolar complementary feedback pair (curve 2) and a normal complementary FET stage (curve 4) are shown for comparison.

References

1. Olsson, B. 'Better audio from non-complements?' *EW + WW*, December 1994, pp 988–992.

2. Self, D. 'Distortion in power amplifiers: Part 1', *EW + WW*, August 1993.

3. Self, D. 'Distortion in power amplifiers: Part 3', *EW + WW*, October 1993.

4. Self, D. 'Distortion in power amplifiers: Part 4', *EW + WW*, November 1993.

5. Hood, J.L. 'Simple Class-A amplifier, *Wireless World*, April 1969.

6. Self, D. 'Distortion in power amplifiers: Part 7, *EW + WW*, March 1994.

Chapter 37

Loudspeaker undercurrents

(Electronics World, February 1998)

Intended title: *Enhanced Loudspeaker Currents*

I must admit that when I first read that exotic waveforms that increased the current demands of speakers by several times could be thought up, I was a bit sceptical. Well, actually, I was very sceptical. However, I was wrong to be so untrusting. Simulating a very simple analogue of a loudspeaker showed that transient currents of considerable size can be made to flow if careful adjustments are made to the stimulus waveform. I should emphasise that things need to be set up quite precisely to get the effect, and the results are rather sensitive to the slew-rate of the transitions between the step levels. Such transitions do not occur in real audio, which gives an immediate hint that what we are studying is a sort of trick rather than a legitimate concern.

While the effect can be demonstrated, the implications for practical amplifier design are much less clear. The square-edged waveforms required to induce current-enhancement do not resemble speech or music at all, and in the intervening years since the article was published it has gradually become clearer that they do not exist in real life speech and music. It therefore seems that the extent to which such alarming current-enhancement occurs is negligible, and it need not be considered in the design process. Timely support for this view came from an article called 'Current Affairs' by Keith Howard, in *Hi-Fi News* (Feb. 2006) [1], which found no evidence of current-enhancement at all. Note also my conclusion at the end of the article that even if the effect was occurring now and then at high powers, it would not be audible.

To date (2015), there has been no more interest in this topic, and no hint that it might ever be relevant to practical amplifier design. If it *had* turned out to be highly significant, what would the consequences have been? Essentially, we would have had to design output stages with two or three times the peak current capability. Also, since the peak currents cannot be assumed to occur when there are low voltages across the output devices, as demonstrated at Point E in Figure 37.2 of the article, we would have to design for much larger peak powers as well. This will clearly cost more money than an output stage which is designed to only drive 8Ω resistive loads and so sees no current-enhancement.

However, we don't just design for 8Ω resistive loads. Amplifiers have to drive reactive loads rather than just resistances, so we have to allow for higher average and peak power dissipations; accommodating a 45° load is entirely reasonable and necessary, and that requires average power capability to be doubled, and peak power capability increased by 2.7 times. Coping with a 60° load at full output is rather more ambitious, and it is questionable if it is really necessary; if we decide it is, the peak power capability must be increased by 3.4 times. Therefore, the extra cost of allowing for the enhanced-current effect might not actually have been too severe, though I'm not going to pretend I have thought it worthwhile to do any detailed calculations on this. At any rate, it's one notion I'm sure we don't need to worry about.

Reference

1. Howard, K., Current Affairs, *Hi-Fi News,* Feb. 2006.

LOUDSPEAKER UNDERCURRENTS

February 1998

It is easy to show that the voltage/current phase shifts in reactive loudspeaker loads increase the peak power dissipation in a cycle, using sine wave test signals of varying frequency.[1] The effect of this on device selection in output stages is complicated by the inability to treat power ratings as average power, for as far as safe-operating areas are concerned, low audio frequencies count as d.c.

But sinewave studies do not give insight into what can happen with arbitrary waveforms. When discussing amplifier current capability and loudspeaker loading, it is often said that it is possible to synthesise special waveforms that provoke a loudspeaker into drawing a greater current than would at first appear to be possible. This is usually stated without further explanation. Since I too have become guilty of this,[1] it seemed time to make a quick investigation into just how such waveforms are constructed.

The possibility of unexpectedly big currents was raised by Otala,[2] and expanded on later.[3] But these information sources are not available to everybody. The effect was briefly demonstrated in *EW* by Cordell,[4] but this was a long time ago.

Speaker model

Figure 37.1 is the familiar electrical analogue of a single speaker unit. Component R_c is the resistance and L_c the inductance of the voice coil. In series, L_r and C_r represent

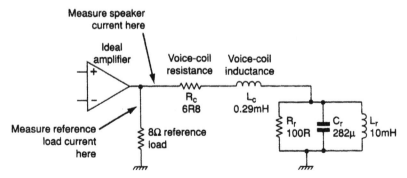

Figure 37.1 Equivalent circuit of single-unit speaker. Although the nominal impedance is 8 Ω, the coil resistance R_c is only 6.8 Ω. This is based on a successful commercial 10 in bass unit. The LCR network on the right simulates the cone mass/suspension-compliance resonance.

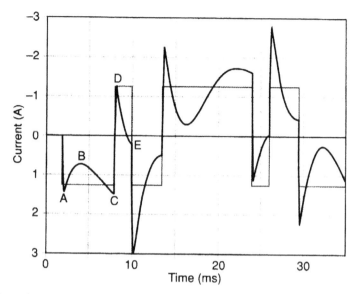

Figure 37.2 Asymmetrical waveforms to generate enhanced speaker currents. The sequence ABCDE generates a negative current spike; to the right, the inverse sequence produces a positive spike. The rectangular waveform is the current through the 8 Ω reference load.

the resonance of cone mass and suspension compliance, while R_r controls the damping. These three components model the impedance characteristics of the real electro-mechanical resonance.

Voice-coil inductance is 0.29 mH, and coil resistance 6.8 Ω. These figures are typical for a 10 in bass unit of 8 Ω nominal impedance. Measurements on this load will never show an impedance below 6.8 Ω at any frequency. This makes it easy to assume that the current demands can never exceed those of a 6.8 Ω resistance. This is not true.

To get unexpectedly high currents moving, the secret is to make use of the energy storage in the circuit reactances, by applying an asymmetrical waveform with transitions carefully matched to the speaker resonance.

Simulating the effects

Figure 37.2 shows PSpice simulation of the currents drawn by the circuit of Figure 37.1. The rectangular waveform is the current in a reference 8 Ω resistance driven with the same waveform. A ± 10 V output limit is used here for simplicity but this will in practice be higher, a little below the rail voltages.

At the start of the waveform at A, current flows freely into C_r but then reduces to B as the capacitance charges. Current is slowly building up in L_r, so the total current drawn increases again to C. A positive transition to the

opposite output voltage then takes us to point D, which is not the same as A because energy has been stored in L_r during the long negative period.

A carefully-timed transition is made at E, at the lowest point in this part of the curve. The current change is the same amplitude as at D, but since it starts off from a point where the current is already negative, the final peak goes much lower to 2.96 A, 2.4 times that for the 8 Ω case. I call this the current timing factor (CTF).

And with multiple speakers?

Otala has shown that the use of multi-way loudspeakers, and more complex electrical models, allows many more degrees of freedom in maximising the peak current, and gives a worst case current timing factor of 6.6 times.[3]

Taking an amplifier designed to give 50 W into 8 Ω, the peak current into an 8 Ω resistance is 3.53 A; amplifiers are usually designed to drive 4 Ω or lower to allow for impedance dips and this doubles the peak current to 7.1 A. In Ref. 3, Otala implies that the peak capability should be at least 23 A, but this need only be delivered for less than a millisecond.

The vital features of the provocative waveform are the fast transitions and their asymmetrical timing, the latter varying with speaker parameters. The waveform in Figure

37.2 uses ramped transitions lasting 10 µs; as the transitions are made slower the peak currents are reduced. Nothing much changes up to 100 µs, but with 500 µs transitions the current timing factor is reduced from 2.4 to 2.1.

Without doing an exhaustive survey, it is impossible to know how many power amplifiers can supply six times the nominal peak current required. I suspect there are not many. Is this therefore a neglected cause of real audible impairment? I think not, because:

- Music signals do not contain high-level rectangular waveforms, nor trapezoidal approximations to them. A useful step would be to statistically evaluate how often—if ever—waveforms giving significant peak current enhancement occur. As an informal test, I spent some time staring at a digital scope connected to general-purpose rock music, and saw nothing resembling the test waveform of Figure 37.2. Whether the asymmetrical timings were present is not easy to say; however the large-amplitude vertical edges were definitely not.
- If an amplifier does not have a huge current-peak capability, then the overload protection circuitry

will hopefully operate. If this is of a non-latching type that works cleanly, the only result will be rare and very brief periods of clipping distortion when the loudspeaker encounters a particularly unlucky waveform. Such transient distortion is known to be inaudible and this may explain why the current enhancement effect has attracted relatively little attention to date.

References

1. Self, D. 'Speaker impedance matters', *Electronics World*, February–November 1987, p 920.

2. Otala, M. *et al.* 'Input current requirements of high-quality loudspeaker systems', *AES* preprint #1987 (D7) for 73rd Convention, March 1983.

3. Otala, M. 'Peak current requirement of commercial Huttunen loudspeaker systems', *JAES* June 1987, p 455.

4. Cordell, R. 'Interface intermodulation in amplifiers', *Wireless World*, February 1983, p 32.

Chapter 38

Class distinction

(Electronics World, March 1999)

Intended title: *A Modest Proposal for Power Amplifier Classification*

Another title change that made me wince; a limp attempt at a pun (a punette?) and virtually no clue given as to what the article is about. My own title was a reference to Jonathan Swift's famous 1729 satirical essay: *A Modest Proposal for Preventing the Children of Poor People From Being a Burthen to Their Parents or Country* [1]. (No, I'm not going to tell you what his proposal was, look it up.) My writing often contains stealth references like that, some of them very obscure indeed. I think this is the first one I have actually outed.

Some ideas lie waiting for a long time before they see the public light of day, especially if I am involved. This one had been at the back of my mind for some 10 years; the simple thought that complex output stages could be analysed into combinations of a few elementary classes, much as atoms are made of elementary particles. I also had the Whyte locomotive classification in mind, where for example 2–6–2 tells you with extreme economy that there are two leading wheels, six driving wheels coupled together, and two trailing wheels.

The goal was not classification alone, but setting up a formal system whereby possible new combinations of operating principles would become evident, and some of these are dealt with—rather briefly, I'm afraid, but space was running out—at the end of the article.

The article was well received at the time, and I received some nice letters, but whether it would have any real impact on the naming of amplifiers seemed highly dubious. My own prediction was no, it would not, so it has been a pleasant surprise to see that it has had some use.

The classification system presented in the article here has a major limitation in that it assumes that the output stage is symmetrical. This is usually the case in solid-state power amplifiers, including those with quasi-complementary output stages, where the circuitry is not visually symmetrical but is essentially symmetrical in operation—for example, both top and bottom output devices working in optimal Class-B. Some Class-A amplifiers are essentially built as an emitter-follower for the top half, with a driven current source for the bottom half (see Chapter 29 on Class-A amplifiers for more details on this), and this again gives a visually asymmetric circuit, but both top and bottom output devices are working in Class-A and in terms of current flow operation is once more symmetrical.

A notable exception of mine own devising is Class-XD (Crossover Displacement) where current flow in the top and bottom halves of the output stage is deliberately made asymmetrical by the use of three output devices so the crossover region is displaced away from the zero voltage point. See Chapter 42 for a full explanation of Crossover Displacement.

Towards the end of the article, Class-G is dealt with by treating it as an inner output stage in Class-B, with outer power devices in Class-C. Class-G does not often get a great deal of publicity, but there is a striking exception in the Arcam A49 power amplifier introduced in 2014. According to Arcam's published data, the amplifier uses ±35V for the lower rails and ± 65V for the upper ones. The transition reportedly occurs at the voltage equivalent of 50W/8Ω, i.e., 28 Vpeak, which leaves a safety margin between the transition voltage and the rail voltage.

As concern for energy efficiency increases, I think we will be seeing more Class-G amplifiers around. Signal Transfer has been selling high-quality Class-G amplifiers as PCBs, kits, or built and tested units for many years [2].

References

1. http://en.wikipedia.org/wiki/A_Modest_Proposal (accessed Oct. 2014)

2. http://www.signaltransfer.freeuk.com/classg.htm (accessed Jan. 2015)

CLASS DISTINCTION

March 1999

Power amplifiers are usually distinguished by their operating class—A, AB or B, and so on. Unfortunately this classification scheme only begins to address the problem, as amplifiers come in many more than three kinds. There is current-dumping, Class-G, error-correction, and so on. Amplifiers that work in quite different ways are all called 'B' or 'AB', and there is still confusion between B and AB in many quarters.

Traditionally, further letters such as G, H, and S have been used to describe more complex configurations. It occurred to me that rather than proliferating amplifier classes on through the alphabet, it might be better to classify amplifiers as combinations of the most basic classes of device operation.

It may be optimistic to think that this proposal will be adopted overnight, or indeed ever. Nevertheless, it should at least stimulate thought on the many different kinds of power amplifier and the relationships between them.

Class structure

At the most elementary level, there are five classes of device operation, as outlined in the panel. More sophisticated amplifier types such as Class-G, Glass-S, etc., are combinations of these basic classes. Class-E remains an rf-only technology,[7] while Class-F does not apparently exist.

All the operating classes above work synchronously with the signal. The rare exceptions are amplifiers that have part of their operation driven by the signal envelope rather than the signal itself.

Krell[8] has produced Class-A amplifiers with a quiescent current that is rapidly increased by a sort of noise-gate side-chain, but slowly decays. An interesting study of a syllabic Class-G amplifier with envelope-controlled rail switching was presented in Ref. 9.

Combinations of classes

The basic classes mentioned in the panel on the right have been combined in many ways to produce the amplifier innovations that have appeared since 1970. Since the standard output stage could hardly be simplified, all of these involve extra power devices that modify how the voltage or current is distributed.

Assuming the output stage is symmetrical about the central output rail, then above and below it there will be at least two output devices connected together, in either a series or parallel format. Since these two devices may operate in different classes, two letters are required for a description, with punctuation—a dot or plus sign—between them to indicate parallel or series connection.

Parallel or series connection

In parallel, i.e., shunt, connection, output currents are summed, the intention being either to increase power capability, which does not affect basic operation, or to improve linearity.

Five basic classes

Class-A

The device conducts 100% of the cycle. This includes Class-A push-pull, where at full output, device current varies from twice the quiescent current to almost zero in a cycle, and Class-A constant-current mode, also known as single-ended Class-A. Any intermediate amount of current swing clearly also qualifies as Class-A, so unlike Class-B there exists an infinite range of variations on Class-A operation.

Class-AB

Conducts less than 100% but more than 50% of the cycle. This is essentially over-biased Class-B, giving Class-A operation up to a certain power level, but above that at least twice as much distortion as optimal Class-B. Once more there is a range of variations on Class-AB, depending on the amount of overbias chosen.

Class-B

The device conducts very nearly 50% of the cycle. The exactness of the 50% depends on the definition of 'conducts' because with Class-B optimally adjusted for minimum crossover distortion, there is always some conduction overlap at crossover, otherwise there would be no quiescent current. This will be 10 mA or so for a complementary feedback pair stage, or about 100 mA for an emitter follower version. With bipolar transistors, collector current tails off exponentially as V_{be} is reduced, and so the conduction period is rather arguable.

So-called 'non-switching' Class-B amplifiers, which maintain a small current in the output devices when they would otherwise be off, such as the Blomley[1] and Tanaka approaches[2] are treated as essentially Class-B.

Class-C

The device conducts less than 50% of the cycle. It is frequently written—indeed I have written it myself—that Class-C is inapplicable to audio and never used therein. A little more thought showed me that this is untrue. The best-known example is Quad current-dumping, a scheme specifically intended to allow the high-power output stage—the 'current-dumpers'—to be run at zero quiescent.[3]

An emitter-follower output stage with no bias has a fixed dead-band of approximately ±1.2 V, so clearly the exact conduction period varies with supply voltage; ±40 V rails and a 1 mA criterion for conduction give 48.5% of the cycle. This looks like a trivial deviation from 50%, but crossover distortion prevents direct audio use

Class-D

The device conducts for any percentage of the cycle but is either fully on or off. Class-D usually refers to a pulse-width modulation scheme where the mark/space ratio of an ultrasonic squarewave is modulated by the audio signal.[4,5,6] However, in this case I am concerned only with the on-off nature of operation, which can be of use at audio frequencies, though not of course for directly driving the load. The conduction period during a cycle is not specified in this definition of Class-D.

A subordinate aim is often the elimination of the Class-B bias adjustment. The basic idea is usually a small high-quality amplifier correcting the output of a larger and less linear amplifier. For a parallel connection the two class letters are separated by a dot, i.e. '•'.

In a series connection the voltage drop between supply rail and output is split up between two or more devices, or voltages are otherwise summed to produce the output signal. Since the collectors or drains of active devices are not very sensitive to voltage, such configurations are usually aimed at reducing overall power dissipation rather than enhancing linearity. Series connection is denoted by a plus sign between the two Class letters.

The order of the letters is significant. The first letter denotes the class of that section of the amplifier

Table 38.1 Sub-class definitions

Parallel		
A•B	Sandman Class-S scheme	Figure 38.1
A•C	Quad current-dumping	
B•B	Self Load-Invariant amplifier	
B•C	Crown and Edwin types	Figure 38.2
B•C	Class-G shunt. (Commutating) 2 rail voltages	Figure 38.3
B•C•C	Class-G shunt. (Commutating) 3 rail voltages	Figure 38.4
Series		
A+B	'Super Class-A'	Figure 38.5
A+B	Stochino error correction	Figure 38.6
A+D	A possible approach for cooler Class-A	
B+B	Totem-pole or cascade output. No extra rails	Figure 38.7
B+C	Classical series Class-G, 2 rail voltages	Figure 38.8
B+C+C	Classical series Class-G, 3 rail voltages	
B+D	Class-G with outer devices in D	Figure 38.9
B+D	Class-H	Figure 38.10

that actually controls the output voltage. Such a section must exist—if only because the global negative feedback must be taken from one specific point—and the voltage at this point is the controlled quantity. The shunt configurations are dealt with first; see Table 38.1.

Class A • B

Class A • B describes an output stage in which the circuitry that actually controls the output is in Class-A, while a second Class-B stage is connected in parallel to provide the muscle.

The best-known example is probably the Sandman output configuration, in which the high-power amplifier A_2 is controlled by its own negative feedback loop so as to increase the effective load impedance until it is high enough for the Class-A stage to drive it with low distortion.[10]

In Figure 38.1, A_1 is the Class-A controlling amplifier while A_2 is the Class-B heavyweight stage. As far as the load is concerned, these two stages are delivering current in parallel. The aim was improved linearity, with the elimination of the bias preset of the Class-B stage as a secondary goal.

If A_2 is unbiased and therefore working in Class-C, A_1 has much greater errors to correct. This would put the amplifier into the next category, Class A • C.

Class A • C

The power stage A_2 is now working in Class-C, the usual motivation being the reduction of power dissipation because current is flowing for less of the cycle. The absence of any bias for a Class-B-type output stage puts it in into Class-C, as conduction is less than 50%— though probably not much less.

If the bias voltage is dispensed with then a number of problems with setting and maintaining accurate quiescent conditions are eliminated. A good example of such use of Class-C is the Quad current-dumping concept. Here, the use of feed-forward error-correction allows the substantial crossover distortion from a heavyweight Class-C—i.e. underbiased Class-B—stage to be effectively corrected by a much smaller Class-A amplifier.[3]

Class B • B

At first there seems little point in using one Class-B stage to help another, as they both have inherent crossover distortion. However, since reducing the current handled by an output stage reduces both crossover and large-signal distortion, the concept can be useful.

An example is my Load-Invariant amplifier, which can be considered as two Class-B output stages collapsed into one.[12]

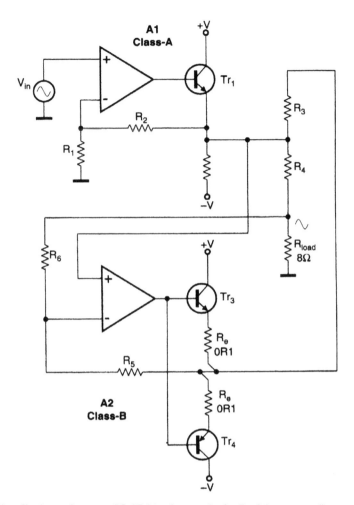

Figure 38.1 Sandman 'Class-S' scheme. Resistors R3,4,5,6 implement the feedback loop controlling amplifier A2 so as to raise the load impedance seen by A2.

Class B • C

Here, the controlling stage A_1 is Class-B, accepting that some crossover distortion in the output will be inevitable. This approach appears to have been introduced by Crown (Amcron) around 1970.[13]

Once more two stages are combined; the drivers—usually compound are required to deliver significant power in Class-B, while the main power devices only turn on when the output is some way from the crossover point, and are in Class-C.

Similarly, the 'Edwin' type of amplifier, Figure 38.2, was promoted by Elektor in 1975.[14] It was claimed to have the advantage of zero quiescent current in the main output devices-though why this might be an advantage was not stated; in simulation linearity appears worse than usual.

Another instance of B • C is Class-G-shunt.[11] Figure 38.3 shows the principle; at low outputs only $Tr_{3,4}$ conduct, delivering power from the low-voltage rails. Above a threshold determined by V_{bias3} and V_{bias1}, D_1 or D_2 conducts and $Tr_{6,8}$ turn on, drawing from the high-voltage rails.

Diodes $D_{3,4}$ protect $Tr_{3,4}$ against reverse bias. The conduction periods of the Class-C devices are variable, but much less than 50%. Class-G-Shunt schemes usually have A1 running in Class-B to minimise dissipation, giving B • C; such arrangements are often called 'commutating amplifiers'.

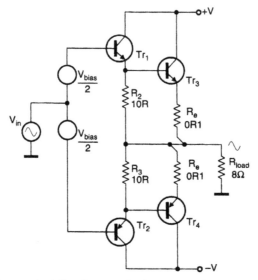

Figure 38.2 Edwin type amplifier; standard Class-B except for the unusually low driver emitter resistors. Effectively B • C.

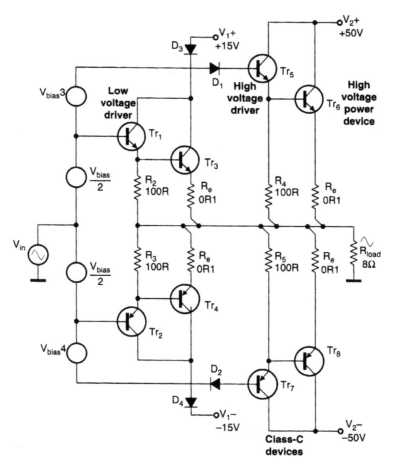

Figure 38.3 A Class-G-shunt output stage, composed of two emitter-follower output stages with the usual drivers. Voltages Vbias3,4 set the output level at which power is drawn from the higher rails. B • C.

Class B • C • C

Some of the more powerful Class-G-shunt public-address amplifiers have three sets of supply rails to further reduce the average voltage-drop between rail and output.

The extra complexity is significant, as there are now six supply rails and at least six power devices. It seems most unlikely that this further reduction in power

consumption could ever be worthwhile for domestic hifi, but it is very useful in large PA amplifiers, such as those made by BSS. Three letters with intervening dots are required to denote this mode, Figure 38.4.

Series connection category

In the second group of configurations, voltages are summed by series connection. The intention is usually the reduction of total power dissipation, rather than better linearity.

Since the devices are not usually operating in the same class, two letters are again required for a description, and I have used a plus-sign between them to indicate the series connection.

Class A + B

Figure 38.5 shows the so-called 'Super-Class-A' introduced by Technics in 1978.[15] The intention is to combine the linearity of Class-A with the efficiency of Class-B.

The Class-A controlling section A_1 is powered by two floating supplies of relatively low voltage, around ± 15 V, but handles the full load current. The floating supplies are driven up and down by a Class-B amplifier A_2. This amplifier must sustain much more dissipation as the same current is drawn from much higher rails, but it need not be very linear as in principle its distortion will have no effect on the output of A_1.

The circuit is complex and costs more than twice that of a conventional amplifier. In addition, the floating supplies are awkward. This seems to have limited its popularity.

Another A + B concept is the error-correction system of Stochino.[16] The voltage summation—the difficult bit—can be performed by a small transformer, as only the flux due to the correction signal exists in the core. This flux cancellation is enforced by the correcting amplifier feedback loop. Complexity and cost are at least twice that of a normal amplifier; Figure 38.6.

Class A + D

The 'Super-Class-A' concept mentioned above can be extended to A + D by running the heavyweight amplifier in the usual high-frequency pwm Class-D configuration.[17] Alternatively, an A + D amplifier can be made by retaining the Class-A stage but powering it from rails that switch at audio frequency between two discrete

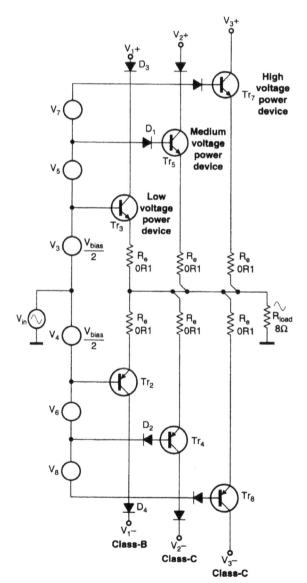

Figure 38.4 Simplified diagram of a three-rail 'commutating' series-Class-G power amplifier, denoted as Class B • C • C.

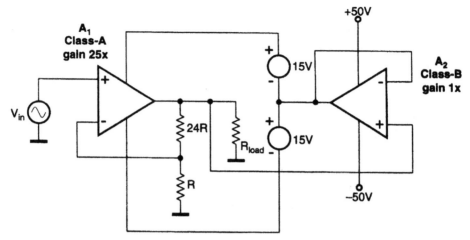

Figure 38.5 The 'Super-Class-A' concept. Amplifier A1 runs in Class-A, while high-power Class-B amp A2 drives the two floating supplies up and down. Denoted as A + B.

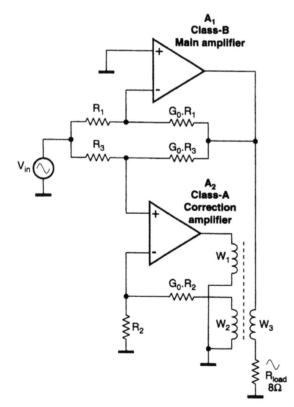

Figure 38.6 The Stochino error-correcting system voltage-sums the outputs of the two amplifiers using transformer W1,2,3, A + B.

voltages. Recall that this definition of Class-D does not mean high-frequency pwm.

Class B + B

Sometimes called a totem-pole stage to emphasise the vertical stacking of output devices, this arrangement shares the power dissipation between two devices. However, a parallel connection does the same thing more simply and with lower voltage losses.

Class B + B has been used to permit high power outputs from transistors with limited V_{ceo}, but this is rarely necessary with modern devices. The concept is usually regarded as obsolete, Figure 38.7.

Class B + C

The basic series Class-G with two rail voltages, i.e., four supply rails, as both voltages are positive and negative, is shown in Figure 38.8. This configuration was introduced by Hitachi in 1976 with the aim of reducing amplifier power dissipation.[18,19]

Musical signals spend most of their time at low levels, and have a high peak/mean ratio, so dissipation is greatly reduced by running from the lower $\pm V_1$ supply rails when possible.

When the instantaneous signal level exceeds $\pm V_1$, Tr_6 conducts and D_3 turns off, so the output current is

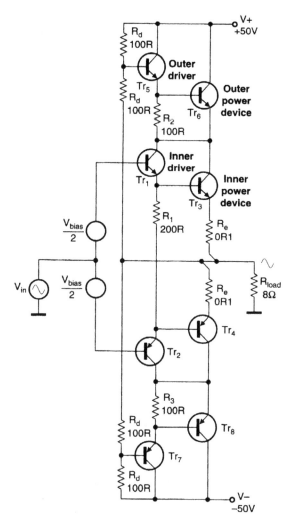

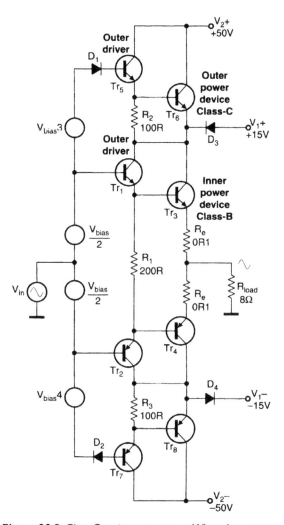

Figure 38.7 A totem-pole or cascade series output. Resistors Rd divide the voltage between rail and output in half, and drive the outer power devices. Inner and outer devices turn on and off together. B + B.

Figure 38.8 Class-G-series output stage. When the output voltage exceeds the transition level, D3 or D4 turns off and power is drawn from the higher rails through the outer power devices. B + C.

now being drawn from the higher $\pm V_2$ rails, with the dissipation shared between Tr_3 and Tr_6. The inner stage $Tr_{3,4}$ normally operates in Class-B, though AB or A are equally possible if the output stage bias is increased.

In principle movements of the collector voltage on the inner device collectors should not affect the output voltage, but practical Class-G is often considered to have worse linearity than Class-B because of glitching due to diode commutation. However, glitches if present occur at moderate power, well away from the crossover region.

Class B + C + C An obvious extension of the Class-G principle is to increase the number of supply voltages, typically to three. Dissipation is reduced and efficiency increased, as the average voltage from which the output current is drawn is kept closer to the minimum.

The inner devices will operate in Class B/AB as before, the middle devices will be in Class-C, conducting for significantly less than 50% of the time. The outer devices are also in Class-C, conducting for even less of the time. Three letters with intervening plus signs are required to denote this.

To the best of my knowledge three-level Class-G amplifiers have only been made in shunt mode. This is probably because in series mode the cumulative voltage drops become too great. If it exists, such an amplifier would be described as operating in B + C + C.

Class B + D

Since the outer power devices in a Class-G-series amplifier are not directly connected to the load, they need not be driven with waveforms that mimic the output signal. In fact, they can be banged hard on and off so long as they are always on when the output voltage is about to hit the lower supply rail.

The outer devices may be simply driven by comparators, rather than via a nest of extra bias generators as in Figure 38.8. Thus the inner devices are in B with the outer in D. Some of the more powerful amplifiers made by NAD—like the *Model 340*—use this approach, shown in Figure 38.9.

The technique known as Class-H is similar but uses a charge-pump for short-term boosting of the supply voltage. In Figure 38.10, at low outputs Tr_6 is on, keeping C charged from the rail via D.

During large output excursions, Tr_6 is off and Tr_5 turns on, boosting the supply to Tr_3. The only known implementation is by Philips[20, 21] which is a single-rail car audio system that requires a bridged configuration and some clever floating-feed-back to function.

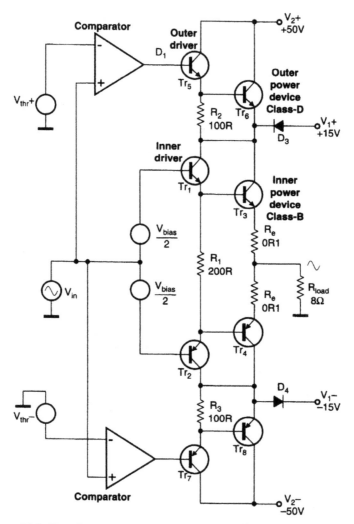

Figure 38.9 Class-G output stage with outer devices in Class-D. Described as B + D.

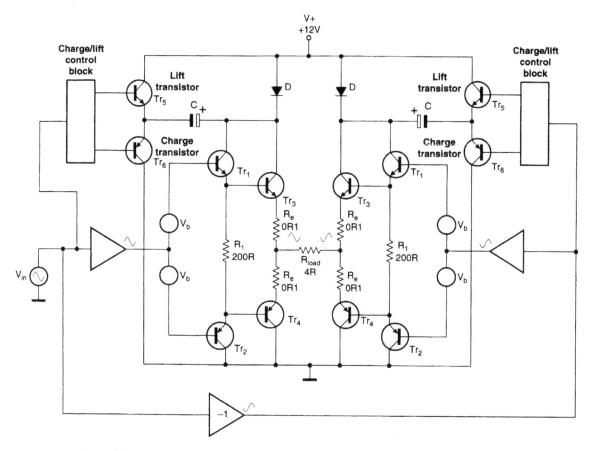

Figure 38.10 The Class-H principle applied to a bridged output stage for automotive use. B + D.

Full circuitry has not been released, but it appears the charge-pump is an on/off subsystem, i.e. Class-D.

In summary

The test of any classification system is its gaps. When the periodic table of elements was evolved, the obvious gaps spurred the discovery of new elements; convincing proof the table was valid.

Table 38.1 is restricted to combinations that are, or were, in actual use, but a full matrix showing all the possibilities has several intriguing gaps; some, such as C • C and C + C are of no obvious use, but others like A + C are more promising—a form of Class-G with a push-pull Class-A inner stage. Glitches permitting, this might save a lot of heat.

The amplifier table really gets interesting when it becomes clear that there are gaps in the entries—things that could exist but are not currently known.

References

1. Blomley, P. 'A new approach to class-B', *Wireless World*, February 1971, p 57.

2. Tanaka, S. 'A new biasing circuit for class-B operation', *Journal of Audio Engineering Society*, January/February 1981, p 27 (Non-switching Class-B).

3. Walker, PJ. 'Current dumping audio amplifier', *Wireless World*, December 1975, p 560.

4. Attwood, B. 'Design parameters important for the optimisation of PWM (Class-D) amplifiers', *JAES*, Vol. 31 (11), November 1983, p 842.

5. Goldberg and Sandler, 'Noise shaping and pulse-width modulation for an all-digital audio power amplifier', *JAES*, Vol. 39 (6), February 1991, p 449.

6. Hancock, J. 'A class-D amplifier using MOSFETS with reduced minority carrier lifetime', *Journal of*

Audio Engineering Society, Vol. 39 (9), September 1991, p 650.

7. Peters, A. 'Class-E RF amplifiers', *IEEE Journal of Solid-State Circuits*, June 1975, p 168.

8. Atkinson, J. 'Krell KSA-50S amplifier review', *Stereophile*, August 1995, p 165.

9. Funada and Akiya, 'A study of high-efficiency audio power amplifiers using a voltage switching method', *Journal of Audio Engineering Society*, Vol. 32 (10), October 1984, p 755 (Syllabic Class-G).

10. Sandman, A. 'Class-S: A novel approach to amplifier distortion', *Wireless World*, September 1982, p 38.

11. Raab, F. 'Average efficiency of class-G amplifiers', *IEEE Trans on Consumer Electronics*, April 1986, p 145.

12. Self, D. 'Load invariant audio power', *Electronics World*, January 1997, p 16.

13. Linsley-Hood, J. 'The straight wire with gain?', *Studio Sound*, April 1975, p 22. (Crown).

14. Unknown, 'Edwin Amplifier', *Elektor*, September 1975, p 910.

15. Sano *et al.* 'A high-efficiency class-A audio amplifier', preprint #1382 for 61st AES convention, November 1978.

16. Stochino, G. 'Audio design leaps forward?', *Electronics World*, October 1994, p 818. (Error correction).

17. Jeong *et al.* 'A high-efficiency Class-A amplifier with variable power supply', (ClassA+Class-D), preprint #4257 for 100th AES convention, May 1996.

18. Sampei, *et al.* 'Highest efficiency & super quality audio amplifier using MOS power FETs in Class-G operation', *IEEE Trans on Consumer Electronics*, Vol. CE-24, #3 August 1978, p 300. (Class-G).

19. Feldman, 'Class-G high efficiency Hi-Fi amplifier', *Radio-Electronics*, August 1976, p 47.

20. Buitendijk, P. 'A 40 W integrated car radio audio amplifier', *IEEE Conf on Consumer Electronics*, 1991 Session THAM 12.4, p 174. (Class-H).

21. Philips, 'TDA1560Q 40 W car radio high power amplifier', Philips Semiconductors' data sheet, April 1993.

Chapter 39

Muting relays

(Electronics World, July 1999)

Intended title: *Relay Control & Muting for Amplifiers*

What could be simpler than switching a relay on and off? This depends on how closely you want to control it. Any design problem, when looked at the right way, can be made more challenging. It is usually possible to identify a parameter that can be focused on and made the subject of some creative improvement. So it is here, where the speed of relay switch-off, and hence its ability to mute turn-off transients, is examined into in depth. This article gives amongst other advice, details on how to make relays operate faster than at first appears to be possible. This is done by recognising that the simple and almost universal practice of protecting relay driver transistors from damage at turn-off by putting a reverse-biased diode across the relay coil slows down its opening drastically, as it permits current to keep circulating and hold the relay in. This can be prevented in a delightfully simple way by putting a high-voltage Zener diode in series with the ordinary diode, clamping the voltage across the transistor to tens of volts rather than less than one, but still fully protecting it. As Table 39.1b in the article shows, the drop-out time with a 27V Zener is reduced by at least two-thirds. This is a robust effect that has been proven for many types of relay and, as far as I can see, should work with any type at all. With big relays, it is vital to ensure that both the ordinary diode and the Zener diode have enough current capability.

Since this article was published, further experience leads me to sound a note of warning about trying to be *too* quick in sensing the disappearance of the mains supply. Doubtful electrical appliances (and I am thinking of a particular veteran refrigerator here, with a vicious motor start-up transient) combined with antique mains wiring can put short-duration dips in the supply that trigger the AC-loss circuit, when in fact the hold-up time of the power supply in the equipment with the relay is perfectly capable of keeping the show going. If you run into trouble with this, increasing the value of R5 or C1 (Figure 39.3) should fix things.

In general, audio equipment does not need to rely on very rapid detection of mains supply loss. If the mains switch generates a spark and a burst of RF when it opens, that may be demodulated and reach the output faster than any electromechanical relay can react. The cure for this is obviously proper switch suppression with appropriately rated components rather than blindingly fast muting. The next possibility is that the equipment generates thumps and bangs the moment the supply rails start to subside; this is not very common and really should not be allowed to happen. It does not normally occur with IC-based small-signal circuitry like preamplifiers, or dual-rail DC-coupled amplifiers like the Blameless design, but I have encountered it in one discrete-transistor preamplifier. Usually there is a little time available while the supply rails are held up by the reservoir capacitors, and so a simpler mains-loss approach can be used, typically a half-wave rectifier charging a small capacitor, sized and resistively loaded so that it discharges much faster than the main supply reservoirs, and so triggers the opening of the mute relay. This method has been employed innumerable times, including by me in the Cambridge Audio 340A, 840A, and 840W products. It has proved thoroughly dependable. Details are given in [1]; once you've got the timing right, the only thing to watch out for is the ripple current in the small capacitor, which will have a big ripple voltage across it. If the heating effect is excessive, you could find yourself with a nasty little electronic trap where the capacitor fails after a few months of use.

It takes much more voltage to pull a relay in than to hold it in, so you can save power by dropping the voltage once the relay has had time to operate. Given increasing concern for energy efficiency, I think we will be seeing more of efficiency circuits like that described in the last section of the article.

Reference

1. Self, D, *Audio Power Amplifier Design*, 6th edition, pp. 587–588. Focal Press 2013. ISBN 978-0-240-52613-3.

MUTING RELAYS

July 1999

Most power amplifiers incorporate an output relay that not only provides muting to prevent transients reaching the loudspeakers, but also protection against destructive DC faults.

Loudspeakers are expensive, and no amplifier should ever be connected to one without proper DC-offset protection. This applies with particular force to experimental amplifiers.

Sensible preamplifiers—i.e. those with AC-coupled outputs—do not require DC protection, but the muting of thumps is no less important. Electronic switching at preamp outputs is feasible, but still presents technical challenges if high standards of linearity are to be combined with a reasonably low output impedance.

Electronic output switching is impracticable at power amplifier signal levels; however, if the amplifier is powered by a switch-mode supply, then turning it off is an option if positive and negative rails can be relied upon to collapse quickly and symmetrically.

Protection circuit operation

Basic functions of a power-on thump elimination and DC protection circuit are as follows:

- Delay relay pull-in until amplifier turn-on transients are over.
- Drop out relay as fast as possible when AC power is removed.
- Drop out relay as fast as possible when DC fault occurs.
- Drop out relay on excess temperature, etc. Speed non-critical.

Figure 39.1 is a block-diagram of a system to perform these functions. Since this is in part a protection system, simplicity and bullet-proof reliability are essential.

The main dynamic parameters of a relay are the pull-in and drop-out time. For this kind of application, the pull-in time is more or less irrelevant, as it is milliseconds compared with the seconds of the turn-on delay.

Relay contacts bounce when they close, but the duration of pull-in contact bounce is not important for this application.

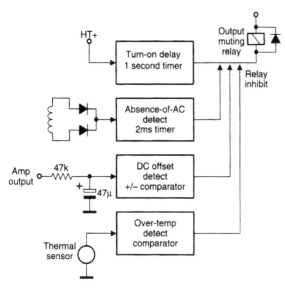

Figure 39.1 Block diagram of a relay control system.

Table 39.1a Relay specifications as presented by their manufacturers

	P&B	Oko	Schrack	Fujitsu
Nominal voltage	24 V	18 V	12 V	12 V
Must-operate voltage	18 V			8.4 V
Drop-out voltage	2.4 V			1.2 V
Coil resistance	660 Ω			320 Ω
Coil inductance	0.55 H			
Pull-in time maximum	15 ms			5 ms
Pull-in time typical	9 ms			
Drop-out time maximum	10 ms			3 ms
Drop-out time typical	7 ms			

All the relays I examined showed clean contact-breaking on drop-out, and this is essential for fast muting. Table 39.1a gives details of three poweramp relays and the Fujitsu relay used in the Precision Preamp '96 article.[1]

The specifications for the P&B relay are very conservative. The example measured pulled-in at 72% of the must-operate voltage, and dropped out at 350% of the must-drop-out voltage. Likewise the real operating times are much less than those specified.

The critical parameter for audio muting is the drop-out time, for this puts a limit on the speed with which turn-off transients can be suppressed. It seems at first that the drop-out time must be solely a function of the

Table 39.1b Measured relay specifications

	P&B	Oko	Schrack	Fujitsu
Operate voltage	13 V	13 V	7 V	6 V
Drop-out voltage	8.5 V	6.5 V	2.5 V	2 V
Pull-in time	14 ms	10 ms	10 ms	2.7 ms
Drop-out time	1.0 ms	1.3 ms	2.4 ms	1.2 ms
Diode drop-out time	5.4 ms	6.9 ms	11 ms	4.2 ms
27 V-clamp drop-out time	1.8 ms	2.4 ms	2.7 ms	1.3 ms

* P&B is Potter and Brumfield.

relay design, depending on the force in the bent contact spring and the inertia of the moving parts. This is partly true, as mechanical factors set a minimum time, but that time is greatly extended by the normal relay-driving circuits.

Relay-on timing

The delay required at amplifier turn-on depends on the amplifier characteristics.

If there are long time-constants, and voltages that take a while to settle, then the muting period will have to be extended to prevent clicks and thumps. Five seconds is probably the upper limit before the delay gets irritating; one second is long enough for a silent start-up with most conventional amplifiers.

This delay function can be performed in many ways, but there are a few points to consider. The tolerance on the length of the turn-on delay is not critical, and an RC time-constant is quite adequate to define it.

It is convenient—and significantly cheaper—to run the relay control circuitry directly from the main HT rails rather than creating regulated sub-rails or extra windings on the mains transformer. The emphasis is therefore on discrete transistor circuitry.

Figure 39.2 shows the relay control system I used in the Precision Preamplifier '96. Note that there was an error in the original diagram that is corrected here.

Capacitor C_{224} charges through R_{211} until D_{207} is forward-biased and Tr_{205} turns on. This turns on Tr_{206} and energises the relay; the extra current-gain of Tr_{206} enables the timing circuitry to run at low power. The on-timing delay here is 2 s.

A series dropper resistor for the relay is usually required; here it is R_{218}. The highest voltage relay-coil available is usually 48 V, though 24 V is more common, and power amplifier rails are often much higher than this.

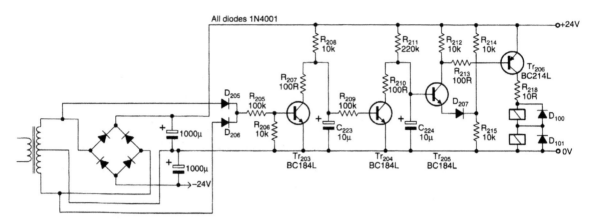

Figure 39.2 Relay-control circuit as used in the Precision Preamplifier '96—effective, but it can be improved. Component numbers retained from original design.

This reverse diodes across the relay coils prevent Tr_{206} being damaged by the inductive spike created when the coil is suddenly de-energised. For relays of the size used in power amplifiers, signal diodes cannot cope with the stored energy and the 1N4001 type should be used.

Off timing criteria

The relay drop-out must be as fast as possible. If a relay is powered directly from the supply rails, then it will drop out eventually as the rails collapse, but this will be far too slow to catch turn-off noises.

The drop-out voltage may well be less than a third of the pull-in voltage, and this slows things down even more. A specific fast off circuit is required, and there are several ways to achieve this.

Mechanical detection

This is a mains switch that closes or opens a control circuit before the mains power contacts are opened. It could give perfect relay operation, but I am not aware that any such mains switch has ever been produced.

Detecting the loss of DC supply

This technique involves a subsidiary supply rail with a small reservoir capacitor. When the mains is switched off, the capacitor discharges quickly and either removes the relay power directly, or resets the turn-on delay timer. The latter is usually easier to implement.

This method is inherently slow, because the relay-off threshold must be below the ripple troughs. Therefore in the worst case, an entire half-cycle of mains must pass before the capacitor becomes fully discharged, so the delay may be 10 ms.

In practice there are component tolerances to be allowed for, and the threshold must be set low enough to prevent spurious operation if the mains voltage is below normal. It is usually prudent to ensure circuitry works with mains down to at least −20%. This extends the minimum delay to about 16 ms. The reservoir capacitor will have a large ripple voltage across it, and its ripple-current rating must be carefully observed.

Detection of loss of AC

Detecting the loss of AC supply, as opposed to the rectified DC, is potentially quicker as there is no reservoir capacitor to discharge before the circuit operates. An AC waveform is effectively appearing and disappearing every half-cycle, so the circuit must distinguish between the zero-crossings that occur every 10 ms, and genuine loss of power.

AC loss detection

The most straightforward method of AC-loss detection exploits the fact that properly-defined zero-crossings are very brief; all that is required is a timer that will not complete and drop out the relay until a period greater than the width of the zero-crossing has expired. This delay can readily be reduced to less than 1 ms.

Referring to Figure 39.2, Tr_{203} is normally held firmly on by the incoming AC, via D_{205} on positive half-cycles and D_{206} on negative ones, and thus keeps C_{223} fully discharged.

At the zero-crossings, Tr_{203} has no base drive and turns off, allowing C_{223} to begin charging through R_{208}. If the absence of base drive persists beyond the preset period, which means the AC has been interrupted, then C_{223} charges until Tr_{204} turns on and rapidly discharges the main timing capacitor C_{210} through R_{207}, dropping out the relay.

When a relay is driven by a transistor, it is standard to put a reversed diode across the coil. Without it, abrupt turn-off of current causes the coil voltage to reverse, driving the collector more negative. For the relays here, the worst spike measured was -120 V, which is enough to destructively exceed the V_{ceo} of most transistors.

This apparently innocent, and indeed laudable practice of diode protection conceals a lurking snag; drop-out time is hugely increased by the reversed diode. It is roughly five times longer, which is very unwelcome in this particular application. This is because the diode gives a path for current to circulate while the magnetic field decays.

This is a good point to stop and consider exactly what we are trying to do: the aim is not to totally suppress the back-EMF but rather to protect the transistor.

If the back-EMF is clamped to about -27 V by a suitable Zener diode in series with the reverse diode, the circulating current stops much sooner, and the drop-out is almost as fast as for the non-suppressed relay.

In general, drop-out is speeded up by a factor of about four on moving from conventional protection to Zener clamping. For the relays examined here, a 500 mW Zener appeared to be adequate.

Preamp enhancement

The preamp relay controller can be improved upon; it works well under most circumstances, but it could be faster. Testing showed that the delay between loss of AC and the relay power being removed could be as long as 17 ms, depending slightly on the phase of the mains when it was cut. The relay drop-out time was 5 ms giving a total of 22 ms before the preamp output is muted.

The following circuit improvements were made to speed up relay drop out.

The on-timing reference divider $R_{214,215}$ is replaced with a 15 V Zener diode. This sharpens up the relay pull-in, making a more 'precise' click. It also prevents the voltage on C_{224} rising beyond that required to turn

Table 39.2 Component revisions for the preamplifier

	Old value	New value
R_{205}	100 kΩ	22 kΩ
R_{208}	10 kΩ	100 kΩ
C_{223}	10 μF	470 nF
R_{209}	100 kΩ	100 Ω
R_{215}	10 kΩ	15 V 400 mW Zener
Relay	1N4001	1N4001 + Suppression27 V, 500 mW Zener

on Tr_{205}; discharging it when the time comes is therefore quicker.

Base drive to Tr_{203} is increased by reducing R_{205} to 22 kΩ. This defines the zero-crossing as twice as narrow, allowing the time-constant R_{208}–C_{223} which bridges this period to be made shorter. Capacitor C_{224} therefore starts discharging sooner after AC is lost.

Impedance of the zero-crossing time-constant R_{208}–C_{223} is increased by changing the values from 10 kΩ–10 μF to 100 kΩ–470 nF. This simultaneously reduces the time-constant mentioned in the previous paragraph. It is now possible to use a non-electrolytic timing capacitor, which reduces tolerances and makes the circuit more designable.

Base drive to Tr_{204} is increased to speed up the discharge of C_{224} by reducing R_{205} from 100 kΩ to 100 Ω.

Finally, a 27 V Zener clamp is applied to each relay, as described above.

After these improvements, the electronic delay was reduced from 17 to 5.4 ms; the total delay including contacts opening now was 9.5 ms worst case.

After adding Zener clamping to the relays this fell to 6.3 ms worst-case, the average being 4.5 ms; the improved circuit is four times faster.

These component changes can be simply retro-fitted to existing Preamp 96 circuit boards using Table 39.2.

Other relay functions

The extra protective functions of a power amp relay require OR-ing together several error signals for DC offset, temperature shutdown, etc.

If a DC fault occurs in a power amplifier, this typically means that the output slams hard to one of the rails and stays there. Assuming the loudspeaker does not suffer instantaneous mechanical damage, it will overheat after a relatively short period as the DC flows through it.

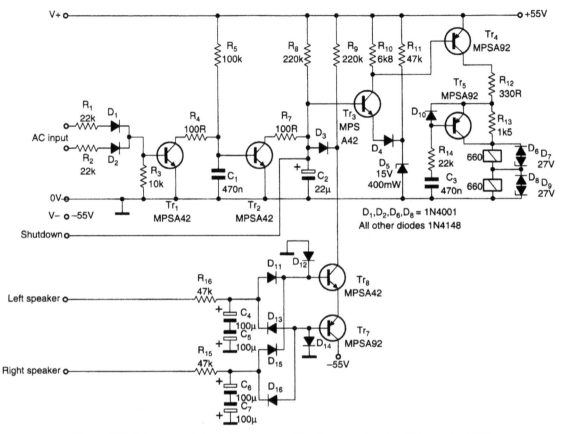

Figure 39.3 Relay control circuit for power amplifier, incorporating the efficiency circuit Tr5.

DC offset protection cannot prevent a loudspeaker hitting its mechanical limits, but it will stop it catching fire if the relay opens promptly. Once more, time is of the essence.

Usually, DC offset is detected by passing the amplifier output through an RC time-constant long enough to remove all audio, followed by a DC-detect circuit that responds to offsets of either polarity.

To allow a safety margin against false triggering on bass signals, I decided that the RC filter must accept full output at 2 Hz without the detector acting. For example, if it triggers at ±2 V, then for supply rails of ±55 V there must be 29 dB of attenuation at 2 Hz; with a single pole this means a -3 db frequency about 0.07 Hz.

This sort of low-pass filtering inevitably introduces a time delay; if the output leaves 0 V and moves promptly to one of the rails, this will be 50 ms with the circuit of Figure 39.3.

Detecting offsets of either polarity requires a little thought. Figure 39.4 shows a common circuit; a positive

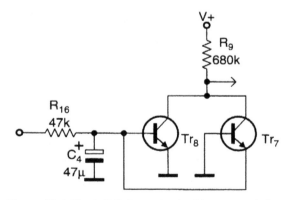

Figure 39.4 Simple DC-detect circuit with asymmetrical thresholds at +1.05 V and −5.5 V.

voltage turns on Tr_8 by forward biasing its base, while a negative voltage turns on Tr_7 by pulling down the emitter. The presence of DC is indicated by the collector voltage falling.

This solution is simple but highly-asymmetrical, requiring either + 1.05 V or −5.5 V to pull the collectors down to 0 V. For positive voltages the stage is common-emitter with high voltage gain, but for negative ones it works in common-base with a lower voltage gain, set by the ratio of R_{16} and R_9. It is difficult to make this ratio large without R_{16} becoming too small and hence C_4 inconveniently big.

If you're unlucky—and chances are you will be—the offset will have the wrong polarity for C_4, which will degrade if left reverse-biased for long periods. Two ordinary electrolytics back-to-back is the cheapest solution.

The improved DC detector in Figure 39.5 is fully symmetrical. Positive voltages turn on D_{11} and Tr_8; Tr_7 also conducts as its emitter is pulled up by Tr_8, while its base is held low by D_{14}. Negative voltages turn on D_{13} and Tr_7; Tr_8 conducts with its base fed by D_{12}. The threshold is now ±2.4 V, as for each polarity there are two diodes and two base-emitter voltages in series. The higher threshold is not a problem as the typical amplifier fault snaps the output hard to one of the rails.

One exception to this statement is HF instability. If an amplifier bursts joyfully into HF oscillation, it will almost certainly show slew-limiting as well. This is unlikely to be very symmetrical so there will be a DC shift at the output.

The magnitude of this is not very predictable, but a 2.4 V threshold will detect most cases. This should save your tweeters, though it may not save the amplifier from internal heating due to conduction overlap in the relatively slow output devices.

Figure 39.5 can be adapted for stereo simply by adding two more diodes, as in Figure 39.3. Note that a positive offset on one channel and a negative one on the other—admittedly highly unlikely—do not cancel out; a fault is still signalled.

Power amplifier relay control

Figure 39.3 shows a power amp relay controller, designed for ±55 V rails and 24 V relays such as the P&B *T*90 type outlined in Table 39.1a. The main differences are the inclusion of DC offset detection and an efficiency circuit to minimise dissipation in the relays, which are now larger than in the preamp, and require more power.

The DC-detect circuit rapidly discharges on-timing capacitor C_2 through D_3 when Tr_8 collector goes low. An extra OR input for thermal shutdown acts via a series diode in the shutdown line.

The circuit now uses *MPSA42/MPSA92* transistors to withstand the higher supply voltages; as usual higher V_{ceo} means lower current-gain, which must be allowed for in the detailed design.

The electronic delay until coil switch-off averages 2 ms, the timing being shown in Figure 39.6. The AC was interrupted at centre screen, and a large positive-going off-transient can be seen just to the right. This is due to the leakage inductance of a large transformer.

The loss of AC cannot be detected until this transient decays to zero, so the delay is slightly extended. This was not a problem with the preamp version as it uses a small toroid with much less leakage inductance.

Figure 39.7 shows the relay coil voltage. At switch off it goes straight down through zero until clamped by the two Zeners at around −50 V. This puts 105 V on Tr_4 collector, which is no problem as it is rated at 300 V V_{ceo}. The relay contacts open just as clamping ceases and the

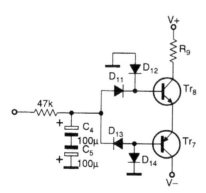

Figure 39.5 Improved DC-detect circuit; fully symmetrical thresholds at ±2.4 V. RC filter can cope with either polarity.

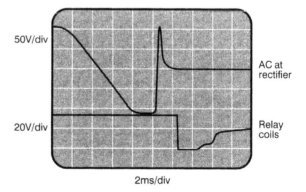

Figure 39.6 Electronic delay for power-amp version. Upper trace is transformer secondary with the usual flat-topped mains waveform, lower is voltage across both relay coils.

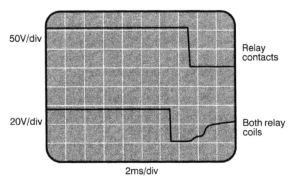

Figure 39.7 Relay drop-out delay for power-amp version. Upper trace is contact timing, lower is voltage across both relay coils.

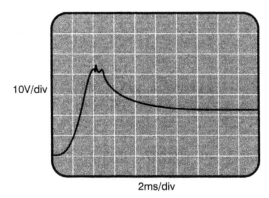

Figure 39.8 Pull-in voltage across both relay coils with efficiency circuit added.

coil voltage returns slowly to zero. Drop-out time is 1.8 ms, giving a total delay of 3.8 ms.

Efficiency circuit

All relays have a pull-in voltage that is greatly in excess of that required to keep them closed. It is therefore possible to save considerable power by applying full voltage only briefly, and then reducing it to a level which is still safely above the maximum drop-out voltage.

From Table 39.1a there is plenty of scope for this. By comparing the specified and measured performance, you will see that the P&B relay can be trusted to pull in at 18 V and not drop out above 8.5 V.

The initial pulse is provided by Tr_5 and R_{13}. At switch-on, Tr_4 is off and Tr_5 does nothing. After the on-timing delay Tr_4 conducts, Tr_5's emitter is pulled up, and its base receives a pulse of current via R_{14} and C_3. Resistor R_{13} is shorted by Tr_5 and the relays get a voltage reduced only by R_{12}; see Figure 39.8.

After 40 ms, C_3 is fully charged and Tr_5 turns off; this is at least four times longer than the minimum pulse to pull-in the T90 relay, but may be adjusted to suit other types by altering C_3. Diode D_{10} protects Tr_5 at switch-off.

In Figure 39.3, the initial voltage is 22 V per relay and the holding voltage 12 V, giving an initial power consumption of 1.85 W, falling to 960 mW long-term. The total power saving is just under a watt.

Running relays at a reduced holding voltage not only avoids the inelegance of consuming power for no good reason, but also speeds dropout time by reducing the magnetic energy stored. It could be argued that such a power saving is negligible. In a big Class-A amplifier it might be, but it makes sense in modest Class-B amplifiers idling for much of the time—which is of course almost all of them.

Reference

1. Self, D. 'Precision Preamplifier '96', *Electronics World*, July/August, September 1996.

Chapter 40

Cool audio power

(Electronics World, August 1999)

Intended title: *Power Amplifiers and Power Dissipation*

Once again, a title to make one cringe. I'm not saying my intended title was a masterpiece of creativity, but it did at least give you some inkling as to what the article was about. The idea for this article took shape several years before it appeared.

I have always thought that some kinds of information cry out to be presented graphically. A case in point is the ultimate destination of the power that an amplifier draws from its supply. Some goes usefully into the load, some is dissipated in the power semiconductors, and a little is lost in drivers, emitter resistors, and so on. All these quantities vary strongly with the percentage power output, and they overlap to give a confused and uninformative picture if plotted on a normal graph. The alternative method of power-partition diagrams, presented here, shows at once how much is power is taken and precisely where it goes; this is particularly important for more complex output stages, such as Class-G types. The idea is not wholly original; my inspiration came from what Victorian engineers called Sankey diagrams [1], a graphical method of showing where the energy went in a steam power-plant.

This chapter only deals with sinusoidal waveforms, for reasons explained at the start of the chapter. I feel bound to point out that I was perfectly aware that sine waves and musical waveforms have little in common, but you have to start somewhere, and a waveform that allows you to easily check your simulator results against pure mathematics has some pretty strong advantages.

For reasons best known to themselves, *Electronics World* chose to put graded shading as a background to the graphs. This didn't look too bad in the original magazine, but as reproduced in the second edition of *Self on Audio,* it made some of the traces disappear at the top of the illustration (for example Figure 40.6), which was not exactly helpful.

The 3D graphs in Figure 40.10 and Figure 40.11 looked cool at the time, but by the time they had been redrawn by EW, were not as informative as they might have been, as most of the Z-axis labelling had disappeared. To be honest, in terms of readability 3D graphs might not have been my brightest idea. So, to make up for it, the most important information from them is displayed in a rather more useful form in Preface Figure 40.1.

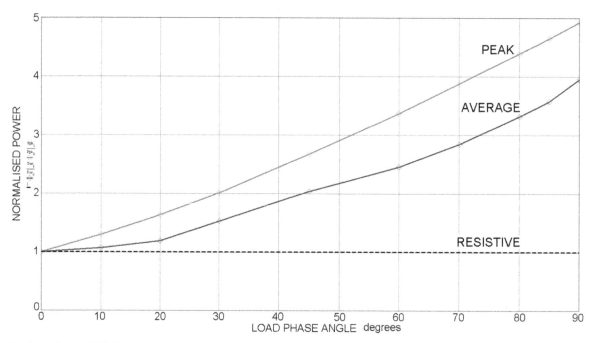

Preface Figure 40.1 The increase of average and peak power with the phase angle of the load. Power out is maximum (Voltage fraction = 0.95), which is the worst case for both average and peak power except for load angles near zero. ±50V rails.

This shows the most demanding cases from Figures 40.10 and 40.11 in the article, and show the average and peak powers in the output stage increase as the load moves away from pure resistance and becomes more reactive, until the phase angle reaches 90° and the load is a pure reactance. The lower trace is from the rear right wall of Figure 40.10, while the upper trace is from the rear right wall of Figure 40.11. It is notable that the peak power increases faster than the average power, as the load becomes more reactive, and so a greater proportion of the input power simply circulates in the amplifier instead of being dissipated by the load.

One of the major decision to be taken in designing a power amplifier is how reactive a load you want to drive at full power. A 45° load is a reasonable target; anything less being rather suspect, and this requires the average power capability to be increased by a factor of 2, and the peak power capability by a factor of 2.7 times. A maximum load angle of 60° is preferable for a quality design, requiring average power to be increased by 2.5 times, and peak power by 3.4 times.

I think this graph will be more useful as a design tool than the original 3D illustrations.

Reference

1. http://en.wikipedia.org/wiki/Sankey_diagram (accessed Feb. 2015)

COOL AUDIO POWER

August 1999

There are several important power relationships in designing an output stage. Both the average and peak power dissipated in the output devices must be considered when determining their type and number. The average power dissipated controls the heat-sink design.

In most amplifier types the power dissipation varies strongly with output signal amplitude as it goes from zero to maximum, so the information is best presented as a graph of dissipation against the fraction of the available rail-to-rail output swing—i.e., the output voltage fraction.

Consideration of average power allows the output devices to be made thermally safe; but it is also essential to consider the peak instantaneous dissipation in them. Audio waveforms have large low-frequency components, too slow for peak currents and powers to be allowed to exceed the DC limits on the data sheet.

For a resistive load the peak power is fixed and easily calculable. With a reactive load the peak power excursions are less easy to determine but highly important because they are increased by the changed voltage/phase relationships in the output device. Thus for a given load impedance modulus the peak power would need to be plotted against load phase angle as well as output fraction to give a complete picture.

Average power drawn from the rails is also a vital prerequisite for the power-supply design; since the rail voltage is substantially constant this can be easily converted into a current demand, which must be known when sizing reservoir capacitors, choosing rectifiers, and so on.

The voltage rating of these components is a much simpler business, requiring simply that they withstand the off-load voltages at the maximum mains voltage, which is usually taken as 10% above nominal. The only thing to decide is how big a safety margin is required.

Power drawn depends on signal-level and is again conveniently displayed with voltage-fraction as the X-axis.

The mathematical approach

When dealing with power amplifier efficiency, most textbooks use a purely mathematical method as shown

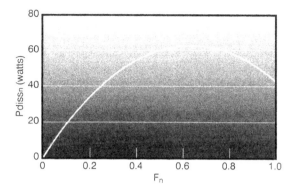

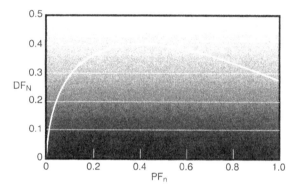

Figure 40.1 The standard mathematical derivation for Class-B. Maximum dissipation occurs at 64% voltage output, equivalent to 42% of maximum power output.

in Figure 40.1, which was produced with the aid of *Mathcad.* The calculation gives only the dissipation in the power devices.

Figure 40.1 gives the familiar information that maximum device dissipation occurs at 64% of maximum voltage, equivalent to 42% of maximum power. These specific numbers are a result of the sine waveform chosen and other waveforms give different values.

To make it mathematically tractable, the situation is highly idealised, assuming an exact 50% conduction period, no losses in emitter resistors or $V_{ce(sat)}$s, and so on. Solving the problem for Class AB, where the conduction period varies with signal amplitude, is considerably more complex due to the varying integration limits.

Simulating dissipation

Alternatively, the power variations in real output stages can be simulated and the results plotted; the circuits simulated in this article are shown in Figure 40.2.

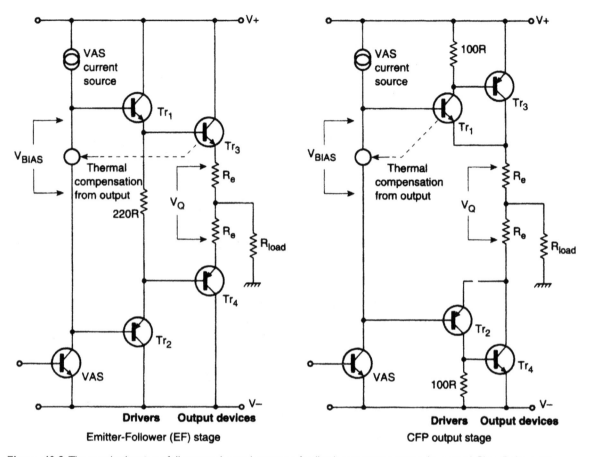

Figure 40.2 The standard emitter follower and complementary feedback pair output stages. In optimal Class-B the emitter follower version takes about 150 mA of quiescent current while the complementary feedback pair draws only 10 mA.

For concision, and by analogy with logic outputs, I have called the upper transistor the source and the lower the sink. In simulation, losses and circuit imperfections are included, and the power dissipations in every part of the circuit, including power drawn from the supply rails, are made available by a single run.

It is an obvious choice—which I duly took—to use a sine waveform in the simulations. This allows a reality-check against the mathematical results. Reactive loads are easily handled, so long as it is appreciated that the simulation often has to be run for ten or more cycles to allow the conditions in the load to reach a steady state.

All simulations were run with ±50 V rails and an 8 Ω resistive or reactive load. The output emitter resistors were 0.1 Ω. The drawback to this approach is that it is

F_n is the fraction of full output swing

$$n = 0 \ldots 10 \quad F_n = \frac{n}{10} \quad \omega = 1 \quad t = 0$$

Instantaneous power dissipation

$$P_n = \frac{V^2}{R}[F_n[\sin(\omega t) - F_n(\sin(\omega t))^2]]$$

Integrate over one half cycle

$$Pdiss_n = \frac{V^2}{R} F_n \frac{1}{\pi} \int_0^x \sin(\omega t) - F_n(\sin(\omega t))^2 \, dt$$

Note the $1/\pi$ due to integration from 0 to π

Since only one device conducts at once, dissipation for one is total dissipation

Rail volts Load resistance R = 8 (± rails)

$$V = 50$$

$$Pout_n = \left[\frac{V}{\sqrt{2}}(F)_n\right]^2 \times \frac{1}{R} \quad \begin{array}{l}\text{Output}\\\text{power}\end{array}$$

$$Pout\,\text{max} = \left[\frac{V}{\sqrt{2}}\right]^2 \times \frac{1}{R} \quad \begin{array}{l}\text{Max. output}\\\text{power}\end{array}$$

Poutmax = 156.25 W

$$PF_n = \frac{Pout_n}{Pout\,\text{max}} \quad \begin{array}{l}\text{Output}\\\text{factor}\end{array}$$

$$DF_n = \frac{Pdiss_n}{Pout\,\text{max}} \quad \begin{array}{l}\text{Dissipation}\\\text{factor}\end{array}$$

rather labour intensive. With my current simulation software, PSpice 6.0 for DOS, the steps are:

- Simulate the output stage over a whole cycle, for each input voltage fraction; 5% steps give enough points for a presentable curve; the •STEP command automates this.
- Display simulation results in the graphical post-processor. (In PSpice this is called PROBE) This assumes it can display computed quantities, e.g., $V_{ce} \times I_c$ to give instantaneous device-power. Peak and average results can be read from the same display as PROBE. There is a function called AVG, which—unsurprisingly—yields the running average over a cycle. This stage can be automated as a macro, which is just as well, since it has to be performed at least 20 times, once for each input fraction value.
- The awkward bit. The computed peak and averaged power dissipations at the end of the cycle are read out from the PROBE cursor and recorded by hand, for each value of input fraction. There seems to be no other way to extract the information.
- The data from the third step is typed into Mathcad, to produce the graphs shown in this article. Once the

data has been entered, Mathcad can manipulate it in almost any way conceivable.

Power-partition diagrams

The graph in Figure 40.1 gives only one quantity, the amplifier dissipation.

I suggest a more informative graph format that I call a power partition diagram, which shows how the input power divides between amplifier dissipation, useful power in the load, and losses in drivers, etc.

Power dissipations are plotted against the input voltage fraction; this is not quite the same as the output voltage fraction as these are real output stages with gain slightly less than one. The input fraction increases in steps of 0.05, stopping at 0.95 to avoid clipping. The X-axis may linear or logarithmic.

Figure 40.3 shows the power-partition diagram for a Class-B complementary feedback pair stage as in Figure 40.2, which has a low quiescent current. Line 1 plots the P_{diss} in the sink (lower) device. Line 2 is source plus sink power. Line 3 is source plus sink plus load power.

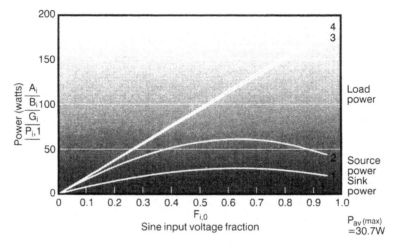

Figure 40.3 Power partition diagram for a Class-B complementary feedback pair driving an 8 Ω resistive load with a sinewave averaged over a cycle.

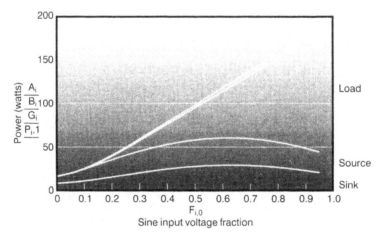

Figure 40.4 Class-B emitter follower power partition into 8 Ω resistive load. Sinewave drive. Significant quiescent dissipation at zero output.

The topmost line 4 is the total power drawn from the power supply, and so the narrow region between 3 and 4 is the power dissipated in the rest of the circuit—mainly the drivers and the output emitter resistors R_e. This power increases with output drive, but remains negligible compared with the other quantities examined.

The diagram shows immediately that the power drawn from the supply increases proportionally to the drive voltage fraction. This is partitioned between the load—represented by the curved region between lines 2 and 3—and the output devices. Note how the peak in their power dissipation accommodates the curve of the load power as it increases with the square of the voltage fraction.

Figure 40.4 shows the same diagram for a Class-B emitter-follower output stage. The quiescent current of an emitter follower output stage is significant—here 150 mA—and pushes up the power dissipation around zero output, but at higher levels the curves are the same. There is no need for extra heatsinking over the complementary feedback pair case.

Effects of increased bias

Figure 40.5 shows Class-AB, with bias increased so that Class-A operation and linearity is maintained up to 5 W r.m.s. output.

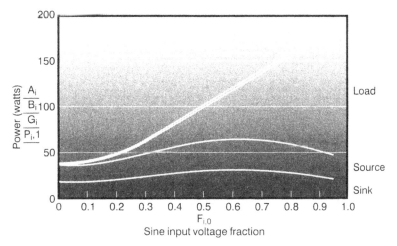

Figure 40.5 Class-AB sinewave drive. If the range of Class-A operation is extended, the area below level = 0.1 advances upwards and to the right.

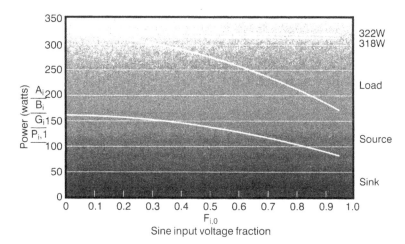

Figure 40.6 Class-A push-pull sinewave drive. Almost all the power drawn is dissipated in the amplifier, except at the largest outputs.

The quiescent current has increased to 370 mA, so quiescent power dissipation is significantly higher for output fractions below 0.1. Device dissipation is still greatest at a drive fraction of around 0.6, so once again no extra cooling is required to deal with the increased quiescent dissipation.

A push-pull Class-A amplifier draws a large standing current, and the picture looks totally different; see Figure 40.6. The power drawn from the supply is constant, but as output increases dissipation transfers from the output devices to the load, so minimum amplifier heating is at maximum output.

The significant point is that amplifier dissipation is only meaningfully reduced at a voltage fraction of 0.5 or more,

i.e. only 6 dB from clipping. Compared with Class-B, an enormous amount of energy is wasted internally.

Single-ended or constant-current versions of Class-A have even lower efficiency, worse linearity, and no corresponding advantages.

Class G

Hitachi introduced the Class-G concept in 1976 with the aim of reducing amplifier power dissipation by exploiting the high peak-mean ratio of music.[1] I have recently explained its operation in Ref. 2.

At low outputs, power is drawn from a pair of low-voltage rails; for the relatively infrequent excursions

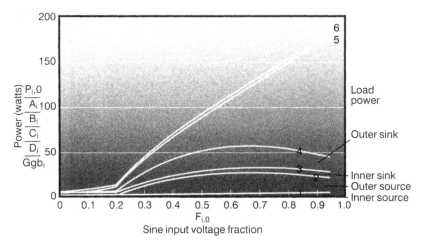

Figure 40.7 Class-G with lower supply rails set at 30% of upper rails, with sinewave drive. Compare Figure 40.4; amplifier dissipation at low levels is much reduced.

into high power, higher rails are drawn from. Here the lower rails are ±15 V, 30% of the higher ±50 V rails, so I call this Class G (30%).

This gives a discontinuous power-partition diagram, as in Figure 40.7. Line 1 is the dissipation in the low-voltage inner source device, which is kept low by the small voltages across it. Line 2 adds the dissipation in the high-voltage outer source; this is zero below the rail-switching threshold.

Above this are added the identical—due to symmetry—dissipations in the inner and outer sink devices, as Lines 3 and 4. Line 5 adds the power in the load, and Line 6 is the total power drawn, as before.

Power consumption and amplifier dissipation at low outputs are much reduced; above the threshold these quantities are only slightly less than for Class-B. Class-G does not show its power-saving abilities well under sinewave drive.

Class-B and reactive loads

The simulation method outlined above is also suitable for reactive loads. It is however necessary to run the simulation not just for one cycle, but sometimes for as many as twenty. This is to ensure that steady-state conditions have been reached.

The diagrams referred to below are for steady-state 200 Hz sinewave drive; the frequency must be defined so the load impedance can be set by suitable component values, but otherwise makes no difference.

Figure 40.8 shows what happens in Class-B emitter follower when driving a 45° capacitive-reactive load

with a modulus of 8 Ω. Comparing it with Figure 40.4, the power drawn from the supply is essentially unchanged, and is still proportional to output voltage fraction.

The larger areas at the bottom show that more power is being dissipated in the output devices and correspondingly less in the load, because the phase shift causes the voltage across and the current through the output devices to overlap more. The amplifier must dispose of 95 W of heat worst-case, rather than 60 W.

Average device dissipation no longer peaks, but increases monotonically up to maximum output. 45° phase angles are common when loudspeakers are driven. It is generally accepted that an amplifier should be able to provide full voltage swing into such a load.

When the load is purely reactive, with a phase angle of 90°, it can dissipate no power and so all that delivered to it is re-absorbed and dissipated in the amplifier. Figure 40.9 shows that the worst-case device dissipation is much greater at 185 W, absorbing all the power drawn from the supply, and therefore necessarily increasing monotonically with output level; there is no maximum at medium levels.

This is a very severe test for a power amplifier. It is also unrealistic, as no assemblage of moving-coil speaker elements can ever present a purely reactive impedance; 60° loads are normally the most reactive catered for. Table 40.1 shows the worst-case cycle-averaged dissipation for various load angles, showing how the position of maximum dissipation moves towards full output as the angle increases.

This is best displayed in 3D, as Figure 40.10, which plots power vertically; the slight hump at the

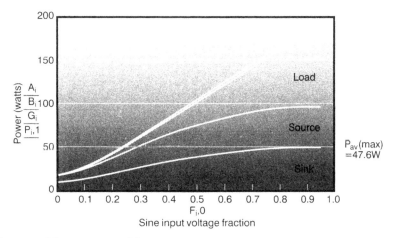

Figure 40.8 Class-B emitter follower reactive 45°, sinewave drive. The load is 11.3 Ω in parallel with 71 μF. Impedance modulus is 8 Ω at 200 Hz. Amplifier dissipation is increased, power delivered to the load decreased.

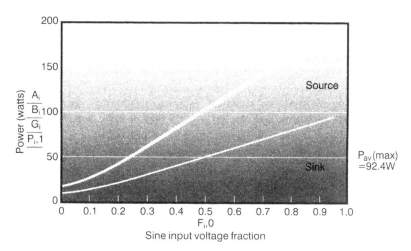

Figure 40.9 Class-B emitter follower reactive 90°, sinewave drive. Load is a 99.5 μF capacitor. Impedance modulus still 8 Ω at 200 Hz. All the supply power is now being absorbed by the amplifier, and none by the load.

Table 40.1

Angle (°)	$P_{diss(max)}$ (W)	Voltage fraction
0	60	0.64
10	63	0.65
20	67	0.70
30	70	0.75
45	95	0.95
60	115	1.00
90	185	1.00

front—non-reactive load—disappearing as the load becomes more reactive.

The dissipation hump is of little practical significance. An audio amplifier will almost certainly be required to drive 45° loads, and these cause higher power dissipations than resistive loads driven at any level.

Figure 40.11 shows the same plot for peak power, which increases monotonically with both output fraction and load angle. Figure 40.12 summarises all this data for design purposes. It shows worst-case peak and average power in one output device against load reactance.

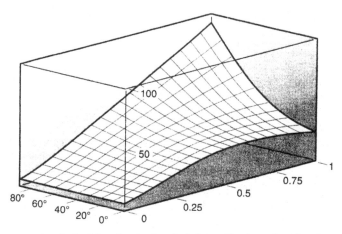

Figure 40.10 The average power (vertical axis) against load angle (left-hand horizontal axis) and output fraction (right-hand horizontal axis).

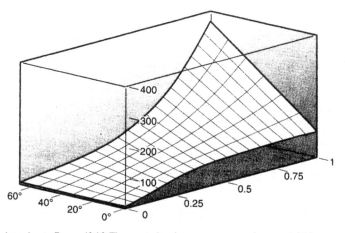

Figure 40.11 Peak power plotted as in Figure 40.10. The vertical scale must accommodate much higher power levels than Figure 40.10.

Peak powers are taken at 0.95 of full output, average power at whatever output fraction gives maximum dissipation. Therefore to design an amplifier to cope with 45° loads, note that average power is increased by 1.4 times, and peak power by 2.7 times, over the resistive case. This can mean that it is necessary to increase the number of output devices simply to cope with the much enhanced peak power.

Considering simple reactive loads like those listed in the panel 'Reactive load observations' gives an essential insight into the extra stresses they impose on semiconductors but is still some way removed from real signals and real loudspeaker loads, where the impedance modulus varies along with the phase, due to electromechanical resonances or crossover dips.

I looked at single and two-unit loudspeaker models in Ref. 3 where the maximum phase angle found was 40°. In brief, the results were:

- Amplifier power consumption and average supply current drawn vary with frequency due to impedance modulus changes.
- The peak device current increased by a maximum of 1.3 times at the modulus minima.
- The average current in the output devices increased by a maximum of 1.3 times.
- Peak device power increased by a maximum factor of 2, mostly due to phase shift rather than impedance dips.
- Average device dissipation increased by a maximum of 1.4 times.

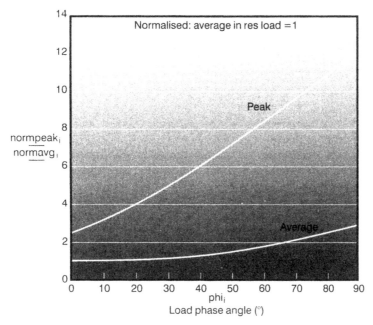

Figure 40.12 Peak power increases faster than worst-case average power as the load becomes more reactive and its phase angle increases. Class-B emitter follower as before.

Reactive load observations

The following conclusions apply to reactive loads.

- Amplifier power consumption and average supply current drawn do not vary with load phase angle if the impedance modulus remains constant.
- Peak device current is not altered so long as the impedance modulus remains constant.
- Average current in the output devices is not altered so long as the impedance modulus remains constant. This follows from the first observation
- Peak device power increases rapidly, as the load becomes more reactive. A 45° load increases power peaks by 2.7 times, and a 60° load by 3.4 times. See Figure 40.12.
- Average device dissipation also increases, but more slowly, as the load angle increases. A 45° load increases average dissipation by 1.4 times and a 60° load by 1.8 times, Figure 40.12.

These numbers come from two specific models that attempted to represent 'average' speakers. Worse conditions could easily have been found.

Ultimately a comprehensive survey of the loudspeakers on the market would be required, but this would be very time-consuming. In Ref. 4, which gives an excellent account of real speaker loading, 21 models were tested and the worst angle found was 67°. Eliminating the two most extreme cases reduced this to 60°.

The most severe effect of reactive loads is the increase in peak power, followed by the increase in average power.

Both are a strong function of load phase, and so the specification of the maximum angle to be driven has a big effect on the devices required, heat-sink design, and hence on amplifier cost.

It is likely that a failure to appreciate just how quickly peak power increases with load angle is the root cause of many amplifier failures.

References

1. Feldman, 'Class-G high efficiency hi-fi amplifier', *Radio-Electronics*, August 1976, p 47.

2. Self, D. 'Classification of power amplifiers', *Electronics World*, March 1999, p 90.

3. Self, D. 'Speaker impedance matters', *Electronics World*, November 1997, p 920.

4. Benjamin, E. 'Audio power amplifiers for loudspeaker loads', *Journal Audio Engineering Society*, Vol. 42(9), September 1994, p 670.

Chapter 41

Audio power analysis

(Electronics World, December 1999)

Intended title: *Power Partition: Part 2*

This article extended the idea of power-partition diagrams to include real waveforms of real music, and therefore required rather a lot of tedious measurement to acquire the probability density data. One of the surprises was that very different musical genres had very similar statistics. It was however reassuring to discover that the demands made on an amplifier by a sinewave were greater than those for music, but not wildly out of line. In other words, sinewave testing builds in a safety margin—if it works on sinewaves it will certainly survive music. Sinewave testing has received a lot of unkind criticism over recent years, but it continues unabated because it works, and this is one of the reasons why.

In preparing the information for this article, it emerged that a push-pull Class-A power amplifier, which is the most efficient kind of Class-A power amplifier there is, has a maximum efficiency of around 1% when handling real-life signals rather than sine waves. This assumes maximum output, defined as occasional clipping. This is as good as it gets—reducing the level quickly makes the efficiency drop off to almost zero. Other versions of Class-A, using constant-current or resistive operation, are worse by factors of two and four respectively. Figure 41.13 shows almost every drop of input power being dissipated in the output devices, the quiescent current sluicing through the transistors from one supply rail to another with the merest trickle being diverted into the load. As with the previous article, the graded shading that was imposed on the graphs by EW obscures the traces at the top, and Figure 41.13 suffered particularly badly. The power into the load is the thin wedge shaped area at the top.

This has put me off designing Class-A amplifiers for a long time, though I'm always happy to do it if someone will pay me. A branch of technology with a practical energy efficiency of less than 1% needs a good justification for its existence, and the distortion from a Blameless Class-B amplifier, especially with the addition of output-inclusive compensation (see Chapter 45) is so low that the extra linearity of Class-A amplifiers is no longer the forceful argument it once was. A Class-A amplifier is now more of a philosophical statement than a technical one.

There is still, and no doubt always will be, interest in trying to bridge the gap between Class-B and Class-A operation, and obtain the efficiency of the first with the linearity of the second. A well-

Waveform PMR	Class-B	Quartic-law Class-A	Cube-law Class-A	Square-law Class-A	Linear Class-A
Sinewave 3 dB	78.5%	77.4%	74.7%	66.7%	50%
Pop music 7 dB	60%	53.9%	46%	33.3%	20%
Classical 12 dB	52.4%	29.3%	19.2%	11.1%	5.9%
Classical 25 dB	35%	2.4%	1.2%	0.6%	0.3%

Note: Table reproduced by kind permission of Ian Hegglun and Jan Didden.

known attempt was Technics 'Super-Class-A' in which a Class-A amplifier ran from floating ±15V rails that were swung up and down by a Class-B amplifier. Both amplifiers have to be able to pass the full load current, and this is clearly a complicated and expensive approach. It did not prosper.

Currently, there is interest in curvilinear operation, in which the output device current changes with voltage according to a square or cube law, for example [1], which reduces the quiescent dissipation. This is ably demonstrated in the table produced by Ian Hegglun for signals with different Peak Mean Ratios (PMRs) at full output level. The efficiency of all versions of Class-A collapses dramatically on going from a sinewave to 25dB PMR classical music, and I am glad to see my finding that linear Class-A has sub-1% efficiency in real life is confirmed. The extra data for quartic law (fourth-power) was kindly provided by Ian Hegglun [2].

As I suggested in the preface to Chapter 35, there is a continuum of curved laws between linear Class-A and optimally biased Class-B, and this idea is reinforced by the steady reduction in sinewave dissipation in the top row as we go from linear Class-A to square-law, cube-law, quartic-law, and finally Class-B, which I suppose you could call 'exponential law'. It's worth pointing out that none of the various curvilinear designs have so far got anywhere near the low distortion of a Blameless Class-B amplifier.

References

1. Hegglun, I., A Cube-law Audio Power Amplifier, *Linear Audio,* Volume 8, Sept. 2014, p. 149.

2. Hegglun, I., Private communication, Feb. 2015.

AUDIO POWER ANALYSIS

December 1999

My last chapter showed how the power consumed by amplifiers of various classes was partitioned between internal dissipation and the power delivered to the load.[1] This was determined for the usual sinewave case.

The snag with this approach is that a sinewave does not remotely resemble real speech or music in its characteristics. In many ways it is almost as far from it as you could get.

In particular, it is well-known that music has a large peak-to-mean ratio, or PMR, though the actual value of this ratio in decibels is a vague quantity. Signal statistics for music appear to be in surprisingly short supply.

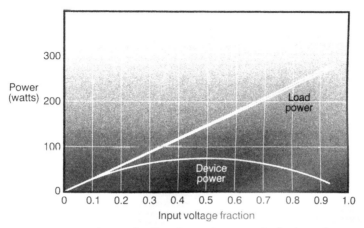

Figure 41.1 Instantaneous power partition diagram for Class-B complementary-feedback pair. Power in the output devices peaks when output is at half the rail voltage.

Very roughly, general-purpose rock music has a PMR of 10 dB to 30 dB, while classical orchestral material—which makes very little use of fuzz boxes and the like—is 20 to 30 dB. The muzak you endure in lifts is limited in PMR to 3 to 10 dB, while compressed bass material in live PA systems is similar.

It is clear that the power dissipation in PA bass amplifiers is going to be radically different from that in hi-fi amplifiers reproducing orchestral material at the same peak level. The PMR of a sinewave if 4.0 dB, so results from this are only relevant to lifts.

Recognising that music actually has a peak-to-mean ratio is a start, but it is actually not much help as it reduces the statistics of signal levels to a single number. This does not give enough information for the estimation of power dissipation with real signals.

To calculate the actual power dissipations, two things are needed; a plot of the instantaneous power dissipations against level, and a description of how much time the signal spends at each level. The latter is formally called the 'probability density function', or PDF, of the signal; more on this later.

The instantaneous power partition diagram, or IPPD, is obtained by running the output stage simulation with a sawtooth input and no per-cycle averaging. Instantaneous power dissipation can therefore be read out for any input voltage fraction simply by running the cursor up the sawtooth.

Figure 41.1 is the instantaneous power partition diagram for the Class-B complementary-feedback pair case, where the quiescent current is very small. This looks very much like the averaged-sinewave power partition diagram in reference,[1] but with the device

dissipation maximum at 50% voltage rather than 64% for the sinewave case.

The instantaneous powers are much higher, as they are not averaged over a cycle. There is only one device-power area at the bottom as only one device conducts at a time. Output device dissipation at the moment when the signal is halfway between rail and ground, input fraction 50%—is 76 W, and the power in the load is 75 W. This total to 151 W, on the lower of the two straight lines, while the power drawn from the supply is shown as 153 W by the upper straight line. The 2 W difference represents losses in the driver transistors and the output emitter resistors.

All the IPPDs for various output stages look very similar in shape to the averaged-sine PPDs in Ref. 1, but the peak values on the Y-axis are higher. The IPPD can be combined with any PDF to give a much more realistic picture of how power dissipation changes as the level of a given type of signal is altered.

The probability density function

The most difficult part of the process above is obtaining the probability density function. For repetitive waveforms the PDF can be calculated, [2] but music and speech need a statistical approach. It is often assumed that musical levels have a Gaussian (normal) probability distribution, as the sum of many random variables.

Positive statements on this are however hard to find. Benjamin [3] says, 'music can be represented accurately as a Gaussian distribution' while Raad [4] states, 'music and mixed sounds typically have Gaussian PDFs'. It appears likely this assumption is true for multi-part music which can be regarded as a summation of many random processes;

whatever the PDF of each component, the result is always Gaussian as indicated by the Central Limit Theorem.

If the distribution is Gaussian, its mean is clearly zero, as there is no DC component, which leaves the variance—i.e. width of the bell-curve—as the only parameter left to determine. The Gaussian distribution tails off to infinity, implying that enormous levels can occur, though very rarely.

In reality the headroom is fixed. I have dealt with this by setting variance so the maximum value, 0 dB. occurs 1% of the time. This is realistic as music very often requires judicious limiting of occasional peaks to optimise the dynamic range.

The PDF presents some conceptual difficulties, as it shows a density rather than a probability. If a signal level ranges between 0 and 100%, then clearly it might be expected to spend some of its time around 50%.

However, the probability that it will be at exactly 50.000% is zero, because a single level value has zero extent. Hence the PDF at x is the probability that the signal variable is in the interval $(x, x + dx)$, where dx is the usual calculus infinitesimal.

The cumulative distribution function

If the probability that the instantaneous voltage will be above—not at—a given level is plotted against that level, a cumulative distribution function, or CDF, results. This is important as it is easier to measure than the PDF.

If the variable is x, then the PDF is often called $P(x)$ and the CDF called $F(x)$. These are related by:

$$P(x) = \frac{d}{dx}F(x)$$

or,

$$F(x) = \int_0^x P(a)\,da$$

where a is a dummy variable needed to perform the integration. The integration starts at zero in this case because signal levels below zero do not occur.

Generating a CDF by integrating a given PDF is straightforward, but going the other way—determining the PDF from the CDF—can be troublesome as the differentiation accentuates noise on the data.

Some probability density functions

Figure 41.2 shows the calculated PDF of a sinewave. As with every PDF, the area under the curve is one, because the signal must be at some level all of the time.

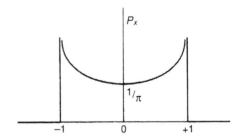

Figure 41.2 Probability density function, or PDF, of a sinewave. Peaks at each end go towards infinity.

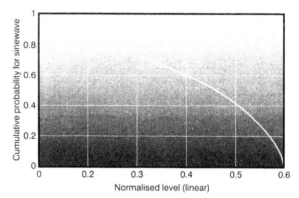

Figure 41.3 Cumulative distribution function, CDF, of a sinewave. Drawn with measured data from the circuit of Figure 41.4 as a reality check.

However, the function blows up—i.e. heads off to infinity—at each end because the peaks of the wave are 'flat', and so the signal dwells there for infinitely longer than on the slopes where things are changing. These 'flat' bits are infinitely small in time extent though, and so the area under the curve is still unity. This shows you why PDFs are not always the easiest things to handle.

The CDF for a sinewave is shown in Figure 41.3; the probability of exceeding the level on the axis falls slowly at first, but then accelerates to zero as the rounded peaks are reached.

Measuring probability density functions

But is all music Gaussian? I was not satisfied that this had been conclusively established from just two brief references.

I decided it was essential to make some attempt to determine musical PDFs. In essence this is simple. The first thing to decide is the length of time over which to examine the signal. For most contemporary music the obvious answer is 'one track', a complete composition lasting typically between three and eight minutes.

Very simple circuitry can be used to determine a CDF, and hence the PDF, though the process is protracted. A variable-threshold comparator is driven by the signal to be measured, and its output applied to a long-period averaging time-constant. Figure 41.4.

A comparator, IC_{1a}, rather than an op-amp, is used to avoid inaccuracies due to slew-rate limiting. Reference IC_2 is an inexpensive 2.56 V bandgap type, while VR_1 sets the comparator threshold. When the signal level is below this threshold, the comparator open-collector output is off, and the voltage seen by the averaging network is zero.

When the signal exceeds threshold, the comparator output is pulled low, so this point carries an irregular rectangular waveform while signal is applied. The average value of this is derived by R_3 and C_2, buffered by IC_{3a}, and drives a moving-coil meter through a suitable resistance R_5.

Switch SW_1 and R_4 enable a quick reset when no signal is present. A moving-coil meter allows much easier reading of a changing signal, though not to any great accuracy.

Potentiometer VR_2 sets the scale so that the meter deflects to full scale for a 100% reading. This is done with no input, so it is essential to check that the circuit offsets have put the comparator in the right state—i.e. output low; if not the inverting input will need to be pulled fractionally negative by a high-value biasing network.

The circuit only measures one polarity of the waveform, in this case the positive half, so signal symmetry is assumed. This is safe unless you plan to do a lot of work with solo instruments or single a cappella voices; the human vocal waveform is notably asymmetrical.

This minimal system is simple, but it only yields one data point at a time. Set the threshold level to say 50%, play the track—I'd pick a short one and as it finishes the reading on the meter shows the percentage of time the signal exceeded the preset level.

Since twenty data points are required for a good graph, this gets pretty tedious. The four comparators of IC_1 could give four points, if the time-constant section was also quadrupled, and some means of freezing the output voltages provided.

The CDF thus obtained for Alannah Myles' 'Black Velvet' is Figure 41.5, and the PDF derived from it is Figure 41.6. It comes complete with some rather implausible ups and downs produced by differentiating data that is accurate to ±1% at best.

I measured several rock tracks, and also short classical works by Albinoni and Bach. The results are surprisingly similar; see the composite CDF in Figure 41.7. This is good news because we can use a single PDF to evaluate amplifiers faced with varying musical styles. However, I decided the method needed a reality check, by deriving the PDF in a completely different way.

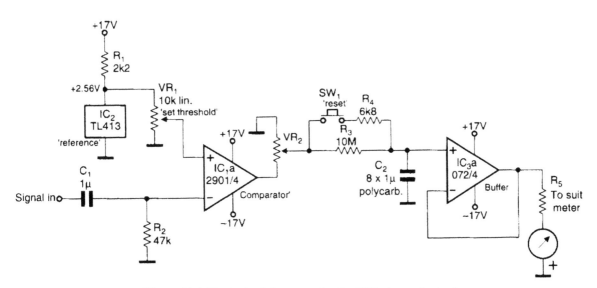

Figure 41.4 Simple circuit for measuring the CDF of an audio signal.

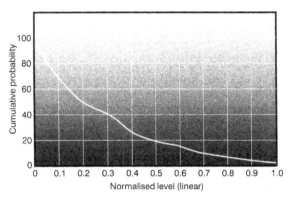

Figure 41.5 The cumulative distribution function obtained from the PDF of Alannah Myles performing 'Black Velvet' by Figure 41.4.

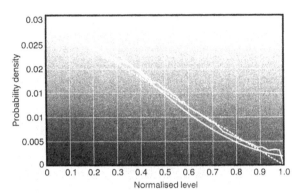

Figure 41.8 The PDF of disco music, sorted into 65 amplitude bins by DSP. Also shown are a Gaussian distribution (smooth curve) and a triangular distribution (dotted line).

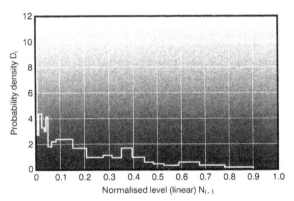

Figure 41.6 Probability density function derived from Alannah Myles performing 'Black Velvet'.

Probability density functions Via DSP

A digital processor offers the possibility of determining as many data points as you want on one playing of the music specimen. In this case a very simple 56001 program sorts the audio samples into 65 amplitude bins.

The result for 30 s of disco music is Figure 41.8, which is somewhere between triangular and Gaussian, if the latter has appropriate variance. The important point is that the difference between them is very small, and either can be used. The triangular PDF simplifies the mathematics, but if like me you use *Mathcad* to do the work, it is easy to plug in whatever distribution seems appropriate.

Deriving actual power

Having found the PDF, it is combined with the power partition diagram. In this case the IPPD is divided into twenty steps of voltage fraction, and each one multiplied by the probability the signal is in that region.

The summation of these products yields a single number—the average power dissipation in watts for a real signal that just reaches clipping for 1% of the time. An obvious extension of the idea is to plot the average power derived as above, against signal level on the X-axis. This gives an immediate insight into how amplifer power varies as the general signal level is reduced, as by turning down the volume control.

Figure 41.9 shows how level changes affect the PDF. Line 1 is maximum volume, just reaching full volume at the right. Line 2 is half volume, –6 dB, and so hits the X-axis at 0.5; it is above Line 1 to the left as the

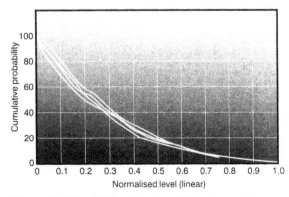

Figure 41.7 The CDFs of 3 rock and 1 classical tracks, showing only small differences.

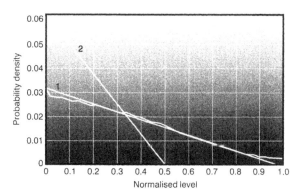

Figure 41.9 The triangular PDF, and how level changes affect it. Line 1 is full volume, and Line 2 half volume.

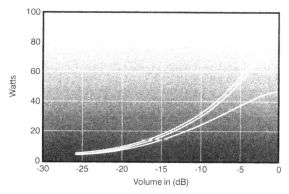

Figure 41.11 Class-B complementary feedback pair plotted with volume on a more useful logarithmic (decibel) X-axis. The shape looks quite different from Figure 41.10.

probability of lower levels must be higher to maintain unity area under it.

This process continues as volume is reduced, until at zero volume the zero-level probability is 1 and all other levels have zero probability. Having generated twenty PDF functions, the powers that result for each one are plotted with the volume setting—not the output fraction—as the X-axis. The results for some common amplifer classes are as follows.

Class-B

The instantaneous power plot for Class-B complementary feedback pair combined with a triangular PDF of Figure 41.10 illustrates how the load and device power varies with volume setting. A signal with triangular PDF spends most of its time at low values, below 0.5 output fraction, and so there is no longer a dissipation

maximum around half output. Device dissipation at bottom increases monotonically with volume. Load power increases with a square-law, which is a reassuring check on all these calculations.

Figure 41.11 is Figure 41.10 replotted with a logarithmic X-axis, which is more applicable to human hearing. Domestic amplifiers are rarely operated on the edge of clipping; a realistic operating point is more like −15 or −10 dB. The plot reveals that here the efficiency is low, with much more power dissipated in the devices than reaches the load.

Class-AB

A decibel plot for Class-AB, biased so Class-A operation is maintained up to 5 W r.m.s. output is shown in Figure 41.12. Quiescent current is now 370 mA, so

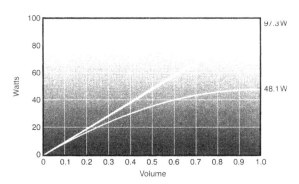

Figure 41.10 Class-B complementary feedback pair power partition versus level. The Class-B IPPD has been combined with the triangular PDF. Device dissipation (lower area) now increases monotonically with volume.

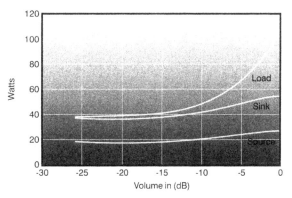

Figure 41.12 Class-AB Power Partition Diagram, stage biased to give Class-A up to 5W. Averaged over whole cycle.

there is greater quiescent dissipation at zero volume. There is also substantial conduction overlap, and so sink and source would be different if the plot only considered voltage excursions in one direction away from 0 V. When positive and negative half-cycles are averaged, symmetry is achieved. The total device dissipation is unchanged but the boundary between the source and sink areas is half way, as in Figure 41.12.

Class-A push-pull

I have stuck with the same ±50 V rails for ease of comparison, and this yields a very powerful Class-A amplifier. The power drawn from the load is constant, and as output increases dissipation transfers from the output devices to the load, giving minimum amplifier heating at maximum output.

The result for sinewave drive is bad enough, [1] but Figure 41.13 reveals that with real signals, almost all the energy supplied is wasted internally—even at maximum volume. Class-A has always been stigmatised as inefficient; this shows that under realistic conditions it is hopelessly inefficient, so much so that it grates on my sense of engineering aesthetics. At typical listening volumes of −15 dB the efficiency barely reaches 1%.

Class-G

This class of amplifiers was introduced by Hitachi in 1976 to reduce amplifier power dissipation by exploiting the high peak-mean ratio of music.[5] Class-G made little headway in the hi-fi market as the power saving does not outweigh the increased circuit complexity, but the rise of five-channel home theatre applications has caused a revival of interest in improved amplifier efficiency.

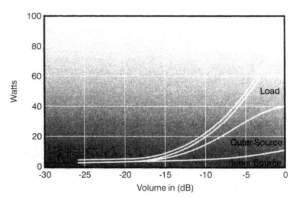

Figure 41.14 Class-G-30%. Low rail voltage is 30% of the high rail. Rail-switching occurs at about −15 dB relative to maximum output.

I recently explained Class-G in Ref. 6. At low outputs, power is drawn from low-voltage rails; for the relatively infrequent excursions into high power, higher rails are switched in.

In Figure 41.14 the lower rails are ±15 V, 30% of the higher ±50 V rails; I call this Class-G-30%. The lower area is the power in the inner devices—i.e. those in all the time. The larger area just above is that in the outer devices, i.e. those only activated when running from the higher rails. This is zero below the rail-switching threshold at a volume of 0.2.

Total device dissipation is reduced from 48 W in Class-B to 40 W, which is not a good return for twice as many power transistors. This is because the lower rail voltage is poorly chosen for signals with a triangular PDF.

If the low rails are increased to ±30 V this become Class-G-60% as in Figure 41.15. Here the low-dissipation region now extends up to a voltage fraction of 0.5, but

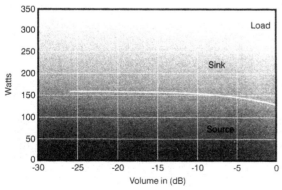

Figure 41.13 Class-A push-pull, for 150 W output. The internal dissipation completely dominates—even at maximum volume.

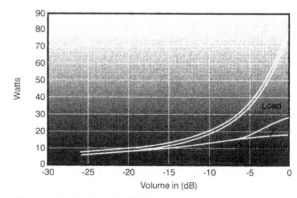

Figure 41.15 Class-G-60%. The low rail voltage is now 60% of the high rail. This reduces both dissipation and power consumed, compared with Figure 41.14.

inner device dissipation is higher due to the increased lower rail voltages.

The overall result is that total device power is reduced from 48 W in Class-B to 34 W, which is a definite improvement. I am not suggesting that 60% is the optimum lower-rail voltage. The efficiency of Class-G amplifiers depends very much on signal statistics.

Reactive loads

The disadvantage of using instantaneous power is that it ignores signal and circuit history, and so cannot give meaningful information with reactive loads. The peak dissipations that these give rise to with real signals are difficult to simulate; it would be necessary to drive the circuit with stored music signals for many cycles; and that would only cover a few seconds of a CD or concert. The anomalous speaker currents examined in Ref. 7 show how significant history effects can be with some waveforms.

In summary

Tables 41.1 and 41.2 summarise how a triangular-PDF signal—rather than a sinewave—reduces average power dissipation, and the power drawn from the supply.

These economies are significant; the power amplifier market is highly competitive, and it is essential to exploit the cost savings in heat-sinks and ower-supply components made possible by designing for real signals rather than sinewaves.

Table 41.1 Device dissipation, worst-case volume

	Sinewave (W)	PDF (W)	Factor
Class-B CFP	64	48	0.75
Class-AB	64	55	0.78
Class-A, push-pull	324	324	–
Class-G-30%	43	40	0.93
Class-G-60%	56	34	0.61

Table 41.2 Power drawn, worst-case. Always maximum output

	Sinewave (W)	PDF (W)	Factor
Class-B CFP	186	97	0.52
Class-AB	188	105	0.58
Class-A, push-pull	324	324	–
Class-G-30%	177	93	0.52
Class-G-60%	169	81	0.48

In particular, Class-G shows valuable economies in device dissipation and ower-supply capacity, though to reduce dissipation, the lower supply voltage must be carefully chosen. This approach is unlikely to reduce the number of power devices required as real signals give no corresponding reduction in peak device power or peak device current.

References

1. Self, D. 'Cool audio power', *Electronics World*, August 1999, p 657.

2. Carlson, A. 'Communication systems', McGraw-Hill, Third Edn 1986, p 144.

3. Benjamin, E. 'Audio power amplifiers for loudspeaker loads', *Journal Audio Engineering Society*, Vol. 42, September 1994, p 670.

4. Raad, 'Average efficiency of class-G amplifiers', *IEEE Trans on Consumer Electronics*, Vol CE-22(2), May 1986.

5. Feldman, 'Class-G high efficiency hi-fi amplifier', *Radio-Electronics*, August 1976, p 47.

6. Self, D. 'Class Distinction: amplifier classes', *Electronics World*, March 1999, p 190.

7. Self, D. 'Loudspeaker undercurrents', *Electronics World*, February 1988, p 98.

Chapter 42

Crossover displacement

A new kind of amplifier
(Electronics World, November 2006)

Now that's more like a proper title. Yes, I chose it myself.

Like several of the ideas described in these articles, this one goes a good deal further back than you might think. The germ of the idea was formed in 1976. At that time, there was a popular quad opamp called the LM324, which was an economical way of deploying four opamps, but which in audio applications gave a fairly pitiful distortion performance due to its markedly asymmetrical load-driving capabilities.

A useful way of easing this situation was to connect a resistor between the opamp output and the +V supply rail, and it was standard design procedure during my time at Electrosonic. Forty years later, this method of reducing opamp distortion has become quite well-known, often being described as 'forcing the output stage into Class-A', which I suppose is reasonably accurate, though I just call it 'output biasing'. It has been applied indiscriminately to many opamps, not just those with questionable load-driving abilities. The only one I have tried it on recently is the 5534/5532, where injecting 5 mA into the output from the V+ rail gives a modest but dependable and easily measurable reduction in distortion; take a unity-gain shunt feedback stage driving 5 Vrms into a 1 kΩ load, and with the 5 mA bias added THD drops from 0.00050% to 0.00040% below 20 kHz. I am not going to pretend that such an improvement is revolutionary, but it is robust and almost cost-free. Switching to the LM4562 opamp will give better THD results, but at considerable increase in cost. Output biasing the 5534/5532 with more than 5 mA causes linearity to start getting worse again. Bizarrely the advocates of this practice almost always connect the resistor *to the wrong supply rail*, i.e., V−, which makes the distortion worse rather than better. Of such are the fruits of blind copying.

The output biasing of the 5534/5532 is dealt with in detail in my book on active crossovers [1]. Regrettably it is *not* included in the 2nd edition of *Small-Signal Audio Design* [2], for simple reasons of space.

This opamp back-story has led the odd ill-disposed person to question the originality of the Crossover Displacement power amplifier. However I do assert, and the British Patent Office agrees with me [3], that I was the first person to think of applying it to power amplifiers. There is no question that I was the first to use a signal-modulated displacement current to improve Crossover Displacement power amplifier efficiency and linearity.

The first Crossover Displacement amplifier to go into production was the integrated Cambridge Audio 840A in 2005, followed by the stand-alone power amplifier 840W in 2006; an 840A V2 with unchanged displacement circuitry was released in 2008. The integrated Cambridge Audio 851A was released in 2014; I do not have definitive information on its internals, but from what I have learnt, the displacement circuitry appears to be still the same.

This seems to indicate that Crossover Displacement technology has not progressed much since it was introduced. So far as Cambridge Audio is concerned, that may be true, though I stress I have incomplete knowledge of the 851A. Developments are in hand at my end that I cannot describe here, but I can tell you about one attempt to improve the technology. In the Prefaces to Chapters 35 and 41, what is usually called 'curvilinear Class-A' is described, in which non-linear laws are used to increase power efficiency. This has not so far led to much in terms of low distortion, but I thought there might be mileage in applying a similar principle to Crossover Displacement. The displacer drives a current into the output of the amplifier, which is controlled by a healthy amount of negative feedback. Therefore, if the modulated displacer current is non-sinusoidal, working on a law that reduces the dissipation due to the displacement current, the effect on the overall distortion should be very small. This pretty much works as described, but I found the increases in efficiency were marginal, especially when compared with the increased complication of the displacement circuitry. There the idea rests for the moment.

I have always thought it slightly regrettable that the first Class-XD amplifier (840A) was a fairly powerful design, using two pairs of output devices apart from the displacer transistor, and the next version (840W) was even bigger using four output pairs. The large number of output devices tended to obscure the intriguing fact that with the displacer device included, there was an *odd number of them*. I have not so far built a low-power Class-XD amplifier with a total of three output devices, but it needs doing because it really gets across the unique method of operation. I will hold it aloft for the photographer with a knowing smile.

References

1. Self, D., *The Design of Active Crossovers,* pp. 389–394. Focal Press 2011. ISBN 978-0-240-81738-5.

2. Self, D., *Small-Signal Audio Design,* 2nd edition. Focal Press (Taylor & Francis) 2014. ISBN: 978-0-415-70974-3 Hardback. 978-0-415-70973-6 Paperback. 978-0-315-88537-7 ebook.

3. Self, D., British patent GB 2,424,137B. Granted and published 13th December, 2006.

CROSSOVER DISPLACEMENT

The state of the art of amplifiers

The great divide in solid-state amplifier technology has always been between the efficient but imperfect Class-B approach and the beautifully linear but dishearteningly inefficient Class A.

It is indisputable that Class A power amplifiers have the potential to give the best linearity, (though a good many designs have thrown this away by using non-linear small-signal circuitry) but they are usually impracticable due to their high power dissipation. A Class-A amplifier can be 50% efficient with a maximum sinewave output, but when reproducing a real music signal this falls to one or two percent.

Class-B linearity can be very good. I have introduced the Blameless amplifier design methodology [Ref 1], and it gives very good linearity- typically below 0.001%-at 1 kHz- especially in its Load-Invariant [Ref 2] form. This approach however, has its limitations as a Class-B amplifier inherently generates crossover distortion, and inconveniently displays this non-linearity at the zero-crossing, where it is always in evidence no matter how low the signal amplitude. At one unique value of quiescent current the distortion produced is a minimum, and this is what characterises optimal Class-B; however at no value can it be made to disappear. It is inherent in the classical Class-B operation of a pair of output transistors.

Fig 42.1 shows a simulation of an output stage that illustrates the heart of the problem. The diagram plots the incremental gain of the output stage against output voltage; in other words the gain for a very small signal. A complementary-feedback pair (CFP) output stage was used. Both 8 and 4 Ohm loads are shown. You can see from the Y-axis that in the 8 Ohm case in particular, the gain variations are very small, with an inoffensive-looking gain ripple around the zero-crossing at 0V output. Unfortunately this gain ripple generates high-order harmonics that are poorly linearised when the negative-feedback factor falls with increasing frequency, as it usually does.

The 4 Ohm case shows lower overall gain due to the increased loading, and a dropoff of gain at each side due to falling transistor beta; this latter distortion mechanism is relatively easy to deal with by the use of negative feedback and other methods, and is not considered further. The awkward gain perturbations at the crossover point are similar to the 8 Ohm case, but larger in size.

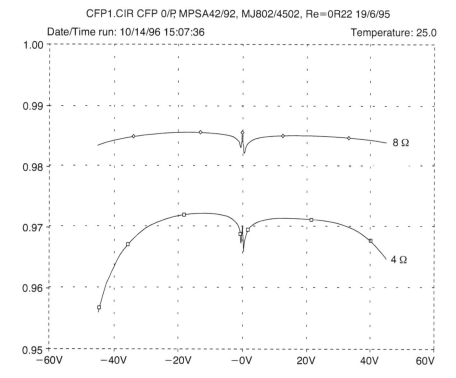

Fig 42.1 The incremental gain of a CFP output stage as output voltage changes. SPICE simulation for 8 and 4 Ohm loads.

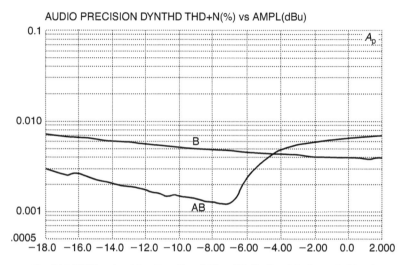

AUDIO PRECISION DYNTHD THD+N(%) vs AMPL(dBu)

Fig 42.2 THD vs level for Class B and Class AB. (0 dB is 30W into 8 Ohms)

There has always been a desire for a compromise between the efficiency of Class-B and the linearity of Class-A, and the most obvious way to make one is to turn up the quiescent current of a Class-B stage. As this is done, an area of Class-A operation, with both output transistors conducting, is created around the zero-crossing. This area widens as the quiescent current increases, until ultimately it encompasses the entire voltage output range of the amplifier, and we have created a pure Class-A design where both output transistors are conducting all the time. There is thus an infinite range of positions between the two extremes of Class-B and Class-A, and this mode of operation is referred to as Class-AB. Unfortunately, while Class-AB is a compromise between Class-A and Class-B operation, it is not a very good compromise.

In my series on power amplifier distortion [Ref 1] I showed that if Class-AB is used to tradeoff between efficiency and linearity, its performance is certainly superior to B below the AB transition level, as it is pure Class A. This can have very low THD indeed, at less than 0.0006% up to 10 kHz, as demonstrated in [Ref 3]. However, once the signal exceeds the limits of the Class-A region, the THD worsens abruptly due to the sudden gain-changes when the output transistors turn on and off, and linearity is inferior not only to Class-A but also to optimally-biased Class-B. This effect is often called "gm-doubling" in the literature. Class-AB distortion can be made very low by proper design, such as using the lowest possible emitter resistors, [Ref 4] but remains at least twice as high as for the equivalent Class-B situation. The bias control of a Class-B amplifier does NOT give a straightforward tradeoff between power dissipation and linearity at all levels, despite the

constant repetition this notion receives in some parts of the audio press. To demonstrate this, Fig 42.2 shows THD plotted against output level for Classes AB and B.

It would be much more desirable to have an amplifier that would give Class-A performance up to the transition level, with Class-B after that, rather than AB. This would abolish the abrupt AB gain changes that cause the extra distortion. This article shows how it can be done.

This design would not have been attempted without the Blameless amplifier concept to provide a defined point of departure for new developments; the distortion performance can be immediately attributed to output stage behaviour, instead of being obscured by non-linearity from the small-signal stages.

The crossover displacement principle

In Class-B it would be better if the crossover region were anywhere else rather than where it is. If the crossover is displaced away from its zero-crossing position, then the amplifier output does not traverse it until the output reaches a certain voltage level. Below this transition point the performance is pure Class-A; above it the performance is normal Class-B, the only difference being that crossover discontinuities on the THD residual are no longer evenly spaced. The harmonic structure of the crossover distortion produced is not significantly changed, as explained in more detail below.

The essence of the Crossover Displacement principle is the injection of an extra current, either fixed or varying with the signal, into the output point of a conventional Class-B amplifier. This is shown in Fig 42.3.

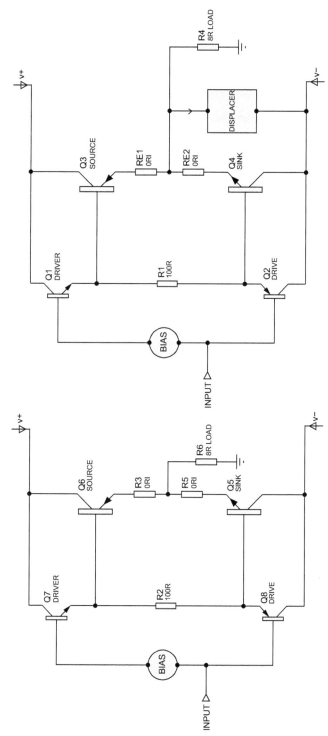

Fig 42.3 On the left is a conventional Class-B output stage, showing drivers and bias voltage source. At the right the stage has been modified by adding a displacer system that draws current from the output and sinks it into the negative rail.

For convenience I have called the current-injection subsystem The Displacer. Similarly the upper transistor is The Source, while the lower is The Sink. The displacement current does not directly alter the voltage at the output- the output stage inherently has a low output impedance, and this is further lowered by the use of global negative feedback. What it does do is alter the pattern of current flowing in the output devices. The displacement current in the version shown here is sunk to V- from the output, rather than sourced from V+, so the crossover region is displaced downward rather than being pulled upwards. This is arbitrary as the direction of displacement makes no difference. The extra current therefore flows through Re1, and the extra voltage drop across it means the output voltage must go negative before the current through Re1 stops and that in Re2 starts. In other words, the crossover point when Q2 hands over to Q4 has been moved to a point negative of the 0V rail; for the rest of this article I refer to this as the "transition point". For output levels below transition no crossover distortion is generated. The resulting change in the incremental gain of the output stage is shown in Fig 42.4. The displacer current may be constant, or vary with the signal.

We now have before us the intriguing prospect of a power amplifier with three output devices, which if nothing else is novel. The operation of the output stage is inherently asymmetrical, and in fact that is its raison d'etre, but this should not cause undue alarm. Circuit symmetry is often touted as being a pre-requisite for either low distortion or healthy operation in general, but this has no real basis in fact. A perfectly symmetrical circuit may have no even-order distortion, but it may still have any amount of odd-order non-linearity, such as a cubic characteristic. Odd-order harmonics are normally considered more dissonant than even-order, so circuit symmetry in itself is not a useful goal.

If the positive-going and negative-going events were identical, all the energy would go into even harmonics. If the events were exact mirror-images, all the energy would go into odd harmonics. In practice both even and odd harmonics appear, though the odd harmonics are usually of higher amplitude, often by 10 dB or more.

In a conventional optimal Class-B amplifier, the crossover events are evenly spaced in time. In the crossover-displacement amplifier, the crossover events are asymmetrical in time and put energy into both even and odd harmonics when operating above the transition

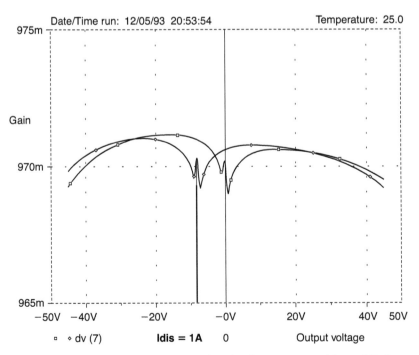

Fig 42.4 SPICE simulation of the output stage gain variation with and without a constant 1Amp of displacement current. The central peak is moved left from 0V to-8V.

point. However, since both even and odd exist already in conventional amplifiers, there is cause for concern. As always, the real answer is to reduce the distortion, of whatever order, to so far below the noise floor that it could not possibly be audible and you never need to fret about it.

Realisation

There are several ways in which a suitable displacement current can be drawn from the main amplifier output node.

Resistive crossover displacement. The most straightforward way to implement crossover displacement is to simply connect a suitable power resistor between the output rail and a supply rail, as shown in Fig 42.5.

This method suffers from poor efficiency, as the resistance acts as another load on the amplifier output, effectively in parallel with the normal load. It also threatens ripple-rejection problems as R is connected directly to a supply rail, which in most cases is unregulated and carrying substantial 100 Hz ripple. A regulated supply to the resistor could be used, but this would be very uneconomic and even less efficient due to the voltage drop in the regulator. The resistive system is inefficient because the displacement of the crossover region occurs when the output is negative of ground, but when the output is positive the resistor is still connected and greater current is drawn from it as the voltage across it

increases. This increasing current is of no use in the displacement process and simply results in increased power dissipation in the positive output half-cycles.

This method has the other drawback that the distortion performance of the basic amplifier will be worsened because of the heavier loading it sees, the resistor being connected to ground as far as AC signals are concerned.

Constant-current displacement

A much better solution is to use constant-current displacement, as in Fig 42.6. A constant-current source is connected between the output and negative rail. Efficiency is better as no output power is dissipated due to the high dynamic impedance of the current source. The output of the current source does not need to be controlled to very fine limits. Long-term variations in the current only affect the degree to which the crossover region is displaced, and this is not a critical parameter. Noise or ripple on the displacement current is greatly attenuated by the very low impedance of the basic power amplifier and its global negative feedback, so complex control circuitry is not required. The efficiency of this configuration is greater, because the output current of the displacer does not increase as the output moves more positive. The voltage across the current source increases, so its dissipation is still increased, but by a lesser amount. Likewise, the source transistor is passing less current on positive excursions so its power dissipation is less.

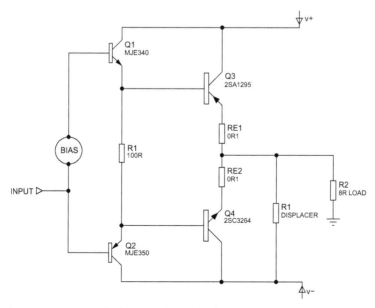

Fig 42.5 The concept of resistive crossover displacement. In and the following cases, the crossover point is displaced positively by sinking a current into the negative rail.

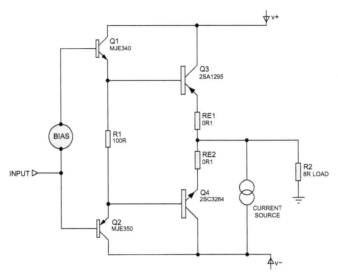

Fig 42.6 The concept of constant-current crossover displacement

Push–pull displacement

Having moved from a simple resistor displacer to a constant-current source, the obvious next step is to move from a constant current to a voltage-controlled current source (VCIS) whose output is modulated by the signal to further improve efficiency. The most straightforward way to do this is to make the displacement current proportional to the output voltage. Thus, if the displacement current is 1 Amp with the output at quiescent at 0V, it is set to increase to 2 Amps with the output fully negative,

and to reduce to zero with the output fully positive. The displacer current is set by the equation:

$$Id = Iq\,(1-Vout/Vrail)$$

where Iq is the quiescent displacement current (ie with the output at 0V) and Vrail is the bottom rail voltage, a constant which must be inserted as a negative number to make the arithmetic work.

Depending on the design of the VCIS, a scaling factor a is required to drive it correctly; see Fig 42.7. Since

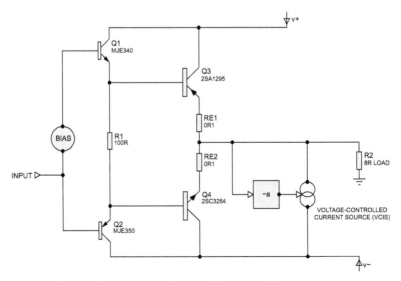

Fig 42.7 The concept of push–pull crossover displacement. The control circuitry implements a scaling factor of-a.

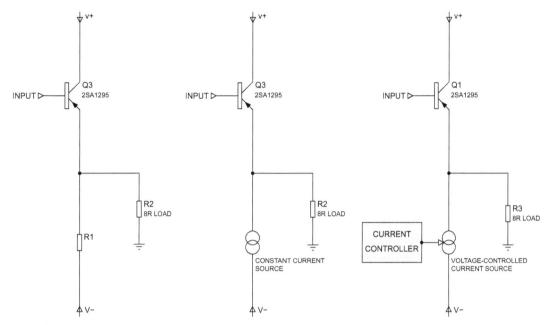

Fig 42.8 Left is a resistive Class-A amplifier giving 12.5% efficiency, while centre shows constant-current Class-A giving 25%. On the right is pushpull Class-A, which achieves; the current source may be controlled by either voltage or current conditions in the output stage.

an inversion is also necessary to get the correct mode of operation, active controlling circuitry is necessary.

The use of push-pull displacement is analogous to the use of pushpull current sources in Class A amplifiers, where there is a well-known canonical sequence of increasing efficiency, illustrated in Fig 42.8. [Ref 5] This begins with a real resistance giving 12.5% efficiency at full power, moves to a constant-current source with effective infinite impedance giving 25%, and finally to a push-pull controlled current-source, giving 50%. In the last case the sink transistor acts in a sense as a negative resistance, though it is more usefully regarded as a driven source (VCIS) than a pure negative resistance, as the current does not depend on rail voltage. In each move the efficiency doubles. Note that these efficiency figures are ideal, ignoring circuit losses, and that Class-A efficiency is seriously reduced at output powers less than the maximum.

Similarly, there is a canonical sequence of efficiency in Crossover Displacers, though the differences are smaller.

The push-pull displacement approach has another benefit; it reduces distortion when operating above transition in the Class-B mode. This is because the push-pull system acts to reduce the current swings in the output devices, as the displacement current varies

in the correct sense for this. This is equivalent to a decrease in output stage loading; this is the exact inverse of what occurs with resistive displacement, which increases output loading. Lighter loading is known to make the current crossover between the output devices more gradual, and so reduces the size of the gain-wobble that causes crossover distortion. [Ref 2] In addition the crossover region is spread over more of the output voltage range, so the distortion harmonics generated are lower-order and receive more linearisation from a negative feedback factor that falls with frequency. In push-pull displacement operation, the accuracy of the current variation does not have to be high to get the full reduction of the distortion, because of the low output impedance of the main amplifier, which maintains control of the output voltage. The global feedback around this amplifier is effective in reducing the inherently low output impedance of the output stage in the usual way, being unaffected by the addition of the displacer.

While the constant-current displacement method is simple and effective, the push-pull version of crossover displacement is to be preferred for the best linearity and efficiency; the extra control circuitry required is simple and works at low power so it adds minimally to total amplifier cost.

Circuit techniques

The constant-current displacer is the simplest practical displacement technique, the resistive version being discarded for the reasons given above.

A practical circuit for a constant-current displacer is shown in Fig 42.9. The displacement current typically chosen will be in the region of 0.5 to 1 A, and so a driver transistor Q15 is added, exactly as drivers are used in the main amplifier, so the control circuitry can work at low power levels. The power device Q14 is going to get hot, so its Vbe must be excluded from having a direct effect on current stability. Therefore the CFP (complementary feedback pair) structure shown is used, so the effect of Vbe variations is reduced by the negative feedback around the local loop Q15—Q14. The bias for the constant current is shown as a Zener diode; if greater accuracy is required an LM385 voltage reference IC may be used instead. One design aim here is that the voltage across R25 should not be large enough to limit the output swing. On the other hand,

if it is small compared with the Vbe of Q15, then the current value may drift excessively with temperature, as Q15 warms up.

Power transistor Q14 dissipates significant heat; clearly the greater the crossover displacement required, the greater the displacement current and the greater the dissipation. Q14 therefore must be mounted on a heatsink in the usual manner. This provides the intriguing sight of a power amplifier with an odd number of output transistors, which might conceivably be exploited for marketing purposes.

The push-pull controller drives the displacer so that as the output rail goes positive, the displacer supplies less current. The basic problem is to apply a scaled and inverted version of the output voltage to the displacer. The signal must also have its reference transferred to the negative rail, which can be assumed to carry mains ripple and distorted signal components. Transferring the reference is done by using the high-impedance (like a current-source) output from a bipolar transistor collector. As before, a driver transistor Q15 is used to drive

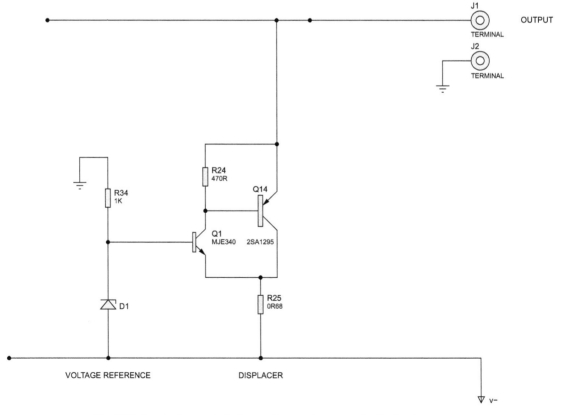

Fig 42.9 Constant-current displacer with complementary feedback pair structure.

the Displacer Q14 so the control circuitry can work at low power levels. This not only minimises total current consumption but also reduces the effect of Vbe changes due to device heating. See Fig 42.10.

The controller is simply a differential pair of transistors with one input grounded and the other driven by the main amplifier output voltage, scaled down appropriately by R34,R35. The differential pair has heavy local

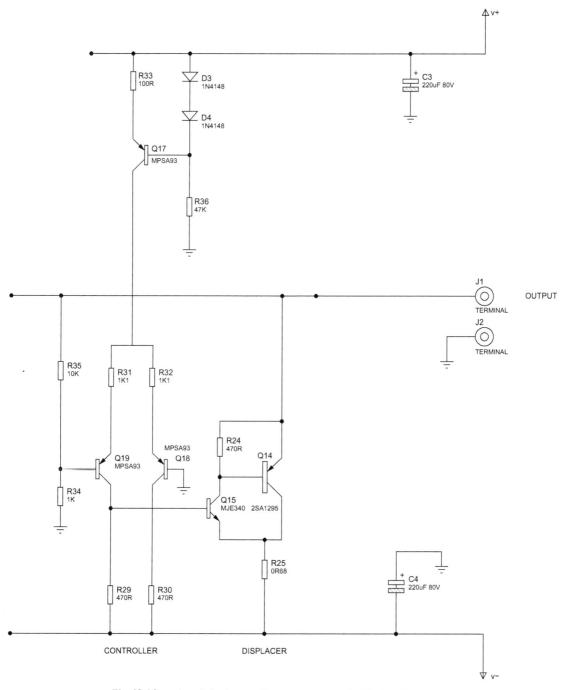

Fig 42.10 push-pull displacer with complementary feedback pair structure.

feedback applied by the addition of the emitter resistors R31,R32, in order to minimise distortion and achieve an accurate gain. The drive to the VCIS displacer is taken from collector load R29, to give the required phase inversion. R30 is present simply to equalise the dissipation in the differential pair transistors to maintain balance.

The tail of the differential pair is fed by constant-current source Q17. This gives good common-mode rejection, which prevents the significant ripple voltages on the supply rails from interfering with the control signal. Since half of the standing current through the differential pair flows through R29, the value of the tail current-source sets the quiescent displacement current. The stability of the current generated by this source therefore sets the stability of the quiescent (no-signal) value of the displacement current. Fig 42.10 shows a simple current-source biased by a pair of silicon diodes, but more sophisticated current-sources using negative feedback can be used when greater stability is required. However, even if the tail current-source is perfect, the value of the displacement current still depends on the temperature of Q15. More sophisticated circuitry could be used to remove this dependency; for example the voltage across R25 could be sensed by an op-amp instead of by Q15. The opamp would however need to be able to work with a common-mode voltage down to the negative rail, or an extra supply-rail provided.

A further possible refinement is the addition of a safety resistor in the differential pair tail to limit the amount of current flowing in the event of component failure. Such a resistor has no effect on normal operation, but it must be employed with care as its presence means that the circuit will not start working until the supply rails have risen to a large fraction of the working value. This is a significant drawback as it is normal to test power amplifiers by slowly raising the rail voltages from zero, and the lower they voltage at which they start working, the safer this procedure is.

A complete crossover displacement power amplifier circuit

Fig 42.11 shows the practical circuit of a push-pull crossover displacement amplifier. The basic Class-B amplifier follows the Blameless design philosophy I

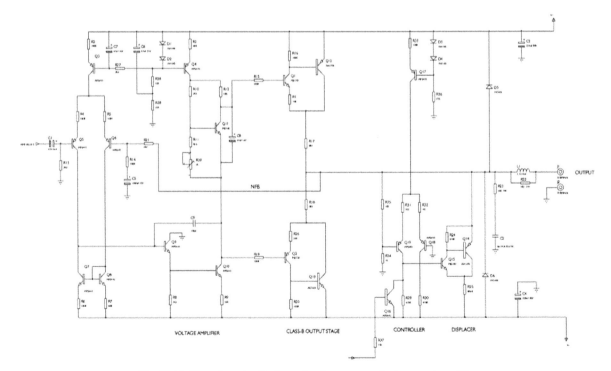

Fig 42.11 Complete circuit of a pushpull crossover displacement amplifier.

introduced in 1993 [Ref 1] and therefore uses the following straightforward techniques to bring the distortion down to the irreducible minimum generated by a Class-B output stage.

1. The local negative feedback in the input differential pair Q5,Q6 is increased by running it at a high collector current and then defining the stage transconductance by adding emitter resistors R4,R5.
2. The crucial collector-current balance between the two halves of the input differential pair is enforced by the use of a degenerated current-mirror.
3. The local negative feedback around the voltage-amplifier transistor Q10 is increased by adding the emitter-follower Q9 inside the Cdom loop.
4. The output stage uses a Complementary Feedback Pair (CFP) configuration to establish local negative feedback around the output devices. This increases linearity and also minimises the effect of output junction temperatures on the bias conditions.

The circuit shown 50W Powers above 100W into 8 Ohms are likely to require two paralleled power transistors in the main amplifier output stage. The displacer does not necessarily require doubling; it depends on the degree of crossover displacement used.

The displacer control circuitry is essentially the same as Fig 42.10. Note that transistor Q16 has been added so that the crossover displacement action can be switched off manually or under microprocessor control. Taking the control terminal high turns on Q16, shunting current away from Q15 and turning the displacer current source off. This facility is useful for testing and fault-finding.

The Measured Performance

Measurements are presented here to demonstrate how the crossover displacement principle reduces distortion in reality. Tests were done with an amplifier similar to that shown in Fig 42.11.

Firstly, the results obtainable from well-implemented Class-B. Fig 42.12 shows THD vs frequency for a standard Blameless Class-B amplifier giving 30W into 8 Ohms. The distortion shown only emerges from the noise floor at 2kHz, and is wholly due to the inherent crossover artefacts; the bias is optimal, so this is essentially as good as Class-B gets. The distortion only gets really clear of the noise at 10 kHz, so this frequency has been chosen for the THD/amplitude tests below. This frequency provides a demanding test for an audio power amplifier. In all these tests the measurement bandwidth was 80 kHz. This filters out ultrasonic harmonics, but is essential to reduce the noise bandwidth; it is also a standard setting on many distortion analysers.

Distortion against amplitude at 10 kHz, over the range 200mW- 20W is plotted in Fig 42.13; this covers the power levels at which most listening is done. (0 dB is 30W into 8 Ohms) Trace B is the result for the Blameless Class-B amplifier; the THD percentage increases as the power is reduced, partly because of the nature of crossover distortion, and partly because the constant noise level becomes proportionally greater as level is reduced. Trace A shows the result for a Class-A amplifier of my design [Ref 6] which is essentially distortion-free at 10 kHz, and simply shows the increasing relative noise level as power reduces. Trace XD demonstrates how a constant-current crossover displacement amplifier has

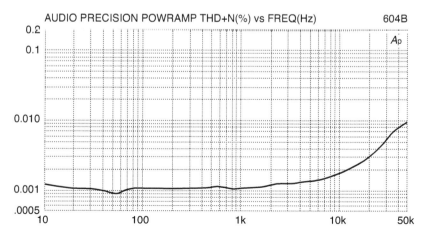

Fig 42.12 THD vs frequency for a standard Blameless Class-B amplifier at 30W/8 Ohm. (604b)

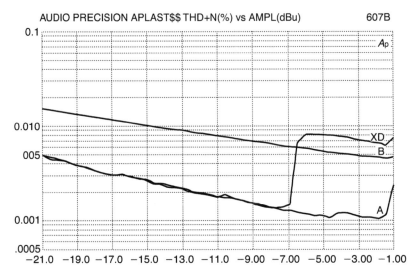

Fig 42.13 THD vs power out for Class A, Class B, and Class B with constant-current crossover displacement. Tested at 10 kHz to get enough distortion to measure; 0 dB is 30W into 8 Ohms. XD stands for Crossover Displacement. (607B)

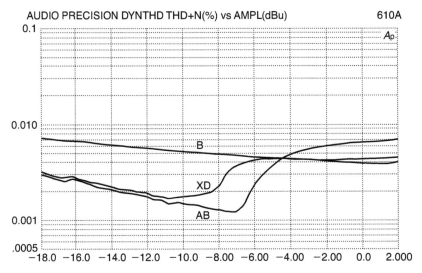

Fig 42.14 (610A) THD vs power out for Class B, Class AB, and Class B with constant-current crossover displacement. (XD) Tested at 10 kHz, power as before.

the same superb linearity as Class A up to an output of-7 dB, but distortion then rises to the Class-B level as the output begins to traverse the displaced crossover region. (In fact it slightly exceeds Class-B in this case; this set of data was acquired before the prototype was fully optimised)

A similar THD/amplitude plot in Fig 42.14 compares Class-B with Class-AB and constant-current crossover displacement. Here the transition point from Class-A to Class-B is at-8 dB and gm-doubling begins in Class-AB at-7 dB. Now the crossover displacement distortion is now slightly below that from Class-B.

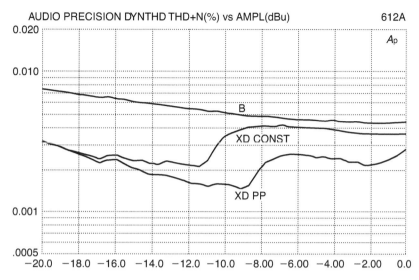

Fig 42.15 shows that push-pull crossover displacement (XD PP) gives much lower distortion than constant-current crossover displacement. (XD CONST) Tested at 10 kHz, power as before. (612A)

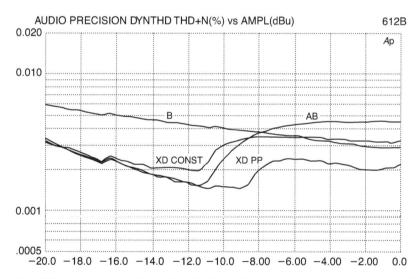

Fig 42.16 (612B) adds a THD vs level plot for Class-AB to the Fig 42.15 diagram, making it very clear that Class-AB gives significantly greater THD above its transition point (say at-4 dB) than Class-B, constant-current crossover displacement gives slightly less, and push-pull crossover displacement gives markedly less.

It is shown in Fig 42.15 that push-pull crossover displacement gives markedly lower distortion than constant-current crossover displacement. The transition points are not quite the same (-8 dB for push-pull versus-11 dB for constant-current) but this has no significant effect on the distortion produced. The salient point is that at-2 dB, THD is very significantly lowered from 0.0036% to 0.0022% by the use of the push-pull method, which reduces the magnitude of the current changes in the output transistors of the main amplifier.

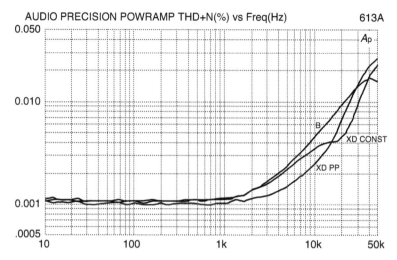

AUDIO PRECISION POWRAMP THD+N(%) vs Freq(Hz)　　　613A

Fig 42.17 (613A) returns to the THD/frequency format, and shows that XD pushpull gives lower THD over the range 1kHz–30kHz.

Effects of loading changes

When a new amplifier concept is considered, it is essential to consider its behaviour into real loads, which deviate significantly from the classical 8 Ohm resistance.

Firstly, what is the effect of changing the load resistance, for example by using a 4-Ohm load? The signal currents in the output stage are doubled, so the voltage by which the crossover region is displaced is halved. Half the voltage across half the resistance means the output power is halved, so the volume at the transition point has been reduced by 3 dB. In terms of SPL and human perception, the reduction is not very significant. One of the concerns facing conventional Class-B amplifiers driving 4 Ohms or less is the onset of Large Signal Nonlinearity [LSN] caused by increased output device currents and consequent fall-off of beta. The use of push-pull displacement reduces LSN in same way that crossover distortion is reduced- by reducing the range of current variation in the output stage.

Secondly, what about reactive loads? In particular we must scrutinise the way that the push-pull displacer is driven by output voltage rather than device currents. In a conventional Class-B amplifier, adding a reactive element to the load alters the phase relationship between the output voltage and the crossover events; this is because voltage and current are now out of phase, and crossover is a current-domain phenomenon. It has never been suggested this presents any sort of problem. Putting reactive loading on a crossover-displacement amplifier moves its crossover events (if the power level is above transition- otherwise they are not generated at all) in time with respect to the voltage output in exactly the

same way, and there is absolutely no reason to suppose that this is any cause for concern.

Stability into reactive loads is not affected by the addition of the displacement system.

Efficiency

The Crossover Displacement technique obviously increases the total power dissipated in the output stage. The dissipation in the Source transistor is increased by the displacement current flowing through it, while that in the sink transistor is unchanged. There is also the additional dissipation in the Displacer itself, which is likely to be mounted on the same heatsink as the main output devices.

When using techniques such as Class AB or Crossover Displacement, that give a limited power output in Class A, you must decide at the start just how much Class A power you are prepared to pay for in terms of extra heat liberated, and what load impedance you intend to drive. For example, assume that 5 Watts of Class-A operation into 8 Ohms is required from an amplifier with a full output of 50W. The crossover point is therefore displaced by the peak voltage corresponding to 5W, which is 8.9V. It is well-established that it takes about a 10 dB increase in sound intensity to double subjective loudness, [Ref 7] which is a power ratio of ten times. Therefore if there is only a doubling in loudness between the transition point into Class-B and full output, the amplifier will be in the Class-A region most of the time; this approach seems reasonable.

Table 42.1 shows the calculated efficiency for the various types of amplifier. "The calculations" were not the

Table 42.1 Efficiency of amplifier types.

	Full output	Half power	1/10 power
Class B	74%	54%	23%
Class XD pushpull	66%	46%	14%
Class XD constant	57%	39%	11%
Class A	43%	23%	4%

usual simple theoretical ones that ignore voltage drops in emitter resistors , transistor saturation voltages and so on, but a lengthy series of SPICE simulations of complete output stage circuits. The effects of transistor non-linearity and so on are fully taken into account. The results are therefore as real as extensive calculation can make them.

For comparison, the "classical" calculations for Class B give a full power efficiency of 78%, but the more detailed simulations show that it is only 73% when typical losses are included.

The output stages were simulated using +/-50V rails, giving a maximum power of about 135W into 8 Ohms. Displacement currents were set to give a transition from Class A to B at 5W. All emitter resistors were 0.1 Ohm.

Here we have demonstrated that there is some penalty in efficiency when crossover displacement is used, but it is far superior to Class A.

Other methods of pushpull displacement control

This article describes in detail pushpull crossover displacement implemented by controlling the displacer from the amplifier voltage output. This has the merit of simplicity, but its design makes assumptions about the load impedance. If the load is of higher impedance than expected, which can often occur in loudspeaker loads because of voice-coil inductance, the displacer current may be increased more than is actually necessary for the desired amount of displacement. This is because the voltage-control method is an open-loop or feedforward system.

This situation could be avoided by using a current-controlled system which senses the current flowing in the main amplifier output devices and turns on enough displacer current to give the amount of crossover displacement desired. This is a negative feedback loop operating at the full signal frequency, and experience has shown that high frequency instability can be a serious problem with this sort of approach. Nonetheless it is worthy of further investigation.

Summary

Crossover displacement aims to provide a genuine way to compromise between the linearity of Class-A and the efficiency of Class-B. I think it is now established that the conventional use Class-AB, by turning the bias, is not such a compromise because it introduces extra distortion.

Advantages

Effective. Crossover distortion has been pushed away from the central point where the amplifier output spends most of its time. Push-pull displacement also reduces distortion when in Class-B operation.

Simple. Only 5 extra transistors are used, of which 3 are small-signal and of very low cost. No extra presets or adjustments. Does not affect HF stability

Patentable. To the best of my knowledge this approach is entirely new.

Straightforward. Easily explained in advertisements.

Versatile. Can act as a bolt-on distortion reducer which may be attached to almost any kind of Class-B amplifier.

Disadvantages

Some extra power dissipation, but far less than the use of Class A.

References

1. Self, D "Distortion In Power Amplifiers" Electronics World, Aug93-Mar94. (the Blameless power amplifier concept)

2. Self, D "Load-Invariant Audio Power" Electronics World, Jan 1997, p16.

3. Self, D "Trimodal Audio Power" Electronics World, June, July 1995.

4. Self, D "Audio Power Amplifier Design Handbook" Newnes 1996. ISBN 0–7506–2788–3. p276. (Low value emitter resistors)

5. Self, D "Distortion In Power Amplifiers" Electronics World, Mar 1994. (Class A efficiency canonical sequence)

6. Self, D "Audio Power Amplifier Design Handbook" as above, p267. (Class A distortion data)

7. Moore, B J "An Introduction to The Psychology of Hearing" Academic Press 1982, pp48–50.

Chapter 43

The 5532 OpAmplifier

Part 1: design philosophy and schematics
(Elektor, October 2010)

For many years, I wondered if the 5532, with its good load-driving capabilities and very low distortion, could be used to make a power amplifier capable of driving 8Ω by simply wiring a bucketful of them in parallel. This is a more economic proposition than it sounds as the 5532 is also usually the cheapest opamp you lay your hands on, probably because large numbers are used in the output stages of CD players. I found that multipath amplifiers with two, three, or four opamps effectively in parallel worked beautifully to give lower noise and increased load-driving ability. The critical feature in this was the averaging of the opamp outputs by connecting them together via 10Ω current-sharing resistors. These are high enough to prevent currents circulating between the opamps, and low enough to have little effect on the load-driving capability. An advantage is that effective overload protection is built into the 5532, so no work to do there; a disadvantage is that the power output is limited by the permissible opamp supply rails to 16W/8Ω.

Scaling this up to make a power amplifier proved to be a totally straightforward business. The amplifier was split into two sections, the first providing a similar voltage gain to a conventional power amplifier, and the second being essentially one big multiple voltage-follower; this is similar to the architecture of discrete power amplifiers, though achieved in a radically different way. An important difference is that there is no global feedback loop enclosing both the voltage-gain stage and the output voltage-follower. The current-sharing resistors were dropped to 1Ω without problems, to reduce voltage losses and the output impedance.

The idea goes back a long way, pretty much from when the 5534 appeared on the market around 1979. This was long before I did the power amplifier research that led to the Distortion In Power Amplifiers series in *Electronics World,* which led in turn to the book *Audio Power Amplifier Design* [1]. The techniques for designing a low-distortion discrete power amplifier were not recognised or simply did not exist, and it looked as though the multiple 5534/5532 approach might be a good way to sidestep these difficulties and get really low distortion. At the time, the high cost of the opamps prevented it from being anything more than a design exercise.

However, times change, and I have never been averse to playing the long game. The 5532 became cheap, the first single-channel prototype amplifier was assembled in (I think) 2002, and proved very capable of driving a small bookshelf loudspeaker. Various other stuff got in the way, as it will, and it was not until 2009 that I had the leisure to work out a fully developed stereo version. It included the option of bridging, which gets the power output up to a more useful 64W into 8Ω.

The voltage-gain section was the most challenging part. Purely as piece of performance art I wanted to make the whole design out of 5532s and no other opamp. The result was a configuration using five opamp sections, plus one more as an inverter to give a balanced output for bridging. This follows the principle of having most of gain in the early stages (IC1a, IC1b) where there is a low signal level, and less gain in the later stages with high signal level (IC2a) to maximise the feedback factor and reduce distortion. In addition, the first stage is doubled to reduce noise by 3 dB where the signal is smallest, so it's quite a nice collection of opamp techniques.

An important feature is the use of zero-impedance outputs [2] built around IC3A, IC3B to drive the output stage; the tracking to connect up 64 opamps is inevitably extensive and has significant stray capacitance to ground, threatening instability. A simple series isolating resistor would have prevented this, but I was worried that all those input currents being drawn by the output opamps might cause a voltage-drop across it that would increase distortion, hence the zero-impedance outputs. The voltage-gain section works pretty well, as you can see from the performance figures in the second half of the article, but it could be done in a sense more elegantly, and probably with lower noise and distortion, by using two or three LM4562 opamp sections.

Using voltage-followers in the output rather than gain stages has two important advantages: firstly the opamps have the maximum possible amount of feedback, which keeps distortion low, and secondly the components required are restricted to 32 dual opamp packages and 64 averaging resistors, plus a few decoupling caps. If each opamp had gain that would have to be defined by two resistors, and that would add another 128 resistors to the circuit; this is clearly well worth avoiding. The 1Ω averaging resistors give an output impedance of 15.6mΩ, more than low enough to drive any loudspeaker. If you turn that into a wholly fallacious 'damping factor' with respect to 8Ω, you get 512, which would be an excellent figure if it meant anything. Which it doesn't.

To put it all into perspective, this amplifier has a conventional series output inductor to ensure stability with capacitive loads; its DC resistance is about 10mΩ. The internal wiring of the unit is not likely to be less than 50 mΩ, so if we take a total of 75 mΩ, we get an equally meaningless but less impressive 107. This ignores the speaker cable resistance and, crucially, the resistance of the loudspeaker voice coils, which at about 7Ω obviously dominates the real damping of the drive units.

Elektor made a few changes, because they were not happy with electrolytics in the signal path. In my original design, the input DC-blocking capacitor C2 was a 2u2 non-polarised electrolytic, and I need hardly say it worked fine. *Elektor* replaced it with a 2u2 250V polypropylene part; the only reservation I have about this is that it is a lot more expensive, and I always design with an

eye to cost. My original design also had a 220uF electrolytic DC-blocking capacitor after IC2b, to prevent offset voltages from being passed on to the power stage. The offsets of the early opamps are amplified because the feedback networks do not have large capacitors at the bottom to reduce the gain to unity at DC. Since the offsets are not large enough to affect headroom, a single DC-block capacitor seemed like the more elegant approach. *Elektor* replaced it with the second-order servo system based on IC6. I have not tested a version with the servo.

In a sense, the power output stage here is the ultimate low voltage-noise amplifier. When multiplying opamps to reduce noise, it is rarely economic to go beyond four sections (two packages of 5532). However, here we have 64 in parallel to get adequate current delivery. 64 parallel amplifiers will reduce the voltage noise by $\sqrt{64} = 8$ times, (-18 dB), and since the output noise of a 5532 voltage-follower is about -119dBu, the noise from the power stage alone will be a subterranean -137 dBu. This is below any means of direct measurement. The current noise is however similarly increased, so a suitably low source impedance would be required to feel the benefit. It would be interesting to see if it would be possible to make a viable MC phono input or a microphone amplifier using multiple 5532s. If the best noise performance is required, it is not likely to be an economic approach.

References

1. Self, D., *Audio Power Amplifier Design*, 6th edition. Focal Press 2013. ISBN 978-0-240-52613-2.

2. Self, D., *Small-Signal Audio Design,* 2nd edition, pp. 538–539. Focal Press (Taylor & Francis) 2014. ISBN: 978-0-415-70974-3 Hardback. 978-0-415-70973-6 Paperback. 978-0-315-88537-7 ebook.

THE 5532 OPAMPLIFIER

The most popular dual opamp in the world of audio is the (NE)5532. An interesting power amplifier can be made by connecting enough 5532s in parallel, how about 32 for a start? This may sound like a radical course of action, but it actually works very well, making it possible to build a very simple amplifier that retains not only the excellent linearity but also the power-supply rejection and the inbuilt overload protection of the 5532, which reduces the external circuitry required to a minimum.

While not exactly a brand-new design, the type (NE)5532 dual operational amplifier (opamp) is a very capable device giving low distortion with good load-driving capabilities, and a remarkably good noise performance. It is only quite recently that better opamps for audio work have become available. While these can give truly outstanding results, the cheapest of them costs ten times more than the 5532, which is available at a remarkably low price—in fact it is one of the cheapest opamps, because it is so widely used in audio applications.

It should be mentioned at once that the obvious limitation with using opamps to drive loudspeakers is that the output voltage swing is limited compared with a conventional power amplifier, and using a single-ended array of 5532s will give about 15 W_{rms} into 8 Ω. This output can be greatly extended by using two such amplifiers in bridge mode; one amplifier is driven with an inverted input signal so the voltage difference between the two amplifier outputs will be doubled, and the power output is quadrupled to about 60 W_{rms} into 8 Ω. This should be enough for most domestic hifi situations.

The other unalterable limit set by the opamps is the maximum output current, set by the internal overload protection. A single 5532 section (one half of the dual package) will drive 500 Ω to the full voltage output, though it is advisable to keep the loading lighter than this to maintain low distortion at high levels. If 4 Ω operation is required, twice as many opamps must be used to supply the doubled current demand. This also applies to bridged operation into 8 Ω. The system is designed so that either single-ended or bridged operation can be used; the basic design described here gives a working stereo amplifier with just three PCBs. The amplifier cards can be paralleled without problems, and facilities

---

Specification—Per Channel, 8 Ohm Load

Supply voltage   ±18.3 V

Input sensitivity	– unbalanced	840 mV (16 W, 1 % THD)

	– balanced	833 mV (16 W, 1 % THD)
Input impedance	– unbalanced	38.8 kΩ
	– balanced	93.6 kΩ
Output power, sinewave	– 0.1 % THD	16 W
	– 1 % THD	16.8 W
Output power bandwidth		1.5 Hz—275 kHz
Slew rate		5 V/μs
Rise time		4 μs
Signal/ noise ratio	(1 W ref.)	110 dBA
Harmonic distortion + noise	– 108 dB (B = 22 Hz—22 kHz linear/ unweighted)	
	– 0.0005% (B = 22 kHz, 1 kHz, 1 W)	
	– 0.0009% (B = 80 kHz, 1 kHz, 1 W)	
	– 0.0004% (B = 22 kHz, 1 kHz, 8 W)	
	– 0.0005% (B = 80 kHz, 1 kHz, 8 W)	
	– 0.003 % (B = 80 kHz, 20 kHz, 8 W)	
Intermodula-tion distortion	– 0.0012% (1 W)	
	– (50 Hz : 7 kHz = 4 : 1) 0.0015% (8 W)	
Dynamic IM distortion	– 0.0011% (1 W)	
	– (3.15 kHz square wave + 15 kHz sine wave)	0.0035% (8 W)
Damping factor	– 194 (1 kHz)	
	– 111 (20 kHz)	
DC-protection	±1.5 V	
Quiescent current	300 mA	

are provided to connect more PCBs in parallel for driving low-impedance speakers.

Overload protection is inherent in the opamps, but output relays are used for on/off muting and to protect loudspeakers against a DC fault.

A tour of the design

The schematic in Figure 43.1 shows one channel of the complete amplifier, which consists of unbalanced and balanced line inputs, and the power amplifier itself, which is divided into a +22.7 dB gain stage and an array of paralleled output opamps configured as voltage-followers, giving the maximum amount of negative feedback around them to minimise distortion. Let's have a look at the various sections of the circuit.

The unbalanced input

This consists simply of RF filter R1, C1 and DC-drain R2, which are directly connected to the gain stage when JP1 is in the 'unbalanced' position.

The balanced input

This amplifier is an innovative design that gives very low noise. The conventional balanced input stage built with four 10 kΩ resistors and a 5532 opamp has a far worse noise performance than a simple unbalanced input, and is also much noisier than most power amplifiers; output noise is approximately −104 dBu. This balanced amplifier here solves this problem partly by the use of a dual balanced stage (IC5A, IC5B) amplifier that partially cancels the uncorrelated noise from each amplifier, giving a 3 dB noise reduction, and in a similar way improves the CMRR; it also uses much lower resistor values than usual (820 Ω instead of 10 kΩ) which produces less Johnson noise in the first place. This is only possible because it is driven by unity-gain buffers IC4A, IC4B, which also allow the input impedances to be much higher than usual, preventing loading of external equipment and further improving the CMRR. The noise output is less than −112 dBu, an 8 dB improvement over conventional technology.

The gain stage

The main input amplifier is another innovative design that achieves very low distortion by spreading the gain required over three stages. +22.7 dB could easily be obtained with one opamp but 5532s are not completely distortion-free, and the THD would be significant.

The first stage (IC1A, IC1B) gives +10.7 dB of gain; the two outputs are combined by R8, R9 to give a 3 dB noise advantage, as in the balanced amplifier. The second stage IC2A gives +6 dB of gain. The gain is less to maximise negative feedback because the signal level is now higher. IC2B is a unity-gain buffer which prevents the 1 kΩ input impedance of final gain stage IC3B from loading the output of IC2A and causing distortion. IC2B is less vulnerable to loading because it has maximal negative-feedback. IC3B gives the final +6 dB of gain; it is used in shunt-feedback mode to avoid the common-mode distortion which would otherwise result from the high signal levels here. It has a 'zero-impedance' output, with HF feedback via C8 but LF feedback via R13, so crosstalk is kept to a minimum while maintaining stability with load capacitance. The output at K3 is phase-inverted and can be used for bridging.

IC3A is a unity-gain inverting stage which corrects the signal phase. The output is also of the 'zero-impedance' type.

The power amplifier

The power amplifier consists of thirty-two 5532 dual opamps (i.e. 64 opamp sections) working as voltage-followers, with their outputs joined by 1 Ω current-sharing resistors. These combining resistors are outside the 5532 negative-feedback loops, and you might wonder what effect they will have on the output impedance of the amplifier. A low output impedance is always a good thing, but not because of the so-called 'damping factor' which is largely meaningless as the speaker coil resistance always dominates the circuit resistance. 'Damping factor' is defined as load impedance divided by output impedance; we have 64 times 1 Ω resistors in parallel, giving an overall output impedance of 0.0156 Ω. This gives a theoretical damping factor of 8 / 0.0156 = 512, very good by any standards. The wiring to the loudspeaker sockets will have more resistance than this!

The output opamps may be directly soldered into the board to save cost and give better conduction of heat from the opamp package to the copper tracks. However, on the prototype built in the Elektor labs, high quality sockets were used. Having a lot of opamps in parallel could make fault-finding difficult—if there is one bad opamp out of 32 then you are likely to have to do a lot of unsoldering (or IC unplugging) to find it. The opamp array is therefore split up into four sections of eight opamps, which are joined together by jumpers K5–K12, so on average you would only need to unsolder (or pull out) four opamps to find a defective one. In my many years of experience

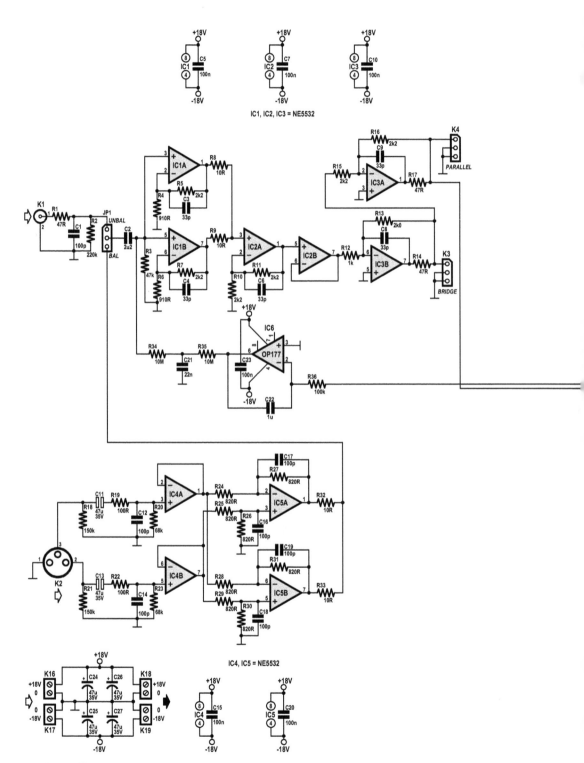

Figure 43.1 Power is indeed in numbers: circuit diagram of the basic NE5532 audio amplifier (one channel shown).

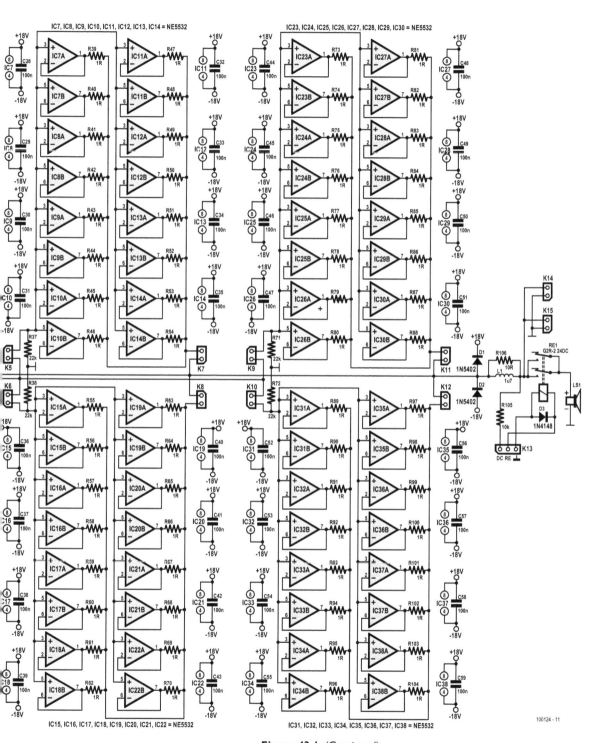

Figure 43.1 (Continued)

100124 - 11

with it the 5532 has proven a very reliable opamp, and I think such failures will be very rare indeed. There is an output choke L1 for stability into capacitive loads, and catching diodes D1–D2 to prevent damage from voltage transients when current-limiting into reactive loads.

The output relay and its control

The output mute relay RE1 protects the loudspeakers against a DC offset fault and gives a slow-on, fast-off action so no transients are passed to the loudspeakers at power-up or power-down. The relay is controlled from the power supply board. With reference to Figure 43.2, at power-up R17 charges C24 slowly to give a turn-on delay. In operation C21 is charged and T3 is on; when the AC power is removed C21 discharges rapidly, T3 turns off, and D8 turns on T4–T5, which discharge C24 and cause the output relay contacts to be opened immediately. Even a brief AC power interruption gives the full turn-on delay.

Normally T4 and T5 are off and D15 non-conducting, but if a DC offset fault applies either a positive or

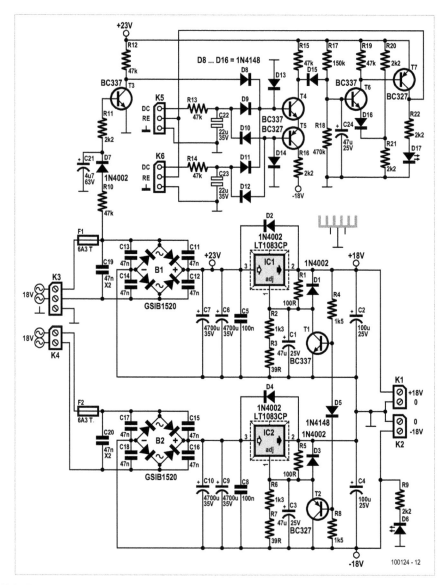

Figure 43.2 The symmetrical power supply is dimensioned for the 2×15-watt, 8 ohm basbic version of the amplifier.

negative voltage via R13 or R14, T4–T5 turn on and the relays are opened at once to protect the loudspeakers.

Power supply

Again referring to the circuit diagram of the power supply unit in Figure 43.2, the ±18 V symmetrical supply is regulated by two type LT1083 TO3-P positive regulators. When a 5532 sees one supply rail disappear, this opamp can get into an abnormal state in which it draws excessive current. This could obviously be catastrophic with this design, so the PSU incorporates a mutual-shutdown facility which shuts off each supply rail if the other has collapsed due to short-circuiting or any other cause. If the positive rail collapses, T2 turns on and disables the negative supply. If the negative rail collapses, T1 turns on and disables the positive supply.

Cost

This project uses quite a lot of 5532s; 37 in each channel, but that does not mean the cost is excessive. In the USA,

NE5532s can be obtained from mail order companies like Mouser at around $0.50 each at 100-off order quantity. This means that the cost of all the opamps in the project would be about $37.00.

To be continued

Next month's closing installment will cover approaches to constructing the amplifier on circuit boards, some performance figures obtained from our high-end test equipment, and an outline of challenges to those of you wishing to modify the amplifier for higher output powers and/or lower output impedance. Meanwhile this month's *E-Labs Inside* section has a page or so on issues with electrolytic capacitors encountered while the first prototype of the amplifier was tested.

(100124)

Internet link

1. www.rapidonline.com

"Observe, record, tabulate, communicate."
- Sir William Osler

Sensor, data storage, RFID, and wireless communication devices for your experiments at **www.parallax.com**.

PARALLAX

www.**parallax**.com
"ParallaxInc" on Twitter, Facebook, and YouTube

To order **Communication, Wireless and RF** accessories visit www.parallax.com or call toll-free at 888-512-1024 (M-F, 7am-5pm, PDT).

Friendly microcontrollers, legendary resources.™

Parallax and the Parallax logo are trademarks of Parallax Inc.

Chapter 44

The 5532 OpAmplifier

Part 2: construction, bridged operation and test results
(Elektor, November 2010)

The second part of the article covers the practical details of construction, and some rather pleasing test measurements. The issues of driving 4Ω loudspeakers and bridging are also discussed. This design approach is not very well suited to low-impedance loudspeakers. The only way to double the output current capability is to double the number of opamps, which naturally doubles the cost. For a 4Ω loudspeaker, there will now be 128 opamps (64 dual packages) per channel. This is catered for in this design simply by using two amplifier PCBs with their outputs connected together. Because of all those 1Ω current-sharing resistors, this is absolutely straightforward and does not cause any increase in distortion. Voltage noise from the output stage would drop by another 3 dB, but it is already below measurement.

The amplifier can be bridged to increase the power output from 16W/8Ω to a more forceful 64W/8Ω. The voltage-gain stages include an inverter to generate the anti-phase output signal. Bridging a power amplifier effectively turns one 8Ω load into two 4Ω loads, so doubled-up output stages are required for the current needed. In addition, two of these doubled-up output stage are necessary to create the two anti-phase outputs that make up the bridging circuit. We therefore have a grand total of 256 opamps (128 packages) per channel; for stereo, we are looking at 512 opamps (256 packages). This really is nudging the envelope of practicality, and demonstrates the ultimate limitations of the multiple-opamp power amp concept. Anything lower than a 4Ω load is probably not worth contemplating. This is in sharp contrast to conventional solid-state power amplifiers, which are virtually all designed from the start to be able to drive a 4Ω load. Adding a 2Ω load capability—for a short period, anyway—does not cost much more.

The Preface to Chapter 42 describes how output biasing, the injection of a steady current into an opamp output, can reduce distortion. For the 5534/5532 the optimal current is about 5 mA, injected from the V+ rail. It occurred to me this might be worth applying to the 5532 power amplifier sections. We have 32 opamp packages and thus 64 opamp sections, so the total current required is $64 \times 5 = 320$ mA, which is a manageable amount. We do not wish to add 64 resistors

to inject this current separately into each opamp output, so I propose a single 320 mA current-source connected to the junction of the 64 1Ω averaging resistors. The current-source is preferred to a resistor as it puts no extra loading on the amplifier output. If a resistor was used, it would have to be 18/0.32 = 56Ω, and it would be better not to waste our precious audio output power in that. A 56Ω resistor would dissipate 5.8W when the amplifier was quiescent, and so would the preferred solution of a simple diode-biased current source, but that could be dealt with by a small vertical heatsink on the PCB. I must stress I have not tried this out, but I can see no reason why it should not work. What could possibly go wrong? At any rate, it is an easily-tried-out way to reduce the already very low distortion to even lower levels, and it is certainly cheaper than converting to LM4562 opamps.

An important feature of this article is the appendix on the linearity of 5532s from different manufacturers, with Texas being dependably the worst. The tests made in the *Elektor* lab replicated my findings, which is always reassuring.

The article has an illustration of the stereo amplifier and power supply built into a suitcase. It was exhibited by Jan Didden at the Burning Amp festival [1] in San Francisco in October 2010, where it was demonstrated with Nelson Pass' open baffle loudspeakers. I understand from Jan that it was well received. You can read discussions about the design on the DIYaudio bulletin board [2].

I do want to emphasise that this was not a 'just-for-fun' project. It was properly designed and engineered, as you can see from the test results. I had, however, no intention of providing serious competition for more conventional solid-state power amplifiers; a technology that it is inherently restricted to 64W into 8Ω obviously has its limitations. But if nothing else, it teaches a lot of important lessons about the application of opamps. It might even trigger some new thinking on power amplifiers. You never know.

References

1. Burning Amp, http://www.burningamp.org/ (accessed Feb. 2015)

2. DIYaudio, http://www.diyaudio.com/forums/chip-amps/174540-doug-selfs-ne5532-power-amp-thoughts-anyone.html (accessed Jan. 2015)

THE 5532 OPAMPLIFIER

In this second and closing installment we get real with two times 32 NE5532 opamps paralleled to form a high-end audio power amplifier with eminent specifications in terms of distortion and general sonic performance. There are also challenges for you: bridged operation for higher output power, and modifying the amplifier for 4-ohm operation.

If you thought that paralleling a few dozen NE5532 opamps is a curious way of designing a high-end audio amplifier and typical of Elektor's off-the-beaten track approach to electronics you are probably right. Last month's design considerations did not fail to trigger responses from you, our readership, in particular from all and sundry aspiring or even claiming to be a high-end audio designer. This month we get real by building

the 5532 OpAmplifier project and putting it through its paces.

Construction—amp board

The amplifier proper is built on a double-sided through plated circuit board of which the component overlay is shown in Figure 44.1. Two of these boards are required for a stereo amplifier. Boards supplied by Elektor come with through plating, a solder mask and holes predrilled. As such they are the best guarantee for success in replicating the project. While on the subject of quality, you are looking at an expensive, high-end amplifier. If you use mishmash components and *ditto* assembly methods and tools, you'll get mishmash results. More on selecting the best NE5532 brand to use in the **inset.**

Construction on the 205 × 84 mm board should not present problems as only standard through-hole parts are used. A properly functioning amp can be expected if you work with care and precision, like Ton Giesberts of Elektor Audio Labs who designed, built and tested all the boards pictured. Some notes and *caveats* deserve mentioning, however.

Most resistors are mounted vertically. Use a uniform method of bending the long terminal twice to obtain right angles. Semicircles indicate where the resistors sit on the board. Where rectangles are printed, the resistor is mounted flat on the board.

Although it's possible to solder all opamps directly on the board, you do so at a risk (see part 1), hence the prototype was built using turned-pin 8-way DIL sockets for all opamps. It is essential to use premium quality sockets here—don't be tempted to use cheap ones with dodgy spring contacts; it is false economy.

The tallest components on the board are capacitors C2, coil L1, relay RE1 and the screw terminal blocks for the loudspeaker and supply connections. The coil consists of 10 turns of 1 mm diameter (AWG18) enamelled copper wire (ECW) with an internal diameter of 20 mm. The coil windings should be spread evenly to obtain an overall length to suit the footprint on the board. Fit the coil first, then R106 at a height of 5–10 mm above the board surface and not touching L1 anywhere.

Although printed on the silk screen PCB overlay, **capacitors C24, C25, C26 and C27 are not fitted.** Instead, a single 1,000 µF, 63 V electrolytic capacitor is used, its terminals being inserted directly into terminal blocks K16/K17 (observe the polarity and provide the leads with sleeving for insulation). This should be a high quality electrolytic capacitor with low ESR. The modification proved necessary to get rid of unexpected distortion levels occurring at about 20 kHz in the initial design and was found to totally cure the problem. The amplifier specifications printed last month apply to the capacitor-modified version only.

Figure 44.1 Perfect sound from perfect construction. Can you match it? Please note that capacitor positions C24–C27 must remain empty.

The four corners of the amplifier board have holes for PCB standoffs—we used 10 mm tall ones.

The finished amplifier board should be given a thorough visual inspection before taking into use. Did you get all the polarized components' orientations right? Are all solder joints beyond reproach? It often helps to have a friend look at the board. Figure 44.1 once again shows the amp board for your reference. See how close you can get to this level of sophistication in building high-end audio circuitry.

Construction—PSU board

This is also a classic board with nothing but standard components, hence should be easy to assemble and get working. It's just the heatsinking of the voltage regulators that requires some mechanical work.

The two bridge rectifiers B1 and B2 should be fitted with 2 mm thick 70 × 35 mm aluminium plates bolted directly to the flat sides. The two voltage regulators IC1 and IC2 are secured to a single black, finned, extruded aluminium heatsink of which the mounting outline is shown on the component overlay. The heatsink is secured to the board with three M4 screws or bolts, for which holes need to be drilled and tapped in the underside. Venting holes are provided in the PCB area under the heatsink to assist cooling. The LT1083CP devices are secured to the heatsink with a heat conducting washer inserted. Two holes have to be drilled for the M3 bolts and nuts that hold the regulators firmly on the heatsink. First, determine the location of the holes for the regulators; then secure the heatsink to the board. The regulator terminals should be 'kinked outward' slightly for thermal relief. The two LT1083 regulators should be mounted, secured and soldered last.

To extend the 5532 amplifier for 4 ohm operation, it is necessary to double the number of 5532 sections driving the output. This is easier than it might sound because facilities are provided for adding one or more power amplifier PCBs in parallel. The connector K4 (see circuit schematic published last month) is used as an output from the main power amplifier PCB; it comes from the 'zero-impedance' stage IC3A, so the output is at a very low impedance (I measured 0.24 ohms at 1 kHz) but is immune to HF instability caused by cable capacitance. Any desired length of cable can be used. The output from K4 drives an identical power amplifier PCB that has the all the output opamps in place, but the redundant input amplifier circuitry omitted. The equivalent connector K4 on this PCB is used as an input, and drives the output opamps in exactly the same way as on the main power amplifier PCB.

The connectors K14, K15 are used to connect together the outputs of the two amplifier PCBs. Note that this connection is made upstream of the output muting relay RE1, to preserve full protection for the loudspeaker in case of a DC fault.

While this conversion for 4 ohm operation is relatively straightforward, it is essential to remember that the power supply requirements are doubled along with the current output. You will need to consider a larger power transformer, more capable rectifiers, larger reservoir capacitors, and increased supply regulator capacity. In view of these various requirements, this 4 ohm version is mentioned here more as an option for the experimenter rather than a fully cut-and-dried design.

Masterclass #1: Extension of Design for 4 Ohm Speakers

It is an inherent property of the 5532 power amplifier that its output current is limited by the internal overload protection of the opamps used. This means that if more current is needed in the load, you need to put more opamps in parallel. The basic design is intended to drive 8 Ohm loudspeakers with a reasonable safety margin, but as it stands it is definitely not recommended for 4 ohm operation.

Figure 44.2 shows the PSU board in a "travel & demo" version of the amplifier. PCB stand-offs are used as with the amplifier board. The PSU can be tested (carefully) by temporarily connecting the secondaries of the toroidal transformer to the AC input terminals and checking for the correct output voltage of 18.0 VDC ±0.7 V on each rail.

The complete amplifier

The OpAmplifier should be built into in a metal case observing all relevant precautions in respect of electrical safety, specifically with respect to earthing and the

Figure 44.2 The power supply board is conventional as far as construction and assembly is concerned, but do make sure you get the mounting of the heatsink and the voltage regulators right.
An ABS suitcase as shown here is not recommended as a permanent housing for the amplifier.

use of wiring carrying the AC grid voltage (230 V or 115 V). The size and shape of the case will depend on the number of amplifier boards used, as well as the associated power supply, see the **insets** on **bridged operation** and **4-ohm conversion.** Allow for quite some heat developing in the amplifier case, from the heatsink as well as from the NE5532s. All those milliwatts add up!

Everything ahead of the bridge rectifier AC terminals should be built, secured and wired with the high

currents and voltages always in mind. If possible, do not extend the transformer wires. The amplifier wiring diagram for the 8 ohm 2 × 15 watts stereo version of the amplifier is shown in Figure 44.3. The rating of the AC power fuse should match the line voltage in your area (115 V/60 Hz or 230 V/50 Hz). Whichever, an approved IEC style appliance socket with integral fuseholder and double-pole rocker switch **must be used.**

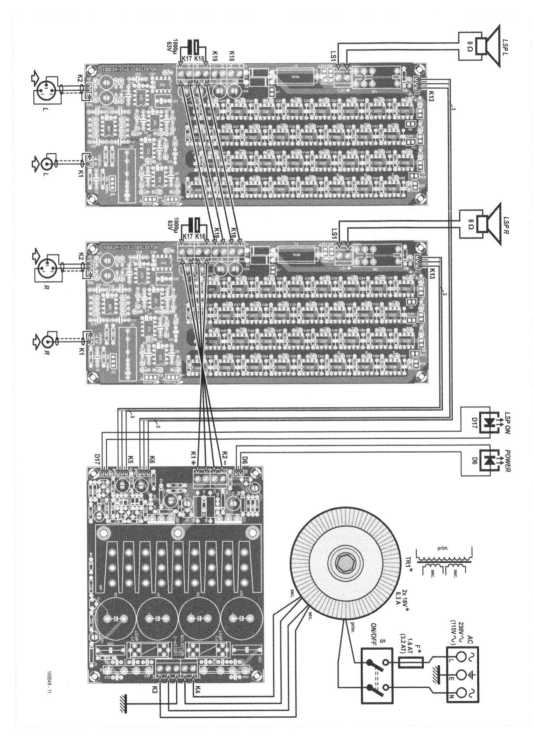

Figure 44.3 OpAmplifier wiring diagram. Note that this is for the 2 × 15 watts, 8-ohm version. The electrical ratings of the toroidal transformer and the fuse should match your local AC line voltage. High amps and loudspeaker connections are shown in slightly thicker lines.

Masterclass #2: Extension of Design for Bridged Operation

Two power amplifiers are bridged when they are driven with anti-phase signals and the load connected between their outputs; the load is not connected to ground in any way. This method of working doubles the signal voltage across the load, which in theory at least quadruples the output power. It is a convenient and inexpensive way to turn a stereo amplifier into a more powerful mono amplifier. Most conventional power amplifiers do not give anything close to power quadrupling- in reality the increase in available power will be considerably less, due to the power supply sagging and extra voltage losses in the two output stages, which are effectively in series. In most cases you will get something like three times the power rather than four times, and it may be less. I have come across many power amplifiers where the bridged mode only gave twice the output power, and it has to be said that in many cases the bridged mode looks like something of an afterthought.

The 5532 power amplifier will give better results than that for two reasons—firstly the power supply rails are regulated, and will not droop significantly under heavy loading. Secondly, the parallel structure minimises voltage losses; for example, all the 1 ohm output combining resistors are in parallel and their effective resistance is therefore very small.

Bridged mode is so called because if you draw the four output transistors of a conventional amplifier with the load connected between them, as in Figure A, it looks something like the four arms of a Wheatstone bridge.

In the drawing, the 8 ohm load has been divided into two 4 ohm halves, to emphasize that the voltage

at their center is zero, and so as far as current output is concerned, both amplifiers are effectively driving 4 Ohms loads to ground. The current capability required is therefore doubled, with all that that implies for increased losses in the output stages. A unity-gain inverting stage is shown generating the anti-phase signal; there are other ways to do it but this is the most straightforward and it simply adds one more 5532 section to a design that already contains quite a few of them. The simple shunt-feedback unity-gain stage shown does the job very nicely, and the 5532 power amplifier incorporates a version of this circuit. The resistors in the inverting stage need to be kept as low in value as possible to reduce their Johnson noise contribution, but not of course made so low that the opamp distortion is increased by driving them. The capacitor C1 across the feedback resistor R2 assures HF stability—with the circuit values shown it gives a roll-off that is −3 dB at 5 MHz, so it does not in any way imbalance the audio frequency response of the two amplifiers.

You sometimes see the glib statement that bridging always reduces the distortion seen across the load because the push-pull action causes cancellation of the distortion products. It is not true. Push-pull systems can cancel even-order distortion products; odd-order harmonics will not be cancelled.

The 5532 power amplifier has facilities for bridging built in. The last stage in the input amplifier is the inverting stage around IC3A, which is needed to get the phase between amplifier input and output correct. If we take the signal off from before this stage, then it is inverted with respect to the amplifier output and can so be used to drive another power amplifier board in anti-phase; this PCB can be built with the input amplifier circuitry omitted, as for the 4 ohm conversion described above. This inverted signal is available at connector K3. The loudspeaker is connected between the two output terminals on the amplifier PCBs and the output ground terminals are not used.

As explained above, driving an 8 ohm load with two bridged power amplifiers means that each amplifier is effectively driving a 4 ohm load to ground, so to take full advantage of the bridging capability requires the output stages to be doubled up, so that there are two power amplifier PCBs in parallel driving in-phase, and two power amplifier PCBs in parallel driving in anti-phase. This is of course quite a serious undertaking, as the number of output opamps has been doubled for bridging, and then doubled again to give adequate output current capability. The power supply capability will also need to be suitably increased.

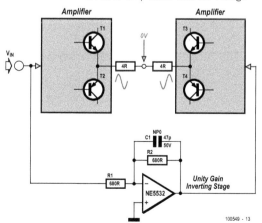

100549 - 13

We should emphasise that the 8.3-amp toroidal transformer indicated in the drawing is advised for the 2 × 15 watt, 8-ohm version of the OpAmplifier. A much beefier transformer is required if you decide to build a 4-ohm or bridged version of the amplifier. As with nearly all audio power amplifiers, it's essential to have a good deal of spare capacity in the 'amps department' so do not skimp on the transformer.

The test results

The basic concept behind the 5532 power amplifier is to create an amplifier which has the very low distortion of the 5532 opamp. Preserving this performance when large currents are flowing to and fro on a circuit board is something of a challenge and requires very careful attention to grounding topology, supply-rail routing, and decoupling arrangements. For this reason it is strongly recommended that you use the Elektor PCB design—better still, ready-made boards supplied by Elektor.

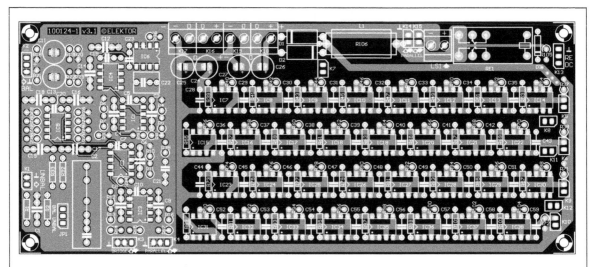

Component list

Amplifier board (one channel)

Resistors

5%, 0.25W, Farnell/Multicomp MCF series, unless otherwise indicated.
R1,R14,R17 = 47Ω
R2 = 220kΩ
R3 = 47kΩ
R4,R6 = 910Ω 1% 0.25W, Multicomp type MF25 910R
R5,R7,R10,R11,R15,R16 = 2.2kΩ
R8,R9,R32,R33 = 10Ω
R12 = 1kΩ
R13 = 2.00kΩ 1% 0.25W, Multicomp type MF25 2K
R18,R21 = 150kΩ
R19,R22 = 100Ω
R20,R23 = 68kΩ
R24-R31 = 820Ω
R34,R35 = 10MΩ
R36 = 100kΩ
R37,R38,R71,R72 = 22kΩ

R39-R70,R73-R104 = 1Ω
R105 = 10kΩ
R106 = 10Ω 5% 3W, Tyco Electronics type ROX3SJ10R

Capacitors

C1,C12,C14,C16-C19 = 100pF, 2.5%, 160V, polystyrene, axial (LCR Components)
    C2 = 2.2μF 5% 250V, polypropylene, Evox Rifa type PHE426HF7220JR06L2
    C3,C4,C6,C8,C9 = 33pF ±1pF 160V, axial, polystyrene (LCR Components)
    C5,C7,C10,C15,C20,C23,C28-C59 = 100nF 10% 100V, lead pitch 7.5mm, Epcos type B32560J1104K
    C11,C13 = 47μF 20% 35V non polarised, Multicomp type NP35V476M8X11.5 (Farnell # 1236671)
    C21 = 22nF 10%, 400V, lead pitch 7.5mm, Epcos type B32560J6223K (Farnell # 9752366)
    C22 = 1μF, 10%, 10V, lead pitch 7.5mm, Epcos type B32560J1105K (Farnell # 9752382)
    C24-C27 = not fitted, see text

Inductors

L1 = 1.7µH; 10 turns 1 mm (18AWG) ECW, diameter 20mm, effective length 20mm

Semiconductors

D1,D2 = 1N5402
D3 = 1N4148
IC1-IC5,IC7-IC38 = NE5532 (see text for notes on device selection)
IC6 = OP177 (Analog Devices)

Miscellaneous

K1,K5-K12,K14,K15 = 2-pin SIL pinheader, lead pitch 0.1 inch

K2,K3,K4,K13 = 3-pin SIL pinheader, lead pitch 0.1 inch
K16-K19,LS1 = 2-way PCB terminal block, lead pitch 5mm
JP1 = 3-pin SIL pinheader and jumper, lead pitch 0.1 inch
RE1 = PCB mount relay, PCB, DPCO, 24VDC/ 1100Ω/5A, Omron type G2R-2 24DC
Phono socket, chassis mount, black, Neutrik type NYS367–0
XLR socket, chassis mount, 3-way, Pro Signal type PSG01588
PCB # 100124–1, Elektor Shop, see www.elektor.com/100549

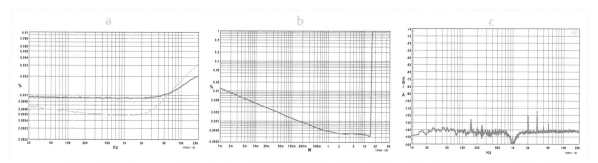

Figure 44.4 The OpAmplifier was duly 'grilled' on our AP System 2 analyzer. Graph (a): THD & noise vs. output power. Graph (b): distortion vs. output power (1 kHz, BW = 22 KHz). Graph (c): FFT graph for 1 kHz, 1 watt, 8 ohms.

The completed prototype was measured using Elektor Labs' **Audio Precision System Two Cascade Plus 2722 Dual Domain** test-set, and Figures 44.4a, 44.4b and 44.4c show the pleasing and impressive results. Figure 44.4a shows the harmonic distortion and noise as a function of signal frequency. Measurements carried out at 1 watt (red) and 8 watts (blue) of output power within 80 kHz bandwidth. The graph in Figure 44.4b shows distortion as a function of output power at 1 kHz and a bandwidth of 22 kHz. From about 3 watts the amplifier actually gets close to the lower measurement threshold and noise floor of our analyser! At 15 watts (roughly) the amplifier veers off into clipping. The curve in Figure 44.4c, finally, is an FFT plot for 1 kHz, 1 watt into 8 ohms. The fundamental frequency is suppressed. The second harmonic is way down at −121 dB and the third harmonic slightly higher at −115 dB. At a bandwidth of 22 kHz, the total level of noise and harmonics is about 0.0005%. If the distortion is measured for the two strongest harmonics only, it is at a level of 0.0002%.

Component list

Power supply board

Resistors

5%, 0.25W, Farnell/Multicomp MCF series, unless otherwise indicated.
R1,R5 = 100Ω
R2,R6 = 1.30kΩ 1% 0.25W, Multicomp MF25 1K3
R3,R7 = 39Ω
R4,R8 = 1.5kΩ 5% 1W, Multicomp MCF 1W 1K5
R9,R11,R16,R20-R22 = 2.2kΩ
R10,R12-R15,R19 = 47kΩ
R17 = 150kΩ
R18 = 470kΩ

Capacitors

C1,C3,C24 = 47µF 20% 25V, lead pitch 2.5mm, Rubycon type 25ZLG47M6.3X7

C2,C4 = 100µF 20% 25V, lead pitch 2.5mm, 1.43A AC, Nichicon type UPM1E101MED

C5,C8 = 100nF 10% 100V, lead pitch 7.5mm, Epcos type B32560J1104K

C6,C7,C9,C10 = 4700µF 20% 35V, lead pitch 10mm, snap in, Panasonic type ECOS1VP472BA (25mm max. diameter)

C11-C18 = 47nF 10% 50V ceramic, lead pitch 5mm

C19,C20 = 47nF 20%, 630VDC X2, lead pitch 15mm, Vishay BCcomponents BFC233620473

C21 = 4.7µF 20% 63V, lead pitch 2.5mm

C22,C23 = 22µF 20% 35V, non polarised, lead pitch 2.5mm, Multicomp NP35V226M6.3X11

Semiconductors

D1-D4,D7 = 1N4002
D5,D8-D16 = 1N4148
D6,D17 = LED, green, 3mm
T1,T3,T4,T6 = BC337
T2,T5,T7 = BC327
IC1,IC2 = LT1083 (Linear Technology)
B1,B2 = GSIB1520 (15A/200V bridge rectifier) (Vishay General Semiconductor)

Miscellaneous

K1,K2,K4 = 2-way PCB terminal block, lead pitch 5mm

K3 = 3-way PCB terminal block, lead pitch 5mm

D6,D17,S1 = 2-pin SIL pinheader, lead pitch 0.1 inch

K5,K6 = 3-pin SIL pinheader, lead pitch 0.1 inch

F1,F2 = fuse, 6.3A antisurge (time lag), with 20×5mm PCB mount fuseholder and cover

Heatsink, Fischer Elektronik type SK92/75SA 1.6 K/W, size: 100×40mm, Farnell # 4621578, Reichelt # V7331G

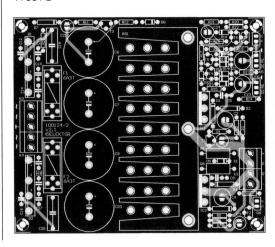

Miscellaneous, mechanical

M3 screws, nuts and washers for mounting IC1 and IC2 to heatsink (see text)

TO-3P thermal pad, (Bergquist type K6–104)

M4×10 screws for mounting heatsink to PCB

PCB # 100124–2, Elektor Shop, see www.elektor.com/100549

Conclusion

To the non-initiated, the OpAmplifier with its case closed may be just another high-end audio amplifier albeit with quite impressive specs relative to the investment. To the discerning electronic engineer, it's a delightful and off-the beaten path approach to getting high-quality audio watts from a dead common component like the NE5532 you normally associate with nano-ampères and microvolts electronics-wise, and pennies in the financial department!

The proof of the pudding is in the eating. That's why a demo version of the OpAmplifier was cased up in a rugged suitcase for easy transporting, packaging and demoing. At the time of writing, it's about to leave Elektor House on a journey along several audiophiles and audio communities around the world for listening tests. Their feedback is awaited eagerly.

(100549)

Spoilt for Choice—Which 5532?

It is an unsettling fact that not all 5532 opamps are created equal. The design is made by a number of manufacturers, and there are definite performance differences. While the noise characteristics appear to show little variation in my experience, the distortion performance does appear to vary significantly from one manufacturer to another. Although, to the best of my knowledge, all versions of the 5532 have the same internal circuitry, they are not necessarily made from the same masks, and even if they were, there would inevitably be process variations between manufacturers

Since the distortion performance of the amplifier is unusually, and possibly uniquely, dependent on only one type of semiconductor, it makes sense to use the best parts you can get. In the course of the development of this project 5532s from several sources were tested. I took as

wide a range of samples as I could, ranging from brand-new devices to parts over twenty years old, and it was reassuring to find that without exception, every part tested gave the good linearity we expect from a 5532. The information here will be of use not only in the 5532 power amplifier project, but should prove very valuable for anyone using 5532s in high-quality applications.

The main sources at present are Texas Instruments, Fairchild Semiconductor, ON Semiconductor, (was Motorola) NJR, (New Japan Radio) and JRC (Japan Radio Company). TI, ON Semi and Fairchild samples were compared in the Elektor labs by Ton Giesberts. The author for his part did THD tests on six samples from Fairchild, JRC, and Texas, plus one old Signetics 5532 for historical interest. The Elektor lab tests were carried out on an actual and very crucial section of the OpAmplifier design: the driver section! See Figure A for the circuit and Figure B for the AP2 plots.

As it turned out, the Texas 5532s (lightest line) proved to be distinctly inferior, both in the Elektor Labs and in the author's tests. We have to admit this surprised us, as we have always thought that the Texas part was one of the best available, but the measurements say otherwise. Distortion at 20 kHz ranges from 0.001% to 0.002%, showing more variation than Fairchild and ON Semi as well as being higher in general level. The low-frequency section of the plot, below 10 kHz, is approaching the measurement floor, as for all the other devices, and distortion is only just visible in the noise.

Compared with other maker's parts, the THD above 20 kHz is much higher—and at least 3 times greater at 30 kHz. Fortunately this should have no effect unless you have very high levels of ultrasonic signals that could cause intermodulation. If you have, then you have bigger problems on your hands than picking the best opamp manufacturer ...

The graphs show beyond doubt that the Fairchild 5532s (darkest line) are top of the bill and true audio purists should select these devices even if they come at a slightly higher price as well as being awkward in terms of type code print on the device. Check with your supplier, it should be worth your while. There is effectively no distortion visible above the noise floor up to about 12 kHz, and distortion at 20 kHz is less than 0.0005%.

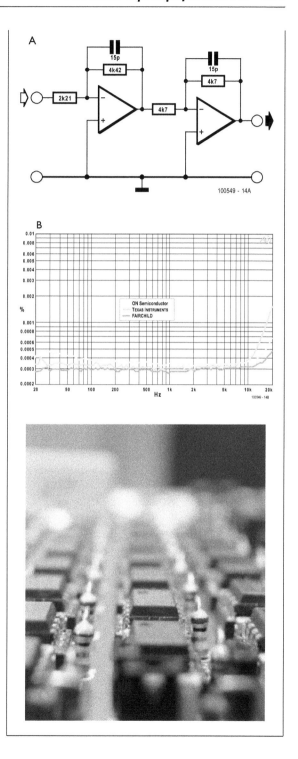

Chapter **45**

Inclusive compensation & ultra-low distortion power amplifiers

(Linear Audio, Volume 0, Sept 2010)

Intended title: *Inclusive Compensation & Ultra-Low Distortion Power Amplifiers*

I wrote this article specially for the first issue of Jan Didden's *Linear Audio,* which has prospered and become the primary journal for innovative audio electronic design [1].

In a standard Blameless power amplifier, all the measurable distortion comes from crossover distortion or large-signal non-linearity (LSN) in the output stage. If we want less distortion, it's the output stage that must be improved in itself, or more negative feedback applied around it—or both. While various methods are available for making an output stage more linear, they tend to be expensive—a good example being the use of multiple output devices in an EF output stage to reduce both crossover distortion and LSN [2]. The use of Class-A is an even more costly example. The application of more negative feedback is much more economical, but also more technically difficult, as the stability of the amplifier must remain satisfactory. Two-pole compensation can effectively reduce distortion [3], but gives you an amplifier that is not unconditionally stable, and so may oscillate during power-down, clipping, or slew-rate limiting. Inclusive compensation does not have this disadvantage.

For many years, it was obvious that if the VAS Miller compensation loop could be configured to include the output stage, there was the possibility of a considerable reduction in distortion. The disabling problem was that the extra phase-shifts in a relatively slow output stage would render that loop unstable, and there would be endemic high-frequency oscillation. The answer is to have global feedback operational around the whole amplifier at low frequencies, transitioning to semi-local feedback around the VAS and output stage at middling frequencies, and transitioning again to local feedback around the VAS alone. This sounds complicated, but as the article demonstrates, only requires the addition of one resistor and one capacitor to a standard Blameless amplifier. This method was suggested to me by Peter Baxandall in a document I received in 1995. This lengthy document, which covers many other aspects of amplifier design, is reproduced in Jan Didden's

book *Baxandall and Self on Audio Power* [4]. Since the article was written, it has been discovered that this form of inclusive compensation is fully described in a 1966 US patent granted to Seki, and assigned to Hitachi [5]. It is not known if they ever exploited it commercially.

My full name for the technique is Output-Inclusive Miller Compensation (OIMC); it could be argued that Output-Inclusive Compensation (OIC) is shorter, but it is less informative as there are other ways of compensating amplifiers apart from using the Miller technique. Transitional Miller Compensation (TMC) has also been suggested as a name but is hardly acceptable because normal Miller compensation already transitions once between global and local feedback, and this is its essence; thus TMC actually makes no distinction between the two methods, and that sort of confusion we can do without. There has been much discussion about naming of parts [6] and inclusive compensation in general on the DIYaudio bulletin board; it is rather disorganised and spread out over many threads, but you should find [7] and [8] to be useful entry points.

The measurements in the article were preliminary tests, and so the output power was limited to 22W/8Ω from a 30W/8Ω design to minimise the size of any explosions that might be provoked. (None were.) This was purely a precaution and does not mean that OIMC is incapable of exploiting the full output of an amplifier. To confirm this, the benefits of OIMC are shown in Preface Figure 45.1 at the rather more convincing power output of 40W into 8Ω, and in addition, Preface

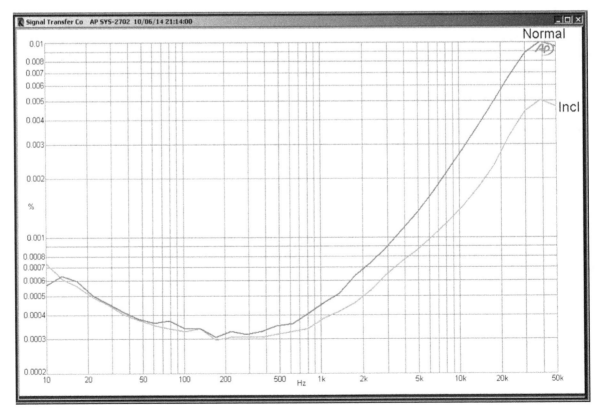

Preface Figure 45.1 THD of Compact Class-B power amp 40W into 8Ω. Normal compensation top trace, inclusive compensation lower trace. Testgear output THD is flat at 0.00025% up to 20 kHz. Bandwidth 10–80kHz.

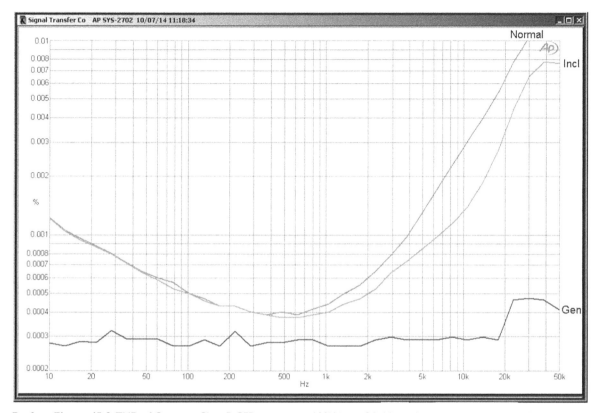

Preface Figure 45.2 THD of Compact Class-B CFP power amp 100W into 8Ω. Normal compensation at top, inclusive compensation middle, testgear output at bottom. Bandwidth 10–80kHz.

Figure 45.2 shows the same test at 100W into 8Ω, which was quite close to clipping with the supply rails used. There are no restrictions on the power level at which OIMC works. The rise in THD at low frequencies is due to thermal distortion in the upper feedback resistor, which is unaffected by the negative feedback. See the Preface to Chapter 31.

References

1. *Linear Audio*, http://www.linearaudio.nl/ (accessed Jan. 2015)

2. Self, D., *Audio Power Amplifier Design*, 6th edition, p. 237, p. 262. Focal Press 2013. ISBN 978-0-240-52613-3.

3. Ibid., p. 444.

4. Baxandall, P., and Self, D., Baxandall and Self on Audio Power, *Linear Audio* 2011, p. 96. Editor: Jan Didden. ISBN 978-94-90929-03-9.

5. Seki, K., Multistage Amplifier Circuit, US Patent 4,145,666, 20th Mar. 1979. Assigned to Hitachi.

6. Reed, H., Naming of Parts. http://www.poetrybyheart.org.uk/poems/naming-of-parts/ (accessed Feb. 2015)

7. http://www.diyaudio.com/forums/solid-state/230492-audio-power-amplifier-design-book-douglas-self-wants-your-opinions-16.html (accessed Jan. 2015)

8. http://www.diyaudio.com/forums/solid-state/94676-bob-cordell-interview-negative-feedback-336.html#post2433940 (accessed Jan. 2015)

INCLUSIVE COMPENSATION & ULTRA-LOW DISTORTION POWER AMPLIFIERS

Two things define the distortion performance of a power amplifier: the open-loop linearity of the circuitry, and the amount of negative feedback that can be safely applied. The latter is determined by the compensation scheme used. The almost-universal Miller dominant-pole method gives excellent and reliable results, but can be significantly improved upon by techniques such as two-pole compensation [1] which allow more feedback at audio frequencies. Inclusive compensation promises even better performance, but as normally conceived it is not workably stable. Here I show how to make it work. The results are dramatic, and the extra cost is trivial.

The vast majority of audio power amplifiers have three stages: a differential input stage that performs voltage-to-current conversion (ie, a transconductance stage), a voltage-amplifier stage (VAS) that performs current-to-voltage conversion, and a unity-gain output stage with a high input impedance but large current-output capability.

A power amplifier must be compensated because its open-loop gain is usually still high at frequencies where the internal phase-shifts add up to 180°. This turns negative feedback into positive at high frequencies, and causes oscillation, which can be very destructive to both the amplifier and attached loudspeakers. This dire scenario is prevented by adding some form of HF roll-off so the loop gain falls to below unity before the phase-shift reaches 180°, so oscillation cannot develop. This process of Compensation makes the amplifier stable, but the way in which it is applied has major effects on its closed-loop distortion performance.

Conventional Miller compensation

Almost all audio power amplifiers use dominant-pole compensation. A "pole" here is a frequency from which the open-loop gain falls of at 6 dB/octave; when a second pole at a higher frequency comes into action, the gain falls off at 12 dB/octave. To implement dominant-pole compensation, you take the lowest frequency pole that exists and make it dominant; in other words so much lower in frequency than the next pole up that the total loop-gain (i.e., the open-loop gain as reduced by the attenuation in the feedback network) falls below unity before enough phase-shift builds up to cause HF oscillation. With a single pole, the gain must fall at 6 dB/octave, corresponding to a constant 90° phase shift. Thus the phase margin, the difference between the actual phase shift and the 180° that will cause oscillation is 90°, which gives good stability.

Since we have caused the open-loop gain to roll-off, the amount of feedback available to linearise the amplifier is reduced at high frequencies, and distortion will be greater. This disadvantage can be offset by applying the dominant-pole in the form of Miller feedback [2] from collector to base of the VAS; this is Ccompen in Figure 45.1. This is by far the commonest method of dominant-pole compensation, but it is also the best. Its action is in fact rather subtle. As the frequency increases, the local feedback around the VAS through the Miller capacitor also increases, so that open-loop gain which was being applied for global negative feedback smoothly transforms itself into gain that linearises the VAS only. Since this stage has to generate the large voltage-swing required to drive the unity output stage, maintaining its linearity in this way is a thoroughly good idea.

I am using the term "local feedback" to refer to feedback that is confined to a single stage only, such as the Miller capacitor around the VAS; other examples are emitter degeneration resistors in an input differential stage and the use of Complementary Feedback Pairs (CFP) in an output stage.

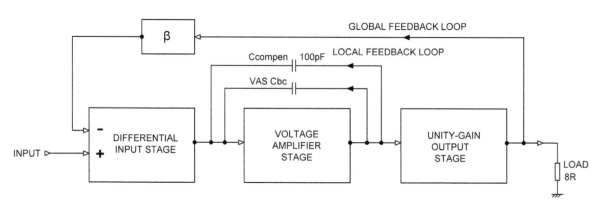

Figure 45.1 The feedback paths in a conventional Miller-compensated amplifier.

There are two other advantages to Miller dominant-pole compensation. Firstly, the input impedance of the VAS is lowered by the local shunt feedback through the Miller capacitor, so the VAS input is effectively a virtual-earth current-input; this prevents unwanted roll-offs occurring at the output of the input stage. Secondly, this local feedback reduces the output impedance of the VAS, enabling it to effectively drive the non-linear input impedance of the output stage.

Figure 45.1 shows the feedback paths in a normal Miller dominant-pole compensated power amplifier. The block marked β represents the attenuation in the global feedback path required to give a closed-loop gain of 1/ β times. Also shown is the collector-base capacitance Cbc, which exists inside the VAS transistor and is non-linear, being a function of the collector voltage Vce. In a typical high-voltage small-signal transistor such as the MPSA42, often used in a VAS, it is quoted as a maximum of 3.0 pF for Vce = 20V, so the normal Miller value of around 100 pF is enough to swamp it. It is included here in every diagram as a reminder that whatever fancy compensation schemes you dream up, this capacitance must be taken into account; if necessary it could be rendered harmless by using a cascoded VAS. The non-linearity of Cbc is not an issue in the input stage transistors as there is only a very low signal voltage on their collectors.

This form of amplifier can give a very good distortion performance, as described below in the section on the Blameless amplifier philosophy, but the great limitation on this is the high-order distortion products generated by crossover distortion in the output stage. These are at a high frequency and so are less effectively linearised by the global feedback factor, which falls with frequency due to the dominant-pole compensation.

This brief description hopefully demonstrates why Miller dominant-pole compensation is near universal; it positively bristles with advantages, is cheap to implement, and gives thoroughly dependable performance. Any alternative compensation scheme has to improve on this. It is a tall order.

Output-inclusive compensation

Looking at Figure 45.1, we note that the output stage has unity gain, and it has occurred to many people that the open-loop gain and the feedback factor at various frequencies would be unchanged if the Miller capacitor Ccompen was driven from the amplifier output, as in Figure 45.2, creating a semi-local feedback loop enclosing the output stage. "Semi-local" means that it encompasses two stages, but not all three- that is the function of the global feedback loop. The inclusion of the output stage in the Miller loop has always been seen as a highly desirable goal because it promises that crossover distortion from the output stage can be much reduced by giving it all the benefit of the open-loop gain inside the Miller loop, which does not fall off with frequency until the much smaller Cbc takes effect.

Figure 45.2 shows the form of output-inclusive compensation that is usually advocated, and it has only one drawback- it does not work. I have tried it many times and the result was always intractable high-frequency instability. On a closer examination this is not surprising.

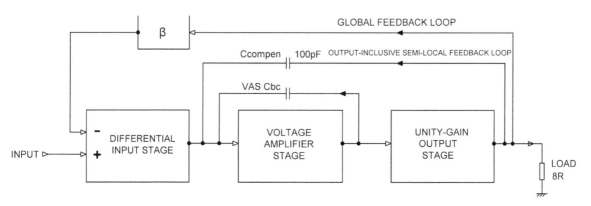

Figure 45.2 The feedback paths in an output-inclusive compensated amplifier. Note that the collector-base capacitance Cbc is still very much present.

Problems of inclusion

Using local feedback to linearise the VAS is reliably successful because it is working in a small local loop with no extra stages that can give extra phase-shift beyond that inherent in the Ccompen dominant pole. Experience shows that you can insert a cascode or a small-signal emitter-follower into this loop, [3] but a slow output stage with all sorts of complexities in its frequency response is a very different matter. Published information on this is very scanty, but Bob Widlar [4] stated in 1988 that output stage behaviour must be well-controlled up to 100 MHz for the technique to be reliable; this would appear to be flat-out impossible for discrete power stages, made up of devices with varying betas, and driving a variety of loads.

Trying to evaluate what sort of output stage behaviour, in particular frequency response, is required to make this form of inclusive compensation workably stable quickly runs into major difficulties. The devices in a typical Class-B output stage work in voltage and current conditions that vary wildly over a cycle that covers the full output voltage swing into a load. Consequently the transconductances and frequency responses of those devices, and the response of the output stage overall, also vary by a large amounts.

The output stage also has to drive loads that vary widely in both impedance modulus and phase angle. To some extent the correct use of Zobel networks and output inductors reduces the phase angle problem, but load modulus still has a direct effect on the magnitude of currents flowing in the output stage devices, and has corresponding effects on their transconductance and frequency response.

Input-inclusive compensation

An alternative form of inclusive compensation that has been proposed is enclosing the input stage and the VAS in an inner feedback loop, leaving just the output stage forlornly outside. This approach, shown in Figure 45.3, has been advocated many times, one of its proponents being the late John Linsley-Hood, for example in [5]. A typical circuit is shown in Figure 45.4. It was his contention that this configuration reduced the likelihood of input-device overload (ie slew-limiting) on fast transients because current flow into and out of the compensation capacitor was no longer limited by the maximum output current of the input pair (essentially the value of the tail current source). This scheme is only going to be stable if the phase-shift through the input stage is very low, and this is now actually less likely because there is less Miller feedback to reduce the input impedance of the VAS, therefore less of a pole-splitting effect, [6] and so there is more likely to be a significant pole at the output of the input stage. There is still some local feedback around the VAS, but what there is goes through the signal-dependent capacitance Cbc.

My experience with this configuration was that it was unstable, and any supposed advantages it might have had were therefore irrelevant. I corresponded with JLH on this matter in 1994, hoping to find exactly how it was supposed to work, but no consensus on the matter could be reached.

A similar input-inclusive compensation configuration was put forward by Marshall Leach [7], the intention being not the reduction of distortion, but to decrease the liability of oscillation provoked by capacitive loading on the output, and so avoid the need for an output inductor.

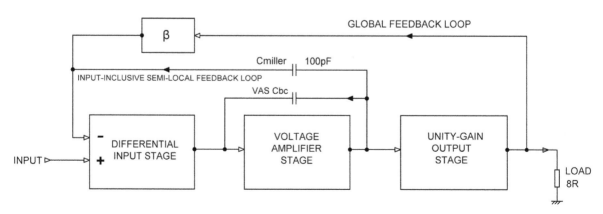

Figure 45.3 The feedback paths in an input-inclusive compensated amplifier.

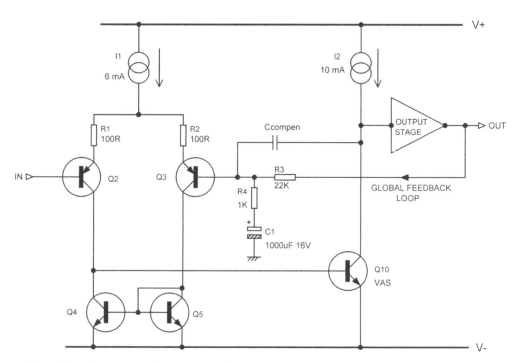

Figure 45.4 Attempt to implement input-inclusive compensation, with some typical circuit values shown.

Attempting to include the input stage in the inner loop to reduce its distortion seems to me to be missing the point. The fact of the matter is that the linearity of the input stage can be improved almost as much as you like, either by further increasing the tail current, and increasing the emitter degeneration resistors to maintain the transconductance, or by using a slightly more complex input stage. Any improvements in slew-rate that might be achieved would be of little importance, as obtaining a more than adequate slew-rate with the three-stage amplifier architecture is completely straightforward.[8]

If any stage needs more feedback around it is the output stage, as this will reduce its intractable crossover distortion. I think that trying to create a semi-local loop around the input stage and VAS is heading off in wholly the wrong direction.

Blameless power amplifiers

Before we proceed to a form of output-inclusive compensation that does work, we need to take a look at the amplifier we are going to apply it to. Since we are

studying only the compensation system rather than the whole amplifier, it makes sense to use a design which is both straightforward and known to have good performance with conventional compensation.

Some time ago I introduced the concept of the Blameless Audio Power Amplifier, [9] following an intensive study of the distortion mechanisms in power amplifiers. The Blameless title is intended to emphasise that although such amplifiers can give very low distortion figures, this is achieved more by avoiding mistakes rather than by using radical and ingenious circuitry. The schematic of such an amplifier is shown in Figure 45.5; the following features make it a Blameless amplifier:

1. The input differential pair Q2, Q3 is run at a higher current than usual, increasing the transconductance of the transistors. This is reduced to a suitable value for a stable amplifier by the emitter degeneration resistors R2, R3 which greatly linearise the transconductance of the input stage;

2. The collector currents of the input differential pair Q2, Q3 are forced into accurate equality by the current-mirror Q4, Q5, which is degenerated by R6, R7

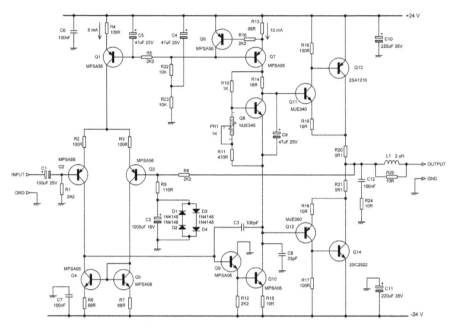

Figure 45.5 Schematic of a Blameless power amplifier with a CFP (complementary feedback pair) emitter-follower) output stage. Overload protection and clamp diodes are omitted for clarity.

for greater accuracy. This equality prevents the generation of second harmonic distortion in the input stage;

3. The VAS (Voltage Amplifier Stage) is linearised by adding the emitter-follower Q9, to increase the open-loop gain inside the local Miller feedback loop;

4. The output stage is in the CFP (Complementary Feedback Pair) configuration which has lower distortion than the EF (Emitter Follower) configuration in this situation;

5. The output stage emitter resistors are set to 0.1Ω, the lowest practicable value that permits acceptable quiescent stability. This reduces the inherent crossover distortion of the output stage with optimal biasing, and also minimises the extra distortion introduced if the output stage strays into Class AB operation due to over-biasing.

There are also topological requirements such as avoiding distortion being introduced by inductive coupling of half-wave currents from the supply connections.

While it is not directly related to distortion, note that the global feedback network R8, R9 has unusually low resistor values in order to improve the noise performance.

My intent was always that the Blameless amplifier was to be a sound and repeatable starting-point for innovative amplifier concepts. To be honest, it proved to be

disconcertingly good, raising the awkward question as to why any further improvement might be necessary. Figure 45.6 shows that the distortion of a $22W/8\Omega$ version is no more than 0.0001% at 400 Hz; it is unfortunately a good deal more at higher frequencies, and a long way short of perfection. Nonetheless, as hoped, this philosophy led to the Load-Invariant amplifier, [10] with unusually low levels of distortion into sub-8Ω loads, the Trimodal Amplifier [11] which was an ultra-low distortion Class-A amplifier that could be switched to Class-B, and was the basis on which the Crossover Displacement (Class-XD) principle was introduced [12]

Figure 45.6 shows the distortion performance of the design in Figure 45.5. The investigations reported here were done at the relatively low power of 22W into 8Ω, as this reduces the amount of damage that occurs if mistakes are made. With ±24 V rails as shown the maximum power output is about 30W into 8Ω,

The 80 kHz trace shows the THD rising at 6 dB/ octave, because the feedback factor is falling at that rate due to the dominant-pole compensation. The 22 kHz trace attempts to show the very low distortion at low frequencies; with this bandwidth the noise floor is approximately 0.00025%. Above 10 kHz this trace falls at 18 dB/octave, the roll-off rate of the Audio Precision bandwidth-definition filter.

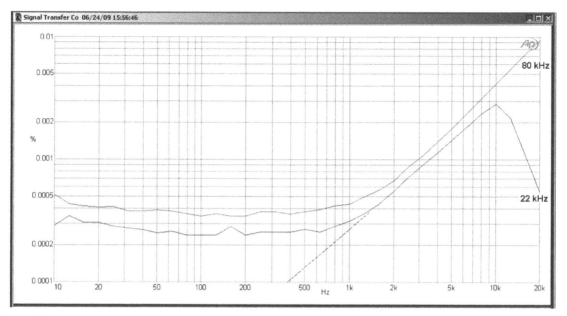

Figure 45.6 Distortion performance of the standard Blameless power amplifier at 22W/8Ω. THD at 10 kHz (80 kHz bandwidth) is 0.0042%. Extrapolating the 22 kHz trace to below the noise floor (dotted line) suggests that the distortion is no more than 0.0001% at 400 Hz.

Stable output-inclusive compensation

Considering the problems described earlier, it is clear that trying to include the output stage in the VAS compensation loop over its full bandwidth looks impractical. What can be done, however, is to include it over the bandwidth that affects audio signals, but revert to purely local VAS compensation at higher frequencies where the extra phase-shifts of the output stage are evident. The method described here was suggested to me by the late Peter Baxandall, in a document I received in 1995, [13] commenting on some work I had done on the subject in 1994. He sent me six pages of theoretical

analysis, but did not make it completely clear if he had personally evaluated it on a real amplifier. However he said that he had "devoted much thought and experiment to the problem" which implies that he had. Whether he invented the technique is not currently clear, but it is a characteristic Baxandall idea—simple but devastatingly effective. The general method was discussed in a DIYaudio forum [14] in March 2007; there appears to have been much simulation but no actual measuring.

The basic technique is shown in Figure 45.7. At low frequencies C1 and C2 have little effect, and the whole of the open-loop gain is available for negative feedback around the global feedback loop. As frequency increases,

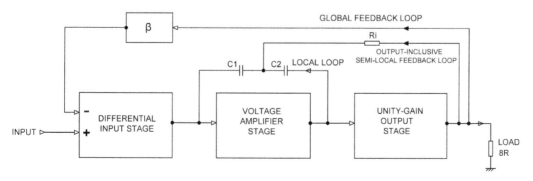

Figure 45.7 The basic principle of the Baxandall inclusive compensation technique.

semi-local feedback through Ri and C1 begins to roll off the open-loop gain, but the output stage is still included in this semi-local loop. At higher frequencies still, where it is not feasible to include the output stage in the semi-local loop, the impedance of C2 is becoming low compared to that of Ri, and the configuration smoothly changes again so that the local Miller loop gives dominant-pole compensation in the usual way. If the series combination of C1 and C2 gives the same capacitance as a normal dominant-pole Miller capacitor, then stability should be unchanged.

Figure 45.8 is a very much simplified version of a VAS compensated in this way, designed for simulation with a minimum of distracting complications. The current feed from the input stage is represented by Rin, which delivers a constant current as the opamp inverting input is at virtual ground. The "opamp" is in fact a VCVS (Voltage-Controlled Voltage Source) with a flat voltage gain of 10,000, once more to keep thing simple. Since there is no global feedback loop as in a complete amplifier, if the input is 1 Volt the output signal will be measured in kilo-Volts, at least at low frequencies; this is not exactly realistic but the magnitude does not alter the basic mechanism being studied.

Figure 45.9 shows how the feedback current through C1 is sourced via Ri at low frequencies, and via C2 at high frequencies. The lower the value of Ri, the higher the frequency at which the transition between the two routes occurs. With C1 and C2 set at 220 pF, and Ri = 1K, this occurs at 723 kHz. At very high frequencies the effective Miller capacitance is 110 pF, a slight

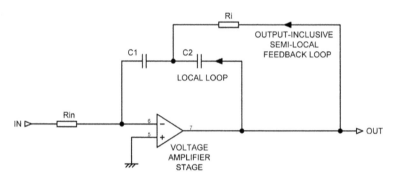

Figure 45.8 Conceptual diagram of the Baxandall output-inclusive compensation scheme for simulation.

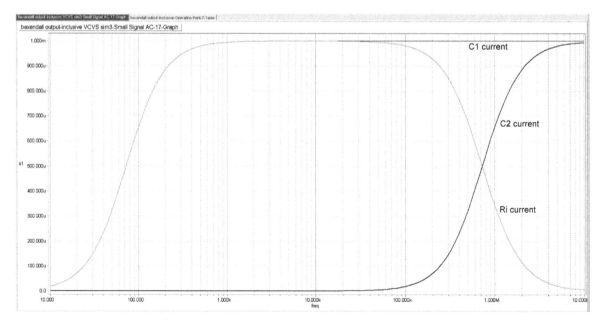

Figure 45.9 Showing how the local feedback loop (C2 current) takes over from the output-inclusive compensation (Ri current) as frequency increases.

increase on the usual value of 100 pF; this simply because 220 pF capacitors are readily available.

At low frequencies the amount of semi-local feedback is controlled by the impedance of C1; at 220 pF it is more than twice the size of the usual 100 pF Miller capacitor, and so the reduced open-loop gain means the feedback factor available is actually 6.5 dB less than normal, at low frequencies only. This sounds like a bad thing, but the fact that the semi-local loop includes the output stage more than makes up for it. At 723 kHz the impedance of C1 has fallen to the point where it is equal to Ri. At 1.45 MHz the impedance of the series combination of C1 and C2 now reaches that of Ri, and the capacitors dominate, giving strictly local Miller compensation, with approximately the normal capacitance value (110pF).

Figure 45.10 attempts to illustrate this; it shows open-loop gain with the closed-loop gain (+27.2 dB) subtracted to give a plot of the feedback factor. Note that the kink in the plot only extends over an octave, and so in reality is a gentle transition between the two straight-line segments. This diagram is for Ri = 10KΩ, as this raises the gain plateau above the X-axis and makes thing a bit clearer.

The thin line shows normal Miller compensation; note that its feedback factor reaches a plateau around 20 Hz as this is the maximum gain of input stage and VAS combined, without compensation.

Figure 45.11 shows the practical implementation of the inclusive technique to the Blameless amplifier of Figure 45.5; the unaltered parts of the circuitry are omitted for greater clarity.

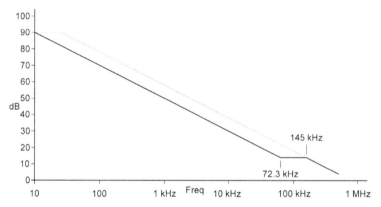

Figure 45.10 How the feedback factor varies with frequency. C1 = C2 = 220 pF, Ri = 10KΩ. The dotted line shows the feedback factor with normal Miller compensation with the capacitor equal to half C1, C2.

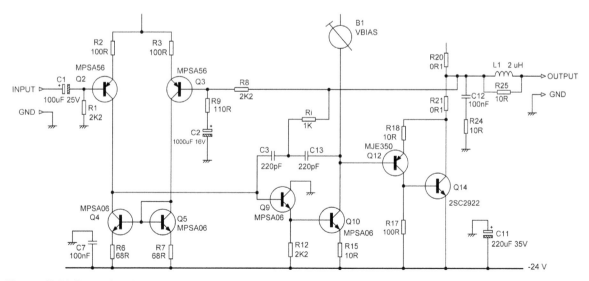

Figure 45.11 Practical implementation of output-inclusive compensation in the Blameless amplifier of Figure 45.5 (circuit simplified for clarity).

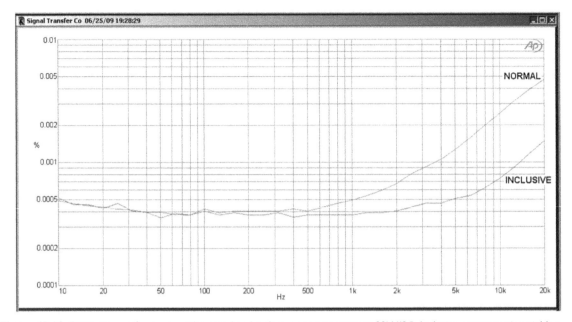

Figure 45.12 Distortion performance with normal and inclusive compensation, at 22W/8Ω. Inclusive compensation yields a THD + Noise figure of 0.00075% at 10 kHz, less than a third of the 0.0026% given by conventional compensation. Measurement bandwidth 80 kHz.

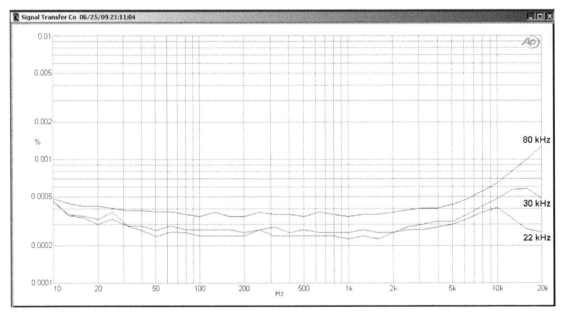

Figure 45.13 Bandwidths are 80 kHz, 30 kHz and 22 kHz.

Figure 45.12 compares the distortion performance of the standard Blameless amplifier with the new output-inclusive version; there may be no official definition of "ultra-low distortion", but I reckon anything less than 0.001% at 10 kHz qualifies. The THD measurement system was an Audio Precision SYS-2702.

Figure 45.13 shows another view of the output-inclusive distortion performance. Because the THD levels are so low, noise is a significant part of the reading. Three measurement bandwidths are therefore shown. The 22 kHz and 30 kHz bandwidths eliminate most of the harmonics when the fundamental is 10 kHz or above, so they do not give meaningful information in this region. It is however clear that a 10 kHz THD of 0.00074% is an overestimate, and 0.0005% is probably more accurate.

This form of compensation is very effective, but it is still necessary to optimise the quiescent bias of the Class-B output stage. This is easier to do without the distortion-suppression of output-inclusion, so you might consider adding a jumper so that the connection via Ri can be broken for bias setting.

As Figure 45.13 shows, this amplifier pretty much takes us to the limits of THD analysis; it is impossible to read any distortion at all below 2 kHz. The 22 kHz trace is only just above the AP distortion output; this varies slightly with output voltage but is below 0.00025% up to 10 kHz.

It is only right that I point out that a slight positional tweak of the output inductor was required to get the best THD figures. I attribute this to small amounts of uncorrected inductive distortion, where the half-wave currents couple into the input or feedback paths; it is an insidious cause of non-linearity.[15] The amplifier I used for the tests was thought to be free from this, but the reduced output stage distortion appears to have exposed some remaining vestiges of it. At 10 kHz the inductive coupling will be ten times greater than at 1 kHz, but in a conventional Blameless amplifier this effect is normally masked by crossover products.

The information here is not claimed to be a fully worked-out design such as you might put into quantity production. As with any unconventional compensation system, it would be highly desirable to check the HF stability at higher powers, with 4Ω loads and below, and with highly reactive loads.

Practical experimentation

One of the great advantages of this approach is that it can be added to an existing amplifier for the cost of a few pence. Possibly one of the best bargains in audio! This does not mean, of course that it can be applied to any old amplifier; it can only work in a three-stage architecture, and the amplifier needs to be Blameless to begin with to get the full benefit.

Another advantage of this arrangement is that if you decide that inclusive compensation is not for you, the amplifier can be instantly converted back to standard Miller compensation by breaking the connection to

Ri. There is in fact a continuum between conventional and output-inclusive compensation. As the value of Ri increases, the local/semi-local transition occurs at lower and lower frequencies.

The quickest and most effective way to experiment with this form of compensation is to start with an amplifier that is known to be Blameless. A very suitable design is the Load-Invariant amplifier produced by The Signal Transfer Company; [16] this has all the circuit features described above for Blameless performance, and the PCB layout is carefully optimised to eliminate inductive distortion. I need to declare an interest here; I am, with my colleague Gareth Connor, the technical management of The Signal Transfer Company.

Conclusion

It is instructive to compare this method with other ways of achieving ultra-low distortion. The Halcro power amplifier is noted for its low distortion, achieved by applying error-correction techniques to the FET output stage; however what appears to be a basic version, as disclosed in a patent, [17] uses 31 transistors. Giovanni Stochino (for whose abilities I have great respect) has also used error-correction in a most ingenious form, [18] but it requires two separate amplifiers to implement; the main amplifier uses 20 transistors and the auxiliary correction amplifier has 17, totalling 37. The output-inclusive compensation amplifier uses only 13 transistors, and gives extraordinary results for such simple circuitry.

Is this the end of history for power amplifiers? Not if I have anything to do with it.

References

1. Douglas Self, Audio Power Amplifier Design Handbook, 5th edition, Newnes, p198 (two-pole compensation).

2. John M Miller, Dependence of the input impedance of a three-electrode vacuum tube upon the load in the plate circuit, Scientific Papers of the Bureau of Standards, 15(351):367–385, 1920.

3. Douglas Self, Audio Power Amplifier Design Handbook, 5th edition, Newnes, p119 (VAS enhancements).

4. Widlar, R A, Monolithic Power Op-Amp, IEEE J Solid-State Circuits, Vol 23, No 2, April 1988.

5. John Linsley-Hood, Solid-State Audio Power, Electronics World +Wireless World, Nov 1989, p1047.

6. Douglas Self, Audio Power Amplifier Design Handbook, 5th edition, Newnes, p198 (pole-splitting).

7. Marshall Leach, Feedforward Compensation of the Amplifier Output Stage for Improved Stability with Capacitive Loads, IEE Trans on Consumer Electronics, Vol 34, No 2, May 1988z.

(So far as I can see, this is a misuse of the word "feedforward". What is actually described is semi-local feedback around the input stage, so that the output stage actually sees less HF feedback. The aim is to avoid using an output inductor; I feel this is a mistake. The amplifier is divided into only two stages for analysis).

8. Douglas Self, Audio Power Amplifier Design Handbook, 5th edition, Newnes, p255 (slew-rates).

9. Douglas Self, Distortion In Power Amplifiers: Parts 1 to 8, Electronics World Aug 1993 to Mar 1994.

10. Douglas Self, Load-invariant audio power, Electronics World Jan 1997 p16.

11. Douglas Self, Trimodal audio power, Part 1: Electronics World June 1995 p462, Part 2: Electronics World July 1995 p584.

12. Douglas Self, Class XD: a New Power Amplifier, Electronics World Nov 2006 p20.

13. Peter Baxandall, Private communication, 1995.

14. http://www.diyaudio.com/forums/solid-state/94676-bob-cordell-interview-negative-feedback-49.html (Pages 49–67).

15. Douglas Self, Audio Power Amplifier Design Handbook, 5th edition, Newnes, p198 (inductive distortion).

16. The Signal Transfer Company: http://www.signal-transfer.freeuk.com/.

17. Bruce Halcro Candy, US patent No. 5,892,398. (1999) Figure 2.

18. Giovanni Stochino, Audio Design Leaps Forward? Electronics World, Oct 1994, p. 818.

Index

The Signal Transfer Company

The Signal Transfer Company supplies PCBs, kits of parts, and fully built and tested modules based on the design principles in this book. All designs are approved by Douglas Self.

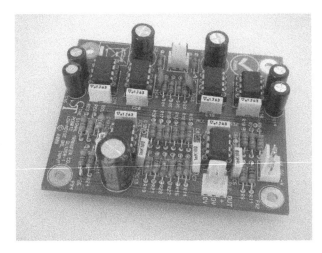

Low Noise Balanced Input PCB

A conventional balanced input stage built with 10K resistors and a 5532 opamp has an output noise level of approximately −104 dBu.

The Signal Transfer low-noise balanced input card uses a multiple amplifier array that causes the noise from each amplifier to partially cancel, and in a similar way improves the common-mode rejection ratio: this array is driven by a multiple-buffer structure that allows the input impedance to be much higher than usual, preventing loading of external equipment and also further improving the CMRR. This elegant design does not require selected or exotic components. The noise output is less than −115 dBu.

Balanced Output Stereo Phono Preamp

This offers both moving-magnet and moving-coil inputs, the latter with two gain options for optimal performance. It gives extremely accurate RIAA equalisation, a noise performance that approaches the theoretical limits, and superb linearity. It has fully balanced XLR outputs.

- Exceptional RIAA accuracy of +/−0.05 dB
- THD better than 0.002% at 6 Vrms. (Forty times the normal operating level)
- 3rd-order Butterworth subsonic filler: −3 dB at 20 Hz
- Separate input connectors for MM and MC inputs, switch selected
- Power indicator LED.

We supply the finest quality double-sided, plated-through hole, fiberglass PCBs. All boards have a full solder mask, gold-plated pads, and a silk-screen component layout. Each PCB is supplied with extensive constructional notes, previously unpublished information about the design, and a detailed parts list to make ordering components simple.

Signal Transfer products include several types of power amplifier, RIAA phono preamplifiers, and an enhanced version of the Precision Preamplifier described in this book. The specialised semiconductors required for some designs are also available.

For prices and more information go to   **http://www.signaltransfer.freeuk.com/**

Or contact:

**The Signal Transfer Company,
Unit 9A Topland Country Business Park,
Cragg Road,
Mytholmroyd,
West Yorkshire,
HX7 5RW, England.**

Tel: 01422 885 196